SOIL MANAGEMENT OF SMALLHOLDER AGRICULTURE

Advances in Soil Science

Series Editors: Rattan Lal and B. A. Stewart

Published Titles

Assessment Methods for Soil Carbon
R. Lal, J. M. Kimble, R. F. Follett, and B. A. Stewart

Food Security and Soil Quality
R. Lal and B. A. Stewart

Global Climate Change: Cold Regions Ecosystems
R. Lal, J. M. Kimble, and B. A. Stewart

Interacting Processes in Soil Science
R. J. Wagenet, P. Baveye, and B. A. Stewart

Methods for Assessment of Soil Degradation
R. Lal, W. H. Blum, C. Valentine, and B. A. Stewart

Principles of Sustainable Soil Management in Agroecosystems
R. Lal and B. A. Stewart

Soil Erosion and Carbon Dynamics
E.J. Roose, R. Lal, C. Feller, B. Barthès, and B. A. Stewart

Soil Management and Greenhouse Effect
R. Lal, J. M. Kimble, E. Levine, and B. A. Stewart

Soil Management: Experimental Basis for Sustainability and Environmental Quality
R. Lal and B. A. Stewart

Soil Management of Smallholder Agriculture
R. Lal and B. A. Stewart

Soil Processes and the Carbon Cycle
R. Lal, J. M. Kimble, R. F. Follett, and B. A. Stewart

Soil Quality and Biofuel Production
R. Lal and B. A. Stewart

Soil Structure: Its Development and Function
B. A. Stewart and K. H. Hartge

Soil Water and Agronomic Productivity
R. Lal and B. A. Stewart

Soils and Global Change
R. Lal, J. M. Kimble, E. Levine, and B. A. Stewart

Structure and Organic Matter Storage in Agricultural Soils
M. R. Carter and B. A. Stewart

World Soil Resources and Food Security
R. Lal and B. A. Stewart

Advances in Soil Science

SOIL MANAGEMENT OF SMALLHOLDER AGRICULTURE

Edited by

Rattan Lal

The Ohio State University
Columbus, USA

B. A. Stewart

West Texas A&M University
Canyon, USA

CRC Press
Taylor & Francis Group
Boca Raton London New York

CRC Press is an imprint of the
Taylor & Francis Group, an **informa** business

CRC Press
Taylor & Francis Group
6000 Broken Sound Parkway NW, Suite 300
Boca Raton, FL 33487-2742

Printed on acid-free paper
Version Date: 20141013

International Standard Book Number-13: 978-1-4665-9858-4 (Hardback)

Visit the Taylor & Francis Web site at
http://www.taylorandfrancis.com

and the CRC Press Web site at
http://www.crcpress.com

Contents

Preface

This volume addresses the challenges and opportunities for hundreds of millions of small landholder farmers, who are the major contributors to global food production, and do so under difficult biophysical and socioeconomic conditions. These resource-poor and small landholders, farming mostly in the tropic and subtropics (e.g., sub-Saharan Africa, Asia, Central America, the Caribbeans), are vulnerable to varieties of climate and related extreme events and depleted/exhausted soils prone to degradation by erosion, nutrient depletion, compaction, and other processes. The major soil-related challenges faced by smallholder farmers, and aggravated by the harsh and uncertain climate, are (i) low soil organic matter content and meager plant-available nutrient reserves, (ii) weak aggregation and poor structure making soils prone to crusting and compaction, (iii) low plant-available water reserves and soils prone to frequent pedological/agronomic/ecologic drought, (iv) shallow effective rooting depth, and (v) elemental imbalance (toxicity of some and deficiency of others) in the root zone. These biophysical problems are exacerbated by unfavorable human dimensions, including issues of land tenure, weak institutions (i.e., extension services), poor infrastructure, and lack of access to credit for purchase of essential inputs.

The Green Revolution of the 1970s, which saved hundreds of millions of smallholders and their dependents from starvation and malnutrition in Asia and South/Central America, has slowed in South Asia since the 1990s and entirely by-passed sub-Saharan Africa. It has become harder to bolster crop yields of smallholders by merely using the seed-based technology because the potential of improved varieties can only be realized if grown under optimal soil conditions. The economically lagging agricultural sector of resource-poor farmers can be revived only through sustainable management of soils, which is the engine of economic development.

The basic inputs of sustainable soil management, for small and large landholders, are enhancing soil structure, improving soil fertility, and increasing plant-available water capacity. Enhancing soil fertility by using chemical fertilizers is a major challenge in Africa because the price of fertilizers can be several times higher than elsewhere, and smallholders cannot afford it. In addition to the high price, resource-poor farmers are not sure about the effectiveness of fertilizers because of frequent droughts and high ambient temperatures. Thus, productivity can be enhanced through the use of biological nitrogen fixation and recycling of by-products (e.g., crop and animal residues, urban waste, compost) in conjunction with water of nutrients through adoption of conservation tillage, water harvesting, and recycling, and other measures of converting blue water to green water.

In the context of biophysical processes, sustainable soil management must replace what is removed, respond wisely to what is changed, and predict what will happen from anthropogenic and natural perturbations. Therefore, this book provides the technological, economic, social, and cultural bases of sustainable management of soils for smallholders. The objective is to promote the adoption of proven technologies of sustainable intensification, producing more from less, both for advancing agronomic production and adapting to changing climate. The strategy is to usher a soil-based Green Revolution by increasing the use efficiency of energy-based inputs (e.g., fertilizers, pesticides, irrigation), restoring soil quality, and sequestering carbon in the terrestrial ecosystems (e.g., soils and vegetation).

The editors thank all the authors for their outstanding contributions to this volume, and for sharing their knowledge and experiences with others. Despite their busy schedules and numerous commitments, authors prepared the manuscripts in a timely manner, and this is greatly appreciated. The editors also thank the editorial staff of Taylor & Francis for their help and support in publishing this volume. The office staff of the Carbon Management and Sequestration Center provided support in the flow of manuscripts between authors and editors, and made valuable contributions; their help and support are greatly appreciated. In this context, special thanks are due to Jennifer Donovan, who formatted the text and prepared the final submission. Help from Laura Hughes is thankfully acknowledged. It is a challenging task to thank, by listing the names, all those who contributed in one way or another in bringing this volume to fruition. Finally, it is important to build on the outstanding contributions of numerous soil scientists and ecologists whose research is cited throughout the book.

Rattan Lal
Bobby A. Stewart

Editors

Rattan Lal, PhD, is a distinguished university professor of soil science and the director of the Carbon Management and Sequestration Center, The Ohio State University, and an adjunct professor at the University of Iceland. After completion of his education from Punjab Agricultural University, Ludhiana (BSc, 1963); Indian Agricultural Research Institute, New Delhi (MSc, 1965); and The Ohio State University, Columbus (PhD, 1968), he served as a senior research fellow with the University of Sydney, Australia (1968–1969); soil physicist at the International Institute of Tropical Agriculture, Ibadan, Nigeria (1969–1987); and professor of soil science at The Ohio State University (1987–to date). His current research focus is on climate-resilient agriculture, soil carbon sequestration, sustainable intensification, enhancing use efficiency of agroecosystems, and sustainable management of soil resources of the tropics.

Dr. Lal is a fellow of the American Society of Agronomy, Soil Science Society of America, Third World Academy of Sciences, American Association for the Advancement of Sciences, Soil and Water Conservation Society, Indian Academy of Agricultural Sciences, and Rothamsted (UK). He received the Hugh Hammond Bennett Award of the Soil and Water Conservation Society, the 2005 Borlaug Award, and the Liebig Award (2006) of the International Union of Soil Science, the M.S. Swaminathan Award (India) of 2009, and the COMLAND Award (Germany) of 2009. He received an honorary doctor of science degrees from Punjab Agricultural University (2001), the Norwegian University of Life Sciences, Aas (2005), and Alecu Russo Balti State University, Moldova (2010). He was president of the World Association of the Soil and Water Conservation (1987–1990), the International Soil Tillage Research Organization (1988–1991), and the Soil Science Society of America (2005–2007). He was a member of the Federal Advisory Committee on National Assessment of Climate Change-NCADAC (2010–2013); member of the Strategic Environmental Research and Development Program Scientific Advisory Board of the Department of Energy (2011–); senior science advisor to the Global Soil Forum of the Institute for Advanced Sustainability Studies, Potsdam, Germany (2010–); member of the advisory board of the Joint Research Programming Initiative on Agriculture, Food Security and Climate Change of the European Council (2013–); and member of the advisory board, UNU-FLORES, Dresden, Germany (2014–2017).

Professor Lal was a lead author of the Intergovernmental Panel on Climate Change (IPCC, 1998–2000), and was awarded the Nobel Peace Prize Certificate by the IPCC in 2007, and Global Dry Land Champion of the United Nations Convention to Combat Desertification and Rothamsted Fellow in 2013. He has mentored 101 graduate students and 54 postdoctoral researchers, and hosted 120 visiting scholars. He has authored/coauthored more than 1700 research publications, including 678 refereed journal articles, has written 12 books, and edited/co-edited 56 books.

Bobby A. Stewart, PhD, is a distinguished professor of soil science at West Texas A&M University, Canyon, Texas. He is also the director of the Dryland Agriculture Institute; a former director of the US Department of Agriculture (USDA) Conservation and Production Laboratory at Bushland, Texas; past president of the Soil Science Society of America; and member of the 1990–1993 Committee on Long-Range Soil and Water Policy, National Research Council, National Academy of Sciences. He is a fellow of the Soil Science Society of America, American Society of Agronomy, and Soil and Water Conservation Society; a recipient of the USDA Superior Service Award; a recipient of the Hugh Hammond Bennett Award of the Soil and Water Conservation Society; and an honorary member of the International Union of Soil Sciences in 2008. Dr. Stewart is very supportive of education and research on dryland agriculture. The B.A. and Jane Anne Stewart Dryland Agriculture Scholarship Fund was established at West Texas A&M University in 1994 to provide scholarships for undergraduate and graduate students with a demonstrated interest in dryland agriculture.

Contributors

Jens B. Aune
Norwegian University of Life Sciences
Aas, Norway

Ermias Aynekulu
World Agroforestry Centre (ICRAF)
Nairobi, Kenya

Youlu Bai
Chinese Academy of Agricultural Sciences
Beijing, China

Roshan M. Bajracharya
Kathmandu University
Kavre, Nepal

Niels H. Batjes
ISRIC, World Soil Information
Wageningen, the Netherlands

Tracy L. Beedy
World Agroforestry Centre (ICRAF)
Lilongwe, Malawi

Prem Bindraban
Virtual Fertility Research Center (VFRC)
Washington, District of Columbia

Boris Boincean
Alecu Russo State University
Selectia Research Institute of Field Crops
Balti, Republic of Moldova

Johan Bouma
Wageningen University
Wageningen, the Netherlands

Corinna Clements
Virginia Polytechnic Institute and State University
Blacksburg, Virginia

Corey Henry Cockerill
Wilmington College
Wilmington, Ohio

Bed Mani Dahal
Kathmandu University
Kavre, Nepal

Jozef A. Deckers
Katholieke Universiteit Leuven
Heverlee, Belgium

Stefaan Dondeyne
Katholieke Universiteit Leuven
Heverlee, Belgium

Tim Ellis
International Water Management Institute
Accra, Ghana

Cornelia B. Flora
Iowa State University
Ames, Iowa

Alan J. Franzluebbers
US Department of Agriculture Agricultural Research Service
Raleigh, North Carolina

Raj Gupta
Borlaug Institute for South Asia
New Delhi, India

Kiros M. Hadgu
World Agroforestry Centre (ICRAF)
Addis Ababa, Ethiopia

Lei Han
China Agricultural University
Beijing, China

Deborah Hellums
International Fertilizer Development Center (IFDC)
Muscle Shoals, Alabama

Andrew Kabanza
Ministry of Agriculture Food Security and Cooperatives (MAFC)
Naliendele Agricultural Research Institute (NARI)
Mtwara, Tanzania

Dong-Gill Kim
Hawassa University
Hawassa, Ethiopia

Didas N. Kimaro
Sokoine University of Agriculture
Morogoro, Tanzania

Xiangbin Kong
China Agricultural University
Beijing, China

Rattan Lal
The Ohio State University
Columbus, Ohio

Hongjun Lei
North China University of Water Conservancy and Electric Power
Zhengzhou, China

Baoguo Li
China Agricultural University
Beijing, China

Kejiang Li
Institute of Dryland Farming
Ministry of Agriculture
Hengshui, China

Eike Luedeling
World Agroforestry Centre (ICRAF)
Nairobi, Kenya

Keith M. Moore
Virginia Polytechnic Institute and State University
Blacksburg, Virginia

Hildelitha B. Msita
Ilonga Agricultural Research Institute
Kilosa, Tanzania

Jerod Myers
Virginia Polytechnic Institute and State University
Blacksburg, Virginia

Ted L. Napier
The Ohio State University
Columbus, Ohio

Constance Neely
World Agroforestry Centre (ICRAF)
Nairobi, Kenya

George Nyamadzawo
Bindura University of Science and Education
Bindura, Zimbabwe

Frank Place
World Agroforestry Centre (ICRAF)
Nairobi, Kenya

Jean Poesen
Katholieke Universiteit Leuven
Heverlee, Belgium

Nani Raut
Kathmandu University
Kavre, Nepal

Amit Roy
International Fertilizer Development Center (IFDC)
Muscle Shoals, Alabama

Dil Prasad Sherchan
Nepal Agricultural Research Council
Kathmandu, Nepal

M.P.W. Sonneveld*
Wageningen University
Wageningen, the Netherlands

Jean L. Steiner
Grazinglands Research Laboratory
US Department of Agriculture Agricultural Research Service
El Reno, Oklahoma

Bobby A. Stewart
West Texas A&M University
Canyon, Texas

Rajbir Yadav
Indian Agricultural Research Institute
New Delhi, India

* Deceased.

1 Small Landholder Farming and Global Food Security

Rattan Lal

CONTENTS

1.1 INTRODUCTION

About 500 million small landholder farms (<2 ha) in the world are critical to achieving global food security and alleviating poverty. Traditionally, these farms are managed by family labor, use manual or animal-driven farm tools, practice mixed farming combining crops and animals, use few external inputs, and are subsistence farms. However, small landholders are now being linked with market, grow modern varieties, use fertilizers and pesticides, and produce high yields. Improved technologies include conservation agriculture, cover cropping, water harvesting and recycling using micro-irrigation, and soil-specific farming based on the principles of precision agriculture. The principal challenges to sustainable intensification of small landholder farmers are restoring soil organic matter content, improving soil fertility, controlling soil erosion, conserving water in the root zone, and abating climate change. Rather than subsidies, payments for ecosystem services are a viable option to promote the adoption of recommended management practices.

The annual yield increase globally was 30% in 1950 and <1% in 2001 (Kotschi 2013). High yields due to adoption of the Green Revolution technology were obtained on fertile soils with liberal use of synthetic fertilizers and pesticides, and by expansion of irrigation. Yet, intensification of agriculture, adopted in the 1960s, has been used by only a small proportion of total farms that are large scale and use commercial farming techniques. It is estimated that small landholders produce food for 70% of the total population and use only 30% of the resources. While these statistics may be debatable, the fact remains that food security and the Millennium Development Goals can only be achieved by increasing the productivity of small landholder farms. Most small landholders, farming <2 ha of arable land, are in Asia and Africa but also in the Caribbean and South and Central America (Von Braun 2005). Globally, the percentage of all small landholder farms is estimated at 85% (Nagayets 2005); however, the total land area farmed by them was 60% in 1980 and ~40% at present (Kotschi 2013). Small landholders have an important role to play in alleviating global hunger and poverty. Adoption of proven and recommended management practices (RMPs) can narrow the yield gap and triple or even quadruple the agronomic yield of cereals, which has stagnated at ~1 Mg/ha since the 1960s. Thus, the objective of this chapter is to deliberate strategies of improving and sustaining the productivity of small landholder farms, and outline potential challenges of implementing these strategies.

1.2 CHARACTERISTICS OF SMALLHOLDER FARMS

There are ~500 million small farms (<2 ha) in the world, and 87% of these are in the Asia and Pacific regions. China and India account for 193 million and 93 million of small farms, respectively (International Fund for Agriculture Development 2010; Trapa and Gaiha 2011). Most small landholders, regardless of the region where they farm, have numerous common attributes (Figure 1.1). Traditionally, several features linking small landholders from around the world included degraded and depleted soils; vulnerability to harsh and changing climate; low-risk and traditional systems based on family labor and low external input; and limited access to services, market, and credit facilities (Figure 1.1). With these characteristics, they have been in the grip of poverty, hunger, and malnutrition. Small landholders are characterized by having a small land area (0.2–2 ha) and low per capita natural resources. Their household income is related to the farm size; poor farmers often cultivate 0.4–1.0 ha of land (Tittonell et al. 2005). Most small farms practice mixed farming but have a lack of a specific forage production system. Thus, there is a strong dependency on using crop residues as fodder and on open grazing leading to a limited cycling of nutrients. Farms closer to homesteads are more fertile than those farther away because ash from kitchen and other household wastes are recycled.

Farm operations (e.g., seedbed preparation, weeding, harvesting) are performed manually or with draft animals. Thus, drudgery and hard work, under harsh climate conditions, are common features. With limited resources, however, small landholders are mostly risk adverse. The goal is to use safe practices that produce a minimum assured yield during the worst season rather than maximize yield during the best season. The food security of families is the major driver in decision making.

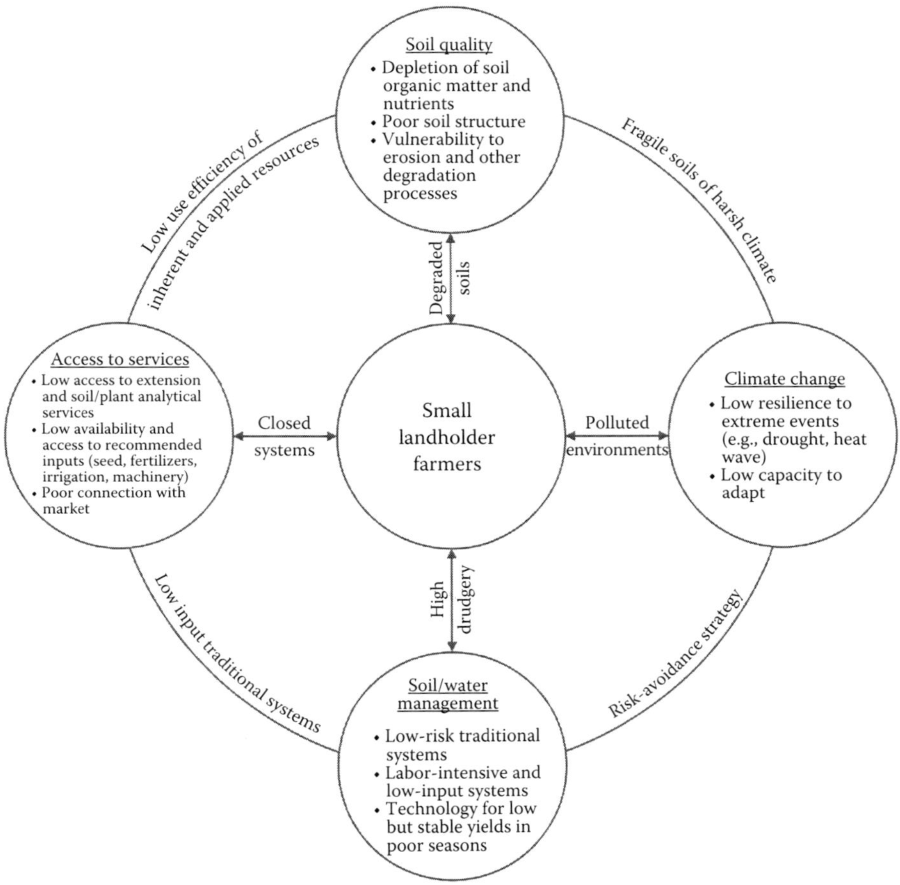

FIGURE 1.1 Characteristics of small landholder farmers of the tropics.

Furthermore, most small landholders are not connected with a market and are subsistence farmers.

Nutrient mining, leading to depletion of soil organic matter (SOM) and plant nutrients, is a widespread problem in South Asia, sub-Saharan Africa (SSA), the Caribbeans, and Andean regions. Negative nutrient budget in SSA has occurred owing to decades of extractive farming (Bekunda et al. 2010). Such nutrient mining practices have exacerbated elemental imbalances, resulting in loss of N and K in soils of southwest Mexico (Flores-Sanchez et al. 2011). On sloping lands (e.g., Andean and Central American hillsides), soil degradation, genetic erosion, and poverty are serious issues with smallholder farmers (Ashby et al. 1999).

There are several factors affecting the agronomic productivity of small landholder farms. These factors, outlined in Table 1.1, are relevant to the characteristics explained in this section, and must be addressed through scientific innovation and prudent governance. Resource-poor farmers generally occupy medium- and low-potential land of marginal productivity (Eswaran et al. 1997). However, small farms

TABLE 1.1
Factors Affecting Low Crop Yields and Agronomic Productivity of Small Landholders in Developing Countries

Factor	Region	Reference
1. Climate change (drought) and land degradation	Northern Ghana	Laube et al. (2012)
2. Soil fertility depletion and nutrient mining	Sub-Saharan Africa	Bekunda et al. (2010)
3. Lack of access and capacity to efficiently apply fertilizers	Sub-Saharan Africa	Bekunda et al. (2010)
4. Spatial variability in soil fertility	Western Kenya	Tittonell et al. (2005)
5. Imbalance of plant nutrients in soil	Southwest Mexico	Flores-Sanchez et al. (2011)
6. Drought risks	Honduras	Nieto et al. (2010)
7. Poverty and low soil quality	Honduras	Ravnborg (2002)
8. Land degradation, genetic erosion, and poverty	Columbian Highlands	Ashby et al. (1999)

are in transition. They are being linked with market and high-value chain, and are adapting to changing climate and market dynamics. Small landholders contribute strongly to agricultural production and global food security, and deserve to receive attention and support for improving production and environment quality. Small farms must not be marginalized and should be the focus of developmental programs (Hazell et al. 2006). Because most smallholders practice mixed farming, pastures can also be improved with good management.

1.3 SEEDBED PREPARATION

Fire has been a widely used tool in land clearance and seedbed preparation. Slash-and-burn agriculture has been practiced for millennia, and is still being practiced by small landholders in the tropics and subtropics. Seedbed preparation even on peatlands in Indonesia involves controlled burning. Destruction of 10 Mha of forests and nonforest lands in 1997 in Indonesia, resulting in more 2 Pg C emission into the atmosphere, was caused by the fire started on agricultural land and forestry plantations (Saharjo and Munoz 2005).

Most smallholder farmers use some tillage for seedbed preparation and believe that tillage increases crop yield. This belief is based on the short-term benefits of tillage through breaking the surface crust and loosening the soil, improving water infiltration, controlling weeds, and accelerating decomposition of SOM, which mineralizes plant nutrients and makes these available to crop. Whereas the merits of no-till (NT) farming have been documented by research experimentation for a range of soils and agroecosystems (Table 1.2), there are numerous constraints to its adoption by resource-poor farmers (Table 1.3). Lack of appropriate seeding and other

TABLE 1.2
Merits of No-Till Farming for Small Landholders of the Tropics

Merits	Region	Reference
1. Crop yields equal to or better than traditional system	South Asia	Johansen et al. (2012)
2. Reduced fuel and labor costs	South Asia	Johansen et al. (2012)
3. Reduced turn-around time between crops	South Asia	Johansen et al. (2012)
4. Increased profitability	South Asia	Johansen et al. (2012)
5. Reduced soil erosion	Vietnam	Affholder et al. (2010)
6. Improved crop nutrient and water balances	Vietnam	Affholder et al. (2010)

TABLE 1.3
Constraints to Adoption of No-Till Farming and Other Recommended Practices by Small Landholders of the Tropics

Constraint	Region	Reference
1. Shortage of mechanized options	South Asia	Johansen et al. (2012)
2. Weed management	South Asia	Johansen et al. (2012)
3. Competing uses of crop residues	South Asia	Johansen et al. (2012)
4. Availability and safe and effective use of herbicides	South Asia	Johansen et al. (2012)
5. Lack of small-scale no-till planters	South Asia	Johansen et al. (2012)
6. Extra labor and input needed during the first year	Vietnam	Affholder et al. (2010)
7. Constraints at the community level	Vietnam	Affholder et al. (2010)
8. Education status of farmers	China	Kong et al. (2002)
9. Lack of support services and facilities	Nepal	Paudel and Thapa (2001)
10. Fragmented small pieces of land	Western Kenya	Tittonell et al. (2005)
11. Limited capacity of resource-poor farmers	Andean Highlands	Fonte et al. (2012)
12. Lack of specific animal fodder systems, animal grazing within communal areas, limited recycling	Southwest Mexico	Flores-Sanchez et al. (2011)
13. Coarse-textured "Terra rossa" soils	Paraguay	Kubota et al. (2005)
14. Acidic soils, low fertility, competition by cover crops	Colombian hillside	Daellenbach et al. (2005)
15. Insufficient extension, economic constraints	Panama	Fischer and Vasseur (2002)
16. Spatially variable fields	Kenya	Booltink et al. (2001)
17. High value of residues of fodder	Mexico	Fischer et al. (2002)

machinery for NT farming is a major obstacle. Use of new equipment being developed can produce crop yields equal to or better than plow tillage (Johansen et al. 2012). Furthermore, appropriate cropping systems are being tested and promoted for NT farming (Affholder et al. 2010). Despite numerous uses of NT farming, it may not be applicable under all soils and cropping systems. Thus, a soil guide is needed to advise small landholders about the site-specific tillage and other methods of seedbed preparation (Lal 1985). In eastern Paraguay, Kubota et al. (2005) observed that the NT system might not be suitable for coarse-textured Terra rossa soils.

1.4 DROUGHT STRESS

Crop yields in rainfed farming is constrained by drought stress. Droughts are disruptive to the development process, and adverse impacts of drought are exacerbated by the reluctance of small landholders to invest in water management options (Nieto et al. 2010). Expensive modern water conservation and management technologies are unlikely to be adopted by the poorest of the poor. Appropriate technologies must be cost-effective, risk reducing, and productivity enhancing over a short term (Ellis-Jones and Mason 1999). Thus, adoption of soil water conservation measures can improve water productivity (Shaheen et al. 2011). Farmer-driven small-scale irrigation can be important in alleviating agronomic/pedologic droughts (Laube et al. 2012). Flexible systems of N fertilization, with variable rate according to the current seasonal rainfall pattern, can improve productivity in semiarid areas (Masvaya et al. 2010). Drought stress can also be managed by adoption of the Zai-type pitting system, which originated in drier parts of West Africa (Bekunda et al. 2010). This traditional technique integrates harvesting of water and applying nutrients/manure through a traditional form of precision agriculture. In addition, affordable drip irrigation system (i.e., bucket kit for home gardens) is being tried to meet the needs of small landholders and poor farmers in SSA, Asia, and Latin America (Postel et al. 2001). In addition to lack of water, agronomic productivity of dryland farming in semiarid tropics is also constrained by low soil fertility (Wani et al. 2007). Thus, judicious strategies of enhancing soil fertility can also improve water productivity.

1.5 SOIL FERTILITY MANAGEMENT

Small landholders use little fertilizers, although they are knowledgeable about soil fertility–enhancing technologies being promoted (Enyong et al. 1999). Some of the factors and techniques are beyond the reach of resource-poor farmers. Furthermore, high levels of risks in drought-prone environments along with high fertilizer prices are important factors precluding poor farmers from using chemical fertilizers (Shiferaw et al. 2004). Therefore, nutrient recycling and the use of biological N fixation (BNF) are important in managing soil fertility in regions such as South and Southeast Asia, SSA, the Caribbeans, and Central America. Thus, efficient management of nutrients in soils and manure is a key factor in crop production. Most small landholders manage soil fertility by recycling nutrients through application of animal and plant wastes. Farmyard manure (FYM) is commonly the only input available (Rufino et al. 2007).

Among several strategies of managing soil fertility are effective erosion control, alternative nutrient sources, and nutrient cycling (Fonte et al. 2012). In general, soil fertility decreases with increase in distance from the homestead (Zingore et al. 2008). All forms of organic wastes are considered as "organic treasures" and recycled as organic fertilizers (Yang 2006). Yet, a judicious use of inorganic fertilizers can be important in improving soil fertility (Masvaya et al. 2010). Soil fertility management technologies on small landholder farms in SSA include farming systems based on the combined use of BNF, recycling of organic wastes, and judicious use of mineral fertilizers (Bekunda et al. 2010; Chikowo et al. 2010; Kanyama-Phiri et al. 1998).

Green manure (with *Sesbania*) has been found useful in improving the lability of soil organic C (SOC) pool compared with FYM and crop residues (Verma et al. 2010). With long-term application and at high rates, use of organic manures can improve soil quality and agronomic productivity (Datta et al. 2010). Cover cropping and mixed cropping (cassava and cover crops) techniques are used to enhance fertility of Colombian hillside soils of low fertility (Daellenbach et al. 2005). In sandy loam soils of the humid tropics, agroforestry (alley cropping) systems could effectively substitute slash-and-burn systems by adding up to 10 Mg/ha of C in the litter layer (Aguiar et al. 2009). Similar to other options of soil management, agroforestry techniques are also site specific in their relevance, performance, and farmer acceptability (Cooper et al. 1996). Use of leguminous trees and cover crops can enhance BNF so that plants can utilize the N_2 available in air (Shiferaw et al. 2004). A study on smallholder perception of agroforestry conducted in Panama by Fischer and Vasseur (2002) indicated that among obstacles to adoption of agroforestry techniques are insufficient extension, inappropriate project design or management (e.g., top–down approach), economic constraints, and policy issues.

Furthermore, the processes of nutrient depletion and soil fertility decline are spatially heterogeneous (Tittonell et al. 2005). This spatial heterogeneity is caused by biophysical and socioeconomic factors. Precision agriculture or soil-specific farming can be used for spatial and temporal optimization of fertilizers and other input (Booltink et al. 2001). Use of precision agriculture enables farmers to use different management practices within a single variable field. Such technology can be used even in small farms such as those in Kenya and Costa Rica (Booltink et al. 2001).

1.6 EROSION CONTROL

Accelerated soil erosion has plagued small landholders, who often cultivate sloping and marginal lands. Up-and-down cultivation and mechanical seedbed preparation exacerbate the erosion hazard. Thus, soil erosion is a serious hazard, especially in erosional hotspots (e.g., the Loess Plateau in Northwest China, the Himalayan region, West Africa/Sahel, East African highlands, Andean region, and the Caribbeans). Accelerated erosion depletes SOC concentration and plant nutrients, truncates the topsoil, and wastes water as surface runoff. Conventional agriculture and extractive farming can exacerbate erosion in erodible lands (Kong et al. 2002). Yet, intensification because of high demographic pressure can reduce erosion (Tiffen et al. 1994; Sahrawat et al. 2010). Integrated use of soil and water conservation practices in conjunction with balanced plant nutrients on a watershed scale can enhance soil quality, improve surface water and groundwater quality, and increase agronomic productivity in the semiarid tropics (Sahrawat et al. 2010). Agroforestry techniques, adding innovation to traditional farming systems (e.g., combination of trees and crops, improved fallow using perennial legumes such as pigeon pea [*Cajanus cajan*], and regeneration of native species) can enhance soil fertility and reduce risks of soil erosion (Rousselet-Gadenne 2004). Conversion of tillage-based systems of seedbed preparation to NT farming is a proven strategy of erosion control (Lal 1976). In conjunction with cover cropping, use of NT techniques is a conservation-effective measure. There are numerous merits and co-benefits of using NT farming (Table 1.1), especially for water conservation and reducing the risks of drought. However, there are also numerous constraints to the adoption of NT farming by small landholders (Table 1.3). These constraints must be objectively addressed.

1.7 INDIGENOUS KNOWLEDGE

Developing on the indigenous knowledge of farmers about soils and farming practices may be critical to the success of agricultural development by small landholders. Thus, strategic combinations of traditional and modern technologies can increase agronomic production (Bekunda et al. 2010). In Laos, most farmers distinguish soils in their field on the basis of color, texture, weed infestation, etc. (Saito et al. 2006). Thus, crops are seeded on the basis of soil characteristics. Black soils, apparently of high fertility, are ranked high in preference and crop yields are generally high in black soils. There are several parameters specifically used by small landholders as indicators of soil quality (Table 1.4). Scientists should understand these parameters and quantify them so that farmers can relate to the scientific date. Examples of traditional technologies include N fixation by indigenous and introduced legumes, and fertilizer trees. Thus, crop allocation on the home fields is most diversified because of a favorable level of SOM and plant nutrients on these soils (Zingore et al. 2007), which receive kitchen ash and other household waste.

Among traditional systems of farming used by small landholders are floating agriculture and soilless systems developed in Meso-America, and South and Southeast Asia (Pantanella et al. 2011). Floating agriculture is a low-technology production system with no use of chemical fertilizer, and with complete recycling of nutrients leached into the lake by fish and other aquatic life.

TABLE 1.4
Perception of Soil Quality Parameters by Traditional Small Landholder Farmers

Soil Quality Parameter	Region/Country	Reference
1. Soil color; black soils are fertile	Laos	Saito et al. (2006)
2. Soil fertility by organic manures	China	Yang (2006)
3. Fat soil vs. thin soil	Haiti	Lal et al. (2013)

1.8 ADAPTATION TO CLIMATE CHANGE

Crop yields on small landholder farms are vulnerable to climate change and the attendant soil degradation. Small landholders have a limited capacity to adapt. Thus, crop production may decline in the tropics even with a moderate increase in temperature (Long et al. 2005). Furthermore, the fertilization effect of CO_2 may be less than projected because of the increasing risks of drought and deficiency of some essential nutrients. Lack of access to irrigation water or the means to use water effectively and efficiently are the key factors (Postel et al. 2001). Furthermore, nations with large natural resources will fare better in adapting to climate change than those in developing countries (Easterling and Apps 2005) where the majority of the resource-poor small landholders are located. For example, small farmers in the high Andes (e.g., Bolivia, Ecuador, and Peru) are prone to climate change and the potential decline in productivity (Fonte et al. 2012). Similarly, farmers in Sahel and South Asia are vulnerable to climate change and uncertainties. In northern Ghana, farmers are diversifying farming to adapt to variable and uncertain climate (Laube et al. 2012). Adaptation measures include use of shallow groundwater irrigation of vegetables for urban market. Adaptation to climate change must be linked to achieving food security in developing countries. For example, transformation of low-productivity croplands to sequential agroforestry can triple system C stocks in 20 years (Sanchez 2000) and offset anthropogenic emissions.

1.9 MANAGEMENT OF SOIL ORGANIC MATTER

SOM is the elixir of all terrestrial life. Yet, soils managed by small landholders are severely depleted of their SOC pool because of the use of extractive farming practices for a long time. Thus, SOC concentration must be restored to above the threshold level of ~1.5% in the rootzone. Indeed, SOC concentration and stock are important determinants of soil quality (Craswell and Lefroy 2001; Palm et al. 2001). Therefore, one of the biggest challenges in warm tropical climates is identification and use of technologies to restore SOM concentration in small landholder farms. Appropriate SOM-enhancing technologies, for small landholders in SSA and elsewhere, must be based on three characteristics (Snapp et al. 1998): (i) residue quality, (ii) importance of deep-rooted species, and (iii) trade-offs between legumes grown for food vs. those established for improving soil quality. In general, species that combine some grain yield with high above- and below-ground biomass (i.e., low N harvest) are useful for the dual goal of achieving food security and improving soil fertility and SOC concentration (Snapp et al. 1998).

In some cases, crop yields can be strongly correlated with SOC concentration in the root zone (Zingore et al. 2008). However, SOC cannot always be used as an index of N supply under all conditions, and the correlation between SOC and N supply may be poor (Cassman et al. 1996). Thus, better understanding of processes governing N dynamic in soils of the tropics is needed toward the development of sustainable crop management practices. Indeed, nutrient input use is the main factor limiting productivity of cereals in SSA (Cobo et al. 2009). In addition to plant nutrients, soil quality indicators must be broad based and include variables such as SOC saturation deficit, compaction, surface scale up, and other biological attributes (Sanchez et al. 2003).

1.10 FOOD SECURITY

Despite impressive gains in crop yields since the 1960s, agronomic production in some regions with predominately small farmers has lagged behind the demand (e.g., SSA). Thus, food insecurity remains to be a problem (Kungu 2007), and about a billion people (one in seven) are food insecure (FAO 2013). Most chronically hungry and malnourished people are poor farm families—small landholders who do not have adequate resources to invest in modern innovations. Resource-poor farmers in SSA have lagged behind in agricultural development (Eswaran et al. 1997).

There is no universal panacea, and there are wide ranges of generic technological options that must be adopted and fine-tuned under site-specific conditions (Table 1.5).

TABLE 1.5
Technological Options for Adaptations to Changing Climate and Alleviation of Soil-Related Constraints by Small Landholders of the Tropics

Technology	Location/Region	Reference
1. Diversification and intensification of small-scale irrigation	Northern Ghana	Laube et al. (2012)
2. Using manure, compost, and cover crops, and varying N input according to seasonal rainfall	Sub-Saharan Africa	Masvaya et al. (2010)
3. Integrated soil fertility management	Sub-Saharan Africa	Bekunda et al. (2010), Cobo et al. (2009), Snapp et al. (1998)
4. Water harvesting by Zai-type pitting	Sub-Saharan Africa	Bekunda et al. (2010), Cobo et al. (2009), Snapp et al. (1998)
5. Manure management	Africa	Rufino et al. (2007)
6. An improved understanding of soil fertility variability/gradient and the nutrient management	Zimbabwe	Zingore et al. (2007)
7. Drip irrigation	Tropics	Postel et al. (2001)
8. Improved management of organic residues with low level or inorganic fertilizer input	Andean Highlands	Fonte et al. (2012)
9. Biological N fixation	Southeast Asia	Shiferaw et al. (2004)
10. Symbiotic N fixation	Tropics	Shiferaw et al. (2004)
11. Preserving soil organic matter	Colombia	Binder and Patzel (2001)
12. N-fixing legumes in agroforestry systems	Brazilian Amazon	McGrath et al. (2000)
13. Risk-reducing options	Andean Bolivia	Ellis-Jones and Mason (1999)
14. Precision farming	Africa	Booltink et al. (2001)
15. Integrated use of soil and water	Semiarid tropics	Zingore et al. (2008)
16. Alley cropping/agroforestry	Humid tropics	Aguiar et al. (2009), Cooper et al. (1996), Sanchez (2000)
17. Soil organic matter management	Tropics	Craswell and Lefroy (2001), Palm et al. (2001), Cassman et al. (1996)

These include conservation agriculture based on NT farming and crop residue mulch, including cover cropping, agroforestry, rainwater harvesting and recycling by drip subirrigation, and integrated nutrient management involving judicious use of biofertilizers and inorganic fertilizers (Winterbottom et al. 2013). Each of these options has merits and trade-offs, and the latter must be critically considered under site-specific situations. Precision farming is an important option to minimize losses and optimize efficiency of inputs.

Any food security initiatives must be integrated with those of improving environment quality. In this context, small-scale farms are important to global food production and well being of a large population, and for improving the environment. Therefore, small landholders must be given the priority they deserve for sustainable development. Through appropriate programs, small landholders can also be symbols of modernity and economic efficiency. Sustainable development of small farms can be realized by linking with market, information, and education network; soil reclamation; and adoption of modern innovations (Bainville et al. 2005). Given opportunities, there is a large resilience and capacity in small landholders to evolve into loci of economic development and environmental improvement. Furthermore, there is a growing consensus that poverty is not necessarily a major cause of environmental degradation (Ravnborg 2003).

1.11 INFORMATION AND COMMUNICATION TECHNOLOGY

Access to essential information (e.g., technology, weather, market) is important to reducing risks and improving productivity of small landholders. The three cornerstones of information are quality, timeliness, and trustworthiness (Mittal and Tripathi 2009). Mobile phones can play a critical role in promoting adoption of BMPs through reduction of transition costs, search costs and saving of time, and travel costs (Mittal and Tripathi 2009).

1.12 PAYMENTS FOR ENVIRONMENTAL SERVICES

Changing fertilizer and poor governance regarding the use of crop residues as animal fodder, animal dung as cooking fuel, and uncontrolled/open grazing are crucial issues of managing soil fertility and restoring depleted SOC pool and nutrient reserves. These constraints have perpetuated the use of extractive farming practices, and hindered the adoption of RMPs. Adoption of RMPs by resource-poor farmers can be promoted through payments for ecosystem/environmental services (PESs) for C sequestration, biodiversity improvement, water quality enhancement, reduction of CH_4 and N_2O emissions from livestock, etc. (Lal et al. 2013). In Costa Rica, Cole (2010) reported that the PES program was effective in overcoming some economic and technical constraints to adoption of agroforestry. Effective policies must be implemented that support efficient distribution of fertilizers along with adoption of RMP (Cobo et al. 2009) through PESs to avoid future soil degradation and improve food security. Furthermore, incentives through PESs can be used to alleviate soil-related and other constraints to realize the potential of small landholders to advance food security while harnessing global environmental services. It is important to

TABLE 1.6
Promoting Adoption of Improved Technology and Rewarding Farmers for Environmental Services

Technology	Impact	Region/County	Reference
Agroforestry	C sequestration, biodiversity	Humid tropics	Cole (2010), Aguiar et al. (2009), Fischer and Vasseur (2002)
Organic matter management	C sequestration	Semiarid tropics (India)	Wani et al. (2007), Datta et al. (2010), Verma et al. (2010)
Effective policy supporting an efficient fertilizer distribution	Enhanced soil fertility	Zimbabwe	Cobo et al. (2009)
Clearer rights to land	Investment in soil restoration	Zimbabwe	Cobo et al. (2009)
Improved tree fellows	Enhance soil fertility	Southern Africa	Ajayi et al. (2009)
BNF	Improved N management	Tropics	Shiferaw et al. (2004)
Improved agronomy	Erosion control and fertility management	Tropics	Fischer et al. (2002)

align smallholder farmer's incentives with those of society (e.g., water quality, global warming, biodiversity) through payments for ecosystem services provisioned by small landholders. Technological options, which can be promoted through PESs, are outlined in Table 1.6. Ecosystem services provisioned through adoption of these technologies affect the world community and are of global relevance. Thus, rewarding farmers through fair and just price is critical to minimize the risks of the tragedy of commons.

1.13 CONCLUSIONS

Increase in agricultural productivity of small landholders is essential in alleviating global hunger and poverty. In rural areas of Asia, Africa, Central America, and the Caribbeans, there are a few alternatives to farming except the agroindustries, which may evolve over time. Agricultural development efforts (e.g., World Bank, United States Agency for International Development, and Gates Foundation) must give a high priority to small farms. Emerging technologies, which require more resources and input, must be promoted among small landholders so that they are not marginalized. Therefore, policy implications must focus on improving small farms. Payments for ecosystem services are one such option. RMPs for small landholders include conservation agriculture based on NT farming and mulching, cover cropping, integrated nutrient management, agroforestry, and water harvesting and recycling. Improving soil fertility and increasing irrigable land area are important strategies of adapting to changing and uncertain climate.

REFERENCES

Affholder, F., D. Jourdain, D. D. Quang, P. Tuong, M. Morize, A. Ricome. 2010. Constraints to farmers' adoption of direct-seeding mulch-based cropping systems: A farm scale modeling approach applied to the mountainous slopes of Vietnam. *Agricultural Systems* 103(1): 51–62.

Aguiar, A. D. F., A. P. Amorim, K. P. Coêlho, E. G. D. Moura. 2009. Environmental and agricultural benefits of a management system designed for sandy loam soils of the humid tropics. *Revista Brasileira De Ciencia Do Solo* 33(5): 1473–1480.

Ajayi, O. C., F. K. Akinnifesi, G. Sileshi, S. Chakeredza, S. Mng'omba. 2009. Integrating food security and agri-environmental quality in Southern Africa: Implications for policy. In *Environment and Health in Sub-Saharan Africa: Managing an Emerging Crisis*, eds. I. N. Luginaah, E. K. Yanful, pp. 39–49. Netherlands: Springer.

Ashby, J. A., J. I. Sanz, E. B. Knapp, A. Imbach. 1999. CIAT's research on hillside environments in Central America. *Mountain Research and Development* 19(3): 241–250.

Bainville, S., F. Affholder, M. Figuié, J. S. M. Netto. 2005. Changes in small-scale farms of Silvânia: A small agrarian revolution in the Brazilian cerrados. *Cahiers Agricultures* 141(1): 103–110.

Bekunda, B., N. Sanginga, P. L. Woomer. 2010. Restoring soil fertility in sub-Sahara Africa. *Advances in Agronomy* 108: 183–236.

Binder, C., N. Patzel. 2001. Preserving tropical soil organic matter at watershed level. A possible contribution of urban organic wastes. *Nutrient Cycling in Agroecosystems* 61(1–2): 171–181.

Booltink, H. W. G., B. J. van Alphen, W. D. Batchelor, J. O. Paz, J. J. Stoorvogel, R. Vargas. 2001. Tools for optimizing management of spatially-variable fields. *Agricultural Systems* 70(2–3): 445–476.

Cassman, K. G., A. Dobermann et al. 1996. Soil organic matter and the indigenous nitrogen supply of intensive irrigated rice systems in the tropics. *Plant and Soil* 182(2): 267–278.

Chikowo, K., M. Corbeels, P. Mapfumo, P. Tittonell, B. Vanlauwe, K. E. Giller. 2010. Nitrogen and phosphorus capture and recovery efficiencies, and crop responses to a range of soil fertility management strategies in sub-Saharan Africa. *Nutrient Cycling in Agroecosystems* 88: 111–120.

Cobo, J. G., G. Dercon et al. 2009. Cropping strategies, soil fertility investment and land management practices by smallholder farmers in communal and resettlement areas in Zimbabwe. *Land Degradation and Development* 20(5): 492–508.

Cole, R. J. 2010. Social and environmental impacts of payments for environmental services for agroforestry on small-scale farms in southern Costa Rica. *International Journal of Sustainable Development and World Ecology* 17(3): 208–216.

Cooper, P. J. M., R. R. B. Leakey, M. R. Rao, L. Reynolds. 1996. Agroforestry and the mitigation of land degradation in the humid and sub-humid tropics of Africa. *Experimental Agriculture* 32(3): 235–290.

Craswell, E. T., R. D. B. Lefroy. 2001. The role and function of organic matter in tropical soils. *Nutrient Cycling in Agroecosystems* 61(1–2): 7–18.

Daellenbach, G. C., P. C. Kerridge, M. S. Wolfe, E. Frossard, M. R. Finckh. 2005. Plant productivity in cassava-based mixed cropping systems in Colombian hillside farms. *Agriculture Ecosystems and Environment* 105(4): 595–614.

Datta, S. P., R. K. Rattan, S. Chandra. 2010. Labile soil organic carbon, soil fertility, and crop productivity as influenced by manure and mineral fertilizers in the tropics. *Journal of Plant Nutrition and Soil Science* 173(5): 715–726.

Easterling, W., M. Apps. 2005. Assessing the consequences of climate change for food and forest resources: A view from the IPCC. *Climate Change* 70(1–2): 165–189.

Ellis-Jones, J., T. Mason. 1999. Livelihood strategies and assets of small farmers in the evaluation of soil and water management practices in the temperate inter-Andean valleys of Bolivia. *Mountain Research and Development* 19(3): 221–234.

Enyong, L. A., S. K. Debrah, A. Bationo. 1999. Farmers' perceptions and attitudes towards introduced soil-fertility enhancing technologies in western Africa. *Nutrient Cycling in Agroecosystems* 53(2): 177–187.

Eswaran, H., R. Almaraz, E. van den Berg, P. Reich. 1997. An assessment of the soil resources of Africa in relation to productivity. *Geoderma* 77(1): 1–18.

FAO. 2013. *The State of Food Insecurity in the World: The Multiple Dimensions of Food Security*. FAO, Rome, Italy, 52 pp.

Fischer, A., L. Vasseur. 2002. Smallholder perceptions of agroforestry projects in Panama. *Agroforestry Systems* 54(2): 103–113.

Fischer, R. A., F. Santiveri, I. R. Vidal. 2002. Crop rotation, tillage and crop residue management for wheat and maize in the sub-humid tropical highlands II. Maize and system performance. *Field Crops Research* 79(2–3): 123–137.

Flores-Sanchez, D., J. K. Koerkamp-Rabelista, H. Navarro-Garza et al. 2011. Diagnosis for ecological intensification of maize-based smallholder farming systems in the Costa Chica, Mexico. *Nutrient Cycling in Agroecosystems* 91(2): 185–205.

Fonte, S. J., S. J. Vanek, P. Oyarzun et al. 2012. Pathways to agroecological intensification of soil fertility management by smallholder farmers in the Andean highlands. In *Advances in Agronomy*, Vol. 116, ed. D. L. Sparks, 125–184. Waltham, MA: Academic Press.

Hazell, P., C. Poulton, S. Wiggins, A. Dorward. 2006. The future of small farming: Synthesis paper. RIMISP 20th Anniversary Paper. IDRC, Canada.

IFAD. 2012. *Sustainable Smallholder Agriculture: Feeding the World, Protecting the Planet*. International Fund for Agricultural Development, Rome, Italy, 12 pp.

Johansen, C., M. E. Haqueb, R. W. Bellc, C. Thierfelderd, R. J. Esdailee. 2012. Conservation agriculture for small holder rainfed farming: Opportunities and constraints of new mechanized seeding systems. *Field Crops Research* 132: 18–32.

Kanyama-Phiri, G., S. Snapp, S. Minae. 1998. Partnership with Malawian farmers to develop organic matter technologies. *Outlook on Agriculture* 273: 167–175.

Kong, Z. H., X. S. Zhang, G. S. Zhou. 2002. Agricultural sustainability in a sensitive environment—A case analysis of Loess Plateau in China. *Journal of Environmental Sciences – China* 14(3): 357–366.

Kotschi, J. 2013. A soiled reputation: Adverse effects of mineral fertilizers on tropical agriculture. Heinrich Böll Stiftung, WWF, 58 pp. Accessed December 11, 2013. Available at http://www.boell.de/sites/default/files/WWF_Mineralduenger_englisch_WEB.pdf.

Kubota, A., J. Bordon, K. Hoshiba, T. Horita, K. Ogawa. 2005. Change in physical properties of "Terra rossa" soils in Paraguay under no-tillage. *Soil Science Society of America Journal* 69(5): 1448–1454.

Kungu, J. B. 2007. Food security in Africa: The challenges of researchers in the 21st century. In *Advances in Integrated Soil Fertility Management in Sub-Saharan Africa: Challenges and Opportunities*, eds. A. Bationo, B. Waswa, J. Kihara, J. Kimetu, 105–113. Netherlands: Springer.

Lal, R. 1976. No-tillage effects on soil properties under different crops in Western Nigeria. *Soil Science Society of America Journal* 40(5):762–768.

Lal, R. 1985. A soil suitability guide for different tillage systems in the tropics. *Soil and Tillage Research* 5:179–196.

Lal, R., A. Spring, R. M. Welch. 2013. *Sustainable Intensification Innovation Lab; Prospects and Opportunities*. US AID, Washington, D.C.

Laube, W., B. Schraven, M. Awo. 2012. Smallholder adaptation to climate change: Dynamics and limits in Northern Ghana. *Climatic Change* 111(3–4): 753–774.

Long, S. P., E. A. Ainsworth, A. D. B. Leakey, P. B. Morgan. 2005. Global food insecurity. Treatment of major food crops with elevated carbon dioxide or ozone under large-scale fully open-air conditions suggests recent models may have overestimated future yields. *Philosophical Transactions of the Royal Society B-Biological Sciences* 360(1463): 2011–2020.

Masvaya, E. N., J. Nyamangara, R. W. Nyawasha, S. Zingore, R. J. Delve, K. E. Giller. 2010. Nitrogen and phosphorus capture and recovery efficiencies, and crop responses to a range of soil fertility management strategies in sub-Saharan Africa. *Nutrient Cycling in Agroecosystems* 88(1): 111–120.

McGrath, D. A., M. L. Duryea, N. B. Comerford, W. P. Cropper. 2000. Nitrogen and phosphorus cycling in an Amazonian agroforest eight years following forest conversion *Ecological Applications* 10(6): 1633–1647.

Mittal, S., G. Tripathi. 2009. Role of mobile phone technology in improving mall farm productivity. *Agriculture Economics Resource Review* 22: 451–459.

Nagayets, O. 2005. Small farms: Current status and key trends. Information Brief. Prepared for the Future of Small Farms Research Workshop, Wye College, June 26–29. Accessed December 11, 2013. Available at http://citeseerx.ist.psu.edu/viewdoc/download?doi=10.1.1.146.4632&rep=rep1&type=pdf.

Nieto, J. D., S. E. Cook, P. LäDerach, M. J. Fisher, P. G. Jones. 2010. Rainfall index insurance to help smallholder farmers manage drought risk. *Climate and Development* 2(3): 233–247.

Palm, C. A., K. E. Giller, P. L. Mafongoya, M. J. Swift. 2001. Management of organic matter in the tropics: Translating theory into practice. *Nutrient Cycling in Agroecosystems* 61(1–2): 63–75.

Pantanella, E., M. Cardarelli, P. P. Danieli, A. McNiven, G. Colla. 2011. Integrated agriculture—Floating agriculture: Is it a valid strategy to raise livelihood? *Acta Horticulture* 921: 79–86.

Paudel, G. S., G. B. Thapa. 2001. Changing farmers' land management practices in the hills of Nepal. *Environmental Management* 28(6): 789–803.

Postel, S., P. Polak, F. Gonzales, J. Keller. 2001. Drip irrigation for small farmers—A new initiative to alleviate hunger and poverty. *Water International* 26(1): 3–13.

Ravnborg, H. M. 2002. Poverty and soil management—Relationships from three Honduran watersheds. *Society and Natural Resources* 15(6): 523–539.

Ravnborg, H. M. 2003. Poverty and environmental degradation in the Nicaraguan hillsides. *World Development* 31(11): 1933–1946.

Rousselet-Gadenne, A. 2004. Adoption of agroforestry innovations in Barra do Turvo São Paulo, Brazil. *Cahiers Agricultures* 13(5): 391–402.

Rufino, M. C., P. Tittonell, M. T. van Wijk et al. 2007. Manure as a key resource within smallholder farming systems: Analyzing farm-scale nutrient cycling efficiencies with the NUANCES framework. *Livestock Science* 112(3): 273–287.

Saharjo, B. H., C. P. Munoz. 2005. Controlled burning in peat lands owned by small farmers: A case study in land preparation. *Wetlands Ecology and Management* 13(1): 105–110.

Sahrawat, K. L., S. P. Wani, P. Pathak, T. J. Rego. 2010. Managing natural resources of watersheds in the semi-arid tropics for improved soil and water quality: A review. *Agricultural Water Management* 97(3): 375–381.

Saito, K., B. Linquist, B. Keobualapha, T. Shiraiwa, T. Horie. 2006. Farmers' knowledge of soils in relation to cropping practices: A case study of farmers in upland rice based slash-and-burn systems of northern Laos. *Geoderma* 136(1–2): 64–74.

Sanchez, P. A. 2000. Linking climate change research with food security and poverty reduction in the tropics. *Agriculture Ecosystems and Environment* 82(1–3): 371–383.

Sanchez, P. A., C. A. Palm, S. W. Buol. 2003. Fertility capability soil classification: A tool to help assess soil quality in the tropics. *Geoderma* 114(3–4): 157–185.

Shaheen, A., M. A. Naeem, G. Jilani, M. Shafiq. 2011. Restoring the land productivity of eroded land through soil water conservation and improved fertilizer application on Pothwar Plateau in Punjab Province, Pakistan. *Plant Production Science* 14(2): 196–201.

Shiferaw, B., M. C. S. Bantilan, R. Serraj. 2004. Harnessing the potentials of BNF for poor farmers: Technological, policy and institutional constraints and research needs. In *Symbiotic Nitrogen Fixation: Prospects for Enhanced Application in Tropical Agriculture*, ed. R. Serraji, 3–27. London: Taylor and Francis.

Snapp, S. S., P. L. Mafongoya, S. Waddington. 1998. Organic matter technologies for integrated nutrient management in smallholder cropping systems of southern Africa. *Agriculture Ecosystems and Environment* 71(1–3): 185–200.

Tiffen, M., M. Mortimore, F. Gichuki. 1994. *More People, Less Erosion: Environmental Recovery in Kenya*, 326 pp. Nairobi, Kenya: ACTS Press.

Tittonell, P., B. Vanlauwe, P. A. Leffelaar, K. D. Shepherd, K. E. Giller. 2005. Exploring diversity in soil fertility management of smallholder farms in western Kenya—II. Within-farm variability in resource allocation, nutrient flows and soil fertility status. *Agriculture Ecosystems and Environment* 110(3–4): 166–184.

Trapa, G., R. Gaiha. 2011. *Smallholder Farming in Asia and the Pacific: Challenges and Opportunities*. Rome, Italy: International Fund for Agriculture Development (IFAD).

Verma, B. C., S. P. Datta, R. K. Rattan, A. K. Singh. 2010. Monitoring changes in soil organic carbon pools, nitrogen, phosphorus, and sulfur under different agricultural management practices in the tropics. *Environmental Monitoring and Assessment* 171(1–4): 579–593.

Von Braun, J. 2005. Small-scale farmers in a liberalized trade environment. In *Small-Scale Farmers in a Liberalized Trade Environment*, eds. H. T. J. Kola, T. Lundström. Proceedings of the Seminar, Haikko, Finland, October 18–19. Department of Economics and Management Publication No. 38. Finland: University of Helsinki. Accessed December 11, 2013. Available at https://helda.helsinki.fi/bitstream/handle/1975/683/Pub38.pdf?sequence=1#page=23.

Wani, S. P., K. L. Sahrawata, T. K. Sreedevia, T. Bhattacharyyab, Ch. Sreenivasa Rao. 2007. Carbon sequestration in the semi-arid tropics for improving livelihoods. *International Journal of Environmental Studies* 64(6): 719–727.

Winterbottom, R., C. Reij, D. Garrity et al. 2013. *Improving Land and Water Management*. World Resources Institute Working Paper, Installment 5 of Creating a Sustainable Food Future, Washington, DC, 43 pp.

Yang, H. S. 2006. Resource management, soil fertility and sustainable crop production: Experiences of China. *Agriculture, Ecosystems and Environment* 116(1–2): 27–33.

Zingore, S., H. K. Murwira, R. J. Delve, K. E. Giller. 2008. Variable grain legume yields, responses to phosphorus and rotational effects on maize across soil fertility gradients on African smallholder farms. *Nutrient Cycling in Agroecosystems* 80(1): 1–18.

Zingore, S., H. K. Murwira, R. J. Delve, K. E. Giller. 2007. Influence of nutrient management strategies on variability of soil fertility, crop yields and nutrient balances on smallholder farms in Zimbabwe. *Agriculture Ecosystems and Environment* 119(1–2): 112–126.

2 Enhancing Soil Security for Smallholder Agriculture

Johan Bouma, Niels H. Batjes, M.P.W. Sonneveld, and Prem Bindraban*

CONTENTS

---

* Deceased.

2.1 INTRODUCTION

2.1.1 Broader Picture

Four-fifths of the developing world's food is produced on about half a billion small farms, supporting more than at least one billion people (International Fund for Agricultural Development [IFAD] 2012). Smallholder farmers live and earn their livelihoods in the world's most ecologically and climatically vulnerable landscapes—hillsides, drylands, and floodplains—and rely on weather-dependent natural resources. They are at the forefront of the world's efforts to deal with climate change and environmental degradation. These women, men, and young people face enormous difficulties. Yet they are themselves among the poorest and least food-secure people on Earth. In developed countries, smallholder farmers cannot compete with large commercial farms producing bulk products and have to focus on "niches" in the urban market, increasingly being framed in terms of ecosystem services that go beyond the production of food.

The term "smallholder agriculture" covers a wide variety of farming systems in different parts of the world. In many developing countries, small farms are often less than a few hectares in size, supporting only the farming family. Only somewhat larger farms may be able to grow a limited amount of cash crops for the local market. In many developed countries, farm sizes have increased strongly during the last decades owing to mechanization and rationalization of production; however, many small farms still exist and increasingly service "niche" markets aimed at urban consumers. Recent reviews (e.g., IFAD 2012) emphasize the enormous diversity of small farming systems in developing countries, requiring a focus on local conditions when formulating desirable future developments, as generalizations may be meaningless. The same diversity applies to small farms in developed countries. Smallholder agriculture will be illustrated in the following by a number of case studies, and their selection has therefore a somewhat arbitrary, subjective character.

Farmers everywhere face three stark challenges over the next four decades. They must (i) contribute to fulfilling an estimated 60% increase in demand for agricultural production by 2050 to feed a growing, more urbanized population; (ii) do so facing growing water scarcity, climate change, and the likelihood that the available area of arable land will hardly increase. At the same time, soil degradation is proceeding in many areas, and they must (iii) ensure that developments are sustainable, continuing to provide a range of ecosystem services to future generations.

Problems and the future potential of smallholder agriculture have been widely analyzed, also considering the future challenges of climate change, water availability, and biodiversity loss (e.g., InterAcademy Council [IAC] 2004; Dorward and Chirwa 2011; IFAD 2012; Food and Agriculture Organization [FAO] 2008, 2009, 2010, 2011; Ncube et al. 2009; Garvelink et al. 2012; Herrero et al. 2014; and many others). The major and well-recognized factors inhibiting future development of smallholder agriculture in developing countries are (i) declining productivity, (ii) insecure land tenure, (iii) insufficient infrastructure, (iv) inadequate education and training, (v) lack of new seeds and technologies, and (vi) inadequate financial services. In addition, many institutional barriers retard development of smallholder practices (e.g., Jiggins

2012). Increasing food production therefore does not only present major technical but also socioeconomic challenges. The Sustainable Livelihoods Approach, inspired by the economist Amartya Sen, reverses the usual reasoning by starting with a basic human right to be fed, followed by exploring various means to realize this right in practice (e.g., Morse et al. 2009). When focusing on the role of soils in smallholder agriculture, the broader socioeconomic and ethical context of the problem should never be ignored. However, to assess the role of soils within this broad context is still relevant because soils are a key resource for farmers everywhere. This broader context is also expressed by the nine planetary boundaries, defining a "safe operating space for humanity" (Rockstrom et al. 2009). One important boundary defines land use, and the authors conclude that resource-poor smallholder agriculture will never be able to feed the demanding inhabitants of the urbanized world of the future where the majority of the 9 billion people in 2050 will live in megacities. Nor can smallholder agriculture provide an adequate income, except, perhaps, when specific "niches" for the urban market can be explored. Upscaling of agricultural enterprises to large, technologically advanced but still sustainable production entities is taking place in many developed countries. This development, which is also bound to occur in developing countries, requires major societal changes if only in terms of providing sufficient employment for unemployed countryside workers. Substantial transition periods are needed, and in the meantime, current conditions need to be improved as much as possible. However, the degree of necessary upscaling is still the object of debate. Recognizing the call for larger, more efficient agricultural production entities, a plea will be made later in this chapter to also consider enterprises of intermediate size, where farming families can still make a living and where the potential for providing a wide range of ecosystem services (thereby increasing soil security) may be larger than in highly industrialized megaproduction facilities that are disconnected from the ecosystems in which they occur.

2.1.2 Role of Soils

The role of soil is not covered specifically in current policy reports on smallholder agriculture, except indirectly when mentioning the importance of conservation agriculture, use of fertilizers, and irrigation, where soils are acknowledged to play a key role (e.g., IFAD 2012). The soil fertility literature is quite extensive but often only considers static chemical data from the topsoil, ignoring the dynamic physical behavior of the entire soil (as described in pedology), which is important in understanding plant growth. Soils will not be considered here as an object, as such, but in terms of soil security, which is defined as "the maintenance or improvement of the world's soil resource so it can provide sufficient food and fiber, fresh water, contribute to energy sustainability and climate stability, maintain biodiversity and overall environmental protection and ecosystem services" (Soil Carbon Initiative 2011, p. 4). As most of the mentioned aspects are covered by the concept of ecosystem services (Millennium Ecosystem Assessment [MA] 2005), the definition can be simplified to "the maintenance or improvement of the world's soil resource so it can continue to contribute to important ecosystem services." Ecosystem services are defined as "benefits people obtain from ecosystems" (MA 2005). Linking soils with soil security and ecosystem services is important as it avoids an inward-looking approach that tends to be common in many disciplines, soil science not excluded, and supports the views of Robinson et al. (2012).

In summary, the objective of this chapter is to (i) describe soil conditions in some selected smallholder farms in both developed and developing countries, expressed in terms of ecosystem services provided and, in turn, soil security, and (ii) explore ways in which soil security can be maintained or increased in the future, focusing in the context of this chapter on use of fertilizers, as mentioned by IFAD (2012) as one of three soil-related measures to alleviate problems of small-scale agriculture.

2.2 MATERIALS AND METHODS

2.2.1 Soil Security Concept

Soil security has been defined in analogy with food security, which aims at the long-term sustainable production of sufficient quantities of food, providing a permanent feeling of security to world citizens. This implies, however, much more than striving for a higher production as such, as many socioeconomic, institutional, and ethical aspects also play a key role. The World Health Organization defines three aspects of food security: food availability, food access, and food use. Food availability refers to having available sufficient quantities of food. Food access refers to having sufficient resources, both economic and physical, to obtain appropriate food for a nutritious diet. Food use is the appropriate use based on knowledge of basic nutrition and care. These elements are also visible in the 1996 World Food Summit definition: "Food security exists when all people, at all times, have physical and economic access to sufficient, safe and nutritious food that meets their dietary needs and food preferences for an active and healthy life."

The security concept is more complicated when applied to soils. Rather than relate to a sustainable, daily need in terms of food intake, soil security relates to what might happen if soils degrade to the extent that sufficient food production is not feasible anymore. Soil degradation is a long-term process, very much related to varying socioeconomic conditions. Except for erosion, its effects are often gradual and difficult to communicate and translate into environmental and economic values. However, when soils degrade to the extent that they cannot anymore provide certain ecosystem services, of which food production is only one provisioning service, the consequences for society are devastating. To mitigate degraded soils is very difficult and even impossible when soil has been removed by erosion. The challenge, therefore, is to create early awareness about the dangers of soil degradation that may, in the end, terminate many ecosystem services the soil can provide (e.g., Vlek et al. 2008; Desire project, www.desire-project.eu).

Considering the relevance of soils only in broad terms of soil conservation practices, fertilization and irrigation (IFAD 2012) ignores the inherent properties and potentials of any given type of soil, which are different in different parts of the world. For example, a relatively nutrient-rich clayey Nitosol in Africa has a higher potential to provide soil-related ecosystem services than a nutrient-poor Ferralsol, let alone a drought sensitive, nutrient-poor Lixisol or Arenosol. The same is true for, for example, poorly drained versus well-drained soils. When discussing existing and potential soil conditions for smallholder agriculture and the associated ecosystem services, attention will therefore be focused on the soil types that occur that may allow extrapolation of results to other sites where the same soils occur. This represents the traditional procedure of soil survey interpretation and land evaluation, where soil types are used as "carriers of information" functioning as class-pedotransfer functions (e.g., Bouma 1989; Bouma et al. 2012). However, a focus on soil type alone is not enough as it is based on permanent properties of soils formed by the soil-forming factors over periods of thousands of years or longer. However, for practical applications, emphasis should also be on soil behavior in terms of the effects of different forms of soil management on the functioning of a given soil type, resulting in a series of characteristic phenoforms. Genoforms are genetic soil types, as distinguished in soil classification systems based on pedogenetic principles that are represented on soil maps, while phenoforms describe different properties as a function of different types of management applied to that particular soil type. A series of phenoforms of a given genoform defines its functionality, not only reflecting its potential but also its limitations and resilience. Each soil type (genoform) has a characteristic range of phenoforms, and extrapolation of data obtained for specific phenoforms is particularly relevant for land evaluation. The phenoform analysis has been made for two major soil series in the Netherlands (Droogers and Bouma 1997; Pulleman et al. 2000; Sonneveld et al. 2002). Thus far, standard soil survey procedures do not include a phenoform analysis but soil surveyors are advised to do so in the future expressing the specific effects of soil management rather than only a focus on static, inherent soil properties.

2.2.2 Ecosystem Services: Role of Soils

2.2.2.1 Introduction

The benefits people obtain from ecosystems are described in terms of ecosystem services, a concept that has proved to be effective as a communication tool (MA 2005). Sixteen services are defined for four categories: supporting, provisioning, regulating, and cultural services. Of these, all have a clear connection with soils. To avoid overly complicated schemes with partly overlapping subcategories, a link can be proposed between ecosystem services and soil functions, as defined by the European Union (EU) Soil Protection Strategy of 2006 (Commission of the European Community [CEC] 2006). Each of these functions represents a soil-provided ecosystem service in terms of producing (i) food and fiber (a provisioning service); (ii) fresh water by its filtering action (a provisioning and regulating service); (iii) biodiversity (a supporting service); (iv) a physical and cultural environment for human activities (a cultural service); (v) raw materials (a provisioning service); (vi) a pool for carbon (a regulating service); and (vii) an archive, expressing our common geological and archeological heritage (a cultural service). In turn, these seven ecosystem services are reflected in the soil security concept because when they are considered to be adequate now and in the future, maintenance and improvement of the soil resource is assured, providing the desired security. Of course, the question as to what is "adequate" has to be defined, representing a major research challenge. Smallholder agriculture represents a very wide range of conditions in different countries. Defining general guidelines for "adequacy" is therefore unrealistic and undesirable. Attention should be focused on local conditions, as presented in this chapter.

2.2.2.2 Selection of Case Studies Focusing on "Use of Fertilizers"

In the limited context of this chapter, a number of case studies have been selected to illustrate the use of fertilizers as mentioned by IFAD (2012) as one of three soil-related means to overcome problems in small-scale agriculture, achieving or maintaining soil security as expressed by the ecosystem services provided. Attention is focused on particular soil types, to allow extrapolation of data obtained, if possible. Case studies will not be restricted to developing countries in Africa but will also include two examples from the Netherlands, demonstrating the universal and unifying power of the soil-related ecosystem services concept.

"Use of fertilizers" is a very broad statement that needs to be specified to become operational both in terms of defining optimal application rates and institutional settings, which will now be discussed.

Determining optimal fertilizer needs of crops is, after pedology, one of the oldest activities in soil science research, and procedures have evolved over time during the past century, summarized as follows:

1. A large number of relatively small plots on experimental stations are fertilized with increasing quantities of chemical fertilizer (N, P, and K), and yields are determined. Soil samples are only taken of the surface soil (e.g., 0–20 cm). Statistical procedures are used to relate application rates to yields, thereby defining optimal rates for farmers. Experiments over many years allow expressions of different weather conditions; however, much scatter is obtained in graphs relating yields to fertilization rates and the soils being used are often not or only broadly characterized in terms of sand, clay, silt, and peat. Nutrient analyses are restricted to surface soil. Still, such graphs are the basis for most current fertilizer recommendations and are the source of the spectacular development of production agriculture in the 20th century (see the two Dutch case studies to be discussed later).
2. In a variant of 1, fertilization rates are adjusted to what is locally feasible (see later examples of Kenya). This relates to the proposal by Tittonell and Giller (2012) to focus on the difference between the actual yields (Ya) and the locally attainable yield (Yl), rather than on the often used difference between Ya and Yw, where Yw is the theoretical water-limited yield where the assumption applies that all growth factors are optimized, except for available water that corresponds to local availability. Yw is beyond reach when soil fertility is as limiting as it often is in smallholder agriculture in developing countries.
3. Considering the major differences between conditions on farm fields and experimental plots on experimental stations, experiments are focused on farmer's fields as part of Farming Systems Research. This involves active participation by farmers and joint learning in interaction with researchers (e.g., Adjei-Nsiah et al. 2008; Sonneveld et al. 2008). Results obtained

are relevant for the particular farms considered but are sometimes difficult to extrapolate because of the peculiarities of the individual farmer's management. This procedure not only includes characterization of existing conditions, as in the Zimbabwe Arenosol case study, but may also involve experimentation with applied manure or inclusion of legumes. This on-farm approach offers opportunities to discover "lighthouse" examples of farmer innovations. When applying organic fertilizers, e.g., manure, it is important to recognize that nutrients are delivered beyond the year of application. This supports the idea that land use history should be taken into account when developing appropriate fertilization strategies within a farm context.

4. Awareness about the possible impact of fertilization on environmental quality has, starting in the 1980s, resulted in a broad system analysis by modeling and monitoring that not only considers nutrient uptake by plants and resulting plant growth, but also leaching of excess nutrients to groundwater and surface waters and production of various gases adversely affecting environmental quality (e.g., Sonneveld et al. 2008). Now, the dynamics of the entire soil are characterized rather than a static fertility sample for surface soil only. This procedure also includes innovative, modern methods to measure soil conditions with, e.g., in situ or proximal sensors.
5. A development of approach 4 is precision agriculture where fertilization in a given farmer's field is "fine-tuned" by applying it differently in space and time, depending on different soil conditions in the field and changing weather and crop demand during the growing season. Dynamic modeling of crop growth, based on the nutrient regime of the soil and weather conditions, allows fine-tuning of fertilization rates and times. A field study in the Netherlands showed that precision procedures resulted in a 30% reduction of fertilizer use as compared with the traditional procedures, while yields did not decrease. This not only represented a considerable savings for the farmer but also a more efficient use of natural resources (Van Alphen and Stoorvogel 2001; Van Alphen 2002; Bouma et al. 2012). Note that only dynamic modeling can provide signals of soil nutrient stocks in the rootzone becoming critically low. The highly promoted use of remote sensing for precision agriculture can indicate nitrogen shortage in crop leaves; however, this signal comes too late as crop growth retardation has already occurred. This is avoided by the dynamic modeling approach. Precision agriculture is the procedure of the future anywhere in the world as it most closely matches the needs of the plants, on the one hand, and fertilizer applications, on the other, taking into account solute fluxes in the soil–plant–atmosphere system. Thus, ecological intensification becomes more feasible, also in developing countries (e.g., Cassman 1999; Tittonell and Giller 2012).

Institutional arrangements have, for a long time, been rather top–down, where fertilizer rates, as determined on experimental stations, were communicated to farmers through extension services. This has been quite successful in both developed and developing countries as long as the focus was on production. As attention shifted since the 1980s to sustainable development and environmental concerns in terms of water and air pollution, knowledge transfer from the research community to practitioners became less effective in the developed world, the more so as extension services were privatized in many countries. Now we see that farmers everywhere become more knowledgeable, certainly as the information revolution proceeds. Also, researchers realize the advantages of involving farmers in a joint learning approach (e.g., Ncube et al. 2007; Adjei-Nsiah et al. 2008; Sonneveld et al. 2008; Bouma et al. 2008, 2011; Rusinamhodzi et al. 2013). This results in a fundamentally different relationship between scientists and users of their information (as will be illustrated in the Northern Frisian Woods [NFW] case study) and this will most likely increasingly apply to developing countries as well, requiring not only an interdisciplinary but also a transdisciplinary research approach (e.g., Bouma et al. 2011). The soil science profession is not quite prepared yet to face this challenge. When describing interdisciplinarity in the *Handbook of Soil Science* (e.g., Levy 2012), attention is focused on technical aspects of soils, ignoring the social sciences while transdisciplinarity is not being covered at all.

An innovative approach to achieving progress in development is being followed in the so-called Millennium Villages by supplying large quantities of improved seeds and fertilizers to allow a "jump-start" of the agricultural production system (www.millenniumvillages.com). Denning et al. (2009) and Dorward and Chirwa (2011) report positive results for Malawi, illustrating that direct assistance to farmers works better than top–down national programs, although long-duration strategic and institutional support at national policy level is crucial.

2.2.2.3 Risk of Gloom

Barriers, as cited above, to raising smallholder agriculture to sustainable levels providing a stable range of ecosystem services are severe. Rather than trying to solve all problems at once, every scientific discipline would be well advised to first explore the potential of its possible contributions to solving the immense problems at hand. As stated, we will therefore first focus on soil-related ecosystem services, expecting that they are quite relevant in an essential broader analysis at a later date, including socioeconomic, institutional, and ethical aspects and consideration of the planetary boundaries of Rockstrom et al. (2009).

Also, rather than being discouraged about all barriers, there are also favorable and encouraging developments that deserve to be recognized as "lighthouses" forming a source of inspiration (IFAD 2012). Worldwide, farmers are demonstrating the benefits of managing natural assets sustainably and in harmony with local ecosystems. They, and their farmers' organizations, developed succesful enterprises in areas where common wisdom considered this to be impossible. The research community would be well advised to be more on the lookout for such succes stories rather than being restricted by classical experimentation. For example, IFAD (2012) reports research on rice intensification undertaken by >100 farmers in Cambodia, resulting in yields increasing by 60% while use of inorganic fertilizers decreased by >70%. In India, the Mahatma Gandhi National Rural Employment Guarantee Act guarantees 100 days of work creating durable assets to help farmers improve productivity more sustainably, such as water harvesting structures. On a broader scale, Brazilian farmers have implemented minimum-till agriculture on 60% of the country's cultivable land. The government of the Philippines has stopped its fertilizer subsidy program, replacing it with a balanced policy that promotes location-specific combinations of organic and inorganic fertilizers. Turkey has tripled agricultural productivity in the past 8 years using sustainable methods. In addition, extension workers are being deployed to live in villages alongside the farmers who need their services. Thus far, with 7500 technicians hired, 80% of the country's 35,000 villages are covered. Open-channel irrigation is being converted to drip or sprinkler irrigation systems through interest-free loans and 50% grants to farmers. Subsidies for fertilizer are conditioned on soil analysis. Once the soil is analyzed and the farmer knows what nutrients it needs, the government pays for the soil analysis and for a certain amount of fertilizer.

Looking for innovative farmers and the "lighthouses" they create, and linking their results to the type of soil where they work, is a key element of the analysis provided in this chapter. The International Union of Soil Sciences World Reference Base (IUSS-WRB 2007) system of soil classification will be used throughout to define soil types being analyzed. Soil classifications are often not provided in the soil fertility literature, and soil types in this chapter are therefore characterized at the highest level of classification, the Reference Soil Groups (RSG). Only when possible are qualifiers added, such as in the Dutch case studies.

2.3 RESULTS

2.3.1 Semiarid Arenosols and Lixisols in Zimbabwe

Ncube at al. (2009) described farming systems in semiarid (av. 590 mm/yr rainfall) southwest Zimbabwe. The soils were Ferralitic Arenosols, locally interspersed by Aridic Arenosols. Farm sizes were 5 ha and smaller. Three types of farmers were distinguished: better, medium, and poorly resourced. Drought-resistant pearl millet (*Pennisetum glaucum*) was grown on 80% of the land. This 3-year study showed that all farmers could, on average, not satisfy the nutritional demands of their households. Total cereal production varied widely as a function of precipitation. All nutrient balances were strongly negative. Clearly, adequate ecosystem services are not delivered: the food and fiber service is inadequate, while negative nutrient balances illustrate worsening future conditions. Soil security is not assured, a conclusion also implicitly presented by Tittonell and Giller (2012) in a broad review of smallholder agriculture in Africa, stating that successful ecological intensification is only feasible when soil nutrient stocks would be strongly increased. Current socioeconomic and institutional conditions are not favorable to achieve this. Incorporation of legumes in the cropping system can increase production levels (Giller and Cadisch 1995), particularly when P is added as well. However, additional input by manure and chemical fertilizer is indispensable (Adjei-Nsiah et al. [2008] working on Lixisols in Ghana, Ojiem et al. [2007] working in West Kenya without mentioning the type of soil they covered, and Rusinamhodzi et al. [2012, 2013] working on Lixisols and Luvisols in Mozambique). Ncube et al. (2007) suggest improving current conditions by applying a series of small doses of manure and fertilizers rather than a single application or double applications. Also, concentrating the little manure and fertilizer there is on fields near the farmstead could produce satisfying yields on at least a small part of the farm. This method has been widely used in Europe before the arrival of chemical fertilizers. Mtambanengwe and Mapfumo (2005) illustrate this approach, based on 120 farm field sites in Zimbabwe, reporting that manifestation of within-field soil fertility gradients on Lixisols and Arenosols under smallholder agriculture is primarily a function of differential capacities by farmers to manage organic matter, which, in turn, is driven by their resource endowment. Designated rich fields consistently contain higher levels of organic matter than corresponding poor fields (or field sections), apparently owing to cumulative effects of applying substantial amounts of organic matter to such specific areas on a regular basis (e.g., Rusinamhodzi et al. 2013). This has conceivably resulted in the formation of specific soil phenoforms that could act as a "lighthouse"; however, no attention was paid to this type of interpretation. Tittonell and Giller (2012) reach comparable conclusions when stating that differential management of the various fields of the farm led to the establishment of gradients of soil fertility, notably decreasing with distance from the homestead. Farmers tended to allocate their scarce nutrient and labor resources in the fields they perceived as most fertile or less risky, or in fields around the homestead where high value crops were better protected from marauding livestock or theft. A close interaction was also found between soil fertility gradients and topography in these highly dissected landscapes, with homesteads located on the upper positions of the slope. Such interactions between inherent soil–landscape variability, historical and current management, nutrient balances, and current soil fertility were documented for smallholder systems in different parts of Africa, e.g., on Lixisols and Luvisols in Zimbabwe (Zingore et al. 2008; Rusinamhodzi et al. 2013), where significantly different yields were obtained between the two types of soil with comparable fertilization rates, and Lixisols in Ghana (Adjei-Nsiah et al. 2008). Zingore et al. (2011) reported for highly degraded Lixisols in Zimbabwe that restoring soil fertility required the application of 10 tons of manure per hectare for 10 years before a maintenance rate of 5 tons/ha could sustain productivity. This illustrates the major investment needed to reestablish soil fertility of highly degraded soils.

2.3.2 Nitosols and Vertisols in the Central Highlands of Ethiopia

This case study is based on an experiment at Holetta Research Centre in the tropical highlands of central Ethiopia (Assefa and Ledin 2001). The study serves to show that, again, the soil type, fertilizer application, and crop variety are factors that determine the yield and quality of the produce. The site is located at an elevation of 2390 m above sea level; the mean air minimum is 6.1°C and the maximum is 21.9°C. Mean annual rainfall is 1100 mm. Trials were conducted on a red Nitosol and a black Vertisol. Nitosols comprise among the most inherently productive tropical soils owing to their high nutrient status and deep, permeable structure, whereas Vertisols show marked shrinking-and-swelling upon drying–wetting and poor physical properties, such as impeded drainage and poor workability when wet (Bridges et al. 1998).

Three oats varieties (*Avena sativa* L.) and two vetch species (*Vicia villosa* and *Vicia dasycarpa*) were sown in monoculture and in mixtures in a randomized complete block under two fertilizer regimes (F1 = 18 kg N and 46 kg Pe_2O_5 and 23 kg N ha^{-1} at tillering; F0 = without application of fertilizer) with three replications. The fertilizer levels considered here reflect conditions in Ethiopia (Assefa and Ledin 2001).

Significantly higher average dry matter forage yields of mixtures, pure oats, and pure vetch were obtained on the red Nitosol (9.7, 9.4, and 4.5 tons/ha, respectively) than on the black Vertisol (5.6, 5.5, and 2.8 tons/ha, respectively). The example illustrates differences in yield for two different soil types that have a similar soil reaction (pH [1:1 H_2O] of 4.6–5.0), organic carbon content (1.8%–2.5%), and % clay (51%–65%) in the study area, yet with markedly different clay mineralogy. The reported experiments were made under controlled conditions, which are likely to differ on farms in the area. Be that as it may, crop yields on the Nitosols are above subsistence level and provided that the amount of fertilizer can again be applied every year, ecosystem service 1 (biomass production) is marginally assured, providing a certain degree of soil security. The picture is different for the Vertisol, for which production levels were approximately 40% lower, leading to lower soil security.

2.3.3 Ferralsols in Mozambique

Materechera and Mkhabela (2001) reported changes in the properties of strongly weathered Ferralsols under conditions of low external input farming systems associated with differences in land use and management practices. Their study was conducted on a small farm at Malindza, Swaziland. The area falls under the eastern part of the low veld ("bushveld"). The 5-ha farm is on a gently undulating terrain at an altitude of about 700 m. The climate is subtropical with a mean annual rainfall of 760 mm (range, 650–1050 mm). For over a decade before 1987, the whole farm was under natural vegetation and used for grazing by about 100 head of cattle. A kraal and holding pen for handling the animals was located on a 0.5-ha piece of land in the center of the farm next to the homestead. In 1987, all animals were removed from the farm and crop production was introduced. Three treatments were compared: (i) "kraal" areas where before 1987 the cattle used to stay during the night had the highest maize yields, followed by (ii) fallow fields after clearing and (iii) by continuous maize. The latter two had negative nutrient balances and yields below subsistence levels. If former "kraal" areas are not fertilized, they are bound to degrade in time.

2.3.4 Nitosols, Ferralsols, and Acrisols in Western Kenya

The highlands of western Kenya support one of the densest rural populations in the world, with >1500 inhabitants/km^2 in the Vihiga district with fertile soils. Population growth, however, has gradually led to depletion of nutrients through crop harvest removal, leaching, and soil erosion. Farmers have largely been unable to compensate such losses via crop residues, manure, and mineral fertilizers (Shepherd and Soule 1998), adversely affecting ecosystem services and soil security. The observed patterns of soil deterioration are spatially heterogeneous in the area. Spatial and temporal niches for targeting soil fertility management and technologies were studied by Tittonnell et al. (2005). The dominant soil types are Humic Ferralsols, Dystric and

Humic Nitosols, and Orthic Acrisols. Sixty farms, located in the three sublocations, were selected by Tittonell et al. (2005). Together, they represent much of the biophysical and socioeconomic variability observed in the highlands of western Kenya where cultural backgrounds differ widely (e.g., tribes). This led to the differentiation of five farm types, referred to here as T1 to T5, where T1 farms have the highest assets and T5 the lowest. A consistent trend of decreasing input use from farm types T1 to T5 was generally observed by Tittonell et al. (2005); however, nutrient resources and land management practices (e.g., fallow) differed markedly between sublocations. Maize (*Zea mays* L.) yields on Nitisols were slightly higher than those on Acrisols and Ferralsols; however, nutrient balances were dominantly negative and self-sufficiency in maize production was achieved by <40% of farmers in all sublocations—ecosystem service 1, thus, is only partially met in the area.

The case for western Kenya and the earlier cases reviewed here illustrate the very high variability in soil fertility management that is associated with the "soilscape," such as the location of land along hillslopes, and with differences in soil fertility management as a function of household wealth. This will be reflected in different gradations of achievable soil security; however, usually, soil security is low. Many studies report a high variability of fertility within farmer's fields, which is based on samples for chemical analyses of surface soil. However, variability of soil types within farms is usually much smaller as they are correlated with landscape position, and this is one reason to use soil type classifications as "carriers" of information.

2.3.5 Sapric Histosols in the Western Part of the Netherlands

Sapric Histosols, located in the lower reaches of the European Rhine–Meuse Delta, contain large amounts of nitrogen (N), and peat decomposition is a substantial contributor to the mineral N supply of crops. For grasslands on nonfertilized poorly drained peat soils in the Netherlands, an average N uptake of 252 kg/ha was reported (Vellinga and André 1999). Despite the large supply of N from the soil, application of fertilizer N was about 205 kg N/ha per year for dairy farms on peat soils in the west of the Netherlands in the 1990s (Reijneveld et al. 2000). This largely corresponded with common fertilizer guidelines: 195–230 kg N/ha per year for well-drained and 235–275 kg N/ha per year for poorly drained peat soils. Together with relatively large inputs from concentrates (102 kg N/ha per year on average) and relatively low exports of N through milk (68 kg N/ha per year) and animals (13 kg N/ha per year), this resulted in N surpluses at farm level of 270 kg N/ha per year on average for a dairy farm on peat soil at the end of the 1990s (Reijneveld et al. 2000). These surpluses are distributed over various environmental fluxes, such as gaseous losses to the atmosphere, because of denitrification and export of N to surface waters (Van Beek et al. 2004).

Following EU Directives, environmental policies were introduced to reduce N losses and increase farm nutrient efficiencies. One farmer in the area, owning a farm of 37 ha, was able to remain economically viable without the input of fertilizer N (Sonneveld et al. 2008) by large and long-term inputs of organic N sources, such as dung, ditch sludge, farmyard manure, cow slurry, and nonharvested herbage. Thus, average N uptake under nonfertilized conditions increased to 342 kg/ha, with only a limited part being derived from peat mineralization and with small losses to the environment (Sonneveld and Lantinga 2011). The example of this pioneering farmer acted as a "lighthouse," inspiring other farmers and scientists. Procedures resulted in the formation of an anthropogenic A horizon in these Histosols, which may be regarded as a separate phenoform. Soil security has improved following this form of innovative management that had a positive effect on several ecosystem functions. Ecosystem services 1 and 3 have improved. Reducing the N surplus has reduced N flow into surface waters and has thus improved service 2. The landscape in which this particular farm is located is the National Landscape "Green Hart" of Holland, which is highly valued for its historic reclamation patterns and dominance of dairy farming. Maintaining farming in this area, which is enhanced by the described innovative new approach to management, contributes to services 4 and 7. Yet, a continuous problem is land subsidence because of drainage, resulting in net carbon loss, giving problems with ecosystem service 6 although the organic matter content of surface soil has increased following innovative management.

2.3.6 Gleyic Podzols in the Netherlands

Gleyic Podzols are the most common sandy soils in the Netherlands. The official fertilizer recommendations for these soils were relatively high at the end of the 1980s, up to 400 kg N/ha per year. They decreased thereafter; however, a group of 100 dairy farmers in the National Landscape Northern Frisian Woodlands reported in 1995 an average N fertilizer use of 290 kg N/ha, ranging from 150 to 478 kg N/ha (Verhoeven et al. 2003). As a result, N surpluses at farm level ranged from 162 kg N/ha per year to 560 kg N/ha per year, with a mean farm N surplus of 326 kg N/ha per year. These high environmental losses resulted in a poor environmental quality, for example, reflected in high levels of nitrate in the upper groundwater (Sonneveld and Bouma 2003).

Ecosystem services 2 and 3 were not satisfied and services 4 and 7 were threatened because rationalization of dairy farming called for large fields and cutting hedgerows, which are considered to be a key quality of the landscape. Concerned farmers organized into an environmental cooperative to improve conditions, taking a proactive approach that also attracted researchers. Thus, a "lighthouse" was created that has received continued national and international attention. An intensive participatory research program from 1997 to 2003, focused on increasing nutrient use efficiencies, resulted in a reduction of fertilizer input to 136 kg N/ha per year in 2001/2002 and a corresponding decrease of N surpluses to 172 kg N/ha on average. More emphasis was placed on organic manuring. Farm N uptake efficiencies increased from 21% in 1997/1998 to 33% in 2001/2002, and organic matter contents of surface soils increased significantly forming a recognizable phenoform: old grassland (Sonneveld et al. 2002). In addition, this study defined two other phenoforms of the genoform Gleyic Podzols: reseeded grassland and continuous maize cropping. After studying 40 fields on different farms, these forms of management could be related to soil C contents:

$$\%\text{C} = 3.40 - (1.54 \times \text{Maize}) + (0.19 \times \text{Old}) + (0.55 \times \text{GWC})(r^2 = 0.75)$$

where Maize = 1 for continuous maize cropping and 0 otherwise; Old = 1 for old grassland and 0 otherwise; and GWC describes groundwater levels with value 1 for class V and 0 otherwise.

Thus, organic matter contents of a given genoform are predicted as a function of past management based on field data. Additionally, the phenoform approach provided farmers with a more holistic view on soil functioning as compared with standard chemical soil analyses. Up until 2000, soil samples for acquiring fertilizer guidelines were taken from the upper 0–5 cm of the soil, which increased to 0–10 cm from 2000 onward, although topsoil depth for Gleyic Podzols is always 30 cm at a minimum. More important, the Gleyic prefix of the genoform indicates shallow groundwater, which has a major effect on plant growth by upward capillary flow of water that is, of course, not expressed by the chemical analyses of surface soil. Six of the seven ecosystem services, with the exception of providing raw materials, therefore had a positive value, thus contributing to soil security. But how about the future?

A group of farmers was not yet satisfied, and they proceeded to refine management to a form of cradle-to-cradle dairy farming, further reducing the amount of chemical fertilizers, increasing organic manuring, and restricting input from outside feed sources and contractors. They were generally more extensive and smaller in size, as compared with other conventional dairy farms in the region. The proportion of arable crops on the farm (silage maize) was significantly lower and farmers were reluctant to apply grassland renovation since this has a negative impact on soil organic matter contents. Application of a life cycle analysis showed that they significantly increased the organic matter content of the soil as compared with comparable farms on sandy soils, thereby increasing the biomass production potential, the filtering capacity of the soil, and its biodiversity (ecosystem services 1, 2, and 3). Energy use was reduced, while the specific character of the landscape was maintained, supporting services 4 and 7 (Dolman et al. 2013). Looking at the future, soil security was not only maintained here, but it is being improved significantly. Note that farmers were the driving force in pursuing these developments, in close cooperation with soil scientists, illustrating a favorable modern tendency for participatory approaches and joint learning.

2.4 MESSAGE FROM DUTCH FARMERS TO SMALLHOLDERS OF DEVELOPING COUNTRIES

Farmers in both National Landscapes in the Netherlands, discussed here, do not qualify as smallholders as distinguished in developing countries, as their farms have an average size of approximately 50 ha. Only one hundred years ago, farm sizes in the Netherlands were as small as they are now in many developing countries. Industrial development and growth of cities providing employment to former agricultural workers has allowed farms to increase in size, making them commercially viable. The same development is likely in developing countries; however, sufficient time should be allowed to transform existing socioeconomic and institutional conditions. The farmers in the two Dutch case studies perform important services to society that go way beyond producing food and fiber (ecosystem service 1). They also have a message for farmers in developing countries by showing that modern agricultural development does not necessarily imply the exclusive need to establish dairy megafarms with 1000 cattle or more or arable farms of thousands of hectares with monocultures. Intermediate sizes of family farms represent, next to megafarms, a viable business proposition for the future that is particularly suitable to satisfy most recognized ecosystem services and that, therefore, presents a major contribution to soil security in the future. Whether such security can also be assured when implementing impersonal and highly industrialized megafarms remains to be seen and should be considered by Rockstrom et al. (2009) when further developing the important planetary boundaries paradigm.

2.5 DISCUSSION

2.5.1 Smallholder Agriculture and the Planetary Boundary of Land Use

Rockstrom et al. (2009) have defined land use as one of nine planetary boundaries and conclude, in a fascinating long-term vision, that feeding 9 billion people in 2050 will require a drastic increase of agricultural production, in roughly the same area that is currently used. Enlarging the area is not only impossible because of lack of suitable land but also undesirable because it would imply a corresponding decrease of nature areas. The reviewed case studies in this chapter, covering smallholder agriculture in Zimbabwe, Kenya, Uganda, Mozambique, Swaziland, and Ghana, show that production levels on small farms of just a few hectares are, in many cases, too low to even support farming families, let alone provide food for fellow citizens. The ecosystem service of producing food and biomass is therefore inadequate and so is therefore soil security, which declines because of continued nutrient depletion. However, experiments show that yields can be improved with proper fertilization practices, preferably to be combined with legumes, although socioeconomic and institutional barriers remain severe. Rockstrom et al. (2009) express a long-term need for establishing larger production units near megacities of the future, using modern management techniques that combine relatively high yields with proper environmental stewardship. This transformation has been and is taking place in the developed world during the past 100 years, and it is hard to see why a similar development would not occur in the developing world. But how large is large? Such transformation should be gradual, avoiding social disruption, and this can be achieved by creating job opportunities in an urban environment. In the short term, the case studies show that smallholder agriculture and soil security can and should be improved. The two Dutch examples intend to show that developing very large industrial megafarms may not be the only future option and that intermediate-sized family farms may offer an additional possibility that could be particularly attractive to ensure future soil security.

2.5.2 Role of Soil Science

The case studies presented in this chapter, focusing on use of fertilizers, show that different soil types have different potentials requiring different management measures to achieve them. Results obtained on Nitosols and Luvisols were significantly better than those on Arenosols, Lixisols, Acrisols, and Ferralsols. Taking this into account when extrapolating favorable results (acting as "lighthouses") to identical soil types elsewhere will improve the transfer of knowledge and may, thus, in principle, contribute to higher soil security. The case studies show that researchers are increasingly investigating conditions on real-life farms in close interaction with farmers and are on the lookout for "lighthouse" examples where inventive farmers have achieved successes on a particular soil type, sometimes to be identified as distinct phenoforms. This clearly happened in the two Dutch case studies.

The soil type, formed by a unique set of interacting soil-forming factors, is represented by mapping units on soil maps, and such units have traditionally been used as "carriers" of information for land evaluation purposes, extrapolating experiences obtained (Bouma et al. 2012). Bouma et al. (1998) and Bouma (2002) illustrated in this context the significantly different behavior of seven soil types from China, Zambia, Nigeria, Colombia, and Indonesia in terms of their Yw values (the water limited yield). If, however, soil mapping units that are, according to the legend of the soil map, supposed to contain certain soil types are quite heterogenous or ill defined, the procedure is questionable and, indeed, many questions have been raised about this procedure. Recent attempts to use sensors to directly obtain soil data (such as the organic matter content, the cation exchange capacity, and pH of surface horizons) (e.g., Kweon 2012) without using soil maps is potentially valuable but cannot explain dynamic physical, chemical, and biological soil behavior the way a well-characterized soil type can, be it in an often qualitative manner. Extrapolation of soil behavior from one well-characterized location to a new location on the basis of CEC, organic matter content, and pH of surface horizons is, of course, impossible. Such sensing data would therefore be most useful when coupled with the type of soil being observed.

Also, when only data obtained at the soil surface are available, important subsurface properties that are essential for soil behavior remain unknown. Soil chemical analyses used in the African case studies, reviewed in this chapter, were made in samples from 0 to 20 cm depth. However, plant roots always go much deeper. Also, many soils have subsurface horizons that strongly affect soil behavior: Lixisols have, for example, a clayey subsoil in contrast to Arenosols. The clay layer often ponds water, causing anaerobic conditions that strongly impair root development and crop growth. This is not observed when only a sample of surface soil is taken for chemical analyses. Soil type data therefore provide a valuable context for separate chemical, physical, and biological measurements.

2.5.3 Other Factors Determining Soil Security

Aside from a set of socioeconomic, institutional, and ethical challenges for smallholder agriculture as mentioned in Section 2.1, IFAD (2012) distinguishes three important soil-related issues: use of fertilizers, conservation agriculture, and irrigation. The restricted size of this chapter only allowed discussion of the soil fertility issue. However, recent reviews of conservation agriculture (e.g., Vlek et al. 2008 and the Desire project, www.desire-project.eu) show that problems are severe but that efforts in many countries to combat soil degradation and erosion are succesful, particularly when approached comprehensively in large-scale watersheds, requiring cooperation among farmers and effective institutional arrangements. Dealing with individual smallholders is much more difficult. Some agronomists, however, challenge the potential of conservation agriculture if only because of the limited amount of crop residues left on the land (e.g., Giller et al. 2009). When applied correctly, irrigation can also strongly improve yields (e.g., IAC 2004). However, it requires a relatively high level of technical expertise that can be applied most effectively in larger irrigation schemes. Again, introducing irrigation on 2-acre fields has little potential. This supports the long-range vision of Rockstrom et al. (2009) that large and modern production systems are needed to satisfy the needs of an ever-wealthier population in the decades to come. The Dutch case studies suggest that there may also be local potential for intermediate-size family farms.

2.6 CONCLUSIONS

Five major conclusions are derived from this study:

1. Soil security is expressed in terms of the degree to which ecosystem services can be provided. The sixteen ecosystem services defined by MA (2005) can, for simplicity, be related to the seven basic soil functions defined by CEC (2006). Large differences among smallholder conditions in different countries, as reviewed in this chapter, make unified expressions of soil security less meaningful and require a local approach taking into account socioeconomic and institutional issues.
2. Recent reports on smallholder agriculture pay little attention to soils. IFAD (2012) mentions conservation agriculture, use of fertilizers, and irrigation as possible soil-related activities that may improve ecosystem services and, in turn, soil security. In this chapter, the very general term "use of fertilizers" is explored by analyzing the various ways in which optimal use of fertilizers has been investigated during the last hundred years, as well as ways in which this knowledge has been communicated to farmers. The traditional top–down approach is now being replaced by interactive joint learning in both developed and developing countries.
3. Case studies in Zimbabwe (Arenosols) and Kenya (Vertisols) demonstrate conditions where ecosystem services are inadequate, as production levels remain below that required by the farming families to meet their food need. This applies to studies in several other countries. Current socioeconomic and institutional conditions do not offer obvious future perspectives for improvement. However, Nitosols in Kenya and Luvisols in Zimbabwe presented an adequate, be it marginal, level of ecosystem services. Rather than only perform classical soil fertility experiments, focusing on surface soil, the suggestion is made to (i) also consider subsoil conditions as defined in soil classification in terms of genoforms, and (ii) look for "lighthouse" examples where farmers, working with a given soil type, have achieved unexpected success. Then, genoforms can be used as a "carriers" of valuable management information.
4. A study on Gleyic Podzols and Sapric Histosols in the Netherlands demonstrated bottom–up actions of farmers applying cradle-to-cradle techniques that resulted in six clearly improved ecosystem services and thus improved

soil security. This improved soil management resulted in distinctly different phenoforms of the given genoform, associated with different forms of management. Distinction of phenoforms offers a more specific possibility to extrapolate management expertise to nonstudied locations with the same soil type: the classic procedure in land evaluation.

5. Smallholder agriculture has to be considered in a broad societal context, with a focus to the future. Rockstrom et al. (2009) have defined planetary boundaries for nine issues, among them land use. Feeding 9 billion people in 2050 will require a 50% increase of agricultural production. The marginal character of many smallholders at this time does not allow such an increase, although case studies reviewed show that significant improvements can be made by improving management practices. Larger production units, using modern production techniques, are needed and this development occurs in the developed world. The two Dutch case studies show the potential of intermediate sizes of farms, providing satisfactory incomes and a range of ecosystem services, creating soil security.

ACKNOWLEDGMENT

The authors gratefully acknowledge valuable comments by Dr. K.E. Giller, professor of plant production systems at Wageningen University, the Netherlands.

REFERENCES

Adjei-Nsiah, S., T. W. Kuyper, C. Leeuwis et al. 2008. Farmers' agronomic and social evaluation of productivity, yield and N-2-fixation in different cowpea varieties and their subsequent residual N effects on a succeeding maize crop. *Nutr. Cycl. Agroecosyst.* 80(3): 199–209.

Assefa, G. and I. Ledin. 2001. Effect of variety, soil type and fertiliser on the establishment, growth, forage yield, quality and voluntary intake by cattle of oats and vetches cultivated in pure stands and mixtures. *Animal Feed Sci. Technol.* 92: 95–111.

Bouma, J. 1989. Using soil survey data for quantitative land evaluation. In *Advances in Soil Science*, Vol. 9, B. A. Stewart (Ed.). Springer Verlag, New York: 177–213.

Bouma, J. 2002. Land quality indicators of sustainable land management across scales. *Agric. Ecosyst. Environ.* 88(2): 129–136.

Bouma, J., N. Batjes and J. J. H. Groot. 1998. Exploring soil quality effects on world food supply. *Geoderma* 86: 43–61.

Bouma, J., J. A. de Vos, M. P. W. Sonneveld, G. B. M. Heuvelink and J. J. Stoorvogel. 2008. The role of scientists in multiscale land use analysis: Lessons learned from Dutch communities of practice. *Adv. Agron.* 97: 177–239.

Bouma, J., J. J. Stoorvogel and W. M. P. Sonneveld. 2012. Land evaluation for landscape units, Chapter 34. In *Handbook of Soil Science*, Second Edition, P. M. Huang, Y. Li and M. Summer (Eds.). CRC Press, Boca Raton, FL: 34-1–34-22.

Bouma, J., A. C. van Altvorst, R. Eweg, P. J. A. M. Smeets and H. C. van Latesteijn. 2011. The role of knowledge when studying innovation and the associated wicked sustainability problems in agriculture. *Adv. Agron.* 113: 285–314.

Bridges, E. M., N. H. Batjes and F. O. Nachtergaele (Eds.). 1998. *World Reference Base for Soil Resources: Atlas*. ISSS Working Group RB, ISRIC-FAO-ISSS-Acco. Acco Press, Leuven.

Cassman, K. G. 1999. Ecological intensification of cereal production systems: Yield potential, soil quality, and precision agriculture. *Proc. Natl. Acad. Sci. U. S. A.* 96: 5952–5959.

Commission of the European Community (CEC). 2006. Directive of the European Parliament and of the Council, establishing a framework for the protection of soil and Amending. Directive 2004/35/EC. Brussels. September 22.

Denning, G., P. Kabambe, P. Sanchez et al. 2009. Input subsidies to improve smallholder maize productivity in Malawi: Toward an African Green Revolution. *PLoS Biol.* 7(1): e23.

Dolman, M. A., M. P. W. Sonneveld, H. Mollenhorst and I. J. M. deBoer. 2013. Benchmarking the economic, environmental and societal performance of Dutch dairy farms aiming at internal recycling of nutrients. *J. Cleaner Prod.* 13: 245–252.

Dorward, A. and E. Chirwa. 2011. The Malawi agricultural input subsidy programme: 2005/06 to 2008/09. *Int. J. Agric. Sustain.* 9(1): 232–247.

Droogers, P. and J. Bouma. 1997. Soil survey input in exploratory modeling of sustainable soil management practices. *Soil Sci. Soc. Am. J.* 61: 1704–1710.

FAO. 2008. *Climate Change, Water and Food Security*. Technical Background Document from the Expert Consultation Held on February 26 to 28, 2008. FAO, Rome, Italy.

FAO. 2009. *Food Security and Agricultural Mitigation in Developing Countries: Options for Capturing Synergies*. FAO, Rome, Italy. Available at http://www.fao.org/docrep/012/i1318e/i1318e00.pdf.

FAO. 2010. *"Climate-Smart" Agriculture: Policies, Practices and Financing for Food Security, Adaptation and Mitigation*. FAO, Rome. Available at http://www.fao.org/docrep/013/i1881e/i1881e00.htm.

FAO. 2011. *Climate Change Mitigation Finance for Smallholder Agriculture. A Guidebook to Harvesting Soil Carbon Sequestration*. FAO, Rome.

Garvelink, W. J., K. Wedding and S. Hanson. 2012. *Smallholder Agriculture: A Critical Factor in Poverty Reduction and Food Security in Africa*. CSIS. Center for Strategic and International Studies, Washington, DC. Available at https://csis.org/publication/smallholder-agriculture-critical-factor-poverty-reduction-and-food-security-africa.

Giller, K. E. and G. Cadisch. 1995. Future benefits from biological nitrogen-fixation—An ecological apporach to agriculture. *Plant Soil* 174(1–2): 255–277.

Giller, K. E., E. Witter, M. Corbeels and P. Tittonell. 2009. Conservation agriculture and smallholder farming in Africa: The heretics' view. *Field Crop Res.* 114(1): 23–34.

Herrero, M., P. K. Thornton, A. Bernues et al. 2014. Exploring future changes in smallholder farming systems by linking socio-econmic scenarios with regional and household models. *Global Environ. Change* 24: 165–182.

InterAcademy Council (IAC). 2004. Realizing the Promise and Potential of African Agriculture. *IAC*. KNAW, Amsterdam, the Netherlands.

International Fund for Agricultural Development (IFAD). 2012. *Sustainable Smallholder Agriculture: Feeding the World, Protecting the Planet*. Proc. of Gov. Council. IFAD, Rome.

IUSS Working Group WRB. 2007. *World Reference Base for Soil Resources 2006*. First update 2007. Report 103. FAO, Rome.

Jiggins, J. (Ed.). 2012. Diagnosing the scope for innovation: Linking smallholder practices and institutional context. Special issue NJAS. *Wageningen J. Life Sci.* 60–63: 1–120.

Kweon, G. 2012. Toward the ultimate soil survey: Sensing multiple soil and landscape properties in one pass. *Agric. J.* 104: 1547–1557.

Levy, G. J. 2012. Introduction to: Interdisciplinary aspects of soil science. In *The Handbook of Soil Science*, Second Volume, Part III, P. M. Huang, Y. Li and M. E. Summer (Eds.). CRC Press, Taylor & Francis Group, Boca Raton, FL: 17–26.

Materechera, S. A. and T. S. Mkhabela. 2001. Influence of land-use on properties of a ferralitic soil under low external input farming in southeastern Swaziland. *Soil Tillage Res.* 62: 15–25.

Millennium Ecosystem Assessment (MA). 2005. *Millennium Ecosystem Assessment—Strengthening the Capacity to Manage Ecosystems Sustainability for Human Well-Being*. United Nations, New York.

Morse, S., N. McNamara and M. Achelo. 2009. *Sustainable Livelyhood Approach (SLA). A Critical Analysis of Theory and Practice*. Geographical Paper 189. Reading University, UK.

Mtambanengwe, F. and P. Mapfumo. 2005. Organic matter management as an underlying cause for soil fertility gradients on smallholder farming in Zimbabwe. *Nutr. Cycl. Agroecosyst.* 73: 227–243.

Ncube, B., J. P. Dimes, S. J. Twomlow, W. Mupangwa and K. E. Giller. 2007. Raising the productivity of smallholder farms under semi-arid conditions by use of small doses of manure and nitrogen: A case of participatory research. *Nutr. Cycl. Agroecosyst.* 77(1): 53–67.

Ncube, B., S. J. Twomlow, J. P. Dimes, M. T. van Wijk and K. E. Giller. 2009. Resource flows, crops and soil fertility management in smallholder farming systems in semi-arid Zimbabwe. *Soil Use Manage.* 25: 78–90.

Ojiem, J. O., B. Vanlauwe, N. de Ridder and K. E. Giller. 2007. Niche-based assessment of contributions of legumes to the nitrogen economy of Western Kenya smallholder farms. *Plant Soil* 292(1–2): 119–135.

Pulleman, M. M., J. Bouma, E. A. van Essen and E. W. Meijles. 2000. Soil organic matter content as a function of different land use history. *Soil Sci. Soc. Am. J.* 64: 689–694.

Reijneveld, J. A., B. Habekotte, H. F. M. Aarts and O. Oenema. 2000. 'Typical Dutch': Zicht op verscheidenheid binnen de Nederlandse melkveehouderij. Report 8, Plant Research International, Wageningen.

Robinson, D.A., N. Hockley, E. Dominati et al. 2012. Natural Capital, Ecosystem services, and soil change: Why soil science must embrace an ecosystems approach. *Vadose J.* 11(1); 1–10.

Rockstrom, J., W. Steffen, K. Noone et al. 2009. A safe operating space for humanity. *Nature* 461: 472–475.

Rusinamhodzi, L., M. Corbeels, J. Nyamangara and K. E. Giller. 2012. Maize–grain legume intercropping is an attractive option for ecological intensification that reduces climatic risk for smallholder farmers in central Mozambique. *Field Crop Res.* 136: 12–22.

Rusinamhodzi, L., M. Corbeels, S. Zingore, J. Nyamangara and K. E. Giller. 2013. Pushing the envelope? Maize production intensification and the role of cattle manure in recovery of degraded soils in smallholder farming areas of Zimbabwe. *Field Crop Res.* 6: 1–16.

Shepherd, K. D. and M. J. Soule. 1998. Soil fertility management in west Kenya: Dynamic simulation of productivity, profitability and sustainability at different resource endowment levels. *Agric. Ecosyst. Environ.* 71: 131–145.

Soil Carbon Initiative. 2011. *Soil carbon summit outcomes document*. Available at http://ussc.edu.au/events-special/soil-carbon-stakeholder-workshop.

Sonneveld, M. P. W. and J. Bouma. 2003. Effects of different combinations of land use history and nitrogen application on nitrate concentration in the groundwater. *Neth. J. Agric. Sci.* 51(1/2): 135–146.

Sonneveld, M. P. W., J. Bouma and A. Veldkamp. 2002. Refining soil survey information for a Dutch soil series using land use history. *Soil Use Manage.* 18: 157–163.

Sonneveld, M. P. W. and E. A. Lantinga. 2011. The contribution of mineralization to grassland N uptake on peatland soils with anthropogenic A horizons. *Plant Soil* 340: 357–368.

Sonneveld, M. P. W., J. J. Schroder, J. A. De Vos et al. 2008. A whole-farm strategy to reduce environmental impacts of nitrogen. *J. Environ. Qual.* 37(3): 333–337.

Tittonell, P. and K. E. Giller. 2012. When yield gaps are poverty traps: The paradigm of ecological intensification in African smallholder agriculture. *Field Crop Res.* 22: 1–15.

Tittonell, P., B. Vanlauwe, P. A. Leffelaar, E. C. Rowe and K. Giller, K. E. 2005. Exploring diversity in soil fertility management of smallholder farms in western Kenya: I. Heterogeneity at region and farm scale. *Agric. Ecosyst. Environ.* 110: 149–165.

Van Alphen, B. J. 2002. A case study on precision nitrogen management in Dutch arable farming. *Nutr. Cycl. Agroecosyst.* 62: 151–161.

Van Alphen, B. J. and J. J. Stoorvogel. 2001. A methodology for precision nitrogen fertilisation in high-input farming systems. *Prec. Agric.* 2: 319–332.

Van Beek, C. L., G. A. H. Van den Eertwegh, F. H. van Schaik, G. L. Velthof and O. Oenema. 2004. The contribution of dairy farming on peat soil to N and P loading of surface water. *Nutr. Cycl. Agroecosyst.* 70: 85–95.

Vellinga, T. V. and G. André. 1999. Sixty years of Dutch nitrogen fertiliser experiments, an overview of the effects of soil type, fertiliser input, management and of developments in time. *Neth. J. Agric. Sci.* 47(3/4): 215–241.

Verhoeven, F. P. M., J. W. Reijs, and J. D. Van der Ploeg. 2003. Re-balancing soil–plant–animal interactions: Towards reduction of nitrogen losses. *Neth. J. Agric. Sci.* 51(1–2): 147–164.

Vlek, P. L. G., L. Quang Bao and T. Lulseged. 2008. *Land Decline in Land-Rich Africa—A Creeping Disaster in the Making*. CGIAR Science Council Secretariat, Rome, Italy: 63 pp.

Zingore, S., R. J. Delve, J. Nyamangara and K. E. Giller. 2008. Multiple benefits of manure: The key to maintenance of soil fertility and restoration of depleted sandy soils on African smallholder farms. *Nutr. Cycl. Agroecosyst.* 80(3): 267–282.

Zingore, S., P. Tittonell, M. Corbeels, M. T. Wijk and K. E. Giller. 2011. Managing soil fertility diversity to enhance resource use efficiencies in smallholder farming systems: A case from Murewa District, Zimbabwe. *Nutr. Cycl. Agroecosyst.* 90(1): 87–103.

3 Smallholder Farmers, Fertilizers, and Increased Food Demand

Deborah Hellums and Amit Roy

CONTENTS

3.1 INTRODUCTION

Since the widespread food deficits in Asia and the initial implementation of agricultural development concepts in the 1960s, the world's population has virtually doubled to 7 billion. Fortunately, timely agriculture interventions based on the adoption of modern inputs (i.e., fertilizers, improved seeds, crop protection products), coupled with rapid expansion of irrigated areas, led to a doubling of world cereal production, enabling billions of additional people to be fed. This scientifically based "Green Revolution," with its significant impacts on cereal crop production, negated the predictions of recurring and widespread famines and provided the foundation for food security.

Subsequently, scientific advancements and improved management practices have continued to increase staple cereal yields. Global cereal production increased from around 1400 million metric tons (Mt) at the start of the 1970s to >2500 Mt in 2011 (Figure 3.1). While recognizing the importance of advances in breeding and crop protection products and the expansion of irrigation, there is widespread acknowledgement that much of the production gains are directly attributable to fertilizer use, which increased from <75 Mt to >150 Mt within the same period. Fertilizers, primarily nitrogen (N), phosphorus (P), and potassium (K), are estimated to have contributed 40%–60% (Figure 3.2). Without fertilizers, Erisman et al. (2008) calculated that global cereal production would have been halved.

Despite widespread adoption, fertilizer consumption by smallholder farmers (the focus of this chapter) varies greatly by region; however, in every case, there is a direct correlation between fertilizer consumption and cereal production (Figure 3.3). In some regions, unbalanced use of mineral fertilizers, typically involving high application rates of N (often influenced by subsidies favoring N fertilizers), exact considerable and increasing environmental costs (Smil 2011). For example, in some areas of high fertilizer consumption in South and East Asia (i.e., China, where the average use rate of NPK fertilizers is 344 kilograms per hectare [kg ha^{-1}]), losses of nutrients through volatilization, leaching, nitrification/denitrification, soil erosion, and surface water runoff are contaminating land, water, and air. Alternately, in much of Sub-Saharan Africa (SSA) (i.e., Ghana, where the average use rate of NPK fertilizers is 7.5 kg ha^{-1}), the underuse of fertilizers, combined with a lack of access to other nutrient inputs (including organic amendments), is degrading soil fertility at the expense of future agricultural productivity.

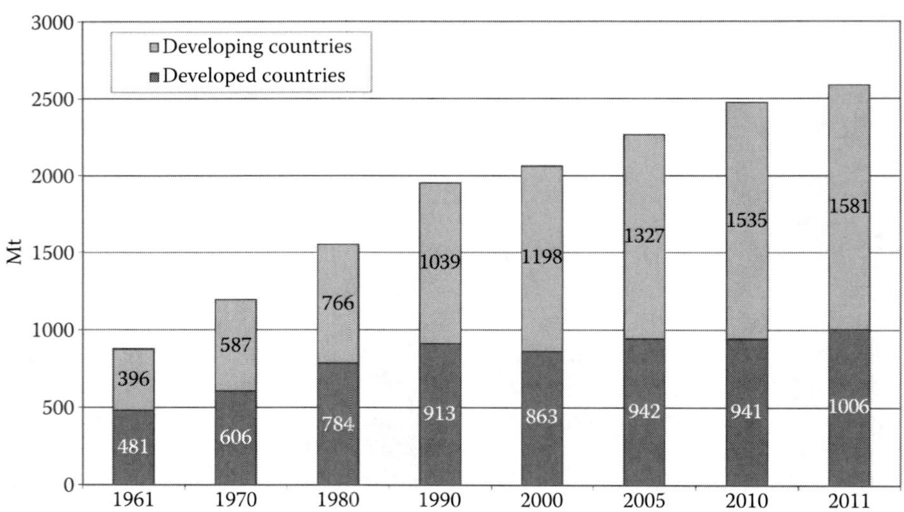

FIGURE 3.1 Global cereal production (Mt). (Adapted from Food and Agriculture Organization of the United Nations. 2012. *FAOSTAT.* Rome, Italy.)

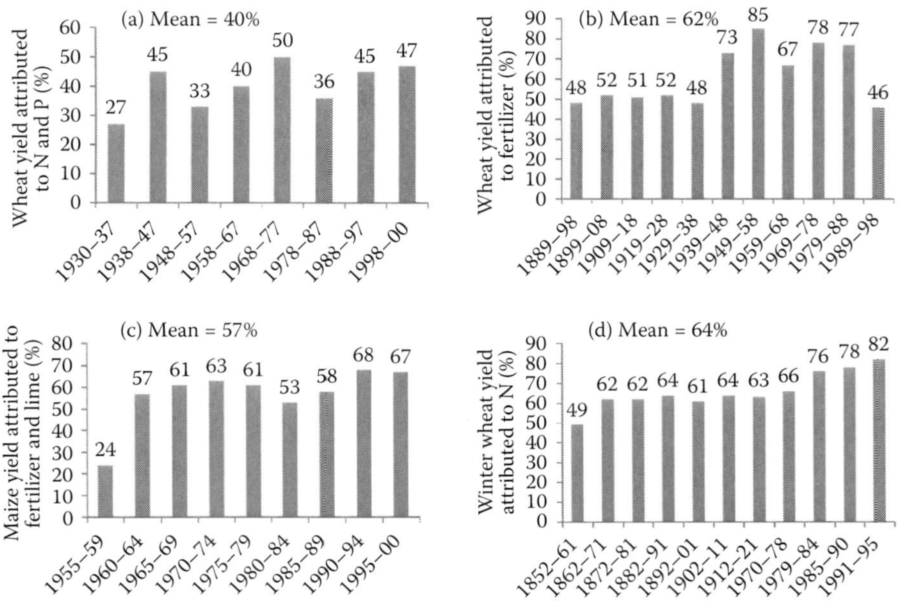

FIGURE 3.2 Yield attributed to fertilizer: (a) N and P from 1930 to 2000, in the Oklahoma State University Magruder plots; (b) N, P, and K from 1889 to 1998 in the University of Missouri Sanborn Field plots; (c) N, P, K, and lime from 1955 to 2000 in the University of Illinois Morrow plots; and (d) N with adequate P and K vs. P and K alone from 1852 to 1995 (years between 1921 and 1969 excluded because part of the experiment was fallowed each year for weed control) in the Broadbalk experiment at Rothamsted, England. (From Stewart, W.M., D.W. Dibb, A.E. Johnston, and T.J. Symth, *Agron. J.*, 97, 1, 2005.)

In many developing regions, including most countries in SSA and South Asia with its significant landless population, most of the rural poor are smallholder farmers. The productivity of these farms is usually very low with large yield gaps, thereby creating opportunities to increase food production (Nin-Pratt et al. 2011; Pinstrup-Andersen, 2013) and to allow smallholders to move from subsistence to commercial agriculture.

To make the transition to commercially oriented intensified agriculture, millions of smallholder farmers must have increased access to inputs and improved technologies, linkages to markets, and information and knowledge on management techniques that improve their production efficiency (Winterbottom et al. 2013). Against this backdrop, these smallholders as an important part of the world's agricultural systems face additional and enormous challenges (land degradation, land use pressures, climatic uncertainties, access to land and water, etc.) to increasing production while reducing the negative environmental impacts of agriculture on the natural resource base. This chapter discusses food demand trends and production requirements in the context of smallholder farmers and the technological tools, particularly mineral fertilizers, needed for them to move from subsistence to commercial agriculture.

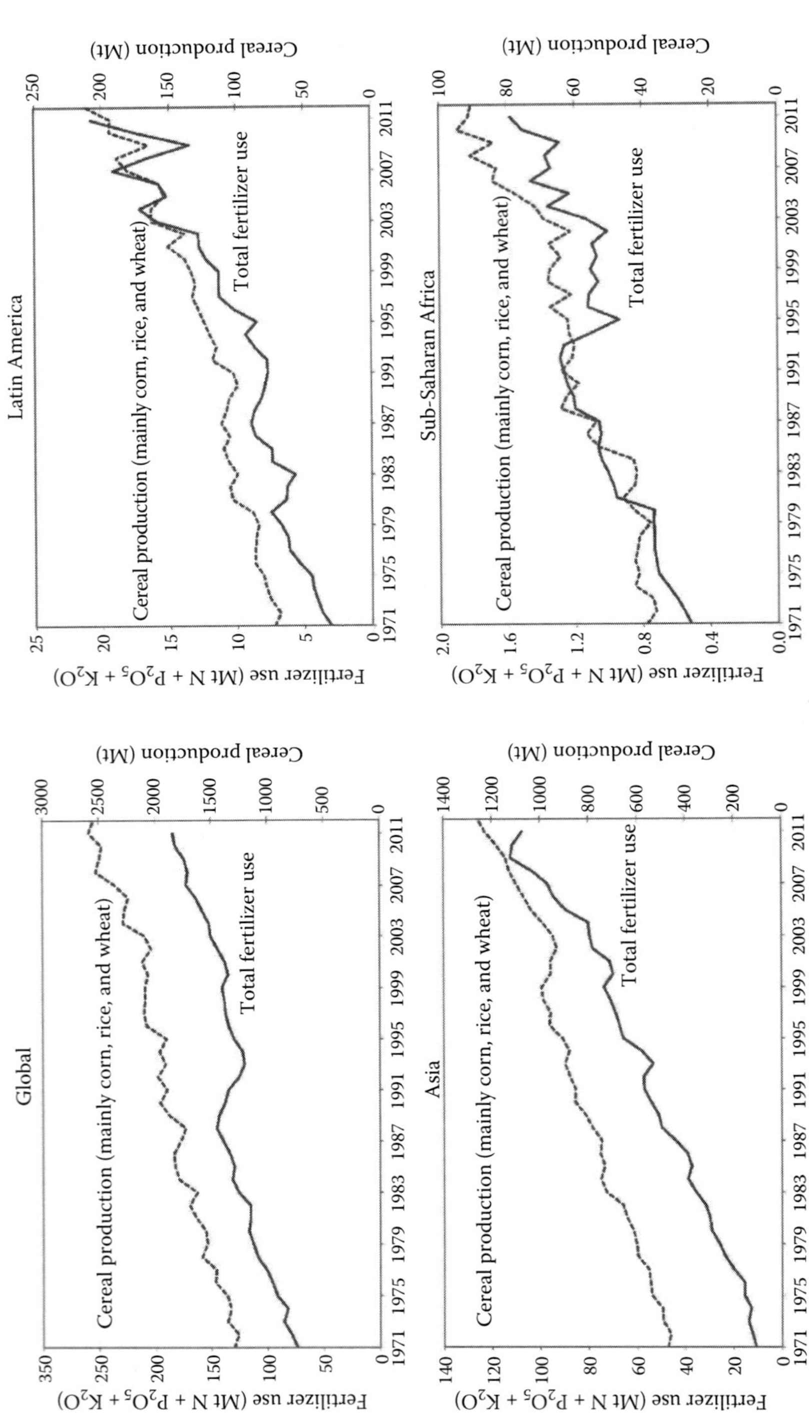

FIGURE 3.3 Global and regional fertilizer consumption. (Adapted from Food and Agriculture Organization of the United Nations. 2012. *FAOSTAT*. Rome, Italy.)

3.2 POPULATION GROWTH AND FOOD PRODUCTION

3.2.1 Demographic Changes

The world population has progressively increased since the beginning of agriculture around 10,000 BC. More recent major breakthroughs in agriculture science, such as the recognition that plants need nutrients in the 19th century and the invention and commercialization of the Haber–Bosch process in the early 20th century, coupled with the beginning of modern medicine, extended longevity and human fertility rates, thereby accelerating the population growth rate (Figure 3.4). Since the 1960s, the population has more than doubled and is estimated to increase up to 9.6 billion in 2050 (Figure 3.5a). While population increases in the developed regions are expected to remain virtually unchanged at 1.3 billion, the developing regions will see increases from the 2010 estimate of 5.7 billion to 7.9 billion. In addition to this significant increase, the regions will also experience massive urbanization, resulting in a shift of the urban/rural ratio from 0.8 to 2.0 (Figure 3.5b) (United Nations [UN] 2013).

The estimated required food production increase of about 60% by 2050 means that, in absolute terms, production increases must be greater than those achieved during the "Green Revolution." As with population growth, most (87%) of the increased food demand will occur in developing countries, with the developed countries

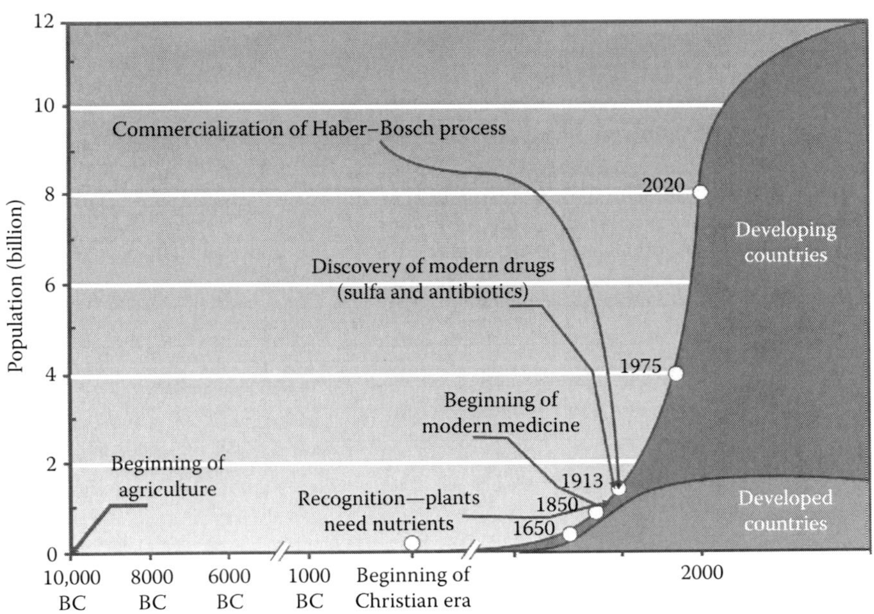

FIGURE 3.4 Impact of scientific advances in agriculture and medicine on global populations. (Adapted from Borlaug, N. 2003. Feeding a world of 10 billion people: The TVA/IFDC legacy. Third Travis P. Hignett Memorial Lecture from International Fertilizer Development Center, Muscle Shoals, AL, March 14, 2003.)

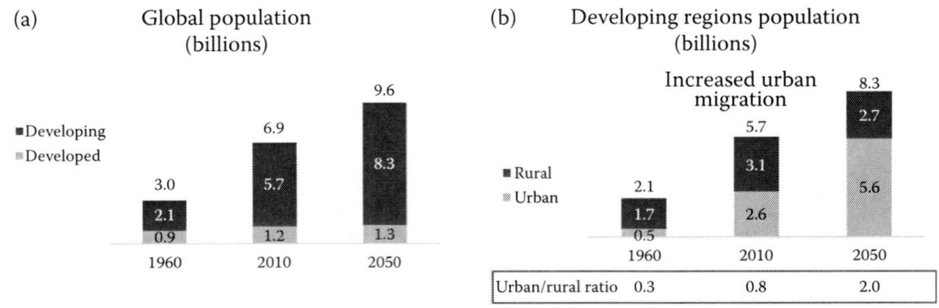

FIGURE 3.5 (a) Population growth in developed and developing regions. (From VFRC. 2012. *Global Research to Nourish the World*. International Fertilizer Development Center.) (b) Population growth in developing regions segregated by rural and urban increases. (Adapted from VFRC. 2012. *Global Research to Nourish the World*. International Fertilizer Development Center.)

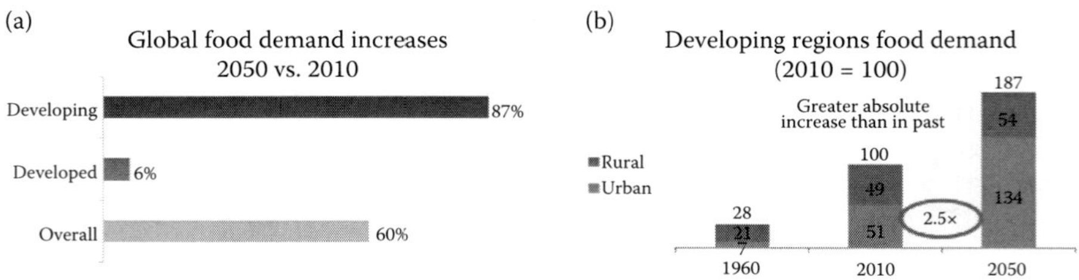

FIGURE 3.6 (a) Overall estimated global food demand increase segregated by developed and developing regions. (Adapted from VFRC. 2012. *Global Research to Nourish the World*. International Fertilizer Development Center.) (b) Estimated food demand increase in developing regions segregated by rural and urban demand. (Adapted from VFRC. 2012. *Global Research to Nourish the World*. International Fertilizer Development Center.)

experiencing only a slight increase (6%). Globally, yields will need to increase at a yearly compound rate of about 1.38%; however, in developing countries, the increase is estimated to be 1.8%. Furthermore, a significant infrastructure expansion will be required to transport up to 2.5 times more food from the rural areas of production to the urban markets (Figure 3.6a and b).

3.2.2 Food Production and the Essential Role of Plant Nutrients

Historical and scientific evidence shows that nutrients removed by harvested crops have to be replenished to maintain soil fertility and agricultural productivity. Before the availability and use of mineral fertilizers, various techniques, including shifting agriculture, use of animal manure and human waste, and crop rotations with legumes, were utilized in attempts to replenish nutrients removed by crops. Most of these attempts were not successful, and as soil fertility declined, increased food production was achieved by bringing additional land into cultivation (extensification). Over time, as the available arable land decreased, these systems were not sustainable, largely because the inputs used were low in nutrient content. Another contributing factor was poor soil conservation practices, with some exceptions such as the dike and pond fields in southeastern China. Even as late as the 1930s in the United States, approximately one-third more land was being cultivated than in the late 1990s; however, yields were only a fraction of what they are today. The damage to the environment from these practices was substantial owing to the large areas of land, including fragile lands, required for such extensive agriculture (Byrnes and Bumb 1998).

The Haber–Bosch process (first commercialized in 1913) revolutionized agriculture and proved to be the major contributor to food security. The chemical process allowed the conversion of nitrogen from air to a form that could be taken up by the plant to produce more grain and biomass. This discovery, combined with the development of phosphate fertilizers from phosphate rock, the commercial mining of potash-bearing minerals, and improved farm management practices, has allowed food production to keep pace with rapidly expanding population growth.

As mentioned earlier, mineral fertilizers are responsible for substantial increases in agricultural and land productivity. While fertilizer use is primarily focused on the major plant nutrients—nitrogen, phosphorus, and potassium—there is increasing recognition of the importance of balanced nutrients, including secondary nutrients (i.e., sulfur) and micronutrients, in yield increases (i.e., zinc, boron, etc.) (Figure 3.7). Recent work in East Africa indicates that deficiencies of sulfur and boron are limiting maize (Zea mays) and wheat (Triticum aestivum) yields in smallholders' fields (Figure 3.8a and b). Micronutrients are also important in human nutrition, and one option to include these nutrients in the human diet may be through crops.

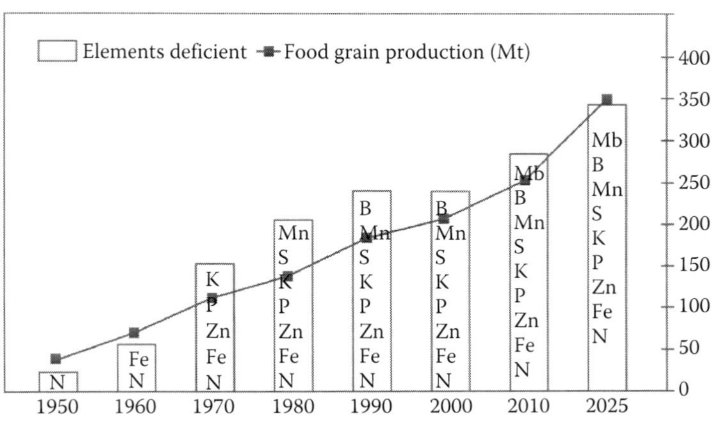

FIGURE 3.7 Relationship between balanced nutrition and yield over time. (From VFRC. 2012. *Global Research to Nourish the World*. International Fertilizer Development Center.)

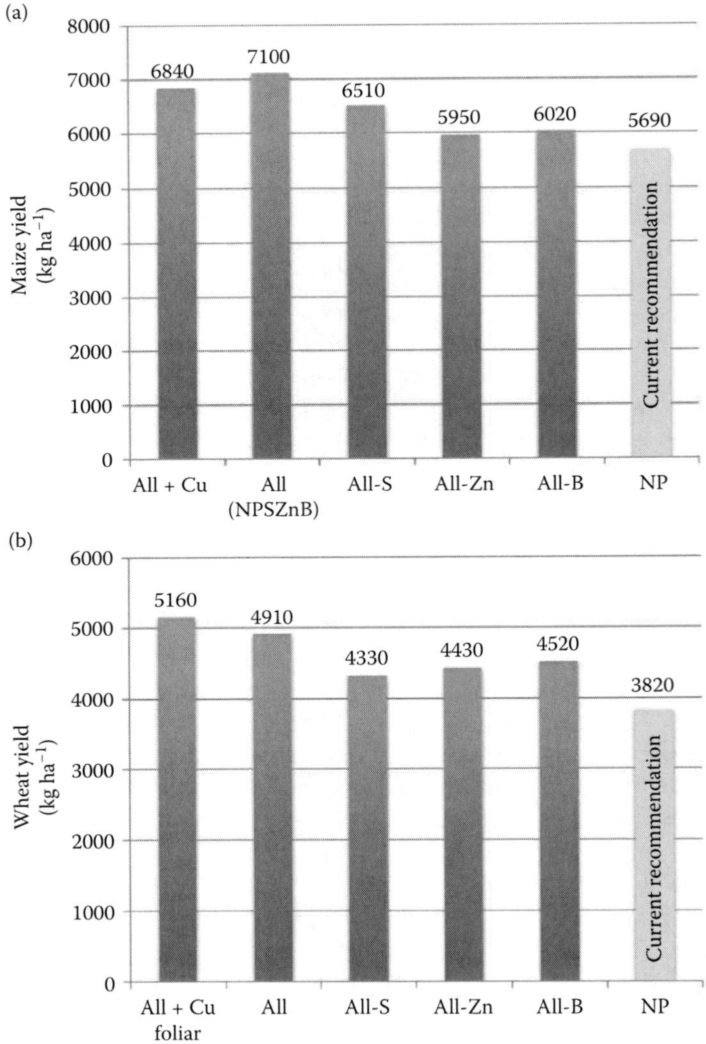

FIGURE 3.8 (a) Impact of applied secondary and micronutrients on yields over current NP fertilizer recommendation on maize in Ethiopia. (From USAID-funded AMDe project, report in preparation.) (b) Impact of applied secondary and micronutrients on yields over current NP recommendation for wheat in Ethiopia. (From USAID-funded AMDe project, report in preparation.)

3.2.3 Food Production and Natural Resources

Proper and judicious use of mineral fertilizers is crucial to maintaining land and water resources. These natural resources provide the foundation for sustainable agricultural systems and increased food production. Unfortunately, natural resources are not evenly distributed, nor are they currently managed in a sustainable manner. Most of the food required by the world's future population will be produced on today's arable lands.

While increases in cultivated areas are possible in some countries (Argentina, Brazil, Democratic Republic of Congo, and Zambia), most countries will experience decreases in arable land area owing to urbanization, land degradation, and population growth. In South and East Asia, arable land per person declined considerably during the 20th century. Therefore, land scarcity will continue to be an important driver for agricultural intensification not only in South and East Asia but also in areas of SSA. Agricultural intensification through the adoption of yield-enhancing technologies, some of which may be more demanding of plant nutrients, is vital to meeting the twin objectives of increased agricultural production and preserving the environment. A combination of efficient use of current mineral fertilizers, new fertilizer products, and recycling of nutrients in organic materials, including human, animal, and food waste, will be needed.

Globally, agriculture accounts for 70% of total water consumption, compared with 20% from industry and 10% from domestic use. While freshwater withdrawals have tripled during the last 50 years, increasing at a rate of 64 billion cubic meters a year, the quantity of freshwater that is continually renewed through the global water cycle is a finite natural resource (Food and Agriculture Organization [FAO] AQUASTAT 2013; Engelman and LeRoy 1993). Currently, irrigated agriculture accounts for 60% of these freshwater withdrawals; however, this percentage is expected to decrease owing to the virtual nonexpansion of irrigated areas since the year 2000 and the increasing availability of cost-effective, efficient irrigation systems.

Smallholder farmers are most vulnerable to water stress and must find practical solutions that increase the efficiency of water use in their farming systems. For rainfed farming systems, rainwater-harvesting practices such as ridge tillage and the use of stone contours, planting pits, and demi-lunes have become fairly common in West Africa (Winterbottom et al. 2013). For irrigated small farms, simple cost-effective microirrigation systems (i.e., Chapin bucket system) are now being utilized to increase efficiency significantly over the furrow row irrigation systems, particularly for high-value fruit and vegetable crops.

The scarcity of arable land and water will be the driver for the development and adoption of new technologies that increase yields and efficiencies of nutrient and water use. Under such conditions, nutrient use strategies must be viewed in the context of an integrated soil fertility management (ISFM) approach, which combines mineral fertilizers with organic amendments, crop rotations, and improved water management strategies that increase land productivity while maintaining soil fertility and water quality (Baanante and Hellums 1998; Syers 1997).

3.3 DOMINANT ROLE OF SMALLHOLDER FARMERS IN THE "PRESSURE POINT" AREAS

Worldwide, about half a billion farms are classified as smallholder farms (<2 ha), with many of these farms becoming smaller with time (Hazell et al. 2007). Smallholder farming dominates the agriculture sector in most developing countries, particularly in the countries classified as "least" developing. Smallholder farmers are estimated to produce four-fifths of the developing world's food; however, the majority of these farmers live in poverty and account for half of the world's undernourished (IFAD 2011).

Smallholders are a very diverse group, ranging from commercially oriented businesses to poor subsistence farmers who are forced to buy food or to seek food aid because their crop production is not sufficient to feed their families. While small farms are getting smaller, they are also becoming more numerous and are increasingly overseen by women. Typically, as countries experience economic growth, small farms gradually decrease in number owing to land consolidation as other sectors (i.e., industrial and service) experience growth and off-farm employment increases. However, many Asian and most African countries have yet to reach that peak and therefore must provide productive employment in smallholder farming for a large number of their workforce. The opportunity to accomplish this is hampered when small farms become too small to provide a reasonable standard of living and the farmers resort to unsustainable practices.

Despite the many challenges, including competition with mechanized and highly capitalized large farms (including domestic farms benefiting from foreign investment), smallholders and their farms have proven to be resilient. While smallholder farms played an important role during the Green Revolution, there is a growing concern that if smallholders continue in their current mode, huge numbers will face unacceptably low incomes and be trapped in poverty.

Ultimately, a market-led approach that results in higher cash incomes is needed to provide the capital necessary for smallholders in SSA to adopt improved technologies (i.e., fertilizers, hybrid seed, crop protection products, efficient irrigation systems) and for smallholders in South and East Asia to adopt management practices, including balanced fertilization that will result in increased agricultural income, productivity, and sustainability.

3.3.1 Characteristics of Smallholder Farmers in Sub-Saharan Africa

The geologic stability of SSA has resulted in a diverse number of primarily low-fertility soils that differ dramatically in their ability to retain and supply nutrients to plants, hold or drain water, withstand erosion and compaction, and allow root penetration. Approximately 45% of the land is classified as suitable for agriculture, including 16% that is classified as high potential or prime land. It is in SSA that one finds the most pressing example of nutrient mining, where current mineral fertilizer use averages <10 kg ha^{-1}, a level that is far below what is needed to replace the nutrients removed in annual harvested crops. In 2006, estimates by the International Fertilizer Development Center (IFDC) (Figure 3.9) indicated that SSA was losing approximately 8 million tons of soil nutrients per year and that >95 million ha of land had been degraded to the point that productivity was greatly reduced. For the 2002–2004 cropping seasons, it was estimated that 80% of the countries in SSA were losing >30 kg of nutrients per year and that 40% of the countries were losing >60 kg/year (Henao and Baanante 2006).

Land degradation is not a choice but the result of many factors, including the previously noted lack of access to mineral nutrients, along with increasing land use pressures resulting from concentrated population growth (i.e., West Africa). The population in the region has constantly increased for the last five decades and is projected to grow from the current 800 million to 1.7 billion by 2050. Of the 800

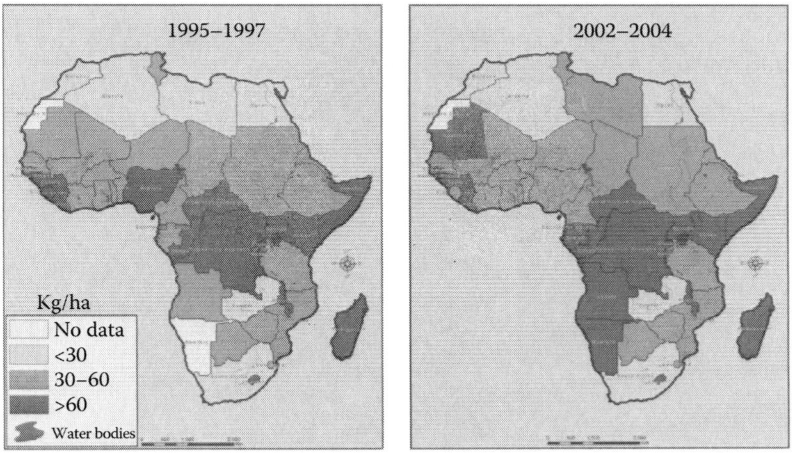

About 75% of the farmland in sub-Saharan Africa is severely degraded by soil nutrient mining. Africa loses $4 billion worth of soil nutrients every year.

FIGURE 3.9 Nutrient mining in agricultural lands of Africa (1995–1997 and 2002–2004). (From Henao, J., and C.A. Baanante. 2006. Agricultural production and soil nutrient mining in Africa: Implications for resource conservation and policy development. Tech. Bull. T-72, International Fertilizer Development Center, Muscle Shoals, AL.)

million, about 63% live in rural areas. The urbanization trend common to all developing countries is progressing more slowly in SSA, and as a result, the tipping point for SSA's transition from a rural to urban majority is predicted around 2050 (IFAD 2011). SSA's population is in the unfortunate position of being the poorest among the developing regions and is seeing its number of poor increasing while other regions have reduced the absolute number of poor despite experiencing population growth.

National economies in SSA are primarily dependent on oil, minerals, and agriculture, with the rural economy being more strongly based on agriculture than in other regions. Smallholder farms make up 80% of all farms in SSA and contribute up to 90% of production (Wiggins 2009). Owing to male urban migration, a large percentage of the smallholders are women who are not only heavily engaged in the production and processing of staple crops but are also the primary producers of vegetables and fruits grown to generate income at local markets.

Agriculture's contribution to SSA's gross domestic product (GDP) has been relatively strong for the last three decades, and in 2005 agriculture generated 27% of the GDP and employed 62% of the population (Staatz and Dembele 2007). Overall, total agricultural production increases have kept pace with population growth (FAOSTAT 2010; IFAD 2011), while cereal production per unit of land area has declined and imports of staple crops have increased. Unlike South Asia where production increases are based on agricultural intensification, SSA's increases are due to increases in cultivated land area or extensification (Figure 3.10). Smallholders who practice extensification to maintain production are, in reality, mining the soil of its nutrients and natural fertility and, in the absence of appropriate fallows, are laying the foundation for land degradation. If smallholders are to benefit from the increased

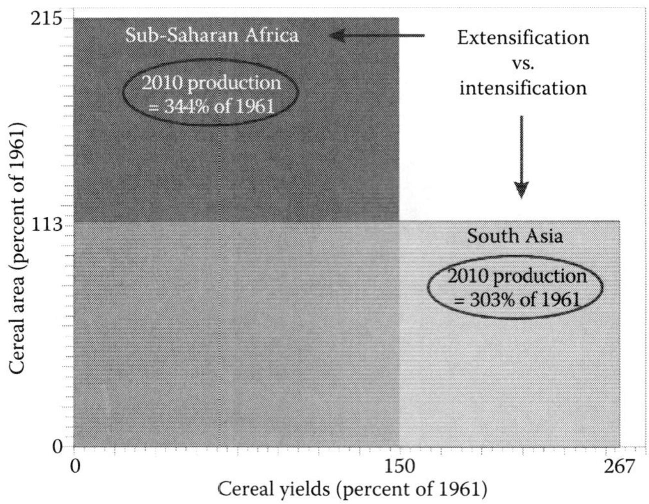

FIGURE 3.10 Cereal production, area and yield increases from 1961 to 2010. (From IFDC. 2012. *Competitive Agricultural Systems and Enterprises.* International Fertilizer Development Center, Muscle Shoals, AL; derived from Henao, J., and C.A. Baanante. 2006. Agricultural production and soil nutrient mining in Africa: Implications for resource conservation and policy development. Tech. Bull. T-72, International Fertilizer Development Center, Muscle Shoals, AL; and Food and Agriculture Organization of the United Nations. 2012. *FAOSTAT.* Rome, Italy.)

food demand and are to be competitive suppliers to local markets, they must adopt a supply response strategy centered on the move from extensification to greater intensification, which will require investments in productive on-farm technologies (IFDC 2000; Breman and Debrah 2003; Gregory and Bumb 2006). However, these investments alone are not sufficient to move smallholders from a dependence on dispersed staple crop markets toward more integrated markets; they must be combined with supportive public sector investments in infrastructure to reduce geographical isolation of SSA smallholders (Livingston et al. 2011).

3.3.2 Characteristics of Smallholder Farmers in South and East Asia

Unlike SSA, which has yet to experience widespread agricultural transformation, most of South and East Asia have experienced two transformations in a relatively short period of time. The first important agricultural transformation was the Green Revolution. The second, more recent, transformation is based on smallholders responding to growth in consumption and production of high-value commodities for domestic urban markets and exports.

As a result, the economic structure of the region has changed dramatically during the last three decades, with agriculture's contribution to GDP declining as the contribution from the service sector increased. Region-wide agriculture now accounts for approximately 25% of the GDP and 60% of employment, down from 75% in 1960 (United Nations Environmental Program 2002, 2007).

Despite agriculture's diminishing impact on the region's GDP, it is estimated that 87% of the world's 500 million small farms are in Asia, with China and India accounting for 193 and 93 million small farms, respectively (Hazell et al. 2007). Three other countries (Bangladesh, Indonesia, and Vietnam) have 10 million or more small farms each, as part of a strong agriculture sector. These small farms (including most <2 ha in size) contribute significantly to agricultural production, food security, and rural poverty reduction. Recent research (UN ESCAP 2012) indicated that improving agricultural productivity in the region could pull an additional 200 million people out of poverty by attracting private sector investment to supporting agribusiness-related industries.

In recent years, the productivity growth of staple food crops, such as rice (*Oryza sativa*) and wheat, has declined significantly. For example, rice yield growth in the irrigated areas declined from 2.31% per annum during 1970–1990 to 0.79% per annum during 1990–2000 (Bhandari et al. 2003; Pingali et al. 1997). A major reason for this decline is the displacement of cereals on better lands by more profitable fruits and vegetables. Other reasons include diminishing returns for modern varieties when fertilizers and irrigation use are already at high levels, low prices of cereals relative to other crops and input costs, pests and disease, and degradation of soil and water (Pingali et al. 1997), all of which make additional intensification less profitable (Thapa and Gaiha 2011).

3.4 IMPROVED NUTRIENT MANAGEMENT PRACTICES—IFDC EXPERIENCES WITH SMALLHOLDERS IN SUB-SAHARAN AFRICA AND SOUTH AND EAST ASIA

As smallholders in developing regions grapple with feeding and providing family livelihoods, any desire to transition to sustainable, intensive agriculture requires that they react to food demand driven by the increasing urban population. In general, urbanization leads to dietary preferences that include more protein, fruits and vegetables, and cereal and grain crops that are more easily prepared food (i.e., rice).

As in the Green Revolution, fertilizers will continue to play a central and vital role in meeting this increased demand; however, this time, there is additional interest beyond yield gains. Fertilizer will need to improve the nutritional value of crops and be used at the proper application rate to avoid underuse of nutrients in most of SSA, and overuse or unbalanced use in many countries of South Asia. Important to meeting this changing demand is the need for identification, promotion, and adoption of technologies that enhance food security, are economically and environmental sustainable, and exhibit resilience in the face of global threats such as climate change, land and water degradation, and increasing competition for finite natural resources. Addressing these challenges will require interventions in many areas, including, most important, maintenance and/or restoration of soil fertility.

3.4.1 Integrated Soil Fertility Management in Sub-Saharan Africa

Soil fertility degradation owing to nutrient mining has been identified as the single most important constraint to food security in SSA. While a large proportion of the soils have inherently low fertility, the major cause of soil fertility decline is the negative nutrient balances that occur when crops remove more nutrients than are supplied by the farmer in the form of external nutrient inputs. ISFM is an important component of efforts to restore soil fertility. Central to ISFM strategies is the combined use of mineral fertilizers and locally available organic amendments (i.e., crop residues, compost, green manures, livestock, and household waste) to provide nutrient inputs to the soil. This approach not only improves the soil's ability to sustain plant and animal productivity but it also increases the efficiency of all inputs, including mineral fertilizers (Bationo et al. 2007; Wopereis et al. 2008). In addition, ISFM promotes the adoption of improved germplasm, balanced fertilization, and improved crop management practices, including efforts to control erosion and leaching and to protect soil organic matter content. Recent work in Central Africa has shown that ISFM relative to farmer practice can more than double yields and significantly increase profits (Table 3.1).

Nonetheless, both productivity and fertilizer use in Africa remain stubbornly low, in part because numerous challenges are hampering widespread ISFM adoption. These include limited access to fertilizers and sufficient amounts of organic inputs, labor constraints, and lack of a whole-farm approach that optimizes resources to meet both commercial and food security objectives while mitigating risks. As a result, the development of profitable, scalable ISFM strategies has been slow, and adoption has been below expectations.

Indeed, for ISFM to be successful, it is increasingly recognized that ISFM must be embedded in a broader strategy that includes increased availability of macro- and micronutrient fertilizers, development of fertilizer and seed input supply chains, output market development, and provision of credit for purchased inputs (primarily seeds and fertilizers).

TABLE 3.1
Summary of Yield Increases Due to ISFM in the Great Lakes Region of Africa

Country	Crop	Farmer Practice (Yield, kg ha^{-1})	Recommended ISFM Practice (Yield, kg ha^{-1})	Increase in Net Returns Due to ISFM (US $ ha^{-1})
Rwanda	Potato (*Solanum tuberosum*)	8000	19,500	1600
	Maize	2200	4100	700
	Wheat	1400	3500	700
Burundi	Potato	3200	15,900	2200
	Rice	1500	3600	400
	Beans (*Phaseolus vulgaris*)	400	1600	300
	Wheat	300	2200	500
Democratic Republic of Congo	Potato	6600	19,100	2200
	Rice	2300	7000	2600
	Beans	200	800	100
	Maize	1000	3600	600

Note: Returns reflect two cropping seasons per year.

3.4.2 Fertilizer Deep Placement for Lowland Rice in South Asia

Working with local and national partners in Bangladesh in the mid-1980s, IFDC further developed the original Japanese concept of fertilizer deep placement (FDP) as an effective alternative to the traditional method of applying N fertilizer by surface broadcasting for lowland rice. The FDP technology involves the one-time placement of large fertilizer granules (1–3 g) in the shape of a briquette below the soil surface (7–10 cm) in close proximity to the root zone of transplanted rice, compared with two to three applications for surface broadcasting. The FDP technology makes N available to the crop throughout its growth cycle while drastically reducing losses to the atmosphere, groundwater, and surface waters. As a result, plants absorb more N, resulting in higher yields per unit of N. In Bangladesh, FDP proved to be an innovative fertilizer application technology that produced 15%–18% more rice with 35% less urea (Figure 3.11).

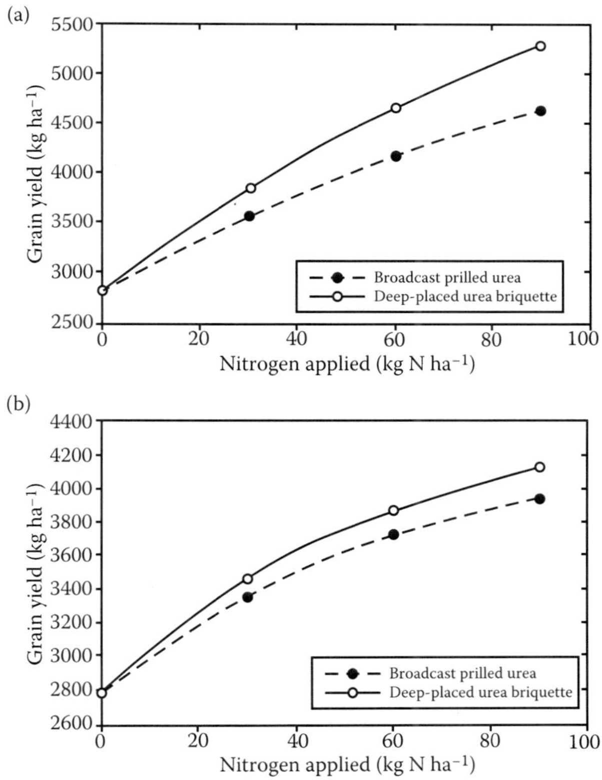

FIGURE 3.11 Field trial results with broadcast and deep-placed urea for paddy rice in Bangladesh. (a) Dry (Boro) season 28 trials; (b) wet season 31 T Aman trials. (From Singh, U., A. Hossain, Md. A. Mazid Miah, and G. Hunter. 2010. Agronomic and environmental benefits of deep-placed multi-nutrient briquettes. Paper presented at the ILSFARM NPK Briquettes Workshop, Dhaka, Bangladesh, July 25, 2010.)

Widespread introduction of FDP in Bangladesh has required significant time. The promotion of the technology consists of two key components that are designed to address issues related to creating a supply of and demand for the briquettes. The first is the production of a fertilizer briquette (Figure 3.12) by compacting commercially available solid fertilizers using a small briquetting machine (Figure 3.13) designed for the rigorous operating conditions sometimes experienced in developing countries. The machine is capable of producing up to 450 kg of 1-g to 3-g briquettes per hour, with the briquette weight being determined by the crop and the desired application rate. FDP briquettes are currently being produced at the village level by

FIGURE 3.12 Briquettes.

FIGURE 3.13 Briquette machines.

>1800 entrepreneurs in Bangladesh. An additional 60 briquette machines have been exported to Africa to produce briquettes for field trials.

The second important component of FDP is the placement of briquettes below the soil surface. When used to fertilize irrigated rice, briquettes are centered between four plants at a depth of 7–10 cm up to 10–14 days after transplanting using a self-loading mechanical applicator (Figure 3.14). Once in the soil, the briquette gradually releases nutrients into the root zone as it dissolves, coinciding with the crop's initial nutrient requirements, unlike broadcast urea applied to flooded rice fields where significant losses of nutrients (especially N) occur via surface water runoff, volatilization, and nitrification/denitrification. When N is deeply placed into the soil, it remains primarily in the form of ammonium, which is much less mobile than the nitrate form of N predominating with broadcast application. Therefore, losses to the atmosphere, groundwater, and waterways are drastically reduced. Only 3%–4% of N from deep-placed urea fertilizer is lost to the environment, compared with about 35% when N as urea is applied via broadcasting. FDP dramatically improves a crop's absorption of N, with two-thirds being absorbed in the grain and straw, compared with one-third when broadcast application is used (Savant and Stangel 1990; Mohanty et al. 1999).

Currently, >2.5 million Bangladeshi farmers have used FDP on >1.5 million ha, and it is being introduced to an additional 1 million farmers across the country. In the last 3 years, the FDP technology has increased rice production by 1.35 Mt, increased farm income by $364 million, and generated savings in government subsidy on urea of about $65 million. In 2009, IFDC initiated its African FDP initiative targeting lowland rice in 13 countries across the continent. As in Bangladesh, the initial results support FDP as a technology with agronomic and economic advantages (Figure 3.15).

FIGURE 3.14 Briquette applicator.

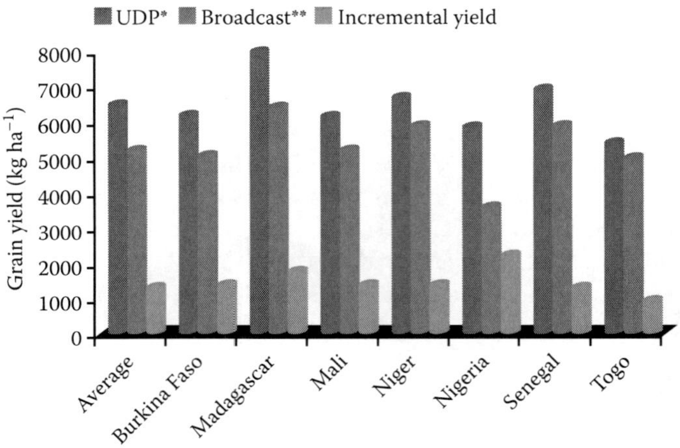

*Selected "best practice" fields using 78 kg N ha^{-1} applied by deep placement method
**Selected "best practice" fields using 115 kg N ha^{-1} applied by broadcast method

FIGURE 3.15 Rice yield increases resulting from FDP technology in Africa. (From IFDC. 2013b. *Fertilizer Deep Placement*. International Fertilizer Development Center, Muscle Shoals, AL.)

While FDP has been used primarily on rice, initial field trials indicate the technology is applicable to upland vegetables. In Bangladesh, field trials with cabbage, cauliflower, tomato, and eggplant show a 20%–30% yield increase with a 10% reduction in N fertilizer use (Figure 3.16).

The potential for rapid diffusion and impact of this technology is significant. For example, in India, the total area under rice cultivation is 43 million ha. If 10% of the

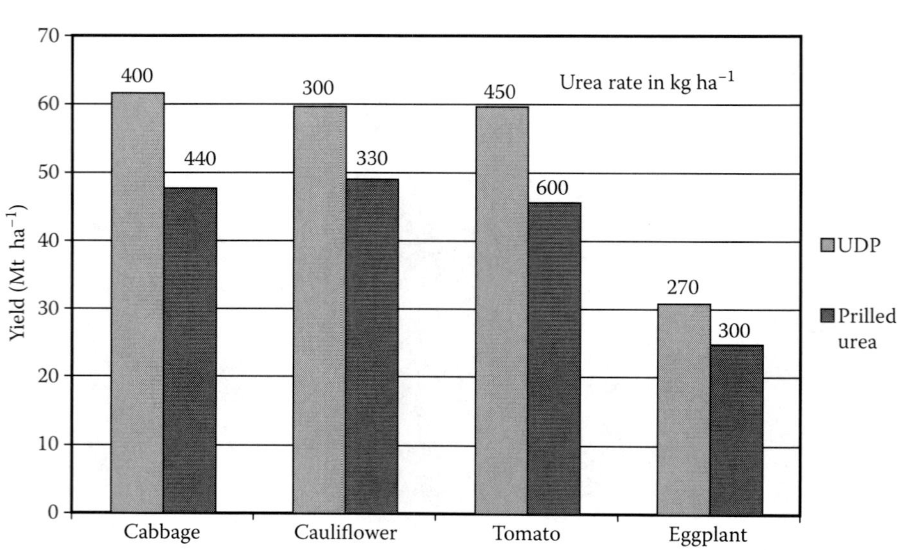

FIGURE 3.16 Comparison of yields and fertilizer use on vegetables. (From IFDC. 2013b. *Fertilizer Deep Placement*. International Fertilizer Development Center, Muscle Shoals AL.)

area could be converted to FDP, it would result in savings of approximately $130 million in fertilizer costs and an increase in rice production by approximately 2 Mt. Similar opportunities exist in Burma, Cambodia, Pakistan, Thailand, and Vietnam.

In Nigeria, rice yields average 1 Mt ha^{-1}, which is 30% less than the global average. Recent FDP trials have enabled farmers to nearly triple their yields using one-third less urea N. This increase in yield could also substantially decrease the nearly 2 Mt of rice imported to augment domestic production of 2.8 Mt, thereby generating significant savings.

3.5 ENABLING ENVIRONMENT—ROLE OF MARKETS AND POLICY IN SUPPORTING SMALLHOLDER FARMERS' ACCESS TO NUTRIENTS

Agriculture offers the best opportunity for generating sustained economic growth among the rural poor. Such growth requires expanded integration of smallholder farms into markets with timely access to information, services, and agro-inputs, such as fertilizers, seeds, and crop protection products. Despite the investment in agricultural development since the 1960s, persistent social, economic, political, and ecological impediments have slowed innovation and market integration of smallholders, particularly in SSA. However, there are numerous and increasing examples of success throughout the developing regions. For example, in SSA, IFDC assists progressive smallholder farmers to transition from subsistence agriculture to commercial farming using the Competitive Agricultural Systems and Enterprises (CASE) solution.

3.5.1 CASE Approach for Market Development

The ultimate goal of CASE is to give smallholder farmers the knowledge and tools they need to intensify the quantity and quality of their crops, and then to link the farmers to profitable markets where they can sell their surplus production. This market-oriented agriculture encourages farmers to invest in their farms to increase yields while decreasing production costs. Once production issues are addressed, CASE integrates farmers into value chains organized around specific crops or commodities, where all value chain stakeholders required to move the product from the farm to the consumer are linked. In these value chains, IFDC not only helps strengthen and professionalize the smallholders but also provides technical support to facilitate agribusiness expansion (i.e., agro-dealer development, processing, and marketing industries), including advocacy by the various private sector stakeholders to work with the government to create an enabling institutional environment for agribusiness development. In summary, CASE is built on agribusiness cluster formation, value chain development, and strengthening the capacities of public and private institutions to produce a favorable policy environment for the growth of agribusiness and trade. In West Africa, CASE facilitated the emergence of agribusiness clusters that linked >700,000 smallholder farmers to markets in Benin, Burkina Faso, Ghana, Mali, Niger, Nigeria, and Togo over a 6-year period (IFDC 2012).

3.5.2 Policy

Inadequate, and sometimes a lack of, agricultural policies are a major reason for the absence of smallholder agricultural transformation in SSA. In general, the policy environment in many SSA countries has undermined or impeded the development of competitive market systems (including credit markets), adoption of improved technologies and management practices, infrastructure development, and capacity building. For this region, the policy environment must be altered to enable smallholder farmers to invest in their soil, and to promote private sector investment in input and output markets (Gregory and Bumb 2006; Bumb et al. 2012; IFDC 2013a).

Because of SSA's generally poor soil resource base and unfavorable socioeconomic conditions, policies that support direct investments by governments and donor agencies in soil amendments (i.e., lime and phosphate rock) that provide long-term benefits to soil fertility and agricultural productivity are needed. Additional policy initiatives that should accompany these efforts include addressing credit problems, tax reforms, and land and water rights; improving rural infrastructure, marketing, and distribution networks; increasing the effectiveness of extension services to provide farmers with technical advice to maximize productivity and profitability (i.e., updated fertilizer recommendations for maximum economic yield); and an overall shifting of government emphasis from consumer support to producer support. Each of these policy initiatives would improve smallholder farmers' access to external inputs, including affordable fertilizers. Table 3.2 provides a matrix of key policy constraints in SSA along with suggestions for ameliorating actions.

Fertilizer played a key role in increasing yields in South and East Asia; however, now, yield increases per unit of fertilizer nutrient input are declining. This slowdown, combined with the degradation of natural resources, soil, and water (partly as a result of the overuse of subsidized urea N at the expense of other multinutrient fertilizers), has raised sustainability and food security concerns. A heavy reliance on N alone results in depletion of soil organic matter, deterioration of soil fertility, nutrient mining, and soil acidification, all of which contribute to the degradation of the soil's physical and chemical properties (Barak et al. 1997; Bothe et al. 2007; Sutton et al. 2011). Policymakers need to address this situation by redesigning and harmonizing current regional policies that led to fertilizer price distortions in favor of urea and unbalanced fertilization practices.

Other important policy initiatives required to ensure that fertilizer use in South and East Asia is efficient and supportive of soil fertility, natural resource protection, and food security include development and/or enforcement of fertilizer law legislation to support accessibility to a diverse mix of quality fertilizers; capacity building for extension services to provide site- and crop-specific fertilizer recommendations and improved management practices that promote balanced plant nutrition and guard against environmental degradation; support for development of robust, accessible market information systems; and promotion of public–private financial services that improve access to credit and initiatives supporting access to complementary inputs (i.e., improved seed, efficient irrigation systems, and quality crop protection products).

In both regions, governments must create an enabling environment for private sector investment in agribusiness growth by becoming facilitators instead of actors. Globally, there are numerous examples in which agro-input accessibility and use were

TABLE 3.2
Policy Constraints and Suggested Actions for Developing Competitive Fertilizer Markets in Sub-Saharan Africa

Policy Constraints	Suggested Action through Country or Regional Efforts
Regulatory architecture	Develop, update, and enact Fertilizer Law by country and also harmonize regionally among and between regional economic communities (RECs) (simultaneously) with the help of national and international experts
	Build enforcement capacity: human, analytical laboratories
Market interventions and price controls	Remove restrictions on import participation
	No import tenders
	No restricted entry to subcountry markets
	Transition to private sector imports
	Remove pan-provincial pricing and price controls
Fiscal issues—tax and tariffs	Remove withholding and value-added tax, etc.
	Zero tariff for RECs external trade
Access to finance	Legalize land property rights/long-term leases
	Mitigate risks (credit guarantees/risk-sharing, contracts, group lending); public–private partnerships framework
Outdated fertilizer recommendations	Soil testing and fertilizer trials
	Regional/mobile laboratories
	Blending services (to meet nutrients)
	Regional information networks
	Knowledge (agro-dealer training, farmer extension)
Port	Replace, repairs
	Twenty-four-hour service
	One-stop window
Inland haulage (rail, road)	Repair, build
	Agree on axle load charges
	Reduce road stops and weighbridges
	Reduce border barriers (delays)

Source: IFDC. 2013a. Developing competitive fertilizer markets in Sub-Saharan Africa: Policy and non-policy solutions. Paper presented at the Technical Convening on Seed and Fertilizer Policy in Africa, Addis Ababa, Ethiopia, December 5–7.

best served by elimination of direct government intervention in agro-input markets. However, market liberalization is not enough. It must be accompanied by a reliance on the private sector to play the leading role in the supply of inputs and the recognition that governments' crucial role is to provide supporting public goods and services and to create conducive policy, legal, and regulatory environments (IFDC 2000).

During the transition period when governments morph from actors to facilitators, farmers and agro-input dealers must organize themselves to create economies of scale in agro-input procurement, gain access to credit and productivity-enhancing technologies, and prepare to advocate for policies that support agricultural development. Consistent efforts must be made to improve the technical and business knowledge of the private sector (farmers and agro-input dealers) so that they are prepared to assume their respective roles and to engage in interactions with the public sector. Ultimately, successful interactions between the public–private stakeholders are critical for increased agricultural productivity; however, increased productivity alone is not sufficient to move the smallholder to profitable farming. Because demand for mineral fertilizers is derived from the demand for agricultural products, these farmers must be integrated into farmer organizations and agriculture value chains to have impact in the marketplace. In summary, the only way forward for smallholders is to intensify agricultural production using productivity-enhancing technologies to boost yields that support increased food availability at the household level and larger marketable surpluses that generate higher incomes.

3.6 CONCLUSIONS

Global agriculture is facing major challenges and changes as it prepares to meet the increased food demands of a growing population that will be characterized by increasing wealth and urbanization. Despite this shift, smallholders in developing countries are expected to continue to play a crucial role in food security for the foreseeable future. Growth in agriculture has and will continue to be an important part of development transformation in many countries. Agricultural growth can stimulate the economy through the provision of capital, labor, and foreign exchange to finance and fuel growth in nonagricultural sectors (Fan et al. 2013). Therefore, development efforts must place greater emphasis on improving smallholders' productivity and sustainability of land and water. For the smallholder farmers, access to and judicious use of current and new fertilizer products will provide the basis for increased productivity and natural resource protection. These fertilizers will not only be expected to increase yields and maintain soil fertility, but they must be designed to improve the nutritional quality of crops and reduce nutrient losses to the environment. Addressing these needs will result in increased and more efficient crop production with multiple benefits. These benefits include the opportunity for some smallholders to transition to intensive commercial farming, significantly increased production in developing regions in response to dramatically increasing food demand, and protection of natural resources resulting from more efficient and sustainable use of land and water.

Public policies and actions that support farmers' access to key inputs at reasonable prices and functioning markets and opportunities for agribusiness growth will be crucial for moving smallholder farmers from subsistence to commercial agriculture and providing an opportunity for smallholders worldwide to escape the poverty trap.

REFERENCES

Baanante, C.A., and D.T. Hellums. 1998. An analysis of the potential demand for phosphate fertilizers: Sources of change and projections to 2025. Paper presented at 11th Annual Workshop of Fertilizer and Lime Center, Palmerston North, New Zealand, February 11–12.

Barak, P., B.O. Jobe, A.R. Krueger, L.A. Peterson, and D.A. Laird. 1997. Effects of long-term soil acidification due to nitrogen inputs in Wisconsin. *Plant Soil* 197: 61–69.

Bationo, A., B. Waswa, J. Kihara, and J. Kimetu. 2007. *Advances in Integrated Soil Fertility Management in Sub-Saharan Africa: Challenges and Opportunities.* Springer, New York.

Bhandari, A.R., R. Amin, C. Yadav et al. 2003. How extensive are yield declines in long-term rice–wheat experiments in Asia? *Field Crop Res.* 81: 159–180.

Borlaug, N. 2003. Feeding a world of 10 billion people: The TVA/IFDC legacy. Third Travis P. Hignett Memorial Lecture from International Fertilizer Development Center, Muscle Shoals, AL, March 14.

Bothe, H., S.J. Ferguson, and W.E. Newton. 2007. *Biology of the Nitrogen Cycle*. Elsevier, Amsterdam, the Netherlands.

Breman, H., and S.K. Debrah. 2003. Improving African food security. *SAS Rev.* 23: 153–170.

Bumb, B.L., M.E. Johnson, and P.A. Fuentes. 2012. Improving Regional Fertilizer Markets in West Africa. IFPRI Policy Brief 20. International Food Policy Research Institute, Washington, DC.

Byrnes, B.H., and B.L. Bumb. 1998. Population growth, food production and nutrient requirements. In *Nutrient Use in Crop Production*, edited by Z. Rengel, 1–27. Food Products Press, New York.

Engelman, R., and P. LeRoy. 1993. *Sustaining Water: Population and the Future of Renewable Water Supplies*. Population Action International, Washington, DC.

Erisman, J.W., M.A. Sutton, J. Galloway, Z. Kilmont, and W. Winiwarter. 2008. How a century of ammonia synthesis changed the world. *Nat. Geosci.* 1: 636–639.

Fan, S., J. Brzeska, M. Keyzer, and A. Halsema. 2013. *From Subsistence to Profit, Transforming Smallholder Farms*. International Food Policy Research Institute, Washington, DC.

Food and Agriculture Organization of the United Nations. 2010. *FAOSTAT*. Food and Agriculture Organization (FAO), Rome, Italy. Available at http://www.faostat.fao.org (accessed February 26, 2014).

Food and Agriculture Organization of the United Nations. 2012. *FAOSTAT*. Food and Agriculture Organization (FAO), Rome, Italy.

Food and Agriculture Organization of the United Nations. 2013. *FAOAQUASTAT*. Food and Agriculture Organization (FAO), Rome, Italy.

Gregory, D.I., and B.L. Bumb. 2006. Factors Affecting the Supply of Fertilizer in Sub-Saharan Africa. Agriculture and Rural Development Discussion Paper 24. World Bank, Washington, DC.

Hazell, P., C. Poulton, S. Wiggins, and A. Dorward. 2007. The Future of Small Farms for Poverty Reduction and Growth. IFPRI Discussion Paper 22. International Food Policy Research Institute, Washington, DC.

Henao, J., and C.A. Baanante. 2006. Agricultural Production and Soil Nutrient Mining in Africa: Implications for Resource Conservation and Policy Development. Tech. Bull. T-72. International Fertilizer Development Center, Muscle Shoals, AL.

IFAD. 2011. New directions for smallholder agriculture. Proc. Conf. Rome, Italy, January 24–25.

IFDC. 2000. *A Strategic Framework for African Agricultural Input Supply System Development*. International Fertilizer Development Center, Muscle Shoals, AL.

IFDC. 2012. *Competitive Agricultural Systems and Enterprises*. International Fertilizer Development Center, Muscle Shoals, AL.

IFDC. 2013a. Developing competitive fertilizer markets in Sub-Saharan Africa: Policy and non-policy solutions. Paper presented at the Technical Convening on Seed and Fertilizer Policy in Africa, Addis Ababa, Ethiopia, December 5–7.

IFDC. 2013b. *Fertilizer Deep Placement*. International Fertilizer Development Center, Muscle Shoals, AL.

IFPRI. 2005. The Future of Small Farms: Proceedings of a Research Workshop, Wye, UK, June 26–29. Available at http://www.ifpri.org/events/seminars/2005/smallfarms/sfproc.asp.

Livingston, G., S. Schonberger, and S. Delaney. 2011. Sub-Saharan Africa: The state of smallholders in agriculture. Paper presented at the New Directions for Smallholder Agriculture, Rome, Italy, January 24–25.

Mohanty, S.K., U. Singh, V. Balasubramanian, and K.P. Jha. 1999. Nitrogen deep-placement technologies for productivity, profitability, and environmental quality of rainfed lowland rice systems. *Nutr. Cycl. Agroecosyst.* 53: 43–57.

Nin-Pratt, A., M. Johnson, E. Magalhaes, L. You, X. Diao, and J. Chamberlin. 2011. Yield gaps and potential agricultural growth in west and Central Africa. *Res. Monogr.* 170. International Food Policy Research Institute, Washington, DC.

Pingali, P., M. Hossain, and R. Gerpacio. 1997. *Asian Rice Bowls: The Returning Crisis.* CABI, Wallingford.

Pinstrup-Andersen, P. 2013. Contemporary food policy challenges and opportunities. *Aust. J. Agric. Res. Econ.* 57: 1–15.

Savant, N.K., and P.J. Stangel. 1990. Deep placement of urea supergranules in transplanted rice: Principles and practices. *Fertilizer Res.* 25: 1–83.

Singh, U., A. Hossain, Md. A. Mazid Miah, and G. Hunter. 2010. Agronomic and environmental benefits of deep-placed multi-nutrient briquettes. Paper presented at the ILSFARM NPK Briquettes Workshop, Dhaka, Bangladesh, July 25.

Smil, V. 2011. Nitrogen cycle and world food production. *World Agric.* 2: 9–13.

Staatz, J., and N. Dembele with A. Mabiso. 2007. Agriculture for development for Sub-Saharan Africa. Background paper for the World Development Report 2008.

Stewart, W.M., D.W. Dibb, A.E. Johnston, and T.J. Symth. 2005. The contribution of commercial fertilizer nutrients to food production. *Agron. J.* 97: 1–6.

Sutton, M.A., C.M. Howard, J.W. Erisman et al. eds. 2011. *The European Nitrogen Assessment: Sources, Effects and Policy Perspectives.* Cambridge University Press, Cambridge, UK.

Syers, J.K. 1997. Managing soils for long-term productivity. *Philos. Trans. R. Soc. Lond. B* 352: 1011–1021.

Thapa, G., and R. Gaiha. 2011. Smallholder farming in Asia and the Pacific: Challenges and opportunities. Paper presented at the New Directions for Smallholder Agriculture, Rome, Italy, January 24–25.

United Nations, Department of Economic and Social Commission for Asia and Pacific. 2012. *Overview of the Agricultural Trade in Asia and the Pacific Region.* Available at http://www.unescap.org/tid/publicatyions/tipub–2616-chap1.pdf.

United Nations, Department of Economic and Social Affairs, Population Division. 2013. *World Population Prospects: The 2012 Revision, Key Findings and Advance Tables. Overview of the Agricultural Trade in Asia and the Pacific Region.* United Nations, Department of Economic and Social Commission for Asia and Pacific, New York. Available at http://www.unescap.org/tid/publications/tipub_2616-chap1.pdf.

United Nations Environmental Program. *GEO-3 Global Environment Outlook.* UNEP, New York, 2002.

United Nations Environmental Program. *GEO-4 Global Environment Outlook.* UNEP, New York, 2007.

Virtual Fertilizer Research Center (VFRC). 2012. *Global Research to Nourish the World.* International Fertilizer Development Center, Muscle Shoals, AL.

Wiggins, S. 2009. Can the smallholder model deliver poverty reduction and food security for a rapidly growing population in Africa? Paper presented at FAO Expert Meeting on How to Feed the World, Rome, Italy, October 12–13.

Winterbottom, R., C. Reij, D. Garrity et al. 2013. Improving land and water management. Working Paper, Installment 4 of *Creating a Sustainable Food Future*, World Resources Institute, Washington, DC. Available at http://www.worldresourcesreport.org.

Wopereis, M.C.S., A. Mando, and B. Vanlauwe. 2008. *Agroecological Principles of Integrated Soil Fertility Management.* Tech. Bull. T-73, International Fertilizer Development Center, Muscle Shoals, AL.

4 Enhancing Soil and Landscape Quality in Smallholder Grazing Systems

Jean L. Steiner, Alan J. Franzluebbers, Constance Neely, Tim Ellis, and Ermias Aynekulu

CONTENTS

4.1 INTRODUCTION

4.1.1 Global Importance of Grazing Systems

Grasslands constitute the largest global land use and are an important part of agricultural and ecological systems on every continent, across a wide range of potential productivity conditions (Figure 4.1). Ruminant livestock grazing constitutes an important, and often the only viable, form of agricultural production on these lands. In most regions of the world, lands suited to some level of grazing (grasslands, woodlands, forestlands, and sparsely vegetated or barren lands) constitute the majority of the land use (Figure 4.2). In the world's lower-income countries, grassland and woodland that support some form of ruminant grazing are proportionally more important than other land uses, compared with middle- and high-income countries (Table 4.1).

It is estimated that 1 billion people depend on livestock, and livestock serves as at least a partial source of income and food security for 70% of the world's 880 million rural poor who live on less than US$ 1 per day (Neely et al. 2009). About 23% of the world's poor are located in sub-Saharan Africa, and the majority of these depend on livestock for some part of their livelihoods (Thornton et al. 2002). About 25 million pastoralists and 240 million agro-pastoralists in sub-Saharan Africa depend on livestock as their primary source of income (IFPRI and ILRI 2000). The Millennium Ecosystem Assessment (2005) found that 10%–20% of drylands were degraded. Additionally, the world's grazing lands are subject to pressures of increasing population, encroachment of cropping onto former grasslands, uncertain land tenure, lack of infrastructure (transportation, markets), lack of capital, and are faced with increasing pressures of variable and changing climate.

Grazing lands provide a wide range of ecosystem services, including provision of food, livelihoods, biodiversity, habitat, carbon storage, water filtration, and others.

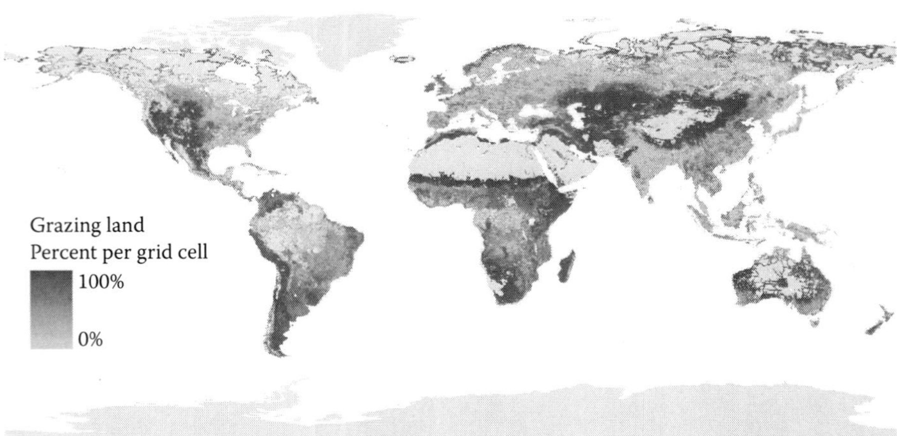

FIGURE 4.1 Global distribution of grazing land in the year 2000. (Adapted from Erb, K.-H., V. Gaube, F. Krausmann, C. Plutzar, A. Bondeau, and H. Haberl, *J. Land Use Sci.*, 2, 191, 2007.)

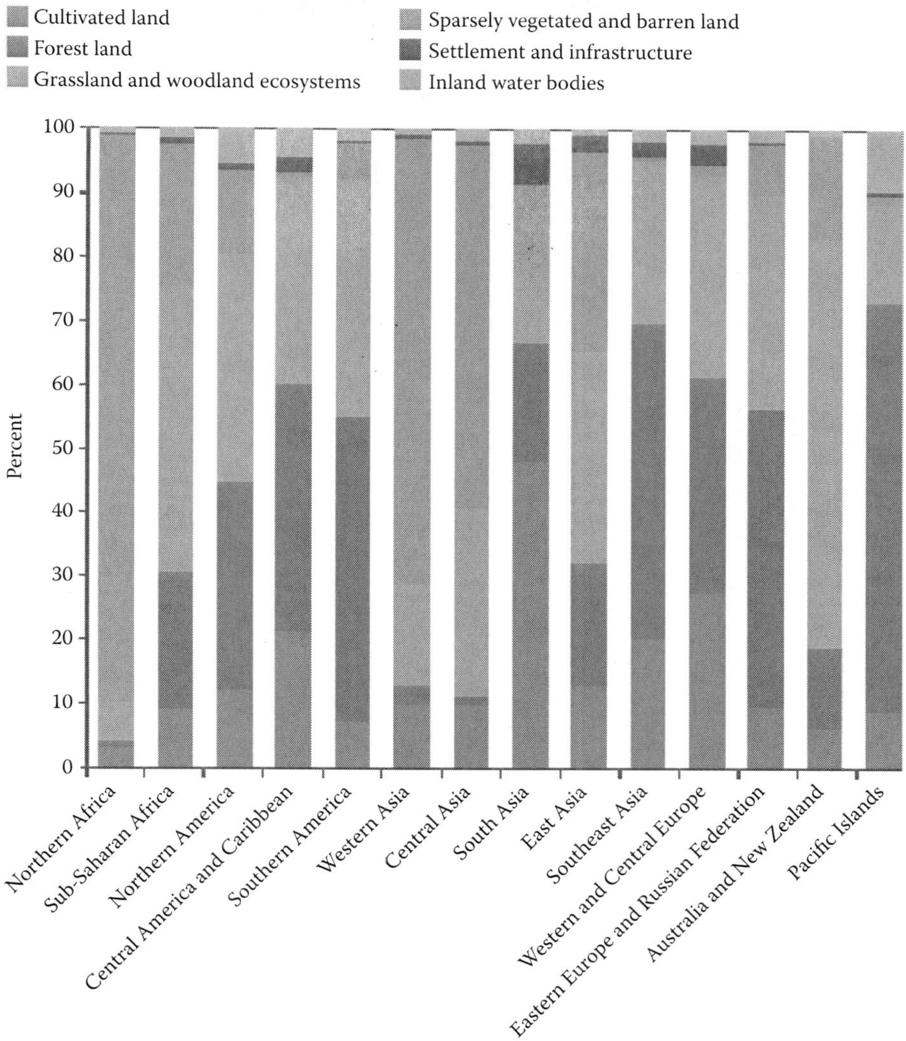

FIGURE 4.2 Land use classification showing the importance of grazing lands (sparsely vegetated and barren, grassland and woodland, and forest land uses) in all agricultural regions. (Adapted from Food and Agriculture Organization [FAO]. 2011. *The State of the World's Land and Water Resources for Food and Agriculture [SOLAW]—Managing Systems at Risk.* Food and Agriculture Organization of the United Nations, Rome and Earthscan, London.)

Grazing lands are particularly important in the world's dryland areas, which are especially sensitive to land degradation. Of the 3.4 billion ha of rangelands worldwide, an estimated 73% are affected by soil degradation (World Overview of Conservation Approaches and Technologies 2009). Additionally, land clearing for expansion of pastures for livestock production is one of the driving forces behind deforestation in the tropics (http://www.fao.org/ag/againfo/themes/en/Environment.html).

TABLE 4.1
Land Use for Low-, Middle-, and High-Income Countries

	Area and Population			Land Use Classification					
Category	**Land Area**	**Global Land Area**	**Global Population**	**Grassland, Woodland**	**Sparsely Vegetated, Barren**	**Cultivated**	**Forest**	**Settlement or Infrastructure**	**Inland Water**
	Mha	%		Mha					
Low income	2862	22	38	1020	744	441	564	52	41
Mid income	6856	53	47	2266	1422	735	2285	69	79
High income	3305	25	15	1299	592	380	880	31	123
Global total	13,023	100	100	4585	2758	1556	3729	152	243

Source: Adapted from Food and Agriculture Organization (FAO). 2011. *The State of the World's Land and Water Resources for Food and Agriculture (SOLAW)—Managing Systems at Risk*. Food and Agriculture Organization of the United Nations, Rome and Earthscan, London.

Rangelands are estimated to hold up to 30% of the world's soil carbon in addition to the carbon biomass stored in trees, bushes, shrubs, and grasses (White et al. 2000; Grace et al. 2006). In view of the vast extent of grasslands and rangelands and the degraded nature of large areas of these systems, the potential to sequester carbon through improved management is significant. Improved management can restore organic matter, reduce erosion, and avoid burning and overgrazing. The capacity to sequester carbon depends on the climatic zone, history, and status of the land resources such as soil and vegetation, and the opportunities available to change management practices (management techniques available, competition with other land uses, economic trade-offs, land tenure, social organization, incentives, and political will).

Walthall et al. (2012) determined that over the next several decades, climate change will increase the volatility and severity of extreme environmental conditions. Climate change directly affects agriculture by rendering local environmental conditions less conducive to crop growth (e.g., by shifting temperature or precipitation patterns), and by increasing the frequency and severity of extreme weather events (Giddings et al. 2013). While increases in temperature, CO_2, and precipitation may increase productivity in some regions, overall, climate variability and extreme weather events are likely to impose significant new constraints on global agriculture, adding to the difficulty of expanding agricultural production to meet increasing demand (Walthall et al. 2012). Herrero and Thornton (2013) synthesized key emerging issues relating to sustainable food systems, and indicated that as the largest land user on Earth, the livestock sector is key to balancing between food production, livelihoods, and environmental goals.

Loss of organic matter from world soils has been a source of atmospheric carbon dioxide since the dawn of settled agriculture, which began about 10 millennia ago. The magnitude of loss is often more in soils prone to accelerated erosion and other degenerative processes. Conversion to restorative land uses (e.g., afforestation, improved pastures) and adoption of recommended management practices can enhance soil organic carbon (SOC) and improve soil quality (Lal et al. 2007). The rate of SOC sequestration with adoption of recommended technologies depends on soil texture and structure, rainfall, temperature, farming system, and soil management. Strategies to increase the SOC pool include soil restoration, perennial pasture and woodland regeneration, no-till farming, cover crops, efficient nutrient management, manure application, reduced grazing pressure, water conservation and harvesting, efficient irrigation, agroforestry, and growing energy crops on spare lands (Lal 2002, 2004).

The magnitude of this potential can be illustrated for the West Asia–North Africa (WANA) region that has a land area of 1.7 billion ha and a population of 600 million. Desertification and soil degradation are severe problems in the region. The historic loss of the SOC pool for the soils of the WANA region may be 6–12 Pg (10^{15} g) compared with the global loss of 66–90 Pg of SOC. Assuming that 60% of the historic SOC loss could be restored, the total SOC sink capacity of the WANA region may be 3–7 Pg (Lal 2002).

Our understanding of the role of grassland ecosystems as net sinks or sources of greenhouse gases (GHGs) is limited by a paucity of information regarding management impacts on the flux of nitrous oxide and methane (Liebig et al. 2010). In an assessment of northern Great Plains grassland systems, the pastures evaluated were significant sinks for SOC and minor sinks for methane. Fertilization of introduced perennial pasture resulted in soil nitrous oxide emissions that were three times greater than unfertilized native pasture with high and medium stocking rates. Liebig et al. (2010) attributed differences in enteric methane emissions from the pastures in their study to differences in stocking rate, although forage and feed quality is a significant factor in enteric methane emission rates from ruminant livestock.

Eswaran et al. (1997b) reported on the development of the Food and Agriculture Organization (FAO) Soil Map of the World, which, together with other data, can be used to make continent-level assessments of land productivity and sustainability. Africa, with a total land mass of about 307 million ha and a population exceeding 746 million, has generally lagged in agricultural development compared with other continents (Eswaran et al. 1997b). Fifty-five percent of the land in Africa is unsuitable for any kind of agriculture except nomadic grazing. These lands, which are largely deserts and steep to very steep lands, have constraints to sustainability; however, about 30% of the population depends on these fragile land resources (Eswaran et al. 1997a). These authors suggested the need for major investments to enhance the productivity of the soil resources available to resource-poor farmers of the African continent.

Abberton et al. (2008) reviewed the potential for genetic improvement of forage species to reduce the environmental footprint of grazing systems through improved nutrient use efficiency of plants by reducing fertilizer requirements and improving forage quality. Such changes would improve livestock nutrient use efficiency and reduce methane emissions from ruminant grazers.

4.1.2 Smallholder Grazing Systems

The term "smallholder" is relative to the size of landholding that prevails in a particular country or region; however, the land area available to a smallholder household is substantially less than to the regional average. Often, smallholders depend on multiple sources of income, of which agricultural production is only one. Although smallholder households produce a relatively lower portion of marketed agricultural products than larger agricultural operations, the land management by smallholder households can constitute a large portion of the landscape of a region or country. In addition, food and goods are often consumed and utilized within the household where they are produced and not typically accounted for in national production indices. Therefore, smallholder production may be undervalued in national and global assessments. In lower-income countries, land, capital, credit, and other resources available to smallholder households are limited. These households must cope with a variety of risks and vulnerabilities (FAO 2013), and livestock provide options for coping with environmental stresses, owing to their mobility, ability to survive on a wide variety of feedstuffs, and provision of food for household consumption or local marketing. In middle- or higher-income countries, the family heritage on a particular piece of land or the rural lifestyle may be equally, or more important than income from smallholder agriculture.

Sere and Steinfeld (1996) classified ruminate production systems on the basis of solely livestock versus mixed crop–livestock systems (Figure 4.3). In this chapter,

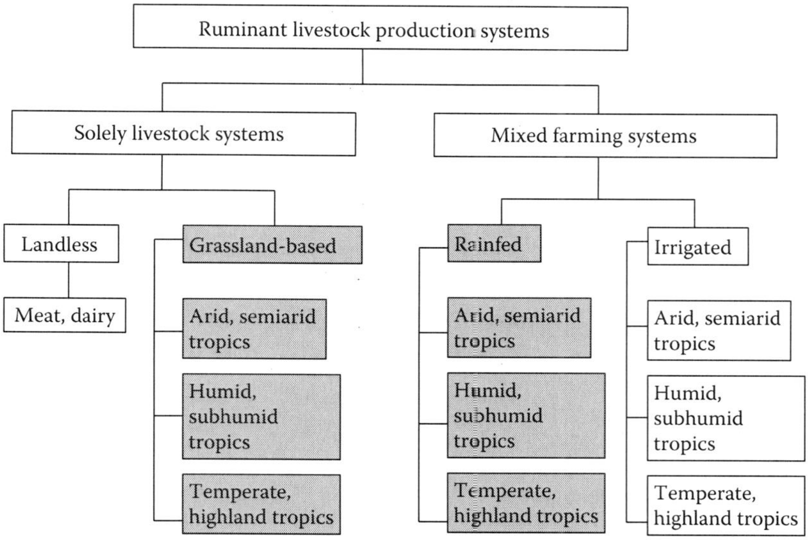

FIGURE 4.3 Livestock production systems classification. Shaded boxes represent ruminant livestock grazing systems that are most important to smallholder households, which are the focus of this chapter. (Adapted from Sere, C. and H. Steinfeld, 1996. World livestock production systems: Current status, issues and trends. Animal production and health paper no. 127. Food and Agriculture Organization of the United Nations, Rome.)

we will focus on grassland-based livestock systems as well as rainfed, mixed crop, forest, and woodland-based ruminant grazing systems, as these are the most dominant systems for smallholder households. The types of grazing systems vary across climate regimes as well as by land tenure (Table 4.2).

4.1.3 Soil–Plant–Animal–Environmental Linkages in Grazing Systems

For annual cropping, soil quality is generally defined in terms of physical, chemical, and biological components of the soil. In grazing lands, soil quality and integrity of the vegetative and faunal communities are intrinsically intertwined, and both are influenced by grazing systems. These are critical relationships in the plant root zone, which provides nature's buffer against precipitation and nutrient deficits. Soil water storage capacity is determined by root depth and plant-available water-holding capacity, and both are likely to be altered by plant and soil faunal activity. Canadell et al. (1996), in a global review of maximum root depth of vegetation types, reported that temperate grasslands had maximum root depths between about 1.2 and 6 m, with a median of about 2.5 m. Reports of maximum root depths for grassland/savannah range between about 1.5 m (presumably the grass components) and >60 m (presumably the tree component). In a study from savannah in the Upper Burdekin of Australia, Williams et al. (1997) estimated total plant-available water capacity in the rootzone to be 140 mm for grass and 380 mm for trees. We have found no reports describing the effect that grazing might have on the rooting depth of pastures;

TABLE 4.2
Characteristics of Key Grazing Systems Suited across a Range of Agroecoregions

	Climate			
System	**Arid**	**Semiarid**	**Subhumid and Humid**	**Temperate and Tropical Highland**
Mobile systems on communal grassland	Nomadic, pastoral societies; camels, sheep, cattle, goats	Transhumance, communal properties, and cropland; cattle, small ruminants	Transhumance, semi-transhumance; may be limited by insect-borne disease (tse-tse fly, ticks)	Cold temperature limits feed availability; may rely on forage conservation or transhumance
Sedentary systems on communal grassland	n/a	Sedentary livestock farmers; feed sources may include browsing, crop residue, and weeds; diverse activities to secure livelihood; mixed livestock species	Sedentary on native pastures; evolves to mixed farming; may include absentee owners with hired herders	May involve transhumance; evolves from sedentarization; fragmented pasture land
Ranching and grassland farming	n/a	Extensive grazing on native plants; private land ownership; cattle and sheep	May involve fencing, pasture improvement, and silviculture; in some regions, deforestation driven by ranching or grass farming expansion	Extensive ranching or livestock farming; requires winter feed source; may involve seasonal transhumance or movement of livestock

Source: Livestock Production Systems Classification, FAO Livestock and Environment Toolbox (http://www.fao.org/ag/againfo/programmes/en/lead/toolbox/index.htm).

however, it is reasonable to assume that closely grazed pastures will have shallower root systems than lightly grazed or intermittently grazed systems.

Infiltration is a critical process of ecosystem water budgets, and grazing can degrade it significantly. Perennial forages often improve soil organic matter (Magdoff and Weil 2004) and link surface and subsurface soil via continuous biopores, e.g., channels made by roots (Elkins et al. 1977) and faunal activity (Shipitalo et al. 2000). These changes in structure typically accompany soil regeneration by perennial pasture species, and basal area and biomass have been used to develop soil infiltration indices in rangeland (Roth 2004; Tongway and Hindley 2004). Bardgett and Ward (2003) and McNaughton et al. (1998) reported no reduction in belowground biomass on an annual timescale owing to "natural" grazing on the African Serengeti. However, heavy grazing by sheep in Mongolia reduced belowground biomass (Zhao et al. 2005), and other authors reported similar biomass interactions and reduced faunal activity (Holt et al. 1996; Bardgett et al. 1998; Gross et al. 1991). If uncontrolled grazing results in excessive removal of living and senesced vegetative cover, a process of soil degradation begins that threatens both the water and nutrient supply to the ecosystem. To sustain a healthy condition, grazing timing, intensity, and duration must be controlled to maintain vegetative cover on the surface and reduce surface soil degradation by animal traffic (Figure 4.4). Herding has been a method of control of stock movement to allow soil and vegetation regeneration. Fencing is also an effective method for doing this but requires capital investment and different management skills.

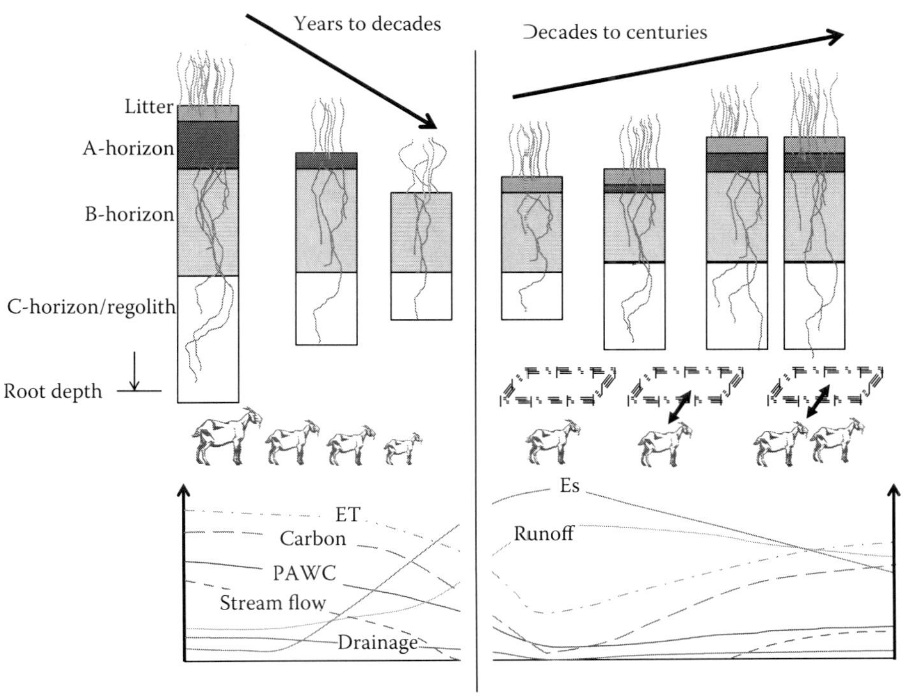

FIGURE 4.4 Schematic of degradation and regeneration of vegetative cover and soil quality with uncontrolled and controlled grazing.

4.2 GRAZING IMPACTS ON SOIL QUALITY

Soil structure and biological health have been directly linked to food production, food security, and environmental quality (i.e., air and water quality, climate stability). Unfortunately, moderate to severe degradation of soils (i.e., loss of soil biodiversity, poor soil structure, and loss of nutrients) has occurred in many agricultural settings as a result of heavy human demands. The pressure to produce as much food and fiber as possible is high, often without sufficient attention to conserving and nurturing resources and processes fundamental to primary production, i.e., soil, water, carbon, and air, and the biophysical processes that control the stocks and flows of these. Reports on the state of land suggest that soil sediment, nutrients, and organic matter stocks are being degraded at rates far exceeding a sustainable level. This has had enormous direct and indirect consequences on the profitability, productivity, and quality of agroecosystems worldwide (National Research Council [NRC] 1993; United States Department of Agriculture–Natural Resources Conservation Service [USDA-NRCS] 1996; FAO 2011).

Scientific assessment of soil quality is essential to monitoring the sustainability of agricultural systems. Soil quality is a complex subject, encompassing the many valuable services humans derive from soil and the many ways soils influence terrestrial ecosystems. Different definitions of soil quality have been proposed, each reflecting a different perspective on the use and value of soils:

- Potential utility of soils in landscapes resulting from the natural combination of soil chemical, physical, and biological attributes (Johnson et al. 1992)
- Capability of soil to produce safe and nutritious crops in a sustained manner over the long term, and to enhance human and animal health, without impairing the natural resource base or harming the environment (Parr et al. 1992)
- Capacity of a soil to function within ecosystem boundaries to sustain biological productivity, maintain environmental quality, and promote plant and animal health (Doran and Parkin 1994)
- Capacity of soil to function (Karlen et al. 1997)
- How well soil does what we want it to do (Schjønning et al. 2003)

Soil quality can be determined from a variety of soil properties or processes (i.e., indicators), the selection of which may be partially dependent on land use. Indicators of soil quality will reflect important soil functions (Magdoff and Weil 2004), including

- Producing vigorous and healthy plants
- Cycling and retaining globally important nutrients, e.g., (i) storing nitrogen in soil and releasing it to roots for efficient plant production and (ii) storing carbon in soil and releasing it to the atmosphere in a dynamic balance that stabilizes atmospheric concentration of CO_2
- Supplying plants with water, nutrients, and plant growth–promoting compounds
- Protecting water quality (both groundwater and surface water) from nutrient and pathogenic contamination
- Providing physical stability and support for vegetation, buildings, and roads

- Enabling animal habitat and serving as a reservoir for biodiversity (microscopic and visible)
- Buffering against toxic accumulation (e.g., salts) and transport of natural and synthetic compounds
- Filtering elements to protect animals, plants, and the environment from undesirable exposure

Land managers and scientists do not have unlimited time and resources to study all of the potential functions served by soil in a region, nor can they predict future needs or demands on soil resources. Soil quality assessments often use a small group of indicators (i.e., a minimum data set) to economically and efficiently characterize selected key soil functions, distinguishing between static and dynamic soil properties. Topography, hydrology, and climate also affect productivity and environmental quality of a site, somewhat independent of management. Static soil properties provide the contextual background for how soil management practices might eventually alter dynamic soil properties and reflect the inherent characteristics of a particular site, e.g., soil texture, mineralogy, and classification, all of which are influenced by geologic history and climatic conditions. Dynamic soil properties are those properties that can change value over relatively short time periods (e.g., months, years, and decades) and are at the leading edge of soil quality assessment because they are sensitive to management. Dynamic soil properties can indicate whether a production system uses agronomically and ecologically sustainable practices. Static as well as dynamic soil properties have been characterized in North America with regional sampling approaches by the USDA-NRCS through the periodic National Resources Inventory (Brejda et al. 2000). Similar efforts have been conducted by Agriculture and Agri-Food Canada (MacDonald et al. 1995). Sanchez et al. (2009) described global soil mapping efforts and resources; however, site-specific assessments of static and dynamic soil properties for monitoring soil quality trends remain limited at all scales.

4.2.1 Soil Organic Matter

One of the most consistent and broadly applied indicators of soil quality is soil organic matter content. Soil organic matter is often defined by carbon and nitrogen contents owing to their overwhelming importance, i.e., carbon is ~58% of soil organic matter, while nitrogen is typically 4%–7% of soil organic matter and is often the most limiting element for plant production. Changing soil organic matter often leads to large changes in soil properties and processes at all scales (Figure 4.5). Perennial forages offer clear benefits to soil quality by producing high carbon inputs and providing vegetative cover to protect the surface from raindrop impact and erosion. From the Sanborn Field in Missouri, topsoil thickness at the end of 100 years of management averaged 20 cm under continuous corn (*Zea mays*), 31 cm under a 6-year rotation of corn–oat (*Avena sativa*)–wheat (*Triticum aestivum*)–red clover (*Trifolium pratense*)–timothy (*Phleum pratense*)–timothy, and 44 cm under continuous timothy (Gantzer et al. 1990). Soil erosion rates predicted from the universal soil loss equation in this study averaged 19, 2.5, and 0.3 Mg ha^{-1} year^{-1} for the three management systems, respectively.

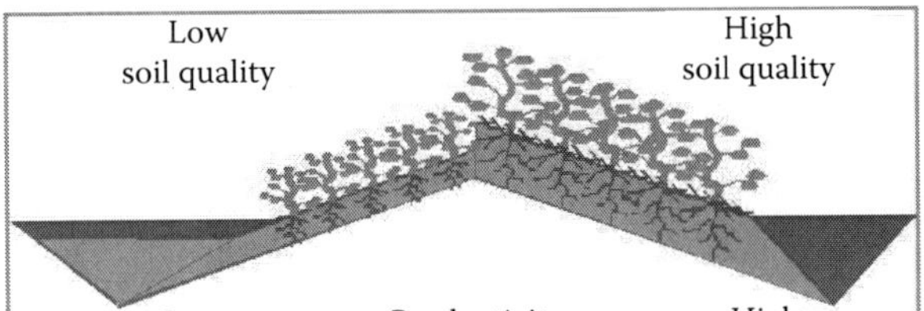

Low soil quality		High soil quality
Low	Productivity	High
?	Healthy food supply	Yes
Low	Environmental protection	High
Low	Land stewardship	High
Low	Soil organic matter	High
High	Erosion/sedimentation	Low
Low	Soil microbial activity	High
Low	Biological diversity	High
Poor	Water quality	Clean
Poor	Air quality	Clean
?	Human condition	Healthy

Soil quality relates directly to the functions performed by soil

High-quality soil is able to produce abundant plant materials, which feed, clothe, and provide shelter to humans. Plant residues not consumed must be returned to the soil to feed soil organisms and provide the organic nutrients for creating a biologically active food web. High-quality soil protects the environment from degradation, by stopping soil erosion and nutrient runoff (i.e., water quality protection) and by storing carbon in soil and reducing greenhouse gas emissions.

Low-quality soil lacks sufficient organic matter to sustain productivity in the long term, leads to excessive soil erosion and poor water quality, has low soil biological activity and diversity, and could lead to an unhealthy food supply and human condition.

FIGURE 4.5 Conceptual depiction of how differences in soil quality of a particular landscape affect various ecosystem forms and functions.

How land is managed has a large impact on the trajectory of soil quality with time (Figure 4.6). Changes in soil properties with time are a key component of dynamic soil quality assessment. Sustainable cropping systems that improve soil quality indicators with time (e.g., increase carbon stocks and flows) will lead to high soil quality, often brought about through diverse crop rotations, minimal use of tillage for weed control and seedbed preparation, and addition of organic amendments such as animal manures, crop residues, and compost. Management systems that cause a decline in soil quality indicators with time (e.g., reduced carbon stock and flows) will lead to low soil quality. This is often induced by overgrazing or cropping systems with low residue retention, intensive tillage, and near monoculture cultivation.

The terrestrial carbon cycle is dominated by two important fluxes, photosynthesis (net ecosystem uptake of CO_2 from the atmosphere) and respiration (release of carbon back to the atmosphere via plant, animal, and soil microbial respiration) (Figure 4.7). Biochemical transformations occur at numerous stages in the carbon cycle, e.g., simple sugars in plants are converted into complex carbon-containing compounds,

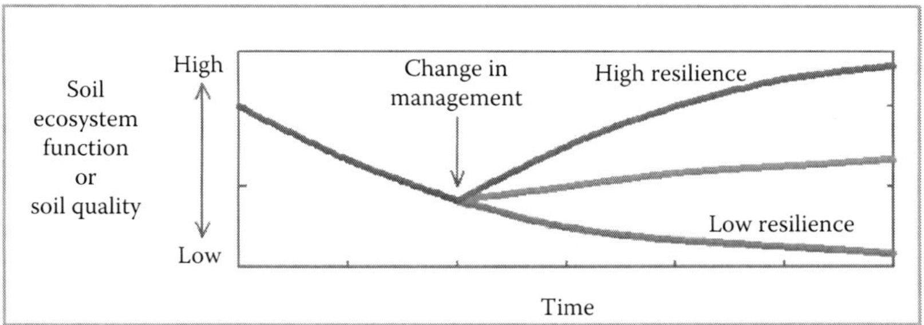

How management affects soil functional capabilities creates the boundaries of soil quality assessment. Farmers making improvements to their operations will find that organic matter inputs, soil disturbance activities, and types and combinations of row cropping and sod-based management scenarios will have some of the largest effects on how soil functions.

Farmers will also find that some soils are resilient to poor management and others are not. Those soils that respond quickly to improved management practices (i.e., high resilience) will function in a sustainable manner relatively quickly and should be targeted for immediate restoration by farmers wanting to transition from conventional to organic agricultural systems. Those soils that do not respond quickly to moderate changes in management approach (i.e., low resilience) may need more intensive management inputs for an extended period of time to restore their functional capabilities within the landscape.

FIGURE 4.6 Scenarios of change in soil quality with time as affected by resilience of a particular soil/ecosystem.

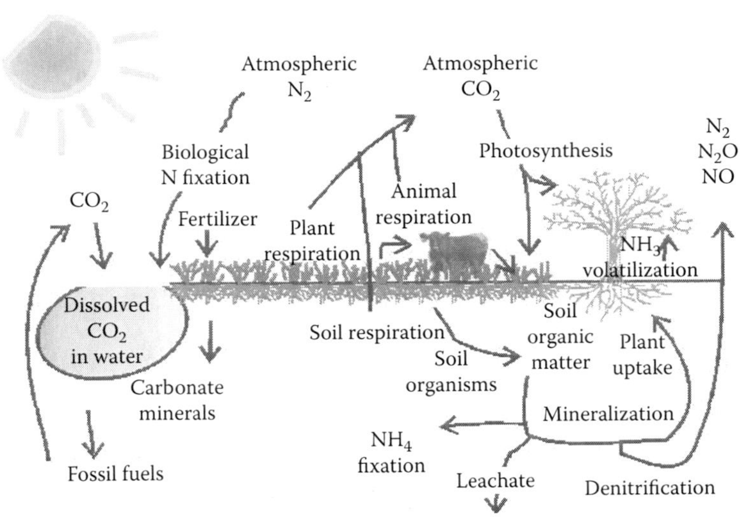

FIGURE 4.7 Pools and fluxes of carbon and nitrogen cycling in agricultural ecosystems. (Reprinted with permission from Taylor & Francis. Franzluebbers, A.J. 2002. Ecology and the cycling of carbon and nitrogen. pp. 374–377. In: Lal, R. (Ed.), *Encyclopedia of Soil Science*, Marcel Dekker, New York.)

animals consuming plants create bioactive proteins, and exposure of plant and animal residues to soil microorganisms and various environmental conditions creates humified soil organic matter complexes. Human intervention often results in harvest of enormous quantities of carbon as food and energy products. Unintended consequences of management can result in significant erosion of soil and leaching of nutrients.

Across a number of studies in different states throughout the southeastern United States, SOC was greater under grasslands than under croplands (Table 4.3). The average difference in SOC between grassland and cropland was 16.3 Mg C ha^{-1}. SOC under grasslands was not different from that under forest. Many of these surveys had single-field estimates of SOC and limited information on the type of management employed, yet pooling the data revealed reasonable conclusions about land use effects on SOC. In a like manner from a survey of soil profiles (ultisols) throughout Georgia in the southeastern United States, SOC stock to a depth of 30 cm was similar between grassland (52.5 Mg C ha^{-1}) and forestland (48.3 Mg C ha^{-1}), both of which were greater than under cropland (39.9 Mg C ha^{-1}) (Franzluebbers 2010).

TABLE 4.3
Soil Organic C Stocks in Different Land Uses in the Southeastern United States

		Carbon Stock (Mg C ha^{-1})			
Study	**Depth (cm)**	**Forest**	**Grass**	**Crop**	**Pr > F**
Eastern Texas[a,b]	30	N.D.	88 ± 18	57 ± 8	<0.01
Ten southeastern states[c]	25	31 ± 12	31 ± 16	23 ± 15	0.04
Maryland[d]	15	N.D.	32 ± 10	20 ± 7	0.01
Alabama[e,f]	25 ± 6	60 ± 21	48 ± 26	34 ± 8	0.03
Mississippi, Georgia[g,h]	25 ± 7	47 ± 2	38	22 ± 6	0.08
Mean	24 ± 6	49.9 z	47.4 z	31.1 y	

Source: Summarized from Franzluebbers, A.J., *Soil Tillage Res.*, 83, 120, 2005.
Note: z and y are mean separation notations, respectively.
[a] Laws, W.D., and D.D. Evans, *Soil Sci. Soc. Am. Proc.* 13, 15, 1949.
[b] Potter, K.N., H.A. Torbert, H.B. Johnson, and C.R. Tischler, *Soil Sci.*, 164, 718, 1999.
[c] McCracken, R.J. 1959. Certain properties of selected southeastern United States soils and mineralogical procedures for their study. *Southern Regional Bull.* 61. Virginia Agric. Expt. Sta., Virginia Polytechnic Inst., Blacksburg, VA, 146 pp.
[d] Islam, K.R., and R.R. Weil, *J. Soil Water Conserv.*, 55, 69, 2000.
[e] Fesha, I.G., J.N. Shaw, D.W. Reeves, C.W. Wood, Y. Feng, M.L. Norfleet et al. 2002. Land use effects on soil quality parameters for identical soil taxa. pp. 233–238. In E. van Santen (Ed.), *Making Conservation Tillage Conventional: Building a Future on 25 Years of Research*, Special Report No. 1, Alabama Agricultural Experiment Station, Auburn University, Auburn, AL.
[f] Torbert, H.A., S.A. Prior, and G.B. Runion, *J. Soil Water Conserv.*, 59, 1, 2004.
[g] Rhoton, F.E., and D.D. Tyler, *Soil Sci. Soc. Am. J.*, 54, 223, 1990.
[h] Franzluebbers, A.J., J.A. Stuedemann, H.H. Schomberg, and S.R. Wilkinson, *Soil Biol. Biochem.*, 32, 469, 2000.

In a survey of agricultural land uses in the Piedmont and Coastal Plain regions of the southeastern United States, SOC under pastures was significantly greater in the 0–5-cm and 5–12.5-cm depths than under conventionally tilled cropland, but no different at 12.5–20 cm depth (Causarano et al. 2008). Although information on pasture age and whether it was hayed or grazed was obtained in this study, more information on specific management practices employed would have been helpful for more insightful interpretation. The average SOC sequestration rate of 0.74 Mg ha^{-1} $year^{-1}$ during 24 ± 11 years was lower than the value of 1.03 Mg ha^{-1} $year^{-1}$ during 15 ± 17 years reported for 12 other pasture vs. crop comparisons in the southeastern United States (Franzluebbers 2005). It is expected that effective SOC sequestration would decrease with longer periods.

On Brazilian farms, Carvalho et al. (2010a) showed nondegraded pasture under fertile soil conditions had greater SOC to 30-cm depth than native vegetation. At another site with degraded pasture, SOC was lower than that of native vegetation. Converting pasture to annual cropping resulted in lower SOC than in nondegraded pasture; however, SOC was similar with cropping as with degraded pasture. Converting sole cropping systems to an integrated crop–livestock system on these sites resulted in an increase in SOC of 7.4 ± 4.6 Mg ha^{-1}. In Chile, cropping systems rotated with 2–5 years of pasture had greater total and macroaggregate-associated organic C than continuous cropping systems (Sandoval et al. 2007).

How SOC sequestration changes with time is illustrated in Figure 4.8. These data suggest that about 50% of the maximum SOC accumulation will have occurred during the first 10 years of pasture establishment, while about 80% of maximum storage could be expected with 25 years of management. The type of forage management had a large effect on the rate of SOC sequestration within the first 25 years, i.e., 0.21 Mg ha^{-1} $year^{-1}$ under hayed Bermuda grass (*Cynodon dactylon*) (Franzluebbers et al. 2000b), 0.33 Mg ha^{-1} $year^{-1}$ under grazed Bermuda grass (Wright et al. 2004), and 0.55 Mg ha^{-1} $year^{-1}$ under grazed tall fescue (*Festuca arundinacea*) (Franzluebbers

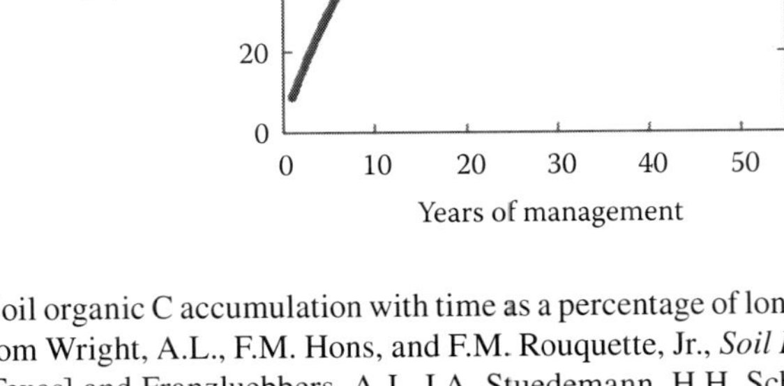

FIGURE 4.8 Soil organic C accumulation with time as a percentage of long-term maximum. (Data derived from Wright, A.L., F.M. Hons, and F.M. Rouquette, Jr., *Soil Biol. Biochem.* 36, 1809, 2004 [in Texas] and Franzluebbers, A.J., J.A. Stuedemann, H.H. Schomberg, and S.R. Wilkinson, *Soil Biol. Biochem.* 32, 469, 2000 [in Georgia].)

et al. 2000b). Grazing increased SOC sequestration relative to haying, likely due to return via animal feces to land. The cool-season tall fescue increased SOC sequestration relative to the warm–warm season Bermuda grass, which may have been in response to differential temporal distribution of moisture for plant growth and soil microbial decomposition.

Management effects on the vertical depth distribution of SOC are a topic of keen interest for scientists and good managers. SOC sequestration could be expected near the soil surface where there are large stocks of plant and animal residues, and where cold and/or dry conditions limit their decomposition. One might also expect carbon stocks in the soil profile where soil remains undisturbed and low available nutrients might limit decomposition. Sequestering SOC deep in the soil profile would likely lead to longer-term storage; however, the time required to achieve this with small downward flux of C is limiting. Data from land use comparisons in the Great Plains of the United States suggest that SOC change with management will be limited to the surface 50 cm of the profile (Figure 4.9). On eroded cropland converted to a pine (*Pinus* spp.) plantation in South Carolina, organic carbon sequestration was 0.95 Mg ha^{-1} $year^{-1}$ in the forest-floor litter layer, 0.04 Mg ha^{-1} $year^{-1}$ in the 0–15-cm soil depth, and unchanged or tending to decline at 15–60-cm depths (Richter et al. 1999). Detailed soil characterization with time would likely be necessary to detect significant changes in SOC with depth in management systems with vigorous deep-rooting capability. To provide some perspective, roots of Bermuda grass at the end of 3 years of experimentation were 3.3 ± 1.7, 0.9 ± 0.5, and 0.9 ± 0.5 Mg ha^{-1} at depths of 0–30, 30–60, and 60–90 cm, respectively (Adams et al. 1966). Assuming half of the root biomass accumulated within a year, 40% of the biomass was organic carbon, and

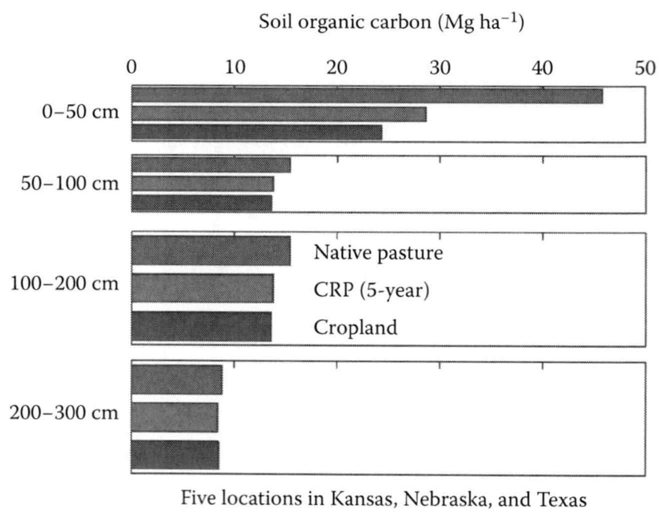

FIGURE 4.9 Soil organic C content as affected by management (native pasture, reestablishment of grassland for 5 years through conservation reserve program, and cultivated cropland) and depth of sampling across five locations in the Great Plains of the United States. (From Gebhart, D.L., H.B. Johnson, H.S. Mayeux, and H.W. Polley, *J. Soil Water Conserv.*, 49, 488, 1994.)

20% of the organic carbon could be retained as SOC following a year of decomposition, then it would have taken 9, 35, and 32 years to achieve an SOC accumulation of 1.5 Mg ha^{-1} (a minimum detectable limit) at depths of 0–30, 30–60, and 60–90 cm, respectively. However, much larger detectable differences can be expected; for example, values of 10.3 ± 3.0 Mg ha^{-1} were reported for three studies in Ontario and Illinois at a depth of 40–50 cm (Yang et al. 2008).

Fertilization of pasture with an organic amendment should increase SOC, given the high carbon concentration of the amendment relative to its nitrogen concentration. However, the evidence available to support this effect is not overwhelmingly strong. Two on-farm surveys of pastures in Alabama and Oklahoma found 5.7 Mg ha^{-1} greater SOC after one to three decades with broiler litter application than without (Sharpley et al. 1993; Kingery et al. 1994). In a 12-year pasture study in Georgia, SOC was statistically greater with broiler litter than without in only one of four management scenarios (Franzluebbers and Stuedemann 2010). The calculated rate of SOC sequestration with broiler litter was 0.21 ± 0.43 Mg ha^{-1} $year^{-1}$ among the four regimes, which was an average retention of 9% of carbon applied as broiler litter. In a review of literature, retention of carbon from manure application was estimated as 7 ± 5% in thermic regions and 23 ± 15% in temperate or frigid regions (Franzluebbers and Doraiswamy 2007), suggesting that manure application could have a more positive impact on SOC accumulation in the northern half of the eastern United States owing to temperature limitation on decomposition.

When animals graze pastures, there is a balance between carbon removal by grazing and deposition via manure that becomes available for storage as SOC. As theorized by Odum et al. (1979), pasture productivity could increase with a moderate level of grazing pressure and decline with time under excessive grazing pressure compared with no grazing (Figure 4.10). In a 5-year evaluation of coastal Bermuda grass in Georgia, the mean annual forage production was 8.6 Mg ha^{-1} under unharvested management, 9.2 Mg ha^{-1} under low grazing pressure, and 7.5 Mg ha^{-1} under high grazing pressure (Franzluebbers et al. 2004). Similar to the response in forage productivity, SOC stock and various other soil biochemical properties at the end of 5 years of management were greatest at a moderate stocking rate (Sollenberger et al. 2012). At the end of 12 years of Bermuda grass/tall fescue management in Georgia, SOC sequestration to a depth of 90 cm followed the order low grazing pressure (1.17 Mg ha^{-1} $year^{-1}$) > unharvested (0.64 Mg ha^{-1} $year^{-1}$) = high grazing pressure (0.51 Mg ha^{-1} $year^{-1}$) > hayed management (−0.22 Mg ha^{-1} $year^{-1}$) (Franzluebbers and Stuedemann 2009).

From a long-term pasture survey in Georgia, SOC was greater when Bermuda grass was grazed than when hayed (Franzluebbers et al. 2000b). The surface residue carbon was 1.8 Mg C ha^{-1} when grazed and 1.2 Mg C ha^{-1} when hayed. SOC to a depth of 20 cm was 38.0 Mg ha^{-1} when grazed and 31.1 Mg C ha^{-1} when hayed. The difference in soil and residue carbon was 7.5 Mg C ha^{-1}, suggesting an SOC sequestration rate in response to grazing vs. haying of 0.46 Mg ha^{-1} $year^{-1}$. In these temperate, humid conditions, moderate grazing levels had beneficial impacts on SOC.

Establishment of perennial grass pastures in the southeastern United States can sequester SOC at rates of 0.25–1.0 Mg ha^{-1} $year^{-1}$. SOC sequestration rate can be affected by forage type, fertilization, forage utilization, animal behavior, and soil

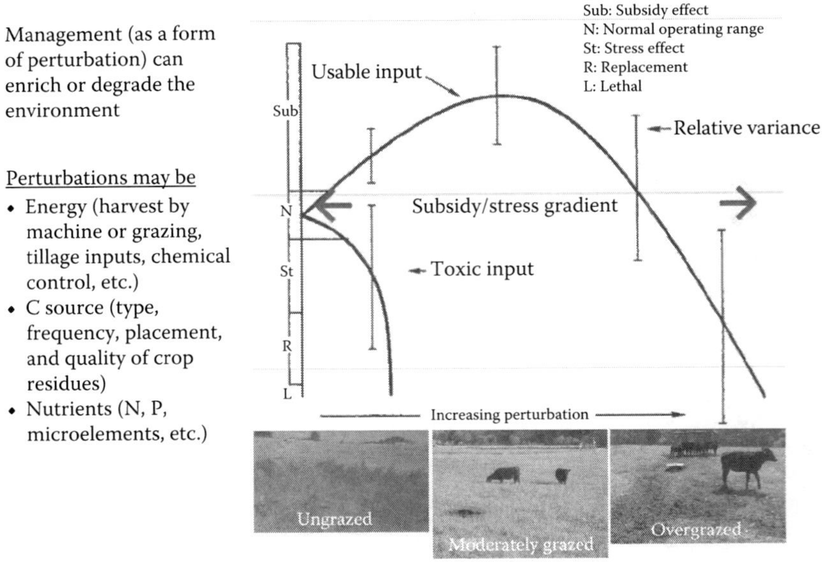

FIGURE 4.10 Effect of grazing as a form of ecosystem perturbation that results in either a subsidy to increase ecosystem productivity (e.g., at moderate grazing intensity) or a stress to decrease ecosystem productivity (e.g., when overgrazed). (Adapted from Odum, E.P., J.T. Finn, and E.H. Franz, *BioScience*, 29, 349, 1979.)

sampling depth. Data from the limited number of studies presented showed that nitrogen fertilization increases SOC storage and emissions, tall fescue pastures store more SOC than Bermuda grass pastures, and grazing returns more carbon to soil than haying or unharvested management. Additionally, SOC can also be spatially affected by animal behavior and soil depth. SOC storage under pastures is not only important for mitigating GHG emissions, but more important on the farm level for improving water relations, fertility, and soil quality (Franzluebbers 2005; Franzluebbers and Doraiswamy 2007).

There have been assessments of land use effects on SOC in many different parts of the world. Cardoso et al. (2009) reported that continuous grazing on native pasture reduced total SOC and microbial carbon contents compared with ungrazed pasture in Brazil. In Inner Mongolia, heavy grazing and continuous grazing increased soil labile organic carbon significantly compared with ungrazed sites (Wu et al. 2012); the authors recommended a grazing system with moderate intensity for improved sustainability. In semiarid grasslands in Inner Mongolia, Zheng et al. (2010) reported that species diversity and ecosystem productivity were related with abiotic factors, such as soil nutrition and precipitation, as well as anthropogenic activities, such as grazing and agriculture. Diversity and community composition provided the best predictors of system productivity in this study. In a study of a degraded steppe *Stipa*-grassland community on the Loess Plateau of China, Cheng et al. (2012) recommended rotational grazing as one of the methods of choice to promote gradual recovery of species diversity. In Kenyan savanna, removal of cattle grazing negatively influenced ecosystem gross primary productivity, while the presence of termitaria and *Acacia* trees facilitated soil water, nitrogen availability, and ecosystem productivity (Otieno et al. 2011). The authors found that heterogeneity related to topographic variations and disturbances critically influenced ecosystem functioning, productivity, and carbon storage.

In long-term land management sites, Li et al. (2007) found that fenced-grazing management was a better option for sustaining SOC and water-stable aggregates than cropping or nonfenced extensive grazing in arid grassland. The effects of cultivation and overgrazing on soil quality in arid regions have been rarely addressed. In a Mongolian meadow steppe, Han et al. (2008) found SOC, total soil nitrogen, and coarse root biomass decreased with grazing intensity. For these grasslands, using judicious herding to distribute livestock might be needed to sustain light to moderate grazing levels. Xiao-Gang et al. (2007) reported that SOC, organic nitrogen, and soil microbial respiration were lower under annual oats and perennial pasture compared with native alpine pasture. The significant decreases in many of the SOC pools in agricultural systems compared with native alpine pasturelands raise concerns about the long-term sustainability of annual pasture involving intensive soil disturbances in this environment.

4.2.2 Soil Biota and Biological Activity

A strong, functioning soil food web is largely dependent on soil organic matter and the continual cycling of plant litter, roots, and animal feces and microbial biomass back to the land (Figure 4.11). Creating this strong food web in pastoral systems is dependent on a balance between plant production and herbivore harvesting of forage. An optimum level of grazing can be hypothesized for each particular soil and ecoregion (Figure 4.10). Lavelle and Spain (2001) described in detail the microbial communities and food chains whereby the litter system is the primary food source for the biota and accessed, redistributed, and incorporated within the soil matrix by roots, worms, and termites. These are the "the ecosystem engineers," and fungi are the primary heterotrophic decomposers that make resources available to them (Colloff et al. 2010). Worms can be of several types, inhabiting only the litter layer, the litter soil interface, and the soil profile to varying depths. Worms can favor a range of diets and can be of "compacting" types, which produce resilient casts, or "decompacting" types, which feed on the casts of others. Together, the various types of earthworms heavily influence nutrient cycling and soil macrostructure (Lavelle et al. 1999). The longevity of earthworm casts and channels can be up to several years. In tropical savannahs, termites play the role of earthworms in concert with other mesofauna (Holt et al. 1996).

Biological crusts (Belnap 2006) form on soil surfaces and are hugely diverse, comprising combinations of algae, cyanobacteria, fungi, mosses, and lichen (Figure 4.12). They can play a huge role in the fine-scale hydrology and stability of soil surfaces. This is especially true in semiarid environments without complete vegetation cover. Organic compounds are also important and must remain protected within microaggregates—bacteria can break down these compounds and lead to structural decay. If organic matter is protected within (hydrophobic) microaggregates, this

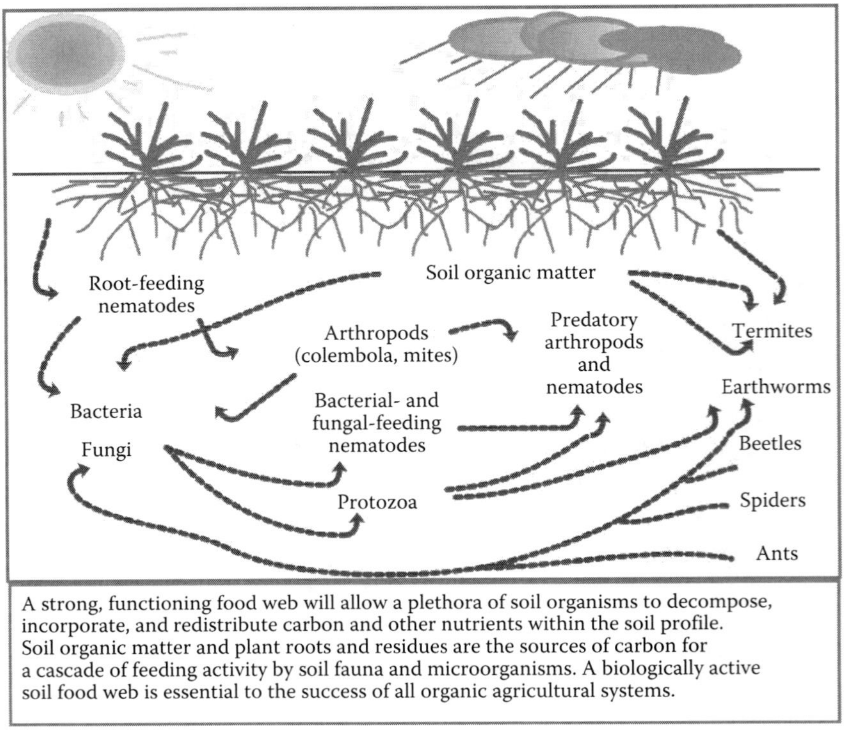

FIGURE 4.11 Soil food web dominated by the influence of plant production on soil organic matter accumulation and decomposition through the activity of various macro- and microorganisms.

FIGURE 4.12 Biological crust composed of moss and other cryptogams in a regenerated site at Canberra, Australia. This surface is resistant to erosion. (Photographer: T.W. Ellis.)

will give time for humification. Biological crusts can be an early casualty when natural rangeland is grazed. Figure 4.13 demonstrates the erosion-resistant effects of both the pasture and the biological crust, protecting the shallow loam topsoil over a gravelly subsoil with massive structure in the Upper East Region of Ghana. With continued grazing, erosion will continue; stock exclusion could help stabilization. Removing vegetative cover leaves the soil fabric exposed to raindrop impact and erosive forces of overland flow (Belnap et al. 2005). During rainfall, litter and biological crusts can protect soil structure from raindrop impact and compaction (Moss 1991a,b; Belnap 2006; Belnap et al. 2005). During overland flow, litter can form microterraces (Figure 4.14), increasing surface hydraulic roughness and producing longer durations for infiltration (Ellis et al. 2006).

Grazing affects surface soil structure, biological crusts, and microtopography by either treading or reduction in litter and soil fauna (Lusby 1970; Bridge et al. 1983; Williams and Bonell 1988; McIvor et al. 1995; Trimble and Mendel 1995; Holt et al. 1996; Mwendera and Saleem 1997; Greenwood and McKenzie 2001; Roth 2004; Belnap 2006; Drewry et al. 2008). These effects lead to reduced infiltration and soil water storage.

In a semiarid South African environment, Materechera and Murovhi (2011) found that bare land had the lowest of all the measured soil biological properties signifying limited biological activity. They found significantly greater SOC and particulate matter under fallow, while microbial biomass carbon and nitrogen were greater under grazing land use practice. The major contributing factor under grazing land was considered to be the large amount of organic matter that is returned to soil, especially that from animal dung and turnover of grass roots. Soil biological properties were further enhanced beneath the canopy of *Acacia erioloba* trees, owing to favorable microclimate and surface deposition of organic materials.

FIGURE 4.13 Degrading biological crust from an overgrazed granitic soil in the Upper East Region of Ghana. This demonstrates the erosion resistant effects of both the pasture and the biological crust, protecting the shallow loam topsoil over a gravelly subsoil with massive structure. With continued grazing, erosion will continue; stock exclusion could help stabilization. (Photographer: T.W. Ellis.)

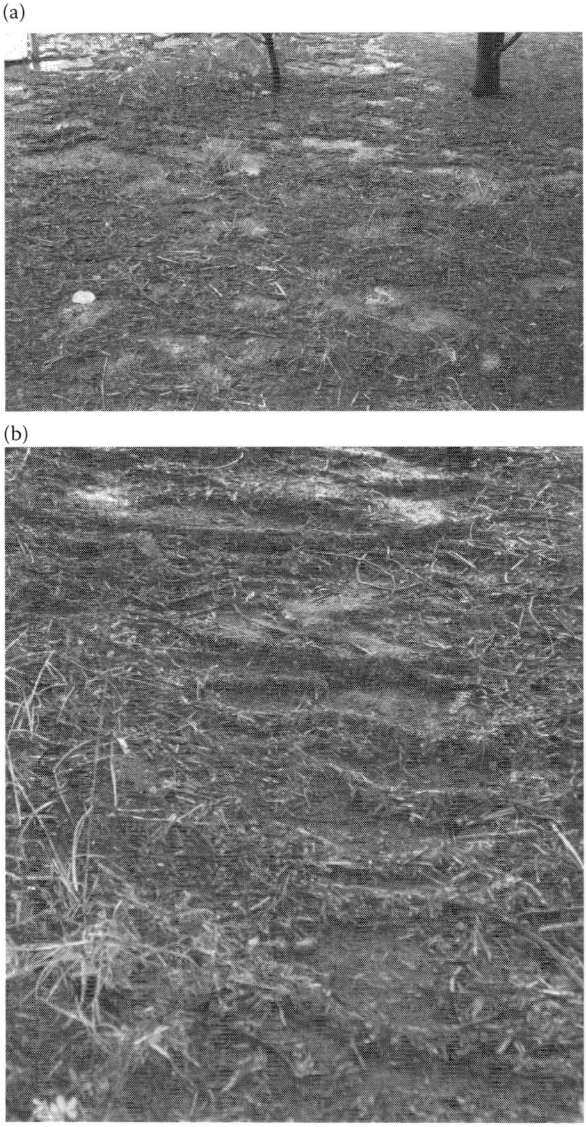

FIGURE 4.14 Microterraces formed from leaf litter of eucalypt and acacia the tree belt following sheet overland flow (a) and concentrated overland flow (b) in a rainfall simulation experiment. (From Ellis, T.W., S. Leguedois, P.B. Hairsine, and D.J. Tongway, *Austr. J. Soil Res.*, 44, 117, 2006.)

The timing of livestock grazing affects soil surface hydrology and physical quality, thus influencing soil hydrologic function (Stavi et al. 2011). Soil properties were measured in a paddock under rotational grazing during the growing season only and compared with those under grazing during the dormant season and rotational grazing during the growing season. In comparison with grazing during the growing season in the midwestern United States, paddocks with dormant season grazing had consistently worse indicators of soil structure, i.e., reduced soil water sorptivity, transmissivity, equilibrium infiltration rate, cumulative infiltration, penetration resistance, and bulk density but larger volumetric field moisture capacity, water-stable aggregates, and coarse root biomass (Stavi et al. 2011). Impairment of the hydrologic function inevitably leads to lower productivity of the system.

4.2.3 Soil Density and Crusting

When grazing animals congregate and frequently tread land, typically at drinking and feeding stations, they poach the land, denuding it of vegetation and making it susceptible to erosive forces and unbalanced nutrient loading, leading to risks of water quality impairment. This visual consequence of animal behavior on perennial pastures has been typically a focus of research documenting the impacts of grazing animals on soil and water quality. The consequences on soil in these heavy-use areas, however, are not typical of what may occur in well-managed integrated crop–livestock systems or throughout a landscape, where managed levels of grazing that maintain vegetative cover maintain good hydrologic function.

Soil compaction reduces porosity, thereby limiting air and water storage and transport, which alter nutrient cycling and the exploration potential for roots. Bulk density is a commonly used measure of soil compaction, and density of uncompacted soil is a function of texture, aggregation, and organic matter content. Coarse-textured soils tend to have greater inherent bulk density than fine-textured soils (Table 4.4), partly because of the greater propensity of fine-textured soils to be more highly aggregated (Franzluebbers et al. 2000c). Soils with high organic matter content have low bulk density, as a result of the low particle density of organic matter and the positive influence of organic matter on aggregation (Weil and Magdoff 2004).

In a review of grazing effects on soil bulk density, Greenwood and McKenzie (2001) cited 22 studies, most of which found an increase in bulk density with increased traffic. The change in bulk density between the extremes of grazing treatments was

TABLE 4.4
Soil Bulk Density as Affected by Soil Texture and Extent of Cattle Trampling in Finland

	Clay ($Mg\ m^{-3}$)			Sandy Loam ($Mg\ m^{-3}$)		
Soil Depth (cm)	Grazed, but No Visible Trampling		Poached	Grazed, but No Visible Trampling		Poached
0–5	0.90		0.88	1.30	<	1.39
10–15	1.06	<	1.14	1.41		1.45
20–25	1.08	<	1.16	1.41		1.40
30–35	1.13		1.09	1.49		1.46

Source: Pietola, L., R. Horn, and M. Yli-Halla, *Soil Tillage Res.* 82, 99, 2005.

0.12 ± 0.12 Mg m^{-3}. Bartley et al. (2010) identified a clear indication of the tendency for grazing practices to reduce saturated hydraulic conductivity by up to 90%. Most notable was a relationship between bulk density and infiltration. The low bulk densities of light grazing (10 years) and no grazing (16 years) were associated with infiltration rates up to 1000 mm h^{-1}, about 10 times greater than in denser soils associated with more intense grazing. Adams (2009) found that removal of vegetation in the Ecuadorian Andes, whether by tillage, grazing, or burning, reduced SOC and increased the tendency to crust, which would result in increased runoff and erosion risk. The authors called for leadership by nongovernmental organizations, research institutions, and the government in supporting intensification projects such as rotational grazing or silviculture to reduce pressure on the forested landscape.

Livestock can exert a significant mechanical load on the soil surface, especially considering the small footprint of large animals, such as mature dairy cows. An adult Friesian cow was determined to exert a pressure of 220 kPa on the soil (Scholefield and Hall 1986). However, the pressure can vary significantly depending on type and age of animal, land slope, and extent of movement. The range of hoof pressures reported in the literature has been 130–350 kPa for cattle (Willatt and Pullar 1983; Scholefield and Hall 1986; Nie et al. 1997), 331 kPa for horses (Cohron 1971), 83–124 kPa for sheep (Cohron 1971; Willatt and Pullar 1983), and 60 kPa for goats (Willatt and Pullar 1983). This compares to a contemporary tractor tire exerting a pressure of 100–200 kPa (Schjønning et al. 2006).

Trampled soil from cattle traffic around drinking stations in Finland had greater bulk density than in grazed pasture with no visible trampling; however, the effect was depth and soil texture dependent (Table 4.4). In Texas, bulk density at a depth of 0–5 cm was greater with cattle grazing (240 kg head^{-1}) than without, and the effect increased with increasing stocking rate (Warren et al. 1986). Soil bulk density averaged 0.91 Mg m^{-3} without grazing, 1.00 Mg m^{-3} with 0.12 animal units ha^{-1}, and 1.04 Mg m^{-3} with 0.24 animal units ha^{-1}. Increasing sheep stocking rate from 0 to 22 head ha^{-1} in Victoria Australia resulted in an increase in soil bulk density from 0.89 to 1.05 Mg m^{-3} (Willatt and Pullar 1983). These studies demonstrated increases in bulk density with animal treading (but remaining <1.3 Mg m^{-3}), which may have caused disruption of aggregation and surface sealing, but should not have been debilitating to root growth and/or air and water storage and transport.

From a survey of farms managed with continuous cropping and pasture–crop rotations (8 years cropping + 4 years grazed pasture) in Argentina, SOC did not differ. However, relative compaction was lower in pasture–crop rotation than continuous cropping in loamy/sandy loam soils (Fernandez et al. 2011). Penetration resistance of the topsoil (0–7.5-cm depth) was greater under pasture–crop rotation, although not necessarily due to surface compaction since bulk density was unaffected. This study showed that the physical condition of soil was minimally affected by conversion of continuous no-till crop management to pasture–crop rotation under no till.

Drewry et al. (2008) reviewed impacts of animal treading on soil physical properties and pasture productivity, and found a shortage of data to develop yield response curves to support decision making on pastures. In plot studies, Miguel et al. (2009) reported a 73% decrease in infiltration rate after 15 passes of cattle trampling. du Toit et al. (2009) found infiltration rate to be inversely related to stocking rate, with greater infiltration rate related to loosening of topsoil by light trampling compared with lower infiltration rate at higher stocking rates. An ungrazed control had greater bulk density and lower infiltration rate compared with the light stocking rate. In dairy systems, where soil pugging may be more intense, Singleton and Addison (1999) cautioned to manage grazing to minimize long-term compaction of the soil profile. Mills and Fey (2003) synthesized South African research and found that removal of vegetative cover by tillage, grazing, or burning reduces SOC and increases soil crusting. Interactions of SOC with soil mineralogy influences clay dispersion and crusting, and the relationship must be understood in finding appropriate soil quality indicators. Francis et al. (2001) showed that, after 6 years under pasture, several soil quality attributes had improved compared with soil cropped annually. However, the improvement during 3 years under perennial pasture was of similar magnitude to the decline under 3 years of cropping. They suggested that similar lengths of pastoral and arable cropping are needed in crop rotations for the long-term maintenance of the crop–livestock systems they evaluated.

In the southeastern United States, stocker cattle (young, weaned animals weighing 70–350 kg head^{-1}) managed for 3 years with a near-continuous grazing system (January to October) on Tift-44 Bermuda grass pasture overseeded with rye (*Secale cereale*) resulted in greater bulk density (1.63 Mg m^{-3}) than under pasture excluded from cattle grazing (1.50 Mg m^{-3}) (Tollner et al. 1990). Another study in the region showed soil bulk density at a depth of 0–6 cm declined with time from 1.53 to 1.26 Mg m^{-3} due to accumulation of surface soil organic matter. Bulk density tended to be greater at the end of summer grazing than before summer, but was not significantly affected by grazing with stocker cattle as compared with ungrazed plots during the first 5 years of management (Franzluebbers et al. 2001). These results suggest that cattle trampling could have both negative and neutral effects on soil compaction, likely depending on soil water content, vigor of plant growth recovery following grazing, stocking rate, and/or landscape features.

In the Midwest United States, soil bulk density was not affected by winter grazing of corn stalks for 1-month periods in the winter (Clark et al. 2004). The estimated corn residue consumption by cows was only 9%, so grazing time and trampling were minimal. In the southeastern United States, winter grazing by cattle of a rye cover crop following soybean (*Glycine max* [L.] Mer.) for 3 years resulted in surface soil bulk density that was not different from the same cropping system that was not grazed when managed with conventional disk tillage (1.50 Mg m^{-3}). With no-till management, the bulk density of grazed plots was greater than that of ungrazed plots (1.60 vs. 1.52 Mg m^{-3}) (Tollner et al. 1990). Disk tillage apparently removed any evidence of compaction. In contrast, grazing of winter and summer cover crops by cow–calf pairs in Georgia did not significantly alter surface bulk density under either conventional tillage or no till (Franzluebbers and Stuedemann 2008b). SOC, total soil nitrogen, and mineralizable carbon and nitrogen were unaffected by grazing of cover crops; however, soil microbial biomass carbon was sometimes enhanced with grazing (Franzluebbers and Stuedemann 2008a). Deep-profile SOC was unaffected by grazing of cover crops; however, greater SOC was retained following termination of pasture with no till than with conventional tillage (Franzluebbers and Stuedemann 2013). The surface 5 cm of a soil in Argentina had greater bulk density with winter grazing of corn and soybean residues than without grazing under conventional tillage (1.34 vs. 1.17 Mg m^{-3}), but not under no till (1.27 vs. 1.25 Mg m^{-3}) (Diaz-Zorita et al. 2002).

Pastures and forage cropping systems can result in higher bulk density than conventional cropping systems, because of heavy traffic combined with general lack of tillage, but can also result in lower bulk density due to SOC accumulation, vigorous rooting, and soil faunal activity. In the southeastern United States, soil bulk density at a depth of 0–20 cm was 1.48 Mg m^{-3} under a 20-year-old tall fescue–common Bermuda grass pasture and 1.57 Mg m^{-3} under a 24-year-old conservation-tillage cropland (Franzluebbers et al. 2000b). In western Canada, soil bulk density at a depth of 0–10 cm was not different among forage and no-till cropping systems managed continuously for 10 years (Arshad et al. 2004).

Taking all of the results on bulk density into consideration, animal grazing generally compacts soil; however, the extent of this compaction may be mitigated by controlling the timing and extent of grazing and whether the soil surface is firm enough to withstand the traffic.

4.2.4 Runoff and Soil Erosion

The loss of nutrient-rich topsoil by erosion can degrade many soil properties, including bulk density, available water-holding capacity, soil pH, and cation exchange capacity, thereby lowering crop productivity even with high fertilizer additions (Gantzer and McCarty 1987; Bauer and Black 1994). During overland flow, litter can increase surface hydraulic roughness and produce deeper overland flow to allow longer duration for infiltration (Ellis et al. 2006). Paudel et al. (2009) reported that about two-thirds of the land in the Nepal Himalaya ecosystem is degraded and that soil erosion could be as high as 87 Mg ha^{-1} $year^{-1}$ on a sloping terrace. High human population density leading to cultivation of marginal lands, livestock grazing, and depletion of biomass cover are accelerating land degradation with associated problems of malnutrition, out migration, and biodiversity loss. Soil conservation in the Inner Mongolia Autonomous Region is required to protect the soil resource base to meet nutritional requirements of the population and reduce the health risks to people and livestock associated with dust storms (Lafond et al. 2009). Appropriate conservation tillage practices were identified for small farms in the region without large requirements for capital and the need to incur a lot of risk. The authors recommend education and training programs focusing not only on farmers but also on custom machinery operators and crop input suppliers.

Ritchie et al. (2009) found that shrub-dominated subwatersheds contributed most of the suspended sediment that was measured at the outlet flume of the Walnut Gulch Experimental Watershed in southern Arizona, United States. These subwatersheds delivered more suspended sediment to the stream systems than grass-dominated landscapes. They recommended management techniques to protect grass-dominated areas from shrub invasion. In a karst topographic region, Stamati et al. (2011) found that soils devegetated by grazing had lower SOC and nitrogen contents and lower biological activity than naturally vegetated soils. Furthermore, they found a linear relationship between dissolved nitrogen export from a watershed and livestock grazing intensity.

4.2.5 Spatial Patterns Related to Animal Behavior

Cattle tend to congregate around shade and water sources, and therefore, can affect the distribution of manure and carbon inputs in pastures. At the end of 12 years of management, SOC was greater nearest shade and water sources at 0–3-, 3–6-, and 6–12-cm depths (Franzluebbers and Stuedemann 2010). The total SOC plus residue was 2–5 Mg ha^{-1} greater near shade compared with farther away in systems where the stock of SOC was 42 Mg ha^{-1} throughout the pasture under low grazing pressure and 39 Mg ha^{-1} under high grazing pressure.

In tall fescue pastures grazed by cattle for 8–15 years, SOC was greatest near shade and water sources, and declined logarithmically with increasing distance. To a depth of 30 cm, SOC was 46.0 Mg ha^{-1} at 1 m from shade, 43.2 Mg ha^{-1} at 10 m from shade, 39.9 Mg ha^{-1} at 30 m from shade, 40.5 Mg ha^{-1} at 50 m from shade, and 39.4 Mg ha^{-1} at 80 m from shade (Franzluebbers et al. 2000a). The zone within a 10-m radius of shade and water sources became enriched in SOC, most likely owing to the high frequency of organic deposition from cattle defecation and urination. To minimize the probability of nitrogen contamination of surface and groundwater supplies (since total nitrogen also increased with increase in SOC), shade/water sources are recommended to be moved periodically, positioned on the landscape to minimize flow of percolate or runoff directly from these areas to water supplies, or avoided during routine fertilization. In livestock congregation sites, such as mineral feeders, water troughs, and shaded areas, Sigua and Coleman (2009) reported soil penetrometer resistance decreased linearly with increase in distance away from the center of mineral feeders and water trough, but that it increased slightly with distance from the center of shade.

Stream channel instability is often attributed to the effects of livestock grazing. Miller et al. (2010) found that cattle exclusion significantly reduced the surface runoff depth of water and nitrogen loads compared with the grazed pasture, suggesting that this fenced pasture may act as a buffer for certain runoff variables. However, in the 3-year study, turbidity, electrical conductivity, pH, concentrations and loads of total suspended solids, and certain nitrogen and phosphorus fractions in the cattle-excluded pasture were generally not improved by stream bank fencing.

Verdoodt et al. (2009) evaluated soil and vegetation properties in communally managed and privately managed enclosures for controlled grazing, compared with uncontrolled extensive grazing in Kenya. The communal enclosures with higher levels of management than the private enclosures exhibited biomass production fully recovered up to its optimal level relative to neighboring nature reserves, while the private enclosures exhibited restricted biomass production. In the communal enclosures, improvements in topsoil bulk density, SOC, total soil nitrogen, and microbial biomass carbon and nitrogen were observed compared with the open rangeland. In private enclosures, only topsoil bulk density and the microbial biomass carbon stock were significantly higher than open rangeland.

Smet and Ward (2006) reported significant negative effects of management type on soil parameters (i.e., soil pH, nitrogen, and SOC) within 0–100 m from the water point of three production systems in South Africa. Commercial livestock ranching had the greatest negative effect on the immediate area around the water point compared with communal grazing or game ranching systems in their assessment. In contrast, Tefera et al. (2007) evaluated semiarid rangelands under communal land, government ranch, and traditional grazing reserve enclosure land use systems in Ethiopia, and found that vegetation distribution and grazing intensity was high in the communal land but more moderate in the government ranch and the traditional grazing reserves. However, there were no significant differences in vegetation along the gradient of distance from water for any of the three land management systems, suggesting that grazing disturbance had exceeded a threshold of degradation. Fterich et al. (2012) studied soil characteristics for *Acacia*-dominated drylands in Tunisia and reported that soil pH, electrical conductivity, SOC, microbial biomass carbon, the microbial-to-organic carbon ratio, and enzyme activities increased as tree size increased, while soil carbon-to-nitrogen ratio and metabolic quotient were lower with increasing tree age. In intensively grazed sites, SOC, microbial biomass, and enzyme activities were lower, while soil carbon-to-nitrogen ratio and metabolic quotient were greater than in ungrazed sites.

4.3 MANAGEMENT PRACTICES AND STRATEGIES FOR SUSTAINABILITY

4.3.1 Extensive Grazing Systems

Management practices such as fire and grazing modify the structure and function of ecosystems, affecting SOC storage (e.g., Piñeiro et al. 2010). In fire-prone ecosystems of arid landscapes, the largest pools of carbon are typically found in soils rather than in the aboveground biomass (e.g., Rau et al. 2010). At a global scale, SOC sequestration represents about 90% of the potential of what is technically feasible (Gattinger et al. 2012). Grasslands have high SOC stocks (~12% of terrestrial SOC); thus, it is important to understand the effects of management on SOC in setting appropriate management intervention and policy. Grass and woody species coexist in savanna ecosystems in which fire, herbivory, and soil types influence the relative dominance of grass and woody cover where moisture is a determining environmental factor (Sankaran et al. 2005).

The potential of African rangelands to sequester SOC is based on the assumption that they are undersaturated in SOC as a result of overgrazing and excessive use of fire (Neely and de Leeuw 2011). Consequently, grazing management and control of fire are considered interventions with the potential to sequester SOC in rangelands (Intergovernmental Panel on Climate Change [IPCC] 2000; Derner and Schuman 2007; FAO 2010). The potential of these two interventions is further supported by a number of models that predict enhanced SOC stocks under moderate grazing intensity (e.g., Conant et al. 2001; Conant 2002, 2012).

4.3.1.1 Fire

Fire is an important management tool in African savanna and grassland systems for both livestock herders and wildlife managers who use it regularly to control bush encroachment and to remove dead and dying vegetation that has low forage quality and is unpalatable to animals (Fynn et al. 2003; Sankaran et al. 2005). Fire is an important driver of ecosystem structure in savanna ecosystems (Beringer et al. 2007; Knicker 2007) in which reducing fire frequency increases carbon stocks in woody biomass (Beringer et al. 2007).

After the combustion of organic matter in the surface mat layer, SOC recuperation follows with decomposition of roots of dead or burned plants, which overtime attains a balance with the unburned plot (Oluwole et al. 2008). However, increases in SOC content have also been reported in response to an increased deposition of dry leaves and charred plant materials in fires that affect the tree canopy (Yan et al. 2012).

The role of fire in managing extensive rangelands remains controversial. Some have found controlled fire to provide positive effects in terms of control of invasive species (Fuhlendorf and Smeins 1997; Winter et al. 2011), while others have reported negative effects on vegetative composition (Nicholas et al. 2009). Teague et al. (2010) reported minimal impacts of burning on soil physical properties of southern Great Plains prairie rangelands. They found instead that rotational grazing provided benefits of less bare ground, greater SOC and nitrogen, and lower soil temperatures compared with continuous grazing on either burned or unburned areas.

Although regular burning has been reported to have a consistent short-term negative effect on SOC in grassland soils (Ansley et al. 2006; Piñeiro et al. 2010), the immediate and long-term effects of burning on SOC remain unclear. For instance, Bird et al. (2000) reported greater SOC in unburned plots in subhumid savanna in Zimbabwe, whereas Oluwole et al. (2008) found lower SOC in unburned plots than in frequently burned plots in South African dry savanna.

Fire not only influences the total biomass of savanna systems but it also markedly influences the vegetation structure of savannas (Higgins et al. 2007). Vegetation structural changes influence the microclimate and distribution of resources such as nutrients and moisture (Ludwig et al. 2002; Higgins et al. 2007). Fire disturbances can lead to ecosystem degradation associated with the disruption of the fundamental environmental cycles such as carbon and other nutrients (Corvalan et al. 2005), although managed fire is also a tool to intentionally manipulate rangeland ecosystems.

Large carbon inputs to the soil occur via photosynthesis, plant growth, and litter decomposition (Fynn et al. 2003), while a disturbance such as fire can alter plant species diversity and dominance by changing microclimate and the availability of limiting resources (light, water, and nutrients). These changes in ecosystem structure and the abiotic environment often modify key functional characteristics of ecosystems (primary productivity, hydrology, and nutrient fluxes) that have the potential to alter the storage and turnover of carbon in plants and soils (Ansley et al. 2006).

Management that enhances SOC generally results in increased protection of the soil from erosion. Conversely, poor management such as intensive burning can result in SOC losses and the land being vulnerable to soil erosion, especially on steeply sloping land (Ansley et al. 2006; Piñeiro et al. 2010). A significant contribution to SOC by the litter is only in the top few centimeters of soil and therefore its removal by fire reduces the organic matter content near the surface. The effect of fire is negligible at deeper levels because most of the organic matter in deep soils originates from root turnover (Fynn et al. 2003).

Understanding the impacts of fire on carbon dynamics is fundamental to sound land management recommendations; however, long-term effects of fire on SOC storage, particularly in the savanna–woodlands, are poorly documented (Gattinger et al. 2012). Nineteen years (1992–2011) of exclosure and annual prescribed burning in two Sudanian savanna ecosystems sites in Burkina Faso resulted in no significant difference in SOC concentration between burned vs. unburned and closed vs. open plots (Aynekulu et al. 2014). Sawadogo et al. (2005) investigated the effect of fire frequency and grazing intensity on vegetation composition at the same sites, and concluded that mean total biomass was reduced by the presence of livestock while it was not significantly affected by prescribed fire. Coetsee et al. (2010) found no significant influence of 50 years of frequent burning on SOC in a southern African savanna. However, Oluwole et al. (2008) reported greater SOC by the decomposition of roots of dead or burnt-off plants.

In addition to the external inputs of organic matter from fire-affected vegetation, it is also necessary to consider that litter (which is usually removed before soil sampling) turns, after fire, into particulate, fine-earth-sized (<2 mm) particles. These mix with the whole soil material in the organic horizon, thus causing a net increase in SOC content with highly friable charred organic matter and particulate charcoal fractions (Yan et al. 2012). Additionally, fire stimulates the turnover of root materials (Fynn et al. 2003).

Apart from these reasons, which could explain the lack of fire influence on SOC concentration and stock, changes in SOC storage occur slowly (Ansley et al. 2006). Consequently, 19-year duration of controlled studies (Aynekulu et al. 2014) could be insufficient to observe changes in SOC concentration in response to a fire regime.

4.3.1.2 Grazing Management

Savory and Butterfield (1998) introduced the concepts of holistic management to address, in a comprehensive way, the reversal of the complex interactive processes that have led to degradation of extensive rangelands in arid to semiarid regions of Africa and beyond. Today, the principles of holistic grazing planning are practiced by tens of thousands of people in many countries and contexts on as many as 12 million ha (Savory and Butterfield 2010; African Center for Holistic Management 2013). The largest impact has been on semiarid to arid rangelands and grasslands where large herds of wild herbivores originated and where they have, for the most part, disappeared. In holistic grazing planning, livestock are managed to mimic the role that wild herds once played in maintaining ecosystem health by removing or trampling senesced vegetation and loosening crusted surfaces and incorporating dung and plant material into the surface. Neely and Butterfield (2004) described the successful application of holistic management in a communal context in the Wange community in Zimbabwe. On the basis of partnerships, the community is applying holistic decision making to restore the natural resource base and empower the community members. Beukes and Cowling (2003) found that short-duration, low-frequency, intensive herbivory by livestock, leading to nonselective grazing, resulted in a greater microbial activity, soil stability, and infiltration compared with control treatments.

Controlled grazing management has been recognized as one of the strategies to enhance SOC sequestration and preservation; however, empirical evidence is sparse. Aynekulu et al. (2014) investigated effects of long-term (14–36 years) livestock exclosures on SOC in the semiarid savanna of southern Ethiopia and found no significant ($P > 0.05$) differences between exclosures and open-grazed rangelands across three age categories and two soil depths. The age chronosequence further suggested no significant changes in SOC content with increasing duration of exclosures (Aynekulu et al. 2014).

Milchunas and Lauenroth (1993) reviewed reports from 34 studies involving grazed and ungrazed sites around the world, and reported both decreased (40%) and increased (60%) SOC as result of grazing exclusion. Other studies of grazed lands worldwide have also shown both increases (e.g., Schuman et al. 1999; Reid et al. 2004) and decreases (Derner et al. 1997; Yong-Zhong et al. 2005) in SOC storage and accumulation compared with adjacent ungrazed soils.

The inconclusive evidence indicates the need to further study the effect of livestock grazing on SOC stocks before deciding to implement broad-scale SOC sequestration strategies in African rangelands. In East African rangelands, semiprivate exclosures are extensively practiced by pastoralists to put aside a fodder bank for use during the dry season when grazing resources are in short supply. Here, exclosures may be defined as an area of land that is enclosed by a fence to prevent grazing and/or browsing by livestock and restore vegetation resources (Coppock 1994; Aerts et al. 2009). Calves and sick/weak animals are allowed to graze inside exclosures for 3–4 months depending on the length of the dry season.

The Borana rangelands in southern Ethiopia comprise extensive grazing lands with indigenous knowledge on natural resource management (Homann et al. 2008). For the last four decades, however, the Borana rangelands have undergone substantial reduction in grassland cover due to bush encroachment, expansion of cultivation, and increased settlements (Dalle et al. 2006; Solomon et al. 2007; Angassa and Oba, 2008), with negative consequences on the livelihood of the local communities.

The conversion of the savanna rangelands of Borana into bushland, exclosures, and cropland began in the 1970s, with a peak expansion of bush encroachment and crop cultivation in the 1980s (Angassa and Oba 2008). During that time, about 40% of the Borana rangelands directly shifted to bush encroachment (Coppock 1994), while the communal rangelands were further shrunk by the expansion of cropland, exclosures, and ranches (Angassa and Oba 2008), leading to large declines in valuable perennial grass species and grass biomass (Angassa et al. 2012).

Widespread bush encroachment and increased grazing pressure around water points and settlements, as well as intensified cultivation, have resulted in a loss of biodiversity (Oba et al. 2000), altered nutrient cycle (Angassa et al. 2012), and reduced resilience of the ecosystem. Although not quantified to date, changes in land use may greatly affect carbon and GHG emissions, as savanna grasslands are believed to have the potential to restore carbon and avoid emissions (Guo and Gifford 2002).

In response to the shifts in patterns of land use, many villages in the Borana plateau have set aside some of their lands that have been degraded by high densities of livestock to manage as enclosures. Angassa et al. (2010) indicated that herbaceous biomass and grass basal cover were significantly greater in the zones protected from grazing than in the open-grazed areas. Enclosures had higher herbaceous species richness than the open-grazed rangelands and promoted the recovery of some herbaceous species. The authors indicated that older enclosures had no superior benefits over younger enclosures in terms of herbaceous production. The exclusion for decades of fire as a diversifying factor in the savannas of southern Ethiopia may have led to the loss of herbaceous species richness and diversity (Angassa and Oba 2008).

The Borana used the communal rangelands for seasonal grazing that involved livestock movements between the wet season and dry season grazing rangelands (Coppock 1994). The Dida-Hara rangelands before the development of semipermanent stock water ponds in the 1980s were part of the traditional wet season rangelands exploited by the mobile fora herds (Oba et al. 2000; Homann and Rischkowsky 2005). The pastoral population has now settled in semipermanent settlements (Olla). The mean livestock holding in Dida-Hara community is estimated at 12.6 cattle, 11.1 small ruminants, and 2.4 camels (Solomon et al. 2007), and the community adopted semiprivate range enclosures to cope with periodic feed shortage for vulnerable animal class such as calves.

Overall, studies have shown a lack of significant difference owing to grazing exclusion alone on SOC between grazed and ungrazed areas even for >30 years (Aynekulu et al. 2014). Kieft (1994) observed similar results with lack of significant differences in SOC between grazed lands and land not grazed for 11 and 16 years. However, Reeder and Schuman (2002) reported greater SOC levels in grazed compared with protected pastures in semiarid grasslands. The same authors noted that under the exclosure system, there is immobilization of carbon in excessive aboveground plant litter, and an increase in annual forbs and grasses that lack dense fibrous rooting systems conducive to SOC formation and accumulation. Descheemaeker et al. (2006) also studied grazing and exclusion areas established in different periods of time (5, 14, and 20 years), and found higher biomass, potassium, phosphorus, SOC, and soil nitrogen in exclusion areas.

African results differed from similar studies in other parts of the world in which SOC was greater in exclusion areas compared with continuous grazing areas (Schuman et al. 1999; Reeder et al. 2004; Mekuria et al. 2011; Sousa et al. 2012). According to Reeder and Schuman (2002), differences in SOC contents in response to grazing varied with climate conditions, soil properties, pasture location, vegetation community composition, and pasture management practices. Ingram et al. (2008) found that SOC levels increased similarly for light and heavy grazed treatments during the first 11 years of a long-term study conducted in Wyoming relative to an ungrazed control. However, SOC levels declined substantially during the subsequent 10-year period (with increased drought) in the heavy grazed treatment, but not the lightly grazed or ungrazed treatments. The SOC increase with moderate grazing was in part the result of more rapid annual turnover and redistribution of carbon within the plant–soil system and changes in plant species composition.

Stohlgren et al. (1999) concluded that for levels of grazing in Rocky Mountain grasslands, (i) grazing probably had little effect on native species richness at landscape scales; (ii) grazing probably had little effect on the accelerated spread of most exotic plant species at landscape scales; (iii) grazing affected local plant species and life-form composition and cover, but spatial variation was considerable; (iv) soil characteristics, climate, and disturbances may have a greater effect on plant species diversity than the studied levels of grazing; and (v) few plant species showed consistent, directional responses to grazing or cessation of grazing.

In communally grazed rangeland in South Africa, Moussa et al. (2007) showed no significant differences in SOC and soil microbial biomass between grazed and ungrazed plots at any of the sites. In diverse degraded landscapes, soil characteristics were enhanced in grazing exclosures compared with grazed pasture, such as in an active sand dune area (Li et al. 2012) in a desertified region (Chen et al. 2012), in degraded land in Ethiopia (Girmay and Singh 2012), and in fescue paddocks in Canada (Dormaar and Willms 1998). In moderately grazed and fertilized pastures, soil characteristics were better in grazed pasture than in exclosure areas (Manley et al. 1995; Wienhold et al. 2001). Stohlgren et al. (1999) found no significant differences in species diversity, evenness, cover of various life forms (grasses, forbs, and shrubs), soil texture, or soil percentage of nitrogen and SOC between grazed and ungrazed sites in Rocky Mountain grasslands. Dormaar et al. (1997) reported that grazing mixed prairie at a low stocking rate had no effect on the vegetation but did alter soil quality.

4.3.2 Pasture Systems

Rotational or management-intensive grazing has increasingly received attention as a potential strategy to rejuvenate degraded pastures, increase productivity, increase profit, and possibly sequester SOC and improve soil quality (Beetz and Rinehart 2010). There are too few research data to fully support claims for greater SOC and improved soil quality with these practices; however, there are some positive results reported. From a field survey of eight pastures in Virginia, Conant et al. (2003) reported 8.2 ± 4.5 Mg ha^{-1} greater SOC under pastures with management-intensive grazing than with extensive grazing. This equated to an SOC sequestration rate of 0.61 Mg ha^{-1} year^{-1} during an evaluation period of 14 ± 11 years. In a survey of pastures in three counties in northern Texas, SOC was greater with rotational than with continuous grazing (Teague et al. 2011). At the end of 4 years of differential grazing method in New South Wales Australia, microarthropod abundance was greater with high-intensity/short-duration grazing than with set stocking (Tom et al. 2006). However, no differences were found in bulk density, earthworm abundance, soil microbial biomass, and respiration.

4.3.3 Crop–Livestock Systems

Integrated crop–livestock systems are usually designed to capture ecological synergies between (i) high cultural requirements and productivity of crops but system susceptibility for nutrient losses, and (ii) relatively low input requirements of animal grazing systems but large potential for utilizing low-value crop residues or cover crops and absorbing nutrients and spreading risk (NRC 2010; Franzluebbers et al. 2011). Some examples of such systems are described to illustrate the impacts of direct grazing on production and environmental quality responses.

In the High Plains of Texas, the issue of water conservation is serious and a study was conducted to compare cotton (*Gossypium hirsutum* L.) grown continuously with cotton integrated with pasture (Allen et al. 2008). Cotton lint yield was similar (1.4 Mg ha^{-1}) between continuous cotton and cotton rotated with grazed wheat, rye (*Secale cereale* L.), and Old World bluestem (*Bothriochloa bladhii* [Retz] S.T. Blake) (Allen et al. 2012). Additional bluestem seed yield and cattle weight gain were achieved in the integrated system. On a farm basis, the integrated system used 25% less irrigation water, required 36% less nitrogen fertilizer, and had fewer other chemical inputs than continuous cotton.

In a 9-year study of integrated crop–livestock systems in North Dakota, an aggregated soil quality index did not differ between integrated annual cropping and perennial grass (Liebig et al. 2012). With careful management, agricultural producers can convert perennial grass pastures to winter-grazed annual cropping systems without adversely affecting near-surface soil quality. Results may have been region specific owing to consistent freeze–thaw and wet–dry cycles typical of the northern Great Plains, use of no-till management, modest fertilizer application rates, and winter grazing that avoided deterioration of soil with exploitation with cattle.

Ryan et al. (2012) made a compelling case for annual vetch (*Vicia* L. [Fabaceae]) for hay or grazing paired with barley (*Hordeum vulgare*) in rotations for the Mediterranean region, suggesting that barley–vetch rotations could potentially enhance barley yields, improve soil quality, and provide valuable fodder for small ruminants. Medic (*Medicago* L. [Fabaceae]) and vetch in barley rotations increased SOC compared with traditional cereal cropping (Ryan et al. 2009). In a 14-year trial that examined numerous legumes in cereal rotations, legumes in rotation and nitrogen fertilization increased SOC levels; however, grazing the cereal stubble reduced SOC (Ryan et al. 2008).

Chan et al. (2011) found that pasture holds the key to maintaining or increasing SOC under long-term crop–pasture studies in New South Wales, Australia. Under pasture converted to continuous cropping, the high initial SOC stock was, at best, maintained with multiple conservation practices but tended to decrease with increased tillage or stubble burning practices. Increases in SOC were observed only in rotations incorporating a pasture phase, with results suggesting that improved soil nutrient and grazing management of permanent pasture can lead to an increase of 0.5–0.7 Mg ha^{-1} $year^{-1}$ where the initial SOC concentrations were below steady-state levels.

Carvalho et al. (2010b) recommended crop–livestock systems as an alternative to unsustainable intensive farming systems to address loss of biodiversity, nutrient pollution, and habitat fragmentation, and illustrated these benefits by focusing on the use of grazing animals integrated with crops under no-till systems characteristic of southern Brazil. With increasing pressure for more cropland in southern Brazil, da Costa et al. (2009) showed that converting from native grassland to an integrated crop–livestock system under no till preserved soil physical conditions better than under conventional tillage, while reduced tillage had a moderate performance. Crop–pasture rotations have been the predominant cropping system in Uruguay since the 1960s (García-Préchac et al. 2004) and compared with continuous cropping with conventional tillage, crop–pasture rotations provide less variable interannual economic results, and soil degradation during the crop phase could be addressed by combining crop–pasture rotations with no tillage management.

4.4 POLICY OPTIONS TO ENHANCE GRASSLAND ECOSYSTEM SUSTAINABILITY

Through a long history, agriculture policy has been crop focused. Livestock and forage-based agriculture have received less attention and resources. Moving forward, the earth's grazing lands require a commitment to regional and national approaches to development that address the intertwined issues of poverty, malnutrition and health challenges, and land degradation in pastoral regions. Increasingly, national and international organizations are adopting policies that aim to develop rural economies and improve livelihoods through increasing agricultural productivity, ensuring food security, and enhancing the sustainable use and management of natural resources. Issues addressed in these policies may include land tenure, credit, market access, education, information dissemination, as well as food and health programs, to programs for monitoring of natural resources.

Global climate change policy can potentially have a large impact on the world's grassland systems. Many of the interventions to improve the productivity of degraded soil and vegetation can offer co-benefits of increased carbon sequestration in soils and woody vegetation (IPCC 2007). To find a way to reward smallholder pastoralists in global carbon markets, methods are required to quantify carbon sequestration in soils and vegetation that is associated with management changes or interventions (Perez et al. 2007). Abberton et al. (2010) summarized the state of knowledge to establish the needed monitoring and reporting systems for carbon sequestration in grasslands. They described the multiple ways in which grassland management practices that enhance SOC sequestration can result in greater biodiversity, improved water resources with respect to both quantity (reduced runoff and evaporation or flood control) and quality (reduced or diffused pollution of waterways), and restoration of degraded land.

Neely et al. (2009) made the case for policies to reward use of practices that sequester carbon in grassland ecosystems and observed the following:

- Climate change, biodiversity loss, drought, and desertification are interrelated symptoms of unsustainable land management resulting in loss of agricultural productivity, reduced capacity to sustain rural livelihoods, and increased risk of, and vulnerability to, natural and human disasters.
- Livestock are an irreplaceable source of livelihoods for the poor.
- Drylands occupy 41% of the earth's land area, and therefore, better management can sustain livelihoods of millions of people.
- Grasslands, by their extensive nature, hold enormous potential to serve as one of the greatest terrestrial sinks for carbon.
- Appropriate grassland management practices contribute to adaptation and mitigation, as well as increasing productivity and food security and reducing risks of drought and flooding.
- Livestock play an important role in carbon sequestration through improved pasture and rangeland management.
- Enabling grassland and livestock stewards to manage the vast grasslands for both productivity and carbon sequestration requires a global coordinated effort to overcome sociopolitical and economic barriers.

- Assessing the biophysical, economic, and institutional potential of supporting pastoralists' access to global carbon markets requires a concerted effort.
- Healthy grasslands, livestock, and associated livelihoods constitute a win–win option for addressing climate change in fragile dryland areas where pastoralism remains the most rational strategy for maintaining the well-being of communities.

Delivery of new practices or modification of systems to enhance sustainability depends on integrating knowledge, altering attitudes toward rational use of resources, and improving institutional support for the capability of those managing the land. Tefera and Sterk (2010) determined that farmers in an Ethiopian watershed were aware of erosion problems and of how the soil loss from steep slopes related to a decline in soil fertility. However, the level of investment in soil and water conservation was related to wealth status of farmers, land tenure, and access to information. High labor demand, lack of short-term benefits from conservation measures, and access to free grazing also reduced adoption rates of the soil and water conservation practices.

Land degradation in the Ethiopian highlands is considered to be one of the major problems threatening agricultural development and food security. Amsalu et al. (2007) described substantial deforestation, introduction of plantations, and expansion of grazing land during a period when population density was rapidly increasing in the Ethiopian highlands. Farmers gradually changed from annual cropping to tree planting and livestock production to cope with the problems of soil degradation, water scarcity, and smaller farm size. Income diversification through the sale of wood and cattle dung is becoming a major livelihood strategy, with little attention to continued loss of soil fertility through soil erosion and removal of dung.

Many areas of policy are required to meet the needs of pastoralists and address the pressing concerns related to grassland agriculture. For example, the World Organization for Animal Health's One Health Initiative (OIE 2013) is a global partnership to address human, animal, and environmental health globally by facilitating and promoting national programs to prevent high public health and animal impact diseases at the human–animal interface. Market and policy incentives that encourage agricultural intensification, such as bioenergy, can contribute to biodiversity decline and habitat fragmentation if they encourage a large-scale conversion of native and seminatural ecosystems. Questad et al. (2011) identified opportunities in the United States where changes to existing management practices could benefit both conservation and bioenergy production, including increased use of hay management or other biomass collection on native grassland remnants and improving the Conservation Reserve Program by adding more native species to seed mixes, and incorporating a periodic biomass collection.

The Great Basin of the western United States is characterized by low annual precipitation, diverse desert plant communities, and local economies that depend on the products and services produced by these lands (e.g., livestock grazing, recreation, and mining). The ecological and economic stability of the Great Basin is increasingly at risk because of the expansion of fire-prone invasive species and increase in wildfires. If restoration in the Great Basin ecosystems is not successful, desertification and the associated loss of economic stability and ecological integrity will continue to threaten the sustainability of natural resources and people in the Great Basin (Pellant et al. 2004).

4.5 SUMMARY

Grasslands constitute the largest global land use and are an important part of agricultural and ecological systems on every continent, across a wide range of potential productivity. Ruminant livestock grazing on these lands constitutes an important form of agricultural production. It is estimated that 1 billion people depend almost exclusively on livestock, and livestock serves as at least a partial source of income and food security for 70% of the world's 880 million rural poor who live on less than US$ 1 per day. Grazing lands provide a wide range of ecosystem services, including provision of food, livelihoods, biodiversity, habitat, carbon storage, water filtration, and others. Grazing lands are particularly important in the world's dry land areas, which are especially sensitive to land degradation.

The Millennium Ecosystem Assessment (2005) found that 10%–20% of dry lands were degraded. Additionally, the world's grazing lands are subject to pressures of increasing population, encroachment of cropping onto former grasslands, uncertain land tenure, lack of infrastructure (transportation, markets), lack of capital, and are faced with increasing pressures of variable and changing climate. More than 70% of the rangelands worldwide are estimated to be affected by soil degradation.

Grazing lands are highly diverse, and there are many opportunities to enhance soil quality; however, results in the literature about impacts of grazing on soil quality have been mixed. In dry regions, rangelands are particularly vulnerable to degradation and it is essential that the vegetative and plant litter cover be maintained to protect the soil from raindrop impact and temperature extremes. In many cases, excessive grazing has resulted in severe soil degradation. In other situations, where grazing timing, duration, and intensity are managed, the effects of grazing have had neutral or positive effects on soil quality characteristics. As demand for cropland increases, there are many opportunities for mixed crop–livestock systems that may provide environmental as well as economic benefits. Science must continue to evaluate grazing systems to develop better predictive understanding of the degradation and regeneration processes of grassland agroecosystems to more effectively develop improved management practices and systems.

In the future, the pressures on fragile grassland ecosystems will only increase with increased human population, increased demand for cropland and bioenergy, increased demand for livestock products, and increasing climate variability and change. Policies to address these challenges must address the intertwined issues of poverty, malnutrition and health challenges, and land degradation in pastoral regions.

ACKNOWLEDGMENT

The authors appreciate the review of the draft versions of the manuscript, and the helpful feedback from Dr. Jan de Leeuw, World Agroforestry Centre (ICRAF), Nairobi, Kenya.

REFERENCES

Abberton, A., R. Conant, and C. Batello (Eds.). 2010. Grassland carbon sequestration: Management, policy and economics. Proc., Workshop on the Role of Grassland Carbon Sequestration in the Mitigation of Climate Change, Food and Agriculture Organization of the United Nations, Rome.

Abberton, M.T., A.H. Marshall, M.W. Humphreys, J.H. Macduff, R.P. Collins, and C.L. Marley. 2008. Genetic improvement of forage species to reduce the environmental impact of temperate livestock grazing systems. *Adv. Agron.* 98:311–355.

Adams, M.L. 2009. Pasture extensification in the Southern Ecuadorian Andes: Appraisal and recommendations. *J. Sustainable Forest.* 28(6–7):866–887.

Adams, W.E., C.B. Elkins Jr., and E.R. Beaty. 1966. Rooting habits and moisture use of coastal and common bermudagrass. *J. Soil Water Conserv.* 21:133–135.

Aerts, R., J. Nyssen, and M. Haile. 2009. On the difference between "exclosures" and "enclosures" in ecology and the environment. *J. Arid Environ.* 73:762–763.

African Center for Holistic Management. 2013. Available at http://achmonline.squarespace.com/holistic-management. Accessed January 28, 2014.

Allen, V.G., C.P. Brown, R. Kellison et al. 2012. Integrating cotton and beef production in the Texas Southern High Plains: I. Water use and measures of productivity. *Agron. J.* 104:1625–1642.

Allen, V.G., C.P. Brown, E. Segarra et al. 2008. In search of sustainable agricultural systems for the Llano Estacado of the U.S., Southern High Plains. *Agric. Ecosyst. Environ.* 124:3–12.

Amsalu, A., L. Stroosnijder, and J.D. Graaff. 2007. Long-term dynamics in land resource use and the driving forces in the Beressa watershed, highlands of Ethiopia. *J. Environ. Manag.* 83(4):448–459.

Angassa, A., and G. Oba, G. 2008. Herder perceptions on impacts of range enclosures, crop farming, fire ban and bush encroachment on the rangelands of Borana, Southern Ethiopia. *Hum. Ecol.* 36:201–215.

Angassa, A., G. Oba, A.C. Treydte, and R.B. Weladji. 2010. Role of traditional enclosures on the diversity of herbaceous vegetation in a semi-arid rangeland, southern Ethiopia. *Livestock Res. Rural Dev.* 22:Article No. 163. Available at http://www.lrrd.org/lrrd22/9/anga22163.htm. Accessed January 30, 2014.

Angassa, A., B. Sheleme, B. Oba, A.C. Treydte, A. Linstädter, and J. Sauerborn. 2012. Savanna land use and its effect on soil characteristics in southern Ethiopia. *J. Arid Environ.* 81:67–76.

Ansley, R.J., T.W. Boutton, and J.O. Skjemstad. 2006. Soil organic carbon and black carbon storage and dynamics under different fire regimes in temperate mixed-grass savanna. *Global Biogeochem. Cycl.* 20:1–11.

Arshad, M.A., A.J. Franzluebbers, and R.H. Azooz. 2004. Surface-soil structural properties under grass and cereal production on a Mollic Cryoboralf in Canada. *Soil Tillage Res.* 77:15–23.

Aynekulu, E., E. Koala, J., Feyisa, L. Sawadogo, and K. Shepherd. 2014. Livelihood Diversifying Potential of Livestock Based Carbon Sequestration Option in Pastoral and Agro Pastoral Systems in Africa Project. Final report. World Agroforestry Centre, Nairobi, Kenya.

Bardgett, R.D., and A.D. Ward. 2003. Herbivore-mediated linkages between aboveground and belowground communities. *Ecology* 84(9):2258–2268.

Bardgett, R.D., D.A. Wardle, and G.W. Yeates. 1998. Linking above-ground and below-ground interactions: How plant responses to foliar herbivory influence soil organisms. *Soil Biol. Biochem.* 30:1867–1878.

Bartley, R., J.P. Corfield, B.N. Abbott, A.A. Hawdon, S.N. Wilkinson, and B. Nelson. 2010. Impacts of improved grazing land management on sediment yields, Part 1: Hillslope processes. *J. Hydrol.* 389:237–248.

Bauer, A., and A.L. Black. 1994. Quantification of the effect of soil organic matter content on soil productivity. *Soil Sci. Soc. Am. J.* 58:185–193.

Beetz, A.E., and L. Rinehart. 2010. *Rotational Grazing*. Nat. Center Appropr. Techn., Fayetteville, AR, 12 pp. Available at www.attra.ncat.org/attra-pub/rotgraze.html.

Belnap, J. 2006. The potential roles of biological soil crusts in dryland hydrologic cycles. *Hydrol. Proc.* 20:3159–3178.

Belnap, J., J.R. Welter, N.B. Grimm, N. Barger, and J.A. Ludwig. 2005. Linkages between microbial and hydrologic processes in arid and semiarid watersheds. *Ecology* 86(2): 298–307.

Beringer, J., L.B. Hutley, N.J. Tapper, and L.A. Cernusak. 2007. Savanna fires and their impact on net ecosystem productivity in North Australia. *Global Change Biol.* 13:990–1004.

Beukes, P.C., and R.M. Cowling. 2003. Non-selective grazing impacts on soil-properties of the Nama Karoo. *J. Range Manag.* 56:547–552.

Bird, M.I., E.M. Veenendaal, C. Moyo, J. Lloyd, and P. Frost. 2000. Effect of fire and soil texture on soil carbon in a sub-humid savanna (Matopos, Zimbabwe). *Geoderma* 94:71–90.

Brejda, J.J., T.B. Moorman, J.L. Smith, D.L. Karlen, D.L. Allan, and T.H. Dao. 2000. Distribution and variability of surface soil properties at a regional scale. *Soil Sci. Soc. Am. J.* 64:974–982.

Bridge, B.J., J.J. Mott, and R.J. Hartigan. 1983. The formation of degraded areas in the dry savanna woodlands of Northern Australia. *Austr. J. Soil Res.* 21:91–104.

Canadell, J., R.B. Jackson, J.B. Ehleringer, H.A. Mooney, O.E. Sala, and E.D. Schulze. 1996. Maximum rooting depth of vegetation types at the global scale. *Oecologia* 108(4): 583–595.

Cardoso, E.L., M.L.N. Silva, F.M.D.S. Moreira, and N. Curi. 2009. Biological indicator attributes of soil quality under cultivated and natural pasture in the Pantanal wetlands. *Pesq. Agropec. Bras.* 44(6):631–637.

Carvalho, J.L.N., G.S. Raucci, C.E.P. Cerri et al. 2010a. Impact of pasture, agriculture and crop–livestock systems on soil C stocks in Brazil. *Soil Tillage Res.* 110:175–186.

Carvalho, P.C., I. Anghinoni, A. de Moraes, E.D. de Souza, R.M. Sulc, and C.R. Lang. 2010b. Managing grazing animals to achieve nutrient cycling and soil improvement in no-till integrated systems. *Nutr. Cycl. Agroecosyst.* 88(2):259–273.

Causarano, H.J., A.J. Franzluebbers, J.N. Shaw, D.W. Reeves, R.L. Raper, and C.W. Wood. 2008. Soil organic carbon fractions and aggregation in the Southern Piedmont and Coastal Plain. *Soil Sci. Soc. Am. J.* 72:221–230.

Chan, K.Y., M.K. Conyers, G.D. Li et al. 2011. Soil carbon dynamics under different cropping and pasture management in temperate Australia: Results of three long-term experiments. *Soil Res.* 49(4):320–328.

Chen, Y., Y. Li, X. Zhao, W. Shang, J. Han, and T. Awada. 2012. Effects of grazing exclusion on soil properties and on ecosystem carbon and nitrogen storage in a sandy rangeland of Inner Mongolia, Northern China. *Environ. Manag.* 50(4):622–632.

Cheng, J., J. Cheng, T. Hu, H. Shao, and J. Zhang. 2012. Dynamic changes of *Stipa bungeana* steppe species diversity as better indicators for soil quality and sustainable utilization mode in Yunwu Mountain Nature Reserve, Ningxia, China. *Clean—Soil Air Water* 40(2): 127–133.

Clark, J.T., J.R. Russell, D.L. Karlen, P.L. Singleton, W.D. Busby, and B.C. Peterson. 2004. Soil surface property and soybean yield response to corn stover grazing. *Agron. J.* 96: 1364–1371.

Coetsee, C., W.J. Bond, and E.C. February. 2010. Frequent fire affects soil nitrogen and carbon in an African savanna by changing woody cover. *Oecologia* 162:1027–1034.

Cohron, G.T. 1971. Forces causing soil compaction. In K.K. Barnes, W.M. Carleton, W.M. Taylor, R.G. Throckmorton, and G.E. Vandenburg (Eds.), *Compaction of Agricultural Soils*. American Society of Agricultural and Biological Engineers, St. Joseph, MI, pp. 106–124.

Colloff, M.J., K.R. Pullen, and S.A. Cunningham. 2010. Restoration of an ecosystem function to revegetation communities: The role of invertebrate macropores in enhancing soil water infiltration. *Restor. Ecol.* 18:65–72.
Conant, R.T. 2002. Potential soil carbon sequestration in overgrazed grassland ecosystems. *Global Biogeochem. Cycl.* 16(4):90.1–90.4. doi:10.1029/2001GB001661.
Conant, R.T. 2012. Grassland soil organic carbon stocks: Status, opportunities, vulnerability. In R. Lal, K. Lorenz, R.F. Hüttl, B.U. Schneider, and J. von Braun (Eds.), *Recarbonization of the Biosphere: Ecosystems and the Global Carbon Cycle*. Springer Dordrecht, Heidelberg, pp. 275–302.
Conant, R.T., K. Paustian, and E.T. Elliott. 2001. Grassland management and conversion into grassland: Effects on soil carbon. *Ecol. Appl.* 11:343–355.
Conant, R.T., J. Six, and K. Paustain. 2003. Land use effects on soil carbon fractions in the southeastern United States. I. Management-intensive versus extensive grazing. *Biol. Fertil. Soils* 38:386–392.
Coppock, L.D. 1994. *The Borana Plateau of Southern Ethiopia: Synthesis of Pastoral Research. Development and Changes, 1980–90*. International Livestock Centre for Africa (ILCA), Addis Ababa, Ethiopia, 393 pp.
Corvalan, C., D. Hales, and A. McMichael. 2005. Ecosystems and Human Well-Being: Health Synthesis. Millennium Ecosystem Assessment Rep. World Health Organization, Geneva Switzerland, 53 pp.
da Costa, A., J.A. Albuquerque, A.L. Mafra, and F.R. da Silva. 2009. Soil physical properties in crop-livestock management systems. *Rev. Bras. Cienc. Solo* 33(2):235–244.
Dalle, G., B.L. Maass, and J. Isselstein. 2006. Encroachment of woody plants and its impact on pastoral livestock production in the Borana lowlands, southern Oromia, Ethiopia. *Afr. J. Ecol.* 44:237–246.
Derner, J.D., D.D. Briske, and T.W. Boutton. 1997. Does grazing mediate soil carbon and nitrogen accumulation beneath C4, perennial grasses along an environmental gradient? *Plant Soil* 191:147–156.
Derner, J.D., and G.E. Schuman. 2007. Carbon sequestration and rangelands: A synthesis of land management and precipitation effects. *J. Soil Water Conserv.* 62:77–85.
Descheemaeker, K., B. Muys, J. Nyssen et al. 2006. Litter production and organic matter accumulation in exclosures of the Tigray highlands, Ethiopia. *Forest Ecol. Manag.* 233:21–35.
Diaz-Zorita, M., G.A. Duarte, and J.H. Grove. 2002. A review of no-till systems and soil management for sustainable production in the subhumid and semiarid Pampas of Argentina. *Soil Tillage Res.* 65:1–18.
Doran, J.W., and T.B. Parkin. 1994. Defining and assessing soil quality. In J.W. Doran, D.C. Coleman, D.F. Bezdicek, and B.A. Stewart (Eds.), *Defining Soil Quality for a Sustainable Environment*. Soil Sci. Soc. Am. Spec. Publ. 35, ASA-SSSA, Madison, WI, pp. 3–21.
Dormaar, J.F., B.W. Adams, and W.D. Willms. 1997. Impacts of rotational grazing on mixed prairie soils and vegetation. *J. Range Manag.* 50(6):647–651.
Dormaar, J.F., and W.D. Willms. 1998. Effect of forty-four years of grazing on fescue grassland soils. *J. Range Manag.* 51(1):122–126.
Drewry, J.J., K.C. Cameron, and G.D. Buchan. 2008. Pasture yield and soil physical property responses to soil compaction from treading and grazing—A review. *Austr. J. Soil Res.* 46(3):237–256.
du Toit, G.v.N., H.A. Snyman, and P.J. Malan. 2009. Physical impact of grazing by sheep on soil parameters in the Nama Karoo subshrub/grass rangeland of South Africa. *J. Arid Environ.* 73(9):804–810.
Elkins, C.B., R.L. Haaland, and C.S. Hoveland. 1977. Grass roots as a tool for penetrating soil hardpans and increasing crop yields. Proc. 34th Southern Pasture Forage Crop Improvement Conf., Auburn, AL, April 12–14, pp. 21–26.

Ellis, T.W., S. Leguedois, P.B. Hairsine, and D.J. Tongway. 2006. Capture of overland flow by a tree belt on a pastured hillslope in south-eastern Australia. *Austr. J. Soil Res.* 44(2): 117–125.

Erb, K.-H., V. Gaube, F. Krausmann, C. Plutzar, A. Bondeau, and H. Haberl. 2007. A comprehensive global 5min resolution land-use dataset for the year 2000 consistent with national census data. *J. Land Use Sci.* 2(3):191–224.

Eswaran, H., R. Almaraz, P. Reich, and P. Zdruli. 1997a. Soil quality and soil productivity in Africa. *J. Sustainable Agric.* 10(4):75–94.

Eswaran, H., R. Almaraz, E. Van Den Berg, and P. Reich. 1997b. An assessment of the soil resources of Africa in relation to productivity. *Geoderma* 77(1):1–8.

Fernandez, P.L., C.R. Alvarez, and M.A. Taboada. 2011. Assessment of topsoil properties in integrated crop–livestock and continuous cropping systems under zero tillage. *Soil Res.* 49:143–151.

Fesha, I.G., J.N. Shaw, D.W. Reeves et al. 2002. Land use effects on soil quality parameters for identical soil taxa. In E. van Santen (Ed.), *Making Conservation Tillage Conventional: Building a Future on 25 Years of Research.* Special Report No. 1. Alabama Agricultural Experiment Station, Auburn University, Auburn, AL, pp. 233–238.

Food and Agriculture Organization (FAO). 2010. Challenges and Opportunities for Carbon Sequestration in Grassland Systems. A Technical Report on Grassland Management and Climate Change Mitigation Compiled by Rich Conant. Integrated Crop Management, Vol. 9, 56, United Nations, Rome, Italy.

Food and Agriculture Organization (FAO). 2011. *The State of the World's Land and Water Resources for Food and Agriculture (SOLAW)—Managing Systems at Risk.* Food and Agriculture Organization of the United Nations, Rome and Earthscan, London.

Food and Agriculture Organization (FAO). 2013. Agriculture and Consumer Protection Department. Livestock and Livelihoods. Available at http://www.fao.org/ag/againfo/themes/en/protecting_livelihoods.html. Accessed January 28, 2014.

Francis, G.S., F.J. Tabley, and K.M. White. 2001. Soil degradation under cropping and its influence on wheat yield on a weakly structured New Zealand silt loam. *Austr. J. Soil Res.* 39(2):291–305.

Franzluebbers, A.J. 2002. Ecology and the cycling of carbon and nitrogen. In Lal, R. (Ed.), *Encyclopedia of Soil Science.* Marcel Dekker, New York, pp. 374–377.

Franzluebbers, A.J. 2005. Soil organic carbon sequestration and agricultural greenhouse gas emissions in the southeastern USA. *Soil Tillage Res.* 83:120–147.

Franzluebbers, A.J. 2010. Depth distribution of SOC as a signature of soil quality. Proc. 19th World Congress of Soil Science, Brisbane, Australia. CD-ROM.

Franzluebbers, A.J., and P.C. Doraiswamy. 2007. Carbon sequestration and land degradation. In M.V.K. Sivakumar and N. Ndiangu'ui (Eds.), *Climate and Land Degradation.* Springer-Verlag, New York, pp. 337–354.

Franzluebbers, A.J., and J.A. Stuedemann. 2008a. Early response of soil organic fractions to tillage and integrated crop–livestock production. *Soil Sci. Soc. Am. J.* 72:613–625.

Franzluebbers, A.J., and J.A. Stuedemann. 2008b. Soil physical responses to cattle grazing cover crops under conventional and no tillage in the Southern Piedmont USA. *Soil Tillage Res.* 100:141–153.

Franzluebbers, A.J., and J.A. Stuedemann. 2009. Soil-profile organic carbon and total nitrogen during 12 years of pasture management in the Southern Piedmont USA. *Agric. Ecosyst. Environ.* 129:28–36.

Franzluebbers, A.J., and J.A. Stuedemann. 2010. Surface soil changes during twelve years of pasture management in the Southern Piedmont USA. *Soil Sci. Soc. Am. J.* 74:2131–2141.

Franzluebbers, A.J., and J.A. Stuedemann. 2013. Soil-profile distribution of organic C and N after 6 years of tillage and grazing management. *Eur. J. Soil Sci.* 64:558–566. doi:10.1111/ejss.12057.

Franzluebbers, A.J., J.A. Stuedemann, and H.H. Schomberg. 2000a. Spatial distribution of soil carbon and nitrogen pools under grazed tall fescue. *Soil Sci. Soc. Am. J.* 64:635–639.

Franzluebbers, A.J., J.A. Stuedemann, H.H. Schomberg, and S.R. Wilkinson. 2000b. Soil organic C and N pools under long-term pasture management in the Southern Piedmont USA. *Soil Biol. Biochem.* 32:469–478.

Franzluebbers, A.J., J.A. Stuedemann, and S.R. Wilkinson. 2001. Bermudagrass management in the Southern Piedmont USA: I. Soil and surface residue carbon and sulfur. *Soil Sci. Soc. Am. J.* 65:834–841.

Franzluebbers, A.J., R.M. Sulc, and M.P. Russelle. 2011. Opportunities and challenges for integrating North-American crop and livestock systems. In G. Lemaire, J. Hodgson, and A. Chabbi (Eds.), *Grassland Productivity and Ecosystem Services*. CAB International, Wallingford, UK, pp. 208–218.

Franzluebbers, A.J., S.R. Wilkinson, and J.A. Stuedemann. 2004. Bermudagrass management in the Southern Piedmont USA: X. Coastal productivity and persistence in response to fertilization and defoliation regimes. *Agron. J.* 96:1400–1411.

Franzluebbers, A.J., S.F. Wright, and J.A. Stuedemann. 2000c. Soil aggregation and glomalin under pastures in the Southern Piedmont USA. *Soil Sci. Soc. Am. J.* 64:1018–1026.

Fterich, A., M. Mahdhi, and M. Mars. 2012. Impact of grazing on soil microbial communities along a chronosequence of *Acacia tortilis* subsp. *raddiana* in arid soils in Tunisia. *Eur. J. Soil Biol.* 50:56–63.

Fuhlendorf, S.D., and F.E. Smeins. 1997. Long-term vegetation dynamics mediated by herbivores, weather and fire in a *Juniperus–Quercus* savanna. *J. Veget. Sci.* 8:819–828.

Fynn, R.W.S., R.J. Haynes, and T.G. O'Connor. 2003. Burning causes long-term changes in soil organic matter content of a South African grassland. *Soil Biol. Biochem.* 35:677–687.

Gantzer, C.J., S.H. Anderson, A.L. Thompson, and J.R. Brown. 1990. Estimating soil erosion after 100 years of cropping on Sanborn Field. *J. Soil Water Conserv.* 45:641–644.

Gantzer, C.J., and T.R. McCarty. 1987. Predicting corn yields on a claypan soil using a soil productivity index. *Trans. ASAE* 30:1347–1352.

García-Préchac, F., O. Ernst, G. Siri-Prieto, and J.A. Terra. 2004. Integrating no-till into crop-pasture rotations in Uruguay. *Soil Tillage Res.* 77(1):1–13.

Gattinger, A., A. Muller, M. Haeni et al. 2012. Enhanced top soil carbon stocks under organic farming. *Proc. Nat. Acad. Sci. U.S.A.* 109(44):18226–18231. doi:10.1073/pnas.1209429109.

Gebhart, D.L., H.B. Johnson, H.S. Mayeux, and H.W. Polley. 1994. The CRP increases soil organic carbon. *J. Soil Water Conserv.* 49:488–492.

Giddings, L.F., M. Stepp, and M. Caine. 2013. *Feeding the Planet in a Warming World: Building Resilient Agriculture through Innovation*. The Information Technology & Innovation Foundation, Washington, D.S. and the London School of Economics and Political Science, London. Available at http://itif.org. Accessed April 12, 2013.

Girmay, G., and B.R. Singh. 2012. Changes in soil organic carbon stocks and soil quality: Land-use system effects in northern Ethiopia. *Acta Agric. Scand. Sec. B: Soil Plant Sci.* 62(6):519–530.

Grace, J., J. San Jose, P. Meir, H. Miranda, and R. Montes. 2006. Productivity and carbon fluxes of tropical savannas. *J. Biogeogr.* 33:387–400.

Greenwood, K.L., and B.M. McKenzie. 2001. Grazing effects on soil physical properties and the consequences for pastures: A review. *Aust. J. Exp. Agric.* 41:1231–1250.

Gross, M., M. Hardisky, P. Wolf, and V. Klemas. 1991. Relationship between aboveground and belowground biomass of *Spartina alterniflora* (smooth cordgrass). *Estuaries* 14(2):180–191.

Guo, L., and R. Gifford. 2002. Soil carbon stocks and land use change: A meta analysis. *Global Change Biol.* 8:345–360.

Han, G., X. Hao, M. Zhao et al. 2008. Effect of grazing intensity on carbon and nitrogen in soil and vegetation in a meadow steppe in Inner Mongolia. *Agric. Ecosyst. Environ.* 125(1–4):21–32.

Herrero, M., and P.K. Thornton. 2013. Livestock and global change: Emerging issues for sustainable food systems. *Proc. Nat. Acad. Sci. U.S.A.* 110(52):20878–20881.

Higgins, S.I., W.J. Bond, E.C. February et al. 2007. Effects of four decades of fire manipulation on woody vegetation structure in savanna. *Ecology* 88:1119–1125.

Holt, J.A., K.L. Bristow, and J.G. McIvor. 1996. The effects of grazing pressure on soil animals and hydraulic properties of two soils in semi-arid tropical Queensland. *Austr. J. Soil Res.* 34(1):69–79.

Homann, S., and B. Rischkowsky. 2005. Integrating the indigenous knowledge of Borana pastoralists into rangeland management strategies in southern Ethiopia. IK Notes, No. 81. June. Available at http://www.worldbank.org/afr/ik/. Accessed February 10, 2014.

Homann, S., B. Rischkowsky, J. Steinbach, M. Kirk, and E. Mathias. 2008. Towards endogenous livestock development: Borana pastoralists' responses to environmental and institutional changes. *Hum. Ecol.* 36:503–520.

IFPRI, and ILRI. 2000. Property Rights, Risk, and Livestock Development in Africa. In N. McCarthy, B. Swallow, M. Kirk, and P. Hazell, (Eds.), *Ethiopian Case Study*. International Food Policy Research Institute (IFPRI) and International Livestock Research Institute (ILRI), Washington, DC, 433 pp.

Ingram, L.J., P.D. Stahl, G.E. Schuman et al. 2008. Grazing impacts on soil carbon and microbial communities in a mixed-grass ecosystem. *Soil Sci. Soc. Am. J.* 72:939–948.

Intergovernmental Panel on Climate Change (IPCC). 2000. Land Use, Land-Use Change, and Forestry. In R.T. Watson, I.R. Noble, B. Bolin, N.H. Ravindranath, D.J. Verardo, and D.J. Dokken (Eds.), *Grazing Hands Management*. Cambridge University Press, UK, p. 375.

Intergovernmental Panel on Climate Change (IPCC). 2007. Contribution of Working Group II to the Fourth Assessment Report. In M.L. Parry, O.F. Canziani, J.P. Palutikof, P.J. van der Linden, and C.E. Hanson (Eds.), *Perspectives on Climate Change and Sustainability*. Cambridge University Press, Cambridge, UK.

Islam, K.R., and R.R. Weil. 2000. Soil quality indicator properties in mid-Atlantic soils as influenced by conservation management. *J. Soil Water Conserv.* 55:69–78.

Johnson, M.B., D.A. Lammers, C.P. Anderson, P.T. Rygiewcz, and J.S. Kern. 1992. Sustaining soil quality by protecting the soil resource. Proc. Soil Quality Standards Symp., San Antonio TX, October 23, 1990, Watershed Air Manage. Rep. No. WO-WSA-2, USDA-Forest Service, Washington, DC, pp. 72–80.

Karlen, D.L., M.J. Mausbach, J.W. Doran, R.G. Cline, R.F. Harris, and G.E. Schuman. 1997. Soil quality: A concept, definition, and framework for evaluation. *Soil Sci. Soc. Am. J.* 61:4–10.

Kieft, T.L. 1994. Grazing and plant canopy effects on semiarid soil microbial biomass and respiration. *Biol. Fertil. Soils* 18:155–162.

Kingery, W.L., C.W. Wood, D.P. Delaney, J.C. Williams, and G.L. Mullins. 1994. Impact of long-term land application of broiler litter on environmentally related soil properties. *J. Environ. Qual.* 23:139–147.

Knicker, H. 2007. How does fire affect the nature and stability of soil organic nitrogen and carbon? A review. *Biogeochemistry*. 85:91–118.

Lafond, G., B.G. McConkey, and M. Stumborg. 2009. Conservation tillage models for small-scale farming: Linking the Canadian experience to the small farms of Inner Mongolia Autonomous Region in China. *Soil Tillage Res.* 104(1):150–155.

Lal, R. 2002. Carbon sequestration in dryland ecosystems of West Asia and North Africa. *Land Degrad. Dev.* 13(1):45–59.

Lal, R. 2004. Soil carbon sequestration impacts on global climate change and food security. *Science* 304:1623–1627.

Lal, R., R.F. Follett, B.A. Stewart, and J.M. Kimble. 2007. Soil carbon sequestration to mitigate climate change and advance food security. *Soil Sci.* 172:(12):943–956.

Lavelle, P., L. Brussaard, and P.F. Hendrix (Eds.). 1999. *Earthworm Management in Tropical Agroecosystems*. CABI Publishing, Wallingford, UK.

Lavelle, P., and A.V. Spain. 2001. *Soil Ecology*. Kluwer Academic, Dordrecht, the Netherlands.

Laws, W.D., and D.D. Evans. 1949. The effects of long-time cultivation on some physical and chemical properties of two rendzina soils. *Soil Sci. Soc. Am. Proc.* 13:15–19.

Li, X.-G., Z.-F. Wang, Q.-F. Ma, and F.-M. Li. 2007. Crop cultivation and intensive grazing affect organic C pools and aggregate stability in arid grassland soil. *Soil Tillage Res.* 95(1–2):172–181.

Li, Y., X. Zhao, Y. Chen, Y. Luo, and S. Wang. 2012. Effects of grazing exclusion on carbon sequestration and the associated vegetation and soil characteristics at a semi-arid desertified sandy site in inner Mongolia, northern China. *Can. J. Soil Sci.* 92:807–819.

Liebig, M.A., J.R. Gross, S.L. Kronberg, R.L. Phillips, and J.D. Hanson. 2010. Grazing management contributions to net global warming potential: A long-term evaluation in the northern Great Plains. *J. Environ. Qual.* 39:799–809.

Liebig, M.A., D.L. Tanaka, S.L. Kronberg, E.J. Scholljegerdes, and J.F. Karn. 2012. Integrated crops and livestock in central North Dakota, USA: Agroecosystem management to buffer soil change. *Renew. Agric. Food Syst.* 27:115–124.

Ludwig, B., P. Khanna, J. Bauhus, and P. Hopmans. 2002. Near infrared spectroscopy of forest soils to determine chemical and biological properties related to soil sustainability. *Forest Ecol. Manag.* 171:121–132.

Lusby, G.C. 1970. Hydrologic and biotic effects of grazing vs non-grazing near Grand Junction, Colorado. *J. Range Manag.* 23:256–260.

MacDonald, K.B., W. Frazer, G. Lelyk, and F. Wang. 1995. A geographic framework for assessing soil quality. In D.F. Acton, and E.G. Gregorich (Eds.), *The Health of Our Soils—Towards Sustainable Agriculture in Canada*. Center Land Biological Resources Res., Agric. Agri-Food Canada, Ottawa, ON, pp. 19–30.

Magdoff, F., and R.R. Weil (Eds.). 2004. *Soil Organic Matter in Sustainable Agriculture*. CRC Press, Boca Raton, FL, 398 pp.

Manley, J.T., G.E. Schuman, J.D. Reeder, and R.H. Hart. 1995. Rangeland soil carbon and nitrogen responses to grazing. *J. Soil Water Conserv.* 50(3):294–298.

Materechera, S.A., and R.N. Murovhi. 2011. Soil biological properties beneath canopies of *Acacia erioloba* (syn. *Acacia giraffae*) trees under different land use practices in a South African semi-arid environment. *Life Sci. J.* 8(Suppl. 2):33–39.

McCracken, R.J. 1959. Certain Properties of Selected Southeastern United States Soils and Mineralogical Procedures for Their Study. Southern Regional Bull. 61. Virginia Agric. Expt. Sta., Virginia Polytechnic Inst., Blacksburg, VA, 146 pp.

McIvor, J.G., J. Williams, and C.J. Gardener. 1995. Pasture management influences runoff and soil movement in the semi-arid tropics. *Austr. J. Exp. Agric.* 35:55–65.

McNaughton, S.J., F.F. Banyikwa, and M.M. McNaughton. 1998. Root biomass and productivity in a grazing ecosystem: The Serengeti. *Ecology* 79(2):587–592.

Mekuria, W., E. Veldkamp, M.D. Corre, and M. Haile. 2011. Restoration of ecosystem carbon stocks following exclosure establishment in communal grazing lands in Tigray, Ethiopia. *Soil Sci. Soc. Am. J.* 75:246–256.

Miguel, F.R.M., S.R. Vieira, C.R. Grego. 2009. Spatial variability of water infiltration rate in soil under pasture as a function of cattle trampling intensity. *Pesq. Agropec. Bras.* 44(11): 1513–1519.

Milchunas, D.G., and W.K. Lauenroth. 1993. Land use, land-use change, and forestry. *Ecol. Monogr.* 63:337–366.

Millennium Ecosystem Assessment. 2005. *Ecosystems and Human Well-being: Synthesis*. Island Press, Washington, DC.

Miller, J.J., D.S. Chanasyk, T., Curtis, and W.D. Willms. 2010. Influence of streambank fencing on the environmental quality of cattle-excluded pastures. *J. Environ. Qual.* 39(3):991–1000.

Mills, A.J., and M.V. Fey. 2003. Declining soil quality in South Africa: Effects of land use on soil organic matter and surface crusting. *South Afr. J. Sci.* 99(9–10):429–436.

Moss, A.J. 1991a. Rain-impact soil crust. 1. Formation on a granite-derived soil. *Austr. J. Soil Res.* 29(2):271–289.

Moss, A.J. 1991b. Rain-impact soil crust. 4. Packing of sand and silt fractions by raindrops. *Austr. J. Soil Res.* 29(2):233–337.

Moussa, A.S., L. Van Rensburg, K. Kellner, and A. Bationo. 2007. Soil microbial biomass in semi-arid communal sandy rangelands in the western Bophirima District, South Africa. *Appl. Ecol. Environ. Res.* 5(1):43–56.

Mwendera, E.J., and M.A.M. Saleem. 1997. Infiltration rates, surface runoff, and soil loss as influenced by grazing pressure in the Ethiopian highlands. *Soil Use Manag.* 13(1):29–35.

National Research Council (NRC). 1993. *Soil and Water Quality: An Agenda for Agriculture*. National Research Council, National Academy Press, Washington, DC.

National Research Council (NRC). 2010. *Toward Sustainable Agricultural Systems in the 21st Century*. Comm. Twenty-First Cent. Syst. Agric., Board Agric. Nat. Resources, National Academies Press, Washington, DC.

Neely, C., S. Bunning, and A. Wilkes. 2009. Review of Evidence on Drylands Pastoral Systems and Climate Change. Land and Water Discussion Paper No. 8. Food and Agriculture Organization of the United Nations, Rome.

Neely, C., and J. Butterfield. 2004. Holistic management of African rangelands. *LEISA* 20(4): 26–29.

Neely, C.L., and J. de Leeuw. 2011. Home on the range: The contribution of rangeland management to climate change mitigation. In E. Wollenberg, A. Nihart, M.-L. Tapio-Biström, and M. Grieg-Gran (Eds.), *Climate Change Mitigation and Agriculture*. Earthscan, London, pp. 333–346.

Nicholas, A.M.M., D.C. Franklin, and D.M.J.S. Bowman. 2009. Coexistence of shrubs and grass in a semi-arid landscape: A case study of mulga (*Acacia aneura*, Mimosaceae) shrublands embedded in fire-prone spinifex (*Triodia pungens*, Poaceae) hummock grasslands. *Austr. J. Bot.* 57(5):396–405.

Nie, Z.N., A.D. MacKay, I. Valentine, D.J. Barker, and J. Hodgson. 1997. Influence of pastoral fallow on plant root growth and soil physical and chemical characteristics in a hill pasture. *Plant Soil* 197:201–208.

Oba, G., E. Post, P.O. Syvertsen, and N.C. Stenseth. 2000. Bush cover and range condition assessments in relation to landscape and grazing in southern Ethiopia. *Landscape. Ecol.* 15(6):535–546.

Odum, E.P., J.T. Finn, and E.H. Franz. 1979. Perturbation theory and the subsidy-stress gradient. *BioScience* 29:349–352.

OIE (World Organization for Animal Health). 2013. *One Health at a Glance*. Paris, France. Available at http://www.oie.int/for-the-media/onehealth. Accessed January 28, 2014.

Oluwole, F.A., J.M. Sambo, and D. Sikhalazo. 2008. Long-term effects of different burning frequencies on the dry savannah grassland in South Africa. *Afr. J. Agric. Res.* 3(2):147–153.

Otieno, D.O., G.O. K'Otuto, B. Jákli et al. 2011. Spatial heterogeneity in ecosystem structure and productivity in a moist Kenyan savanna. *Plant Ecol.* 2129(5):769–783.

Parr, J.F., R.I. Papendick, S.B. Hornick, and R.E. Meyer. 1992. Soil quality: Attributes and relationship to alternative and sustainable agriculture. *Am. J. Altern. Agric.* 7:5–11.

Paudel, P.P., B.D. Devkota, and T. Kubota. 2009. Land degradation in Nepal: A review on its status and consequences. *J. Fac. Agric. Kyushu Univ.* 54(2):477–479.

Pellant, M., B. Abbey, and S. Karl. 2004. Restoring the Great Basin Desert, U.S.A.: Integrating science, management, and people. *Environ. Monit. Assess.* 99(1–3):169–179.

Perez, C., C. Roncoli, C. Neely, and J.L. Steiner. 2007. Can carbon sequestration benefit low-income producers in semi-arid Africa? Potentials and challenges. *Agric. Syst.* 94:2–12.

Pietola, L., R. Horn, and M. Yli-Halla. 2005. Effects of trampling by cattle on the hydraulic and mechanical properties of soil. *Soil Tillage Res.* 82:99–108.

Piñeiro, G., J.M. Paruelo, M. Oesterheld, and E.G. Jobbágy. 2010. Pathways of grazing effects on soil organic carbon and nitrogen. *Rangel. Ecol. Manag.* 63:109–119.

Potter, K.N., H.A. Torbert, H.B. Johnson, and C.R. Tischler. 1999. Carbon storage after long-term grass establishment on degraded soils. *Soil Sci.* 164:718–725.

Questad, E.J., B.L. Foster, S. Jog, K. Kindscher, and H. Loring. 2011. Evaluating patterns of biodiversity in managed grasslands using spatial turnover metrics. *Biol. Conserv.* 144(3):1050–1058.

Rau, B.M., R. Tausch, A. Reiner et al. 2010. Influence of prescribed fire on ecosystem biomass, carbon, and nitrogen in a Pinyon Juniper woodland. *Rangel. Ecol. Manag.* 63:197–202.

Reeder, J.D., and G.E. Schuman. 2002. Influence of livestock grazing on C sequestration in semi-arid mixed-grass and short-grass rangelands. *Environ. Pollut.* 116:457–463.

Reeder, J.D., G.E. Schuman, J.A. Morgan, and D.R. Lecain. 2004. Response of organic and inorganic carbon and nitrogen to long-term grazing of the shortgrass steppe. *Environ. Manag.* 33:485–495.

Reid, R.S., P.K. Thornton, G.J.M.C. Crabb, R.L. Kruska, F. Atieno, and P.G. Jones. 2004. Is it possible to mitigate greenhouse gas emissions in pastoral ecosystems of the tropics? *Environ. Dev. Sustain.* 6:91–109.

Rhoton, F.E., and D.D. Tyler. 1990. Erosion-induced changes in the properties of a fragipan soil. *Soil Sci. Soc. Am. J.* 54:223–228.

Richter, D.D., D. Markewitz, S.E. Trumbore, and C.G. Wells. 1999. Rapid accumulation and turnover of soil carbon in a re-establishing forest. *Nature* 400:56–58.

Ritchie, J.C., M.A. Nearing, and F.E. Rhoton. 2009. Sediment budgets and source determinations using fallout Cesium-137 in a semiarid rangeland watershed, Arizona, USA. *J. Environ. Radioact.* 100(8):637–643.

Roth, C.H. 2004. A framework relating soil surface condition to infiltration and sediment and nutrient mobilization in grazed rangelands of northeastern Queensland, Australia. Earth *Surf. Proc. Landforms* 29(9):1093–1104.

Ryan, J., S. Masri, H. Ibrikçi, M. Singh, M. Pala, and H.C. Harris. 2008. Implications of cereal-based crop rotations, nitrogen fertilization, and stubble grazing on soil organic matter in a Mediterranean-type environment. *Turkish J. Agric. Forestr.* 32(4):289–297.

Ryan, J., S. Masri, and M. Singh. 2009. Seasonal changes in soil organic matter and biomass and labile forms of carbon as influenced by crop rotations. *Comm. Soil Sci. Plant Anal.* 40(1–6):188–199.

Ryan, J., M. Singh, and S. Christiansen. 2012. Assessment of long-term barley–legume rotations in a typical Mediterranean agro-ecosystem: Grain and straw yields. *Arch. Agron. Soil Sci.* 58(3):233–246.

Sanchez, P.A., S. Ahamed, F. Carré et al. 2009. Digital soil map of the world. *Science* 325:680–681. Available at http://www.globalsoilmap.net. Accessed January 28, 2014.

Sandoval, M.A., N.B. Stolpe, E.M. Zagal, and M. Mardones. 2007. The effect of crop–pasture rotations on the C, N and S contents of soil aggregates and structural stability in a volcanic soil of south-central Chile. *Acta Agric. Scand.* 57:255–262.

Sankaran, M., N.P. Hanan, R.J. Scholes et al. 2005. Determinants of woody cover in African savannas. *Nature* 438:846–849.

Savory, A., and J. Butterfield. 1998. *Holistic Management: A New Framework for Decision Making*. Island Press, Washington, DC.

Savory, A., and J. Butterfield. 2010. The Holistic Management framework: Social, economic, and environmental sound development. In O. Ukaga, C. Maser, and M. Reichenbach (Eds.), *Sustainable Development: Principles, Frameworks and Cases*. CRC Press, Boca Raton, FL, pp. 149–175.

Sawadogo, L., E. Tiveau, and R. Nygård. 2005. Influence of selective tree cutting, livestock and prescribed fire on herbaceous biomass in the savannah woodlands of Burkina Faso, West Africa. *Agric. Ecosyst. Environ.* 105:335–345.

Schjønning, P., S. Elmholt, and B.T. Christensen. 2003. *Managing Soil Quality: Challenges in Modern Agriculture.* CABI Publishing, Wallingford, UK, 368 pp.

Schjønning, P., M. Lamande, F.A. Tøgersen, J. Arvidsson, and T. Keller. 2006. Distribution of vertical stress at the soil–tyre interface: Effects of tyre inflation pressure and the impact on stress propagation in the soil profile. *Adv. Geoecol.* 38:38–46.

Scholefield, D., and D.M. Hall. 1986. A recording penetrometer to measure the strength of soil in relation to the stresses exerted by a walking cow. *Eur. J. Soil Sci.* 37:165–176.

Schuman, G.E., J.D. Reeder, J.T. Manley, R.G. Hart, and W.A. Manley. 1999. Impact of grazing management on the carbon and nitrogen balance of a mixed-grass rangeland. *Ecol. Appl.* 9:65–71.

Sere, C., and H. Steinfeld. 1996. World Livestock Production Systems: Current Status, Issues and Trends. Animal Production and Health Paper No. 127. Food and Agriculture Organization of the United Nations, Rome.

Sharpley, A.N., S.J. Smith, and W.R. Bain. 1993. Nitrogen and phosphorus fate from long-term poultry litter applications to Oklahoma soils. *Soil Sci. Soc. Am. J.* 57:1131–1137.

Shipitalo, M.J., W.A. Dick, and W.M. Edwards. 2000. Conservation tillage and macropore factors that affect water movement and the fate of chemicals. *Soil Tillage Res.* 53:167–183.

Sigua, G.C., and S.W. Coleman. 2009. Long-term effect of cow congregation zone on soil penetrometer resistance: Implications for soils and forage quality. *Agron. Sustain. Dev.* 29(4):517–523.

Singleton, P.L., and B. Addison. 1999. Effects of cattle treading on physical properties of three soils used for dairy farming in the Waikato, North Island, New Zealand. *Austr. J. Soil Res.* 37(5):891–902.

Smet, M., and D. Ward. 2006. Soil quality gradients around water-points under different management systems in a semi-arid savanna, South Africa. *J. Arid Environ.* 64:251–269.

Sollenberger, L.E., C.T. Agouridis, E.S. Vanzant, A.J. Franzluebbers, and L.B. Owens. 2012. Prescribed grazing on pasturelands. In C.J. Nelson (Ed.), *Conservation Outcomes from Pastureland and Hayland Practices: Assessment, Recommendations, and Knowledge Gaps*. Allen Press, Lawrence, KS, pp. 113–204.

Solomon, T.B., H.A. Snyman, and G.N. Smit. 2007. Cattle-rangeland management practices and perceptions of pastoralists towards rangeland degradation in the Borana zone of southern Ethiopia. *J. Environ. Manag.* 82:481–494.

Sousa, F.P., T.O. Ferreira, E.S. Mendonça, R.E. Romero, and J.G.B. Oliveira. 2012. Carbon and nitrogen in degraded Brazilian semi-arid soils undergoing desertification. *Agric. Ecosyst. Environ.* 148:11–21.

Stamati, F.E., N.P. Nikolaidis, D. Venieri, E. Psillakis, and N. Kalogerakis. 2011. Dissolved organic nitrogen as an indicator of livestock impacts on soil biochemical quality. *Appl. Geochem.* 26(Suppl.):S340–S343.

Stavi, I., R. Lal, and L.B. Owens. 2011. Effects of cattle grazing during the dormant season on soil surface hydrology and physical quality in a moist-temperate region. *Ecohydrology* 4:106–114.

Stohlgren, T.J., L.D. Schell, and B. Vanden Heuvel. 1999. How grazing and soil quality affect native and exotic plant diversity in Rocky Mountain grasslands. *Ecol. Appl.* 9(1):45–64.

Teague, W.R., S.L. Dowhower, S.A. Baker et al. 2010. Soil and herbaceous plant responses to summer patch burns under continuous and rotational grazing. *Agric. Ecosyst. Environ.* 137(1–2):113–123.

Teague, W.R., S.L. Dowhower, S.A. Baker, N. Haile, P.B. DeLaune, and D.M. Conover. 2011. Grazing management impacts on vegetation, soil biota and soil chemical, physical and hydrological properties in tall grass prairie. *Agric. Ecosyst. Environ.* 141:310–322.

Tefera, B., and G. Sterk. 2010. Land management, erosion problems and soil and water conservation in Fincha'a watershed, western Ethiopia. *Land Use Policy* 27:1027–1037.

Tefera, S., H.A. Snyman, and G.N. Smit. 2007. Rangeland dynamics in southern Ethiopia: (1) Botanical composition of grasses and soil characteristics in relation to land-use and distance from water in semi-arid Borana rangelands. *J. Environ. Manag.* 85:429–442.

Thornton, P.K., R.L. Kruska, N. Henninger et al. 2002. *Mapping Poverty and Livestock in the Developing World.* International Livestock Research Institute, Nairobi, Kenya, 124 pp.

Tollner, E.W., G.V. Calvert, and G. Langdale. 1990. Animal trampling effects on soil physical properties of two southeastern U.S. ultisols. *Agric. Ecosyst. Environ.* 33:75–87.

Tom, N., A. Raman, D.S. Hodgkins, and H. Nicol. 2006. Populations of soil organisms under continuous set stocked and high intensity–short duration rotational grazing practices in the central tablelands of New South Wales (Australia). *N. Z. J. Agric. Res.* 49:261–272.

Tongway, D., and N. Hindley. 2004. Landscape function analysis: A system for monitoring rangeland function. *Afr. J. Range Forage Sci.* 21(2):109–113.

Torbert, H.A., S.A. Prior, and G.B. Runion. 2004. Impact of the return to cultivation on carbon (C) sequestration. *J. Soil Water Conserv.* 59:1–8.

Trimble, S.W., and A.C. Mendel. 1995. The cow as a geomorphic agent—A critical review. *Geomorphology.* 13(1–4):233–253.

United States Department of Agriculture-Natural Resources Conservation Service (USDA-NRCS). 1996. *America's Private Land, A Geography of Hope.* U.S. Government Printing Office, Washington, DC.

Verdoodt, A., S.M. Mureithi, L. Ye, and E. Van Ranst. 2009. Chronosequence analysis of two enclosure management strategies in degraded rangeland of semi-arid Kenya. *Agric. Ecosyst. Environ.* 129(1–3):332–339.

Walthall, C.L., J. Hatfield, P. Backlund et al. 2012. Climate Change and Agriculture in the United States: Effects and Adaptation. USDA Technical Bulletin 1935. U.S. Department of Agriculture, Washington, DC.

Warren, S.D., M.B. Nevill, W.H. Blackburn, and N.E. Garza. 1986. Soil response to trampling under intensive rotation grazing. *Soil Sci. Soc. Am. J.* 50:1336–1341.

Weil, R.R., and F. Magdoff. 2004. Significance of soil organic matter to soil quality and health. In F. Magdoff, and R.R. Weil (Eds.), *Soil Organic Matter in Sustainable Agriculture.* CRC Press, Boca Raton, FL, pp. 1–43.

White, R., S. Murray, and M. Rohweder. 2000. *Pilot Analysis of Global Ecosystems: Grassland Ecosystems.* World Resources Institute, Washington, DC, 112 pp.

Wienhold, B.J., J.R. Hendrickson, and J.F. Karn. 2001. Pasture management influences on soil properties in the Northern Great Plains. *J. Soil Water Conserv.* 56(1):27–31.

Willatt, S.T., and D.M. Pullar. 1983. Changes in soil physical properties under grazed pastures. *Aust. J. Soil Res.* 22:343–348.

Williams, J., and M. Bonell. 1988. The influence of scale of measurement on the spatial and temporal variability of the Philip infiltration parameters—An experimental study in an Australian savannah woodland. *J. Hydrol.* 104(1–4):33–51.

Williams, J., E.N. Bui, E.A. Gardner, M. Littleboy, and M.E. Probert. 1997. Tree clearing and dryland salinity hazard in the upper Burdekin catchment of north Queensland. *Austr. J. Soil Res.* 35(4):785–801.

Winter, S.L., S.D. Fuhlendorf, C.L. Goad, C.A. Davis, and K.R. Hickman. 2011. Topoedaphic variability and patch burning in sand sagebrush shrubland. *Rangel. Ecol. Manag.* 64: 633–640.

World Overview of Conservation Approaches and Technologies. 2009. Benefits of sustainable land management. UNCCD, World Overview of Conservation Approaches and Technologies, Swiss Agency for Development and Cooperation, FAO, Centre for Development and Environment, 15 pp.

Wright, A.L., F.M. Hons, and F.M. Rouquette Jr. 2004. Long-term management impacts on soil carbon and nitrogen dynamics of grazed bermudagrass pastures. *Soil Biol. Biochem.* 36:1809–1816.

Wu, H., M. Wiesmeier, Q. Yu, M. Steffens, X. Han, and I. Kögel-Knabner. 2012. Labile organic C and N mineralization of soil aggregate size classes in semiarid grasslands as affected by grazing management. *Biol. Fertil. Soils* 48(3):305–313.

Xiao-Gang, L., F.-M. Li, Z.R. Bhupinderpal-Singh, and Z. Zheng-Yan. 2007. Soil management changes organic carbon pools in alpine pastureland soils. *Soil Tillage Res.* 93(1): 186–196.

Yan, Y., J. Tian, M. Fan et al. 2012. Soil organic carbon and total nitrogen in intensively managed arable soils. *Agric. Ecosyst. Environ.* 150:102–110.

Yang, X.M., C.F. Drury, M.M. Wander, and B.D. Kay. 2008. Evaluating the effect of tillage on carbon sequestration using the minimum detectable difference concept. *Pedosphere* 18:421–430.

Yong-Zhong, S., L. Yu-Lin, C. Jian-Yuan, and Z. Wen-Zhi. 2005. Influences of continuous grazing and livestock exclusion on soil properties in a degraded sandy grassland, Inner Mongolia, northern China. *Catena* 59:267–278.

Zhao, H.L., X.Y. Zhao, R.L. Zhou, T.H. Zhang, and S. Drake. 2005. Desertification processes due to heavy grazing in sandy rangeland, Inner Mongolia. *J. Arid Environ.* 62(2): 309–319.

Zheng, X.-X., G.-H. Liu, B.-J. Fu, T.-T. Jin, and Z.-F. Liu. 2010. Effects of biodiversity and plant community composition on productivity in semiarid grasslands of Hulunbeir, Inner Mongolia, China. *Ann. N. Y. Acad. Sci.* 1195:(Suppl. 1):E52–E64.

5 Sustainable Food Production in Indo-Gangetic Plains

Role of Improved Cultivars in Cropping System Intensification for Small Farm Holders

Raj Gupta and Rajbir Yadav

CONTENTS

5.1 INTRODUCTION

Before the advent of the Green Revolution in the 1960s, South Asian farmers generally used to cultivate diverse landraces in cereal crops. This allowed the landraces to coevolve with nature and acquire specific adaptive traits. The adaptation not only minimized the risks of crop failures due to aberrant weather but also secured their livelihoods and food security. During the 1970s, these countries achieved substantial yield increase from "Green Revolution" technologies, which comprised short-stature photo-insensitive cultivars of wheat (*Triticum aestivum* L.) and rice (*Oryza sativa* L.) with high harvest index, increased use of fertilizer nutrients, and better irrigation facilities. Genetically homogenous early maturing varieties displaced the traditional landraces and helped the farmers achieve crop intensification with available irrigation facilities. Technologies together with enabling policy support led to the emergence of new cropping systems such as rice–wheat, even in areas nontraditional to rice and/or wheat. Rice–wheat systems now cover a whopping 13 million hectares (Mha) in India, Pakistan, Nepal, Bangladesh, and southern China. This cropping system laid the foundation of South Asian food security (Cassman 1999; Gupta et al. 2003). The development and application of production technologies related to the rice–wheat cropping system became highly relevant. This can be judged if one compares the acreage and productivity of wheat in two Punjabs, i.e., Indian Punjab and Pakistan Punjab (the Punjab territory was partitioned between India and Pakistan in 1947). Up to 1950, wheat acreages, production, and productivity in two Punjabs were almost similar. However, wheat production and productivity in Indian Punjab almost doubled over Pakistan Punjab in the last five decades owing to apparently better land, water, and crop management practices followed in rice–wheat cropping systems (Figure 5.1).

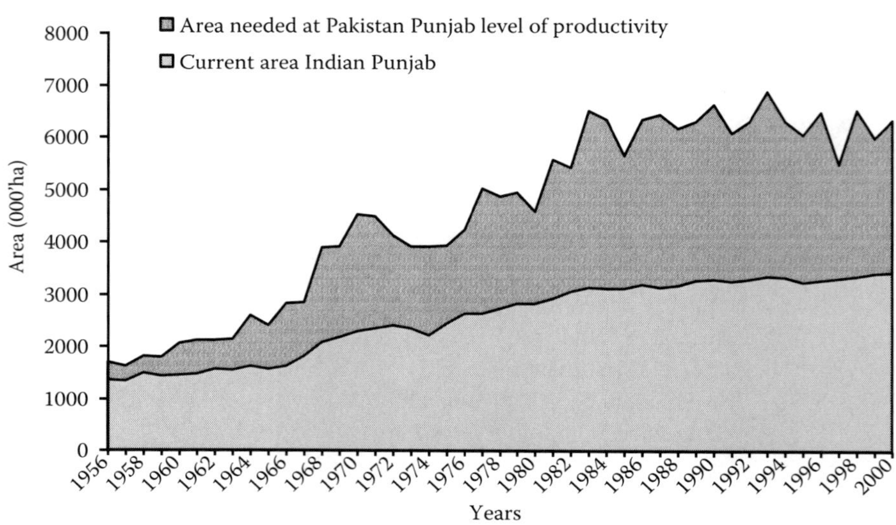

FIGURE 5.1 Comparison of wheat productivity in two Punjabs and saving in land use owing to improved technologies (data beyond 2000 not available). (Personal communication, Sahid Parvez 2001.)

Intensification of the rice–wheat production systems has spared agriculture expansion in marginal areas and replacements in other cropping systems. Data in Figure 5.1 show that achieving the present levels of wheat production in Indian Punjab would have required an additional equal expanse of land acreage in Pakistan Punjab. Prospects of bringing more area under cultivation and increasing irrigation facilities are rather very limited, particularly in South Asia (Poster 1998; The 2030 Water Resources Group 2011), and therefore, crop intensification is inevitable.

Although crop intensification with associated production technologies significantly increased food production, it had a lot of associated land degradation. The problem of land degradation manifests itself in the Indo-Gangetic plains through lower water use efficiency, lower factor productivity, reduced biodiversity, multinutrient deficiencies, declining water tables, groundwater pollution, etc. (Abrol and Gupta 1998; Mehla et al. 2000; Gupta et al. 2003; Ladha et al. 2003; Pathak et al. 2003; Chandna et al. 2010; Rodell et al. 2010). The report of the '2030 Water Resources Group' has projected a deficit of 755 billion m^3 of water in India and an annual depletion of 13–17 billion m^3 of water from the northwest plains in Punjab, Haryana, and western Uttar Pradesh (Rodell et al. 2010). The food situation is further complicated because climate changes have been projected to render almost 40% of the current area under wheat, unsustainable in the future (Ortiz et al. 2008), and each degree Celsius increase in temperature is expected to raise additional water needs by 2% to maintain crop yields at current levels (Report of Subgroup III Soil Health and Water Management. Planning Commission-ICAR-DARE, Govt. of India. 2011. New Delhi) are some of the biggest threats to agriculture in the Indian subcontinent. The current trends and future indication are, therefore, worrisome for rice–wheat systems and require innovative interventions.

The demand and supply gap in irrigation water can be bridged by either increasing its supply or improving its productivity (more crop per drop) by growing more water-efficient crop (substitution/diversification), reduced runoff, increased rainwater infiltration, and reduced evaporation. CA practices encompassing no-till or reduce tillage, along with residue retention and appropriate crop rotation, can improve the water productivity and sustain the cereal production systems at a higher level over time. Keeping in view all these facts and issues, it was felt pertinent to analyze and discuss the productivity gaps experienced in cereal crops of South Asia, and the ability of CA and plant breeding intervention to minimize these gaps.

5.2 AGRICULTURAL PRODUCTIVITY GAPS AND SCOPE FOR IMPROVEMENTS THROUGH MANAGEMENT

The yield gap concept rests on the definition and measurement of the yield potential. Practical yield gap (gap I) can be defined as the difference between the maximum attainable yield (yield obtained in yield maximization trials with best management practices) and the average yields attained at the farm level in the county/districts. This yield gap is due to poor management of the crop at the farmers' level. The other yield gap, designated as yield gap II, is between the theoretical potential yield (simulated yields with no constraints of water and nutrient) and the maximum attainable yield. This yield gap is usually caused by factors beyond management and hence constrains the use of yield-enabling technology components under specific environments. It is therefore very difficult to bridge yield gap II. Biophysical factors such as weather, soil, water, pest, weeds, management practices, variety and seed selection, nutrients, postharvest management etc.; farmers' traditions and knowledge; government policy; seed price; input supply; market; extension; and research are responsible for the yield gap I. Saharawat (2009) carried out a management yield gap analysis in cereal crops across the Indo-Gangetic plains given in Table 5.1 to identify technological interventions for productivity improvements.

TABLE 5.1
Management Yield Gaps in Rice–Wheat and Maize Crops in the Indo-Gangetic Plains

Hub Name	Cereal Crops	Simulated Potential Yield (t/ha), A	Max Attainable Exper. Yield (t/ha), B	Average Yield (t/ha), C	Management Yield Gap I (%), [100(B – C)/B]
Punjab	Rice	8.8	7.0	4.0	43.1
	Wheat	6.5	5.2	4.3	17.3
	Maize	16.6	13.3	3.1	77.0
Haryana	Rice	8.7	7.0	2.8	59.9
	Wheat	5.6	4.5	3.8	14.2
	Maize	8.0	6.4	2.8	56.9
Eastern Uttar Pradesh	Rice	7.2	5.2	3.8	26.9
	Wheat	5.4	4.6	2.4	47.4
	Maize	6.3	4.3	1.2	72.2
Bihar	Rice	8.7	7.0	2.1	70.5
	Wheat	5.5	4.4	2.3	47.7
	Maize	10.7	8.6	2.6	69.6
Nepal	Rice	7.2	5.4	3.3	38.9
	Wheat	5.4	5.0	2.9	42.8
	Maize	6.3	4.3	2.6	39.5
Bangladesh	Rice (boro)	9.0	7.2	5.9	18.1
	Rice (aman)	7.0	5.6	2.7	51.8
	Wheat	6.0	4.8	2.5	47.9
	Maize	16.5	13.2	8.4	36.4

Source: Rainfall: Indian Meteorological Department (http://www.imd.gov.in/). Actual yields: SDDS-DES, Ministry of Agriculture, Government of India and AGRID-NIC, Ministry of Communications and Information Technology, Government of India; Regmi and Ladha (2005); Saharawat (2009); Personal communications from the CSISA hub managers; Chand, R., S. Garg, and L. Pandey. 2009. Regional Variations in Agricultural Productivity: A District Level Study. Discussion Paper No. NPP 01/2009. New Delhi: National Centre for Agricultural Economics and Policy Research.

Note: Boro rice is winter rice and aman rice is rice of the rainy season.

Such large yield gaps (Table 5.1) have also been highlighted in other studies (Aggarwal and Kalra 1994; Aggarwal et al. 2000; Pathak et al. 2003; Saharawat 2009). In an integrated analysis of the long-term trials carried out across 23 Indo-Gangetic plain sites, Tirol-Padre and Ladha (2006) reported that wheat yields had not improved even after 7–23 years, while surprisingly rice yields had declined during the same periods. Despite geographical proximity, the productivity of rice in Bihar is much lower than in those in Nepal and Bangladesh. Late planting, uncontrolled water supplies, monsoon floods, and lack of supplemental irrigation facilities and less fertilizer use are some of the reasons for the low productivity of rice in Bihar. Many of these reasons are surmountable. Boro rice (cultivar Gautam), seeded during receding winter and transplanted in early February in Bihar, West Bengal, and Bangladesh usually yield double than aman (monsoon) rice. A cropping system comprising winter maize (*Zea mays* L.) and boro rice can thus contribute to a significant jump in agricultural productivity in the eastern Gangetic plains of Uttar Pradesh, Bihar, West Bengal, and Terai of Nepal and Bangladesh. This requires further research efforts and shifts in rice production technologies, including the use of modern cultivars, water-efficient technologies like precision leveling, direct dry seeding, intermittent irrigation, and weed and nutrient management. The productivity of scented Basmati cultures is often lower than that of improved nonscented rice. Haryana and Pakistan Punjab farmers grow more Basmati rice than farmers in Indian Punjab. It is for this reason that the rice yield gaps apparently look more in Haryana than in Punjab.

In case of maize, the management yield gaps are more glaring in kharif maize with productivity being comparatively very low in Bihar and West Bengal than in Bangladesh with similar kinds of environments. Lesser adoption of high-yielding single-cross hybrids along with higher yield losses due to biotic and abiotic stresses often results in poor yield realization at the farmers' field in these areas. The productivity of kharif maize is also very low in Punjab, largely because of its restriction to marginal rainfed environments. On the other hand, spring maize after potato, grown with assured irrigation and input supply, gives a very high yield. The development and adoption of waterlogging-tolerant single-cross maize hybrids during the kharif season, or alternatively, the introduction of short-duration rice–potato–winter maize or rice–mustard (*Brassica juncea*)–mung bean (*Vigna radiata*, R. Wilczek) cropping systems along with some groundwater development can provide a much-needed impetus to better yield realization in the eastern Gangetic plains.

Management yield gaps are comparatively very low in northwestern Gangetic plains in comparison with those of the eastern Gangetic plains. Late planting and consequently terminal heat stress have been identified as the biggest reasons for poor yield realization in wheat in the simulation studies reported earlier. The daily losses due to late planting of wheat in the northeastern plains are 60 kg/ha, against 26–32 kg/ha in the northwestern plains (Aggarwal and Kalra 1994; Ortiz-Monasterio et al. 1994; Dhillon et al. 2000; Mehla et al. 2000; Ortiz et al. 2008; Rajaram et al. 2008; Timsina et al. 2008). Late harvesting of rice, delay wheat seeding, and these coupled with lower application of fertilizers and inadequate water supplies result in very large yield gaps (about 50%) in the northeastern plain zone. On the other hand, the issues for management yield gap (20%) in the northwestern plains of India—deteriorating soil health, increased salinity, distorted C-to-N ratio largely because of very low C level (<0.5%), micronutrient deficiency, evolution of herbicide-resistant *Phalaris minor* along with soil compaction, and sharply depleting underground water reserve (Abrol and Gupta 1998; Gupta and Abrol 2000)—are totally different. According to Waddington et al. (2010), nearly half of these yield gaps can be managed by adopting better production techniques. Bridging the remaining yield gaps will require adopting measures in the domain of corrective socioeconomics and a conducive policy environment (Waddington et al. 2010). Gupta (2010) has opined that with appropriate technologies, about 10.7 Tg of additional food can be produced annually from a 28 Mha area covered by rice–wheat, rainfed, mixed production systems in India. An additional 13.4 Tg of food can be added by targeting "rice fallow" area spread over 11.6 Mha in South Asia. The projected deficit of 755 billion m^3 of water (The 2030 Water Resources Group 2011) by 2030 along with groundwater pollution, soil salinization, impact of climate change on water availability, and gradual decline in productivity are some of the major challenges to be faced in the future. Climatic factors such as diminishing radiation due to increasing smogs in winter, increased minimum temperature, although not consistent over the years, are also posing greater difficulty in achieving genetic gain. The productivity gain in the earlier period has also been attributed to greater resistance to multiple stresses (Tollenaar et al. 1994; Sayre and Ramos 1997; Cassman 1999; Gupta and Sayre 2007). Very subtle changes in soil organic chemistry, weather parameters, and production environment make assessable impact on agricultural production, individually and many times synergistically (Cassman and Pingali 1995; Cassman 1999). The subtle changes mentioned previously are complex forms of land degradation that increasingly limit our capacity to produce more with the same level of inputs. These only show that soil quality need to be improved or else additional agri-inputs will be required to offset yield declines due to decrease in soil quality.

5.3 RAINFALL AND AGRICULTURAL PRODUCTIVITY NEXUS IN INDO-GANGETIC PLAINS

The farm productivity and annual rainfall of different districts encompassing various hubs are plotted in Figure 5.2 using latest data from Chand et al. (2009). The two variables are plotted against the longitude of the district headquarters in the respective states. It was observed that agricultural productivity, in spite of low rainfall, was higher in northwest plains than in the middle Gangetic plains. Low productivity has been linked to several lows in rainfall, fertilizer use, and irrigated area (Chand et al. 2009). Data in Figure 5.2 indicate that annual rainfall increases from west to east (with longitudes). Agricultural productivity, high rainfall, and poverty seem to be in parallel in districts located between 78.83° and 86.13° north longitudes. Many of these districts have large tracts of Chaur, Tal, and Diara lands (low-lying lands), which become flooded during monsoon season and vacate the fields very late for wheat planting during winter season. There is a need for a paradigm shift in the approach for early crop establishment, fertilizer application methods, and more area of rice fallow land under wheat cultivation in winter. Relay and surface seeding techniques can prove helpful under such conditions.

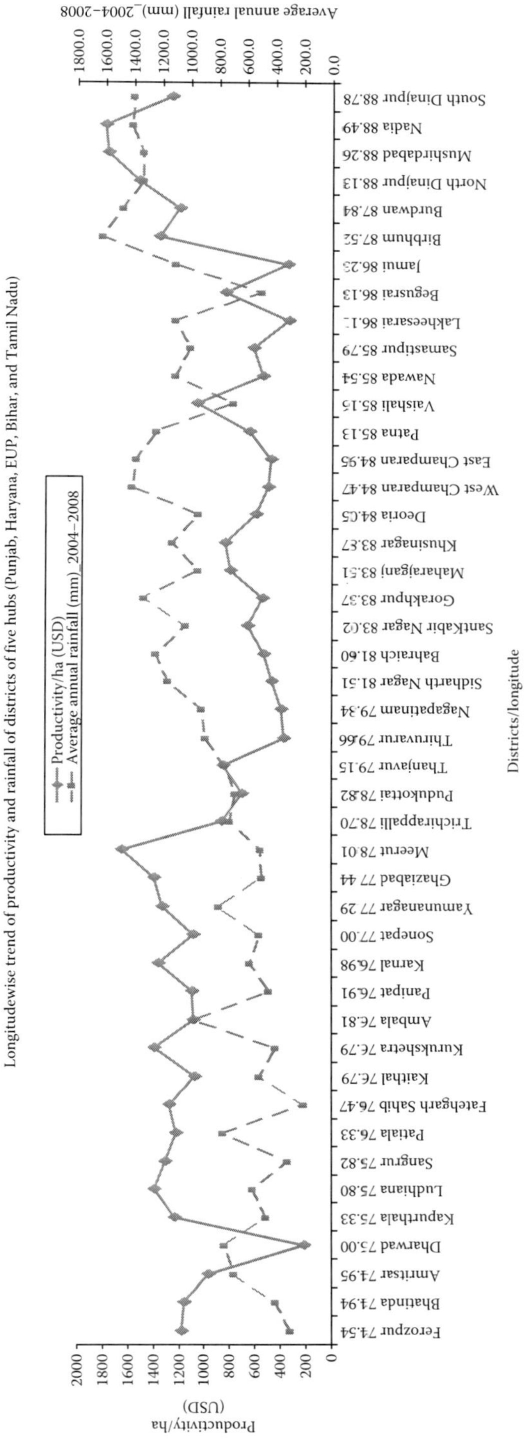

FIGURE 5.2 Productivity and rainfall trends in different districts. (Yield data for the plot taken from Chand et al. 2009.)

Alfisols of Raichur/Dharwad districts in Karnataka receive more rains as compared to Ferozpur (Punjab), but still have low agricultural productivity. This is because Karnataka farmers do not have tube well irrigation which is available to Punjab farmers. The scope for raising the productivity through genetic means in such environments is limited (Cassman 1999). CA practices with residue retention to improve water supply and soil health can be more rewarding. In coarser textured soils found in the southwestern parts of Punjab and Haryana, wheat productivity is low because of soil salinity and saline irrigations, and vacating the fields of cotton late in the season for wheat planting. Growing cotton in kharif in saline areas often delays wheat sowing, and therefore, reduces yield. The recent spurt in cultivation of HD2851, a wheat variety adapted to slight delay in sowing, in these areas is indicative of both the problem and the solution. Relay cropping of wheat in standing cotton can improve productivity of both cotton and wheat crops in the two Punjabs and in Haryana (Buttar et al. 2011). It has been observed that seed-based technologies alone have not done so well in enhancing productivity in the lesser-endowed regions located between 78.83° and 86.13° east longitudes. These areas often have low productivity owing to the presence of Chaur, Tal, and Diara lands; shifting river courses; droughts and floods; >4 Mha of rice fallows; poor-quality seed systems; low fertilizer use; late planting; weak infrastructure and technology fatigue; and little groundwater development during the kharif season. These areas, if put to effective use, can easily produce an additional 10 Tg of food with appropriate technology support. Not only that these risk-prone farmers do not have worthwhile fertilizer recommendations, but also Bihar, with comparatively higher rainfall, has low productivity, largely because of the almost negligible groundwater development work. The northern districts (e.g., Patna, Nalanda, and Vaishali) of Bihar located in the fluvial plains of the major rivers have higher productivity than the southern districts (Lakhisarai, Champaran, Begusarai, Buxur, etc.). With little irrigation water, many districts in south Bihar face drought like situations during the kharif season and largely depend on rainwater to raise puddled transplanted rice crop. It is observed that most farmers waste nearly 40% rainwater in tillage operations before transplanting rice (Singh et al. 2001). With little groundwater development, it is better to establish rice early before the onset of monsoons in these districts, such as to significantly raise rice–wheat productivity in Bihar, West Bengal, and Bangladesh, where there is some parallel between rainfall and productivity trends.

5.4 FOOD REQUIREMENTS AND ALTERNATE SOURCES OF PRODUCTIVITY GROWTH

Easy gains from the original Green Revolution have already been realized; however, to keep pace with the increasing population and demand for food, further increase in production is still required. India alone needs to produce an additional 64 Tg of food over the next decade to achieve the targeted production of 294 Tg by 2020. The important question, therefore, is from where will these future productivity gains come from? Will yield gain be further consolidated through germplasm improvement as done in the past, or will some new avenues need to be exploited? Increasing the efficiency of inputs in irrigated, semiarid, humid, and subhumid tropics through better management of natural resources and improving the productivity in rainfed agroecosystems seem to be more certain methods for further gain. Incidentally, these are also the areas where seed-based technologies alone have not done so well in enhancing productivity. Farmers located in the lesser-endowed regions located between 78.83° and 86.13° east longitudes often have low productivity owing to the presence of Chaur, Tal, and Diara lands (low-lying waterlogged areas); shifting river courses; droughts and floods; rice fallow spread over >4 Mha; poor-quality seed systems; low fertilizer use; late planting; weak infrastructure and technology fatigue; and little groundwater development during the kharif season. These areas if put to effective use can easily produce an additional 10–15 Tg of food with appropriate technological support. It may also be mentioned here that risk-prone (waterlogged/flood-prone/drought-affected) farmers located in eastern Gangetic plains do not have worthwhile fertilizer recommendations.

5.5 CONSERVATION AGRICULTURE AS A CONCEPT FOR INTENSIFICATION OF CROPPING SYSTEMS

Increased food production with better distribution without compromising the environment and degrading the natural resources has been the most sought-after goal worldwide in the last century. Increasingly, CA is being accepted as an important strategy for enhancing productivity, improving environmental quality, preserving natural resources, improving food security, and reducing poverty. The basic components of CA include drastic reduction in tillage, adequate retention of crop residues on the soil surface, use of economically feasible diversified crop rotations, avoidance of freewheeling, and practice of controlled traffic, if possible. These elements of CA are not site specific but represent unvarying objectives that are practiced to extend CA technologies efficiently across all production conditions. The strategic entry points for CA can vary according to locations based on the relative importance of the production constraints. No-till agriculture is considered as a revolutionary step in the direction of preventing land degradation and rehabilitation of the resilient but fragile lands. No-till agriculture together with other associated management practices, such as direct seeding into loose crop residues to provide soil cover and to conserve soil moisture, judicious choice of crop rotations, and agroforestry tree species, constitute CA. Frameworks such as CA, "ecological intensification" (Cassman 1999), and "evergreen revolution" (Swaminathan 2000) share a view of cropping systems as agroecosystems designed to make maximum use of fixed resources (land, light, temperature, etc.) along with optimum use of agri-inputs for attaining sustainable production levels. These systems tap the traditional knowledge of the farmers and add new information relevant to the specific ecologies for the intensification process (Matson et al. 1997).

With limited scope for further expansion of the area under agriculture, decreasing factor productivity, quantum jump in production can only be achieved through agriculture intensification. Can CA lead to agriculture intensification? Agricultural intensification can be accomplished through (i) increasing yields per hectare (e.g., timely planting and with increased inputs such as water and fertilizer nutrients), (ii) increasing cropping intensity per unit of land (e.g., use of short-season crop cultivars, relay and mixed cropping, and growing an additional crop), and (iii) changing land use from low-value crops or commodities to those that receive higher market prices. Sustainable agricultural intensification is defined as producing more output from the same area of land while reducing the negative environmental impacts and at the same time increasing contributions to natural capital and the flow of environmental services (Pretty 2008; Royal Society 2009; Conway and Waage 2010; Godfray et al. 2010). Cropping systems such as rice–wheat in south Asia, which happened to be the backbone of food security in the past, are now being challenged on the basis of their high environmental costs. Thus, our notion of sustainability of agricultural systems is shifting, and alternative practices and systems that reduce negative externalities should be sought. Sustainable agricultural systems, by definition, are less vulnerable to shocks and stresses. It is now widely recognized that yield gaps in cereal crops described previously result from agronomic failings, and that future yield increases depend heavily on this science (Pretty et al. 2011). Many variants of CA have been adopted by farmers in the tropical/subtropical and temperate regions of the world for improved yields. CA has steadily increased worldwide to cover about 7% of the world arable land area (Derpsch and Friedrich 2009). CA is an innovation process of developing appropriate CA implements able to plant in loose and anchored residues, early maturing cultivars for the cropping system, and harnessing appropriate nutrient–water interactions through iterative fine-tuning of crop production technologies. It must, however, be remembered that conservation tillage (CT) and CA are not synonymous. CT refers to reduced tillage with some residues left on the surface and others incorporated using a plough. In CA, tillage is drastically reduced or not practiced to incorporate the residues. Readers are also referred to the use of terms such as resource conservation technologies (RCTs) wherein CA is defined as a practical agricultural production system that strives to achieve acceptable profits, and reduce labor and energy inputs while concurrently conserving and (in time) enhancing environmental quality. All RCT practices may not fit into the elements of CA but may still be useful to the farmers on economic grounds. Thus, there is a need for flexibility in defining CA for operational purposes.

Zero-till/CA systems are not about precision planting without tillage by using appropriate seed drill or planters. It is about management practices (weed, water, nutrient and integrated pest management, etc.) that make zero-till technology successful and provide added advantages to the farmers. The most common CA interventions are reduced (or no) till with residue (straw) retention on the field, resulting in lessened erosion and increased aboveground and belowground biodiversity, improved water penetration and available holding capacity, and greater stores of surface soil carbon. The impact of the interventions/technologies can be maximized by the layering of more than one technology, one above the other, in the crop production chain. However, quite often, circumstances constrain the farmers to overlook or jump some points in the chain such as to suite their resource endowments and the field situations. This facilitates the farmers to adopt CA-based RCTs as per their conveniences. This point is clarified in Figure 5.3 wherein a farmer not having access to a laser-assisted precision land-leveling system and machinery to plant in loose residue can still harness some benefits of CA practices by following the "better-bet" practices shown in Figure 5.3. The "better- and best-bet" management practices for unpuddled transplanted rice–zero-till wheat system has been explained through a schematic view as a specific example. A farmer following some segments of the lower chain (–) and some of the upper chain is only following a set of better-bet rather than a best-bet

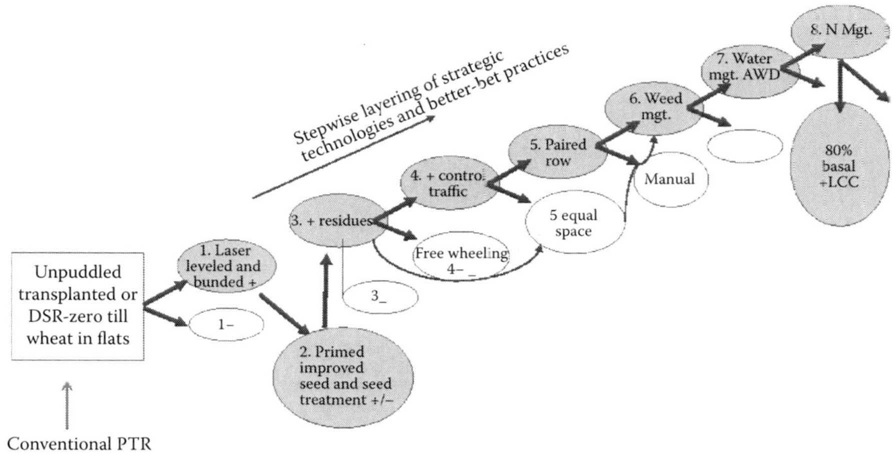

FIGURE 5.3 Example of strategic entry points for layering of the CA components. Notations: laser leveled (L) and traditionally leveled (TL) fields; weed management through chemical control (Hrb) and mechanical means (Mech); compaction-free wheeling of tractor (FW) and movement of tractor in fixed tramlines/controlled traffic (CT); FW, free wheeling; CT, controlled traffic; M, manual weeding; CM, chemical molecules; ES, equal spacing; PR, paired row planting; IC, incorporation; SR, surface retention; BM, brown manuring.

practices. The curved arrows suggest that a farmer can jump from step 2 (+residues) to step 4 (equal spacing), and then to manual weed management (5) to continue the alternate wetting drying (AWD) practice with advantage (a better-bet) than the conventional practice with all (–) in the process chain (business as usual). If the farmer follows the upper chain, it implies that the farmer has opted for the "best-bet" management practices. This illustration enables one to define "better- and best-bet" practices for several other tillage–crop establishment systems.

From the example above, it would appear that stepwise layering of "better-bet" technologies and practices on CA platforms help farmers improve the crop and total system productivity at less costs, by making the best use of appropriate cultivars for no-till systems; avoiding unnecessary use of external inputs; harnessing agro-ecological processes such as nutrient cycling, predation, and parasitism (through mulching); and biological nitrogen fixation (seed treatment). It also minimizes the use of technologies or practices that have adverse impacts on the environment and human health (e.g., preventing residue burning, and fertilizer N broadcasting); carbon sequestration; biodiversity; and dispersal of pests, pathogens, and weeds. Pretty et al. (2011) have earlier suggested that if a technology assists in the efficient conversion of solar energy without adverse ecological consequences, then it is likely to contribute to the system's sustainability.

The example in Figure 5.3 also suggests that RCTs are "divisible" and "flexible" in application under diverse situations, allowing farmers to benefit from them under diverse situations. CA-based RCTs are an "open" approach, easier to mainstream and be adapted even in conventional agriculture systems.

RCTs include a wide range of practices. No-till/minimum tillage approaches lead to a drastic reduction in tillage operations, and hence costs, thereby making it easier for resource-poor and undercapitalized farmers to adopt them. Other technologies include surface seeding, raised-bed planting, skip furrow irrigation in row-planted cropping systems, or with options of complementary crop (intercropping of potato, mung bean, cowpea, chickpea/wheat with maize, sunflower, and sugarcane, etc.), water harvesting and supplemental irrigation, mulching and residue management, live fences and vegetative barriers, agroforestry and horticulture, integrated nutrient management, integrated pest management, integrated tree–crop–livestock farming systems, and the rational use of sloping land (contour farming, upward planting in residues, etc.). The new innovations for direct dry-seeded rice and brown maturing (sowing a green manure crop with the main crop and then knocking down the green manure crop with herbicide molecules), as well as the identification of pre- and postemergence herbicide molecules, have opened the window for practicing CA in irrigated rice–wheat systems. CA-based RCTs have been shown to increase production and improve soil health, make ecosystems more resilient and reduce their vulnerability to climate change. RCTs help produce more at less cost (save labor, fuel/energy, water, and other inputs, and preserve a clean environment), and provide a platform for diversification and intensification of the production systems (Gathala et al. 2013; Jat et al. 2013; Kumar et al. 2013).

5.6 PLANT BREEDING INTERVENTIONS TO MAKE CONSERVATION AGRICULTURE MORE PROFITABLE

Improved cultivars along with external use of fertilizer nutrients, herbicides, and water supplies have tripled the global food production since 1950. Genetic gains in cereal production have been mostly due to breeding × management interactions (Slafer et al. 2005), and are approximately around 0.5% year^{-1} (Calderini et al. 1999; Abeledo et al. 2002). The yield gains were more prominent under environments optimum for crop production. Therefore, the benefits of increased knowledge about genetics and crop production were restricted to a small section of farmers located in favorable environments. According to Pimbert (2008), of 1.3 billion farmers globally, most (96%) are marginal farmers who are unable to apply sufficient external inputs (nutrient, water, etc.) and hence have not been able to realize the full benefits of plant breeding approaches. Increase in food production, either through better-yielding genotypes, better agronomic management, or through their interaction, has become a necessity as the world population is likely to cross the 9 billion mark by 2050 (Tilman et al. 2002). Increased population along with change in dietary habits may result in doubling food demand. With limited scope for expansion in the cropped area, intensification and enhancing crop productivity are the only inevitable routes available. Yield gap at present in many important crops such as wheat and rice results from agronomic failings, and future gains in productivity can come through if we breed the varieties suited to crop management for specific circumstances. Crop production practices that enhance productivity and income sustainably have to be identified. Plant breeders will have to develop varieties that actually complement these systems. It is now generally agreed that climate change has come to stay and is affecting the planting time of food crops, including wheat, particularly in South Asia.

5.6.1 Exploiting Genetic Variability for Better Crop Stand Establishment under Conservation Agriculture Practices

Residue retention for mulching has often been reported to physically hinder shoot emergence and slow down germination, as well as reduce seedling biomass in wheat, rice, corn, and canola (Chastain et al. 1995; Swan et al. 1996; Wuest et al. 2000; Bruce 2003; Mohanty and Painuli 2004). Coleoptile elongation is considered as an important mechanism in overcoming mulch and seed depth barriers to seedling emergence. Coleoptiles in cereals provide protection to subcrown internodes. The elongation capacity of subcrown internodes generally determines the seed placement depth in crops. Deeper placement of seeds under CA, particularly under rainfed conditions, provide uniform stand establishment. Alternately, rainfed farmers have to open "V-shape furrows" with narrow shovels of the cultivator followed by placement of seed 5 cm deep at the sill of the furrows using inverted T-openers of the zero-till drills. Ferguson and Boatwright (1968) have indicated that as surface residues increased, the crown node formed farther from seed—closer to soil surface and, in some cases, above the surface. They noted a positive variety × soil temperature × light intensity interaction mediated via surface mulch. Crown nodes form closer to soil surface as the light available for seedlings or soil temperature decreased. Shorter coleoptiles associated with the presence of the gibberellin-insensitive *RhtB1b* and *Rht-D1b* dwarfing genes in wheat can result in slow seedling emergence, reduced tillering, and poor crop stand establishment when seed is placed deeper in mulched no-till situations (Fick and Qualset 1976; Allan 1989; Matsui et al. 1998; Schillinger et al. 1998; Rebetzke et al. 2001; Rebetzke et al. 2005). Wheat genotypes with the *Rht-B1b* gene for a short coleoptile emerged slower with more nonproductive tillers with less biomass. In contrast, genotypes with longer coleoptile length due to various gibberellin-responsive dwarfing genes, such as *Rht8* and *Rht9*, are less affected by stubble mass and sowing depth. Combining alternative dwarfing genes (*Rht8*, *Rht4*, *Rht12*, and *Rht13*) for reduced plant height with longer coleoptiles can be highly rewarding for a large impact under rainfed CA (Hughes and Mitchell 1987; Schillinger et al. 1998; Rebetzke et al. 1999; Botwright et al. 2001; Ellis et al. 2004). Thus, wheat genotypes well adapted to surface residues coupled with reduced light intensities, higher soil temperature, and higher soil moisture content should produce crown node well below soil surface, and reduce root rot problems (Boosalis et al. 1981).

5.6.2 Reprioritize Emphasis on Resistance Breeding

Wheat is generally planted in the first fortnight of November in South Asia to avoid terminal heat stresses and moisture shortages resulting in shriveled grains and reduced yields. Early planting, however, in October, generally result in stunted growth, poor tillering, and drastic reduction in yields. Thus, heat stresses in early and late planted wheat crop adversely affect grain yields for different reasons. CA is known to help in early planting, and increase cropping intensity by reducing the turn-around time between crops. However, sustainable intensification will need crop varieties having in-built tolerance to both early and late heat stress and drought, and traits for higher resource use efficiency and enhanced pest and disease resistance (http://www.fao

.org/ag/save-and-grow/pdfs/factsheets/en/SG-crops.pdf). Several researchers have indicated that conservation agricultural (CA) practices like zero-tillage and residue retention change the crop production environment significantly (Kuhlman and Steffey 1982; Nyvall 1982; Triplett and Van Doren 1985), and therefore the genotypic requirements under CA are different from the normal tilled production environment (Duvick 1990; Trethowan et al. 2005; Joshi et al. 2007; Yadav et al. 2012). Trethwon et al. (2012) reported significantly higher yield in Berkut × Krichauff population of wheat under zero tillage at Nairobi on two soil types largely because of better availability of water and more days to flower under CA practices. Verhulst et al. (2011) found that water infiltration was significantly improved when tillage was reduced and crop residues are retained. CA practices modulate soil temperature depending on crop growing seasons (mulching keeps the soils cool in summer season and warm in winters) and maintain soil moisture for a longer period. This has a significant influence on crop growth and may also effect disease development. A number of necrotrophic fungal pathogens attacking different parts of crop plants use crop residue retained on surface as a substrate for survival during the off-season and to infect the succeeding crop (Forcella et al. 1994; Bianco 1998; Nazareno et al. 1993). Use of host resistance through plant breeding has been highly rewarding against specialized pathogens because of availability of major genes with large effect within the crop species. However, broad-spectrum host resistance to pathogens like *Pythium* is either lacking or limited within host species. Progress through plant breeding against disease under CA has been comparatively slow largely because of the reluctance of plant breeders to select their genetic stocks under zero-till conditions. In Brazil, where CA has been highly successful, disease control has been recognized as a major weak point (Scopel et al. 2004). In Australia, root diseases caused by *Rhizoctonia solani*, nematodes, and *Pseudomonas* bacteria (Watt et al. 2005, 2006) are reportedly favored in undisturbed soils. The slow growth of wheat during its early stage has been linked to inhibitory *Pseudomonas* on the root tips of wheat in no-till soils. In sandy soil with light texture, *Rhizoctonia* remains a significant problem for no-till cropping (Kirkegaard et al. 2011). Special breeding programs have been established to develop disease-resistant varieties for zero-till condition in soybean, rice, wheat, cotton, and maize. Residue mulch is often considered as the major culprit for increased disease problems under no-till systems. Long-term crop rotation combined with use of moderately resistant cultivars and judicious use of inorganic and organic pesticides may help in breaking the disease and pest cycles (Bolliger et al. 2006).

5.6.3 Use of Landraces for Risk Management and Yield Maximization

In times when our awareness for genetic diversity is growing, large corporations typically rely on releasing few high-yielding, genetically homogenous varieties. This encourages farmers to displace traditional varieties on small farms. However, the fact that large numbers of traditional landraces of maize are still being grown in the tribal belts, in the belly of central India and elsewhere, seems to prove that poorly endowed farmers try to spread their risks in manners that improve their livelihoods and food security. Cultivating landraces with specific traits is a vital gene pool for the high-tech plant breeders who need such traits (earliness in maturity, grain color, spacing between "tassel and silks" and weed competitiveness, tolerance to heat and moisture stresses, etc.). Without a gene pool that is responsive to external use of fertilizer nutrients, a greater input use in agriculture is unlikely to sustain food security for the teeming millions. Thus, we need to cultivate a wide diversity of plants and allow them to coevolve with nature. The success of the smallholder farmers always depend on manipulations of the resources they have on their farms. These small farmers generally have to adapt their crops to the environment, unlike the large/big farmers who use resource/inputs to manipulate the environments (e.g., convert rainfed to irrigated farms, or saline/alkali soils to nonsaline/nonalkali soils). Thus, we have to appreciate the knowledge of the tribal farmers for using specific landraces for overcoming a specific bottleneck of production systems. Landraces when cultivated over many decades consist of plants that have become adapted to specific conditions of the ecosystems/places where such races are grown. These landraces had coevolved with agents influencing the resilience of the landraces. Landraces, evolved in specific environments, respond to specific nutrients, soil moisture, and thermal regimes/dynamics in a manner that make landraces grow more vigorously to compete better with weeds and escape disease and pest attacks. Some landraces do not need much nutrients and water, bind soil particles against erosion, tolerate temporary flooding, and become more resilient after flooding, and others are able to cope with drought and heat, soil moisture, and salinity stresses. Yet others have other traits (e.g., silica hairs), such as to escape insect attack.

5.6.4 Harnessing Genotype–Tillage–Cropping System Interactions for Yield Maximization

Several earlier reports (Newhouse and Crosbie 1986; Kaspar et al. 1987; Hersterman et al. 1988; Duiker et al. 2006) had suggested an absence of genotype × tillage interaction or the presence of non-cross-over-type interactions. Many of the studies where genotype × tillage interaction is either absent, or are small, have been done with very limited number of genotypes, and the genotypes used are the products of breeding programs for conventional tillage. There is a likelihood that genotypes bred for conventional systems may not show specific adaptability for zero-till conditions. However, an absence of genotype × tillage interaction or the presence of a non-cross-over-type interaction does not warrant for a separate breeding program for the zero-till condition. Several recent studies have reported a strong presence of such interactions under no-till situations (Trethowan et al. 2005; Trethowan et al. 2012; Yadav et al. 2012). A strong presence of genotype × tillage interaction, however, necessitates the development of cultivars adapted to CA. The adaptation can be improved by selecting the segregating material as well as by testing of advanced lines under the zero-till condition. Yadav et al. (2012) investigated the genotype × tillage interactions among 720 wheat genotypes selected from advanced breeding lines (F6 generation) grown on permanent untilled raised beds, as well as under conventional tilled conditions in maize–wheat, pearl–millet–wheat, and rice–wheat systems.

These lines were then tested for their specific adaptation for permanent bed or for conventional tillage conditions. It was observed that the top 20 best-performing genotypes on permanent raised beds were different from those in conventionally tilled plots, suggesting the presence of significant genotype × environment interactions. In the same trial, three check varieties of wheat, namely PBW550, HD2967, and DBW17, developed for and bred under conventional tillage conditions, were repeated after every 24 entries and therefore replicated 30 times in the trial to test their adaptability for zero-till conditions.

The average yield of check wheat cultivars planted in a maize–wheat cropping system under conventionally tilled and permanent raised bed systems are given in Table 5.2.

The yield responses of the three check cultivars under two production environments, namely zero-till and conventional till systems, were noticeably interesting (Figure 5.4). The cultivar PBW550 showed specific adaptation for conventional till conditions, and its performance was at par with the other two check varieties under conventional till. However, PBW550 was significantly poor than other two wheat cultivars under permanent zero-till conditions. The cultivar DBW17, on the other hand, performed best under the zero-till condition. It was observed that HD2967 was statistically at par (neutral) to tillage and crop establishment options. In the same study, specific genotypic adaptation was also observed for different cropping systems, as the top-ranking wheat genotypes were different for the three major rice–wheat, maize–wheat, and pearl millet–wheat cropping systems, indicating the existence of a genotype × cropping system interaction also. With accumulating evidences on the effect of CA on the cereal–soil dynamic, there is every likelihood of a prevalence of the genotype × system interaction. Future gains in grain yield, therefore, can be harnessed with more detailed knowledge of the responses of each species to the environment and a precise description of the genotypic variability in promising traits (i.e., a fine-tuned phenotyping for developmental attributes), which seems to be valuable for any cropping system.

5.6.5 Vernalization Genes for Developing Alternate Sustainable Wheat-Based Cropping Systems

Increase in yield in many crops such as wheat has largely been achieved by adjusting the plant phenology as per the requirements of the growing condition in specific regions. Knowledge about genetic factors governing adaptability aids crop breeding for yield potential enhancement. Worldwide, the genetic gain has been limited under dry conditions. More gain can be achieved for these conditions by selecting and combining synergetic traits (Dingkuhn et al. 2006). Phenological development, undoubtedly the most important attribute (Richards 1996; Passioura 1996, 2002; Villegas et al. 2000; Araus et al. 2002; González et al. 2003; Slafer et al. 2005), allows the crop to either escape stresses or avoid the coincidence of the most sensitive phases with the most likely occurrence of stress. Wheat is generally planted in the first fortnight of November in South Asia to avoid terminal heat stresses and moisture shortages resulting in shriveled grains and reduced yields, to provide a more favorable environment for crop growth and avoid early and late heat stress. Genetic gains in wheat yield in the last two decades in the northern plain of India have largely been realized

TABLE 5.2
Top Ten High-Yielding Wheat Genotypes in Three Cropping Systems under Conservation Agriculture

Maize–Wheat (on Raised Beds)			Pearl Millet–Wheat (on Raised Beds)			Rice–Wheat (Zero-Till Flat)		
Rank	Entry No.	Yield, as Percentage of Yield Obtained in Best Check	Rank	Entry No.	Yield, as Percentage of Yield Obtained in Best Check	Rank	Entry No.	Yield, as Percentage of Yield Obtained in Best Check
1	09-103	115	1	09-591	118	1	09-648	127
2	09-107	110	2	09-602	115	2	09-588	124
3	09-39	108	3	09-598	114	3	09-598	118
4	09-117	107	4	09-38	113	4	09-640	117
5	09-33	105	5	09-596	113	5	09-590	113
6	09-93	104	6	09-601	113	6	09-589	112
7	09-282	103	7	09-630	109	7	09-612	112
8	09-289	103	8	09-128	102	8	09-616	112
9	09-489	103	9	09-191	101	9	09-586	109
10	09-321	102	10	09-180	99	10	09-609	108
Mean (kg/ha)		2960			2949			2088
SE		11.87			43.19			44.34

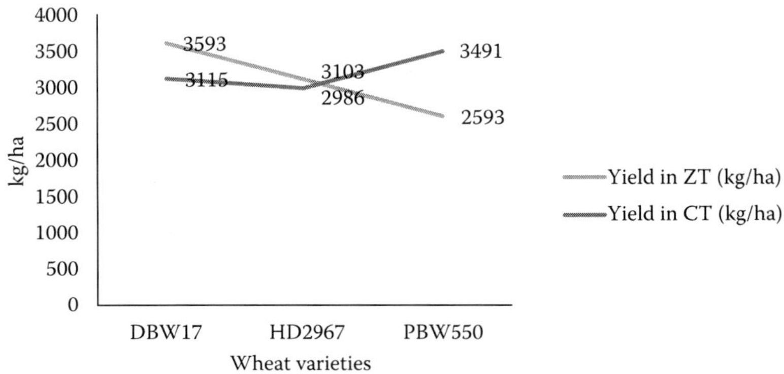

FIGURE 5.4 Genotype × tillage interaction in wheat varieties.

through the development and cultivation of varieties with slightly longer duration (PBW343, UP2338, WH542, HD2733, and DBW17) against the previous predominating varieties such as HD2329. The longer duration of the crop, however, increased the incidence of terminal heat stress and, therefore, led to the recommendation of wheat seeding in last week of October or the first week of November in states such as Punjab and Haryana. Early planting without taking into consideration the vernalization base of the cultivars, however, in October, generally many times results in stunted growth, poor tillering, early flowering, and drastic reduction in yields. Thus, heat stresses in early and late planted wheat crop adversely affect grain yields for different reasons. CA is known to help in early planting, and increase cropping intensity by reducing the turnover time between the crops.

It has been pointed out in preceding sections that heat stresses in early-planted as well as late-season-planted wheat adversely affect grain yields, although with different routes. In most of south Asia, wheat and barley crops are exposed to drought and heat during late reproductive and grain-filling phases. Therefore, crop intensification strategies to breed for earliness to prevent terminal heat and moisture stress can be effective for stabilizing crop yield and increasing food production. In cereals, crop duration may largely depend on vernalization requirements, photoperiod and temperature responses, and presence of genes for earliness per se (Herndl et al. 2008). The knowledge about the major genes influencing vernalization and photoperiod has made phenological manipulations rather easy for breeders (Slafer and Whitechurch 2001) despite sketchy knowledge about earliness genes and their interactions with temperature. In the past, phenological manipulations in wheat and barley have contributed significantly toward yield gains. It is believed that if the duration of stem elongation can be increased, it may result in more grains per spike and harvest index (Miralles et al. 2000; González et al. 2003), without additional need for crop water demands. An extended stem elongation period can be achieved by selecting for slightly higher sensitivity to photoperiod (Slafer et al. 2005). The duration of floret primordial differentiation can be extended through selecting for mild vernalization requirements in spring wheat. In winter wheat, the transition from a vegetative to a generative (reproductive) stage is conditioned by a "vernalization" process—a long action of low temperatures. Vernalization and light requirements largely determine the response of cultivars to sowing date. Exploitation of genetic variations in day-length sensitivity and vernalization requirement result in wider adaptability of wheat crop. Allelic variations at the *VRN-1* locus are one of the main sources of genetic differences in vernalization requirement in both diploid (*Triticum monococcum*) and polyploid wheat (McIntosh et al. 2003). Common wheat is a hexaploid species (2n = 42, genomes AABBDD) and carries three homologous copies of the *VRN-1* gene, one in each of the three genomes, which are designated *VRN-A1*, *VRN-B1*, and *VRN-D1* (McIntosh et al. 2003). A dominant (*Vrn*) allele at any one of these loci is sufficient to confer a spring growth habit (Stelmakh 1998) and spring type flowers without vernalization requirement in the presence of photoperiod-insensitive genes, or once the long day requirement is met. Winter habit is conferred when the recessive alleles are in the homozygous state across the *Vrn-1* locii. However, the different dominant *Vrn* alleles have differential sensitivity to vernalization, with *Vrn-A1* being not only insensitive to vernalization but is also epistatic to *Vrn-B1* and *Vrn-D1*, both of which have residual vernalization requirements (Pugsley 1971; Shindo and Sasakuma 2002). Other vernalization-responsive genes like the *Vrn-2* series, *Vrn-B3* and *Vrn-D5*, have also been identified in wheat (Goncharov 2003; Kato et al. 2003; Yan et al. 2006). The variation in response to vernalization due to the presence or absence of different *Vrn* alleles causes differences in flowering time, where the cultivar with *Vrn-1* allele flowers earliest and those with either *Vrn-D1* and *Vrn-D5* and *Vrn-B1* flower later under nonvernalizing conditions (Iqbal et al. 2011). The presence of two dominant alleles results in early maturity and higher yield, whereas the dominant allele at all the three loci results in earliest maturing but a low yielding cultivar. In South Asia, most of the wheat are planted around mid-November, and the average temperature in November hovers around 20 ± 5°C. Earlier popular wheat varieties in India, such as HD2329, were of 135–140 days duration and therefore were able to complete their life cycle under favorable crop growth conditions and rarely exposed to terminal heat stress, if planted timely. A quantum jump in productivity of the wheat cultivar HD2329 was realized with the development of Veery derivatives such as PBW343, WH542, PBW373, and Raj3765 in India and Inqlab in Pakistan, which are of 140–150 days growth duration. The presence of a rye segment in later cultivars buffers them against abiotic stresses; however, even the timely planted wheat crop is often exposed to terminal heat and drought stresses because of longer durations. Farmers in the northwest of the Indo-Gangetic plains resorted to early sowing of these varieties, such as to maximize wheat productivity. Besides the Veery derivatives, a number of other varieties, such as HD2329, HD2894, and HD2851, are also grown in the northwestern plains of Indo-Gangetic plains. If these cultivars are seeded in October, many times they flower very early, in response to prevailing high temperature owing to the presence of the inappropriate *Vrn-1* allecombinations. This results in a drastic yield reduction of these cultivars (HD2329, HD2894, and HD2851) under early sown condtions. However, variety like PBW343, because of the *vrn-A1* and *Vrn-B1* alleles have partial vernalization requirement, does not flower early even if planted early, as their vernalization requirement is met after mid-December and early January. The *Vrn* allele with partial vernalization requirement therefore needs to be appropriately exploited to develop cultivars that can yield higher even if sown early in the season in October. In central and eastern India, mild winters and water shortages limit crop growth duration. Under such situations, genotypes having *Vrn* allele and partial vernalization requirements can be seeded early, to increase growth duration and yield. Early sowing of such genotypes can help farmers use receding conserved soil moisture for good crop establishment and prelude the need of presowing irrigation. In an experiment conducted for 2 years, the author planted a number of genotypes developed for specific adaptations (zero-till/conventional tilled conditions on two planting dates—early in the last week of September and timely sowing in the mid-November). Under the early-sown condition, days tó flowering were drastically reduced (39.8%–44.7%) in the genotypes HD2329 and CSW2; moderately reduced (30.4%–33.3%) in CSW1, CSW13, and CSW6; and comparatively less reduced in PBW343 and HD2967 (24.7%–25.5%). In CSW18, which has a *Vrn-a* allele similar to many other genotypes, the vernalization requirement seemed to be stronger than all other genotypes, and therefore the percentage reduction in flowering time (12.8%) was least. Therefore, genotypes such as CSW18, having a more vegetative phase (days to flower decreased just by 12.8%), can stabilize wheat yield at a higher level if planted early in the season under conserved soil moisture conditions (Tables 5.3 and 5.4).

5.6.6 Early Wheat Seeding with Appropriate Cultivars for Further Intensification

In the last few years, a series of experiments on early seeding of wheat have been conducted at the Indian Agricultural Research Institute, New Delhi, to reach the conclusion that early seeding in early/mid-October in the Indo-Gangetic plains is feasible without yield penalty only with specially bred cultivars having weak vernalization requirements. Weakly vernalized wheat genotypes are also expected to mature earlier than the normal-sown (November) wheat crop. The development of such new genotypes also provides scope for further intensification (seeding a short-duration crop [green gram, *Vigna radiata* L.] in standing wheat or after early crop harvest; Figure 5.5) and the sustainability of the intensively irrigated rice–wheat cropping systems.

Seeding a green gram crop not only helps improve organic carbon sequestration in cereal systems such as the rice–wheat system but also helps mop up the residual nitrates from the surface soil layers, which otherwise expectedly would leach down during the rainy season. The build up of nitrate-N in aquifers has earlier been reported to be due to deep percolation losses of nitrate-rich rainwater during the monsoon season. Early seeding of wheat in October can be done in residual moisture soils of rice crop, and also obviate the need of irrigation in March/April. The saved water can be used to raise an additional mung crop in tube-well irrigated areas.

In a recent field study, 10 elite, most preferred wheat cultivars by Madhya Pradesh farmers were planted at weekly intervals along with the newly developed genotype CSW18 to explore possibilities for October seeding without yield penalties. Results presented in Figure 5.6 show that weakly vernalized CSW18 and HD2967 genotypes (see Figure 5.6) developed recently at the Indian Agricultural Research Institute, New Delhi, had significant yield advantage at early planting. It was observed that CSW18 outyielded the high-yielding wheat cultivar HD2967 in October planting,

TABLE 5.3
Flowering Response of Wheat Genotypes Planted Early (September) and Timely (November)

Wheat Genotypes	Vrn Alleles	Days to Flower in September Sowing	Days to Flower in November Sowing	Percent Reduction in Days to 50% Flowering (RDF)
HD2329	*Vrn-A1a, Vrn-B1, Vrn-D1*	47^{a}	85^{a}	44.7
PBW343	*vrn-A1, vrn-B1, Vrn-D1*	70^{c}	93^{b}	24.7
HD2967	*vrn-A1, Vrn-B1, Vrn-D1*	70^{c}	94^{b}	25.5
CSW1	*Vrn-A1a, vrn-B1, vrn-D1*	64^{b}	92^{a}	30.4
CSW2	*Vrn-A1a, vrn-B1, Vrn-D1*	53^{a}	88^{a}	39.8
CSW6	*vrn-A1, Vrn-B1, Vrn-D1*	62^{b}	92^{a}	32.6
CSW13	*vrn-A1, Vrn-B1,Vrn-D1*	60^{b}	90^{a}	33.3
CSW18	*A mixture of vrn-A1, Vrn-B1, Vrn-D1 and vrn-A1, Vrn-B1, null for marker of Vrn-D1 and vrn-D1*	82^{d}	94^{b}	12.8
Tukey's HSD at 5%		7.61	7.61	

Note: Values followed by similar superscript are statistically at par whereas those followed by different superscript are statistically significant.

$$\text{Reduction in days to Flowering (RDF\%)} = \frac{(\text{Days to flower in Nov. seeding} - \text{Days to flower in Sept. seeding}) * 100}{\text{Days to flowering in Nov. seeding}}.$$

TABLE 5.4
Analysis of Variation in Wheat for Days to Heading under Different Planting Dates

Source	DF	Sum of Square	Mean of Square	F-Ratio	p-Value	Significant
Replication	1	5.28	5.28	169.00	0.0489	*
Date of sowing	1	6132.78	6132.78	196249	0.0014	**
Error (date of sowing)	1	0.0313	0.0313	–	–	NS
Genotypes	7	1270.47	181.49	38.98	<0.0001	**
Sowing dates × genotypes	7	485.4688	69.3527	14.89	<0.0001	**
Error (genotypes)	14	65.19	4.66			NS
Total	31	7959.22				NS

Note: **, Significant at 1%; *, significant at 5%; NS, nonsignificant.

FIGURE 5.5 Relay seeding of green gram crop with last irrigation of wheat crop.

saving irrigation water and vacating the fields a little early than other timely sown cultivars (not plotted in Figure 5.6) at both Pusa (Bihar) and Jabalpur (Madhya Pradesh). Cultivar GW366 was the best bet for mid-November planting in the central plateau region of India with very mild winter.

Most of the genotypes like CSW18, PBW343, and HD2967 with some degree of vernalization requirement have only a marginal effect on tillering capacity, which is, in most of the cases, overly compensated by increase in grain size. Under canal irrigation, spring maize planted in March can provide a good option for intensification. The profits can be further maximized through a dedicated breeding effort for yield improvement of wheat in early seeding as well as through almost 10–15 days prolongation of moong (green gram) period. The yellow rust of wheat, which is emerging as a challenge to wheat-growing areas in the foothills of Punjab and Haryana, can also be controlled through this intervention to a large extent (Figure 5.7).

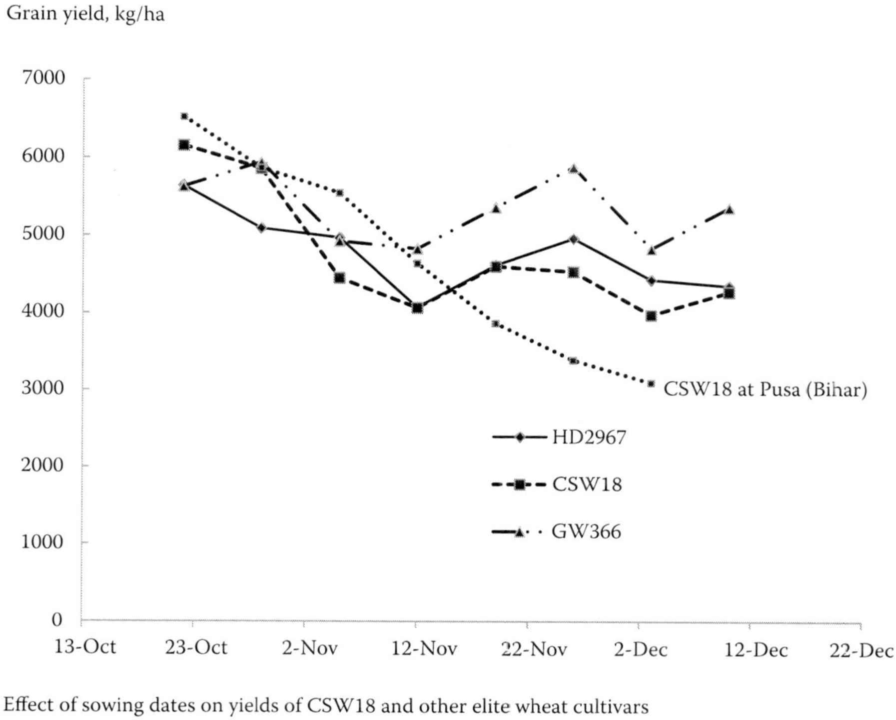

FIGURE 5.6 Performance of weekly vernalized wheat cultivars (CSW18 and HD2967) at Pusa (Bihar) and Lakhanwada (Jabalpur, MP), India (*X*- and *Y*-axis are planting dates and yields in kg/ha, respectively).

FIGURE 5.7 LHS picture shows the differential response to heading in two wheat cultivars. (a) D2967 on first two beds and (b) CSW 18 on next two beds. Both cultivars were planted in early October. Wheat cultivar HD2967 has completed heading in LHS 2 beds, however wheat cultivar CSW18 (next 2 rows) is still in the heading process and will take another 10 days to complete it. Picture on the RHS shows that early-planted CSW18 has improved head size which improves crop productivity.

5.7 CONCLUSION

Enhanced productivity and production of rice and wheat in the Green Revolution era prevented the cultivation of marginal lands and also led to the emergence of the nontraditional rice–wheat cropping system as the primary production system in many irrigated areas. The higher productivity of the rice–wheat system has been apparently due to better land, water, and crop management practices; timely availability of needed agri-inputs; and market support from the local state governments. Long-distance transport of irrigation water through canal networks almost doubled the cropping system intensity in many areas in the northwest Indo-Gangetic plains. However, these developments also led to land degradation as manifested by secondary salinization, reduced biodiversity, lower factor productivity, multinutrient deficiencies, water scarcity, and pollution of groundwater aquifers. Large "management yield gaps" in cereal crops have been observed in the eastern Gangetic plains of Nepal, India, and Bangladesh. In high-rainfall districts located between 78.83° and 86.13° north longitudes, there seem to be some nexus between low agricultural productivity and poverty due to recurring droughts and floods and technology fatigue.

Conservation agriculture (CA) practices, because of their potential to produce more at less costs and ability to bridge "management yield gaps," are increasingly being seen as the panacea of many ills of modern agriculture. It seems that South Asian farmers can produce enough additional food by practicing timely planting, relay cropping, and by adopting appropriate technologies for lowland wet soils and "rice fallows." Yield gain can be further consolidated by developing cultivars able to harness genotype–tillage–cropping system interactions under new microenvironments resulting from the adoption of CA and use of vernalization genes for developing wheat varieties suitable for early seeding in residual soil moisture of rice fields. This calls for some added efforts for setting up special crop-breeding programs. It appears that manipulation of vernalization genes can prove helpful in adjusting the planting times of crops in cropping systems having short turn-around times and the added challenge of climate changes.

Green Revolution technologies improved food production in the developing world through enhanced production and productivity of intensely irrigated intensive cropping systems such as rice–wheat systems in the Indo-Gangetic plains. This, in turn, induced land degradation, low factor productivity, multinutrient deficiencies, and a number of other second-generation problems. Besides providing goods, sustainable land management in agricultural landscapes must focus on key ecosystem services linked to life support (e.g., soil formation, nutrient cycling, flood control, and pollination) and services derived from the regulation of ecosystem processes (e.g., climate regulation, disease control, and detoxification). Crop production systems on CA platforms are closer to natural ecosystems, and hence, if applied properly, can help South Asian farmers to produce enough additional food for the teaming populations. Stupendous gains in crop productivity in South Asian countries were the result of improved crop production environments for high yielding varieties induced by "better-bet" land and crop management practices. Higher yields being realized in different cropping systems through CA can be further consolidated through the development of cultivars appropriate for the conservation agricultural platforms. It is important that the genetic variability present in the germplasm is explored/exploited for designing cultivars for good crop stand establishment under CA environment and use genotype × management interactions. Similarly, agronomy has to be worked for the cropping system as a whole rather than for individual crops. The development of new spring wheat varieties with weak vernalization requirement can be a potential strategy to facilitate early planting in residual soil moisture of rice fields and to overcome adverse terminal heat effects. Timely relay planting of wheat in cotton and of green gram (*Vigna radiata* L.) in wheat can lead to further crop intensification in Indo-Gangetic plains. Disease and insect–pest resistance breeding programs require reorientation, keeping in view prevailing disease scenarios under conservation agriculture.

REFERENCES

Abeledo, L.G., D.F. Calderini, and G.A. Slafer. 2002. Physiological changes associated with breeding progress. In *Barley Science: Recent Advances from Molecular Biology to Agronomy of Yield and Quality*, G.A. Slafer, J.L. Molina-Cano, R. Savin, J.L. Araus and I. Romagosa (Eds.). Food Product Press, New York, pp. 361–386.

Abrol, I.P., and R.K. Gupta. 1998. Indo-Gangetic plains—Issues of changing land use. In *LUCC Newsletter, Proc. GCTE-LUCC, 1998*, No. 1, 5–8 pp.

Aggarwal, P.K., S.K. Bandyopadhyay, H. Pathak, N. Kalra, S. Chander, and S.S. Kumar. 2000. Analysis of the yield trends of rice–wheat system in north-western India. *Outlook Agric.* 29(4):259–268.

Aggarwal, P.K., and N. Kalra. 1994. Analyzing the limitations set by climatic factors, genotype, and water and nitrogen availability on productivity of wheat II. Climatically potential yields and management strategies. *Field Crop. Res.* 38:93–103.

Allan, R.E. 1989. Agronomic comparison between *Rht1* and *Rht2* semi-dwarf genes in winter wheat. *Crop. Sci.* 29:1103–1108.

Araus, J.L., G.A. Slafer, M.P. Reynolds, and C. Royo. 2002. Plant breeding and water relations in C3 cereals: What to breed for? *Ann. Bot.* 89:925–940.

Bianco, R. 1998. Ocorrência e manejo de pragas. In Plantio Direto. Pequena propriedade sustentável. Instituto Agronômico do Paraná (IAPAR) Circular 101, Londrina, PR, Brazil, pp. 159–172.

Bolliger, A., J. Magid, T.J.C. Amado et al. 2006. Taking stock of the Brazilian "zero-till revolution": A review of landmark research and farmers' practice. *Adv. Agron.* 91:47–110.

Boosalis, M.G., B. Doupnik, and G.N. Odvody. 1981. Conservation tillage in relation to plant diseases. In *CRC Handbook of Pest Management in Agriculture*, Vol. 1, D. Pimentel (Ed.). CRC Press Inc., Boca Raton, FL, pp. 445–474.

Botwright, T.L., G.J. Rebetzke, A.G. Condon, and R.A. Richards. 2001. The effect of *rht* genotype and temperature on coleoptile growth and dry matter partitioning in young wheat seedlings. *Aust. J. Plant Physiol.* 28:417–423.

Bruce, S.E. 2003. Poor growth of canola in retained wheat stubble—Causes consequences and control. PhD thesis, Charles Stuart University, Wagga.

Buttar, G.S., H.S. Sidhu, V. Singh, N. Gupta, R. Gupta, M.L., Jat, and S. Baldev. 2013. Innovations for relay planting of wheat in cotton: A breakthrough for enhancing productivity and profitability in cotton-wheat systems of South Asia. *Expl. Agric.* 49(1):19–30.

Calderini, D.F., M.P. Reynolds, and G.A. Slafer. 1999. Genetic gains in wheat yield and main physiological changes associated with them during the 20th century. In *Wheat: Ecology and Physiology of Yield Determination*, E.H. Satorre and G.A. Slafer (Eds.). Food Product Press, New York, pp. 351–377.

Calderini, D.F., and G.A. Slafer. 1999. Has yield stability changed with genetic improvement of wheat yield? *Euphytica* 107:453–460.

Cassman, K.G. 1999. Ecological intensification of cereal production systems: Yield potential, soil quality, and precision agriculture. *PNAS* 96:5952–5959.

Cassman, K.G., and P.L. Pingali. 1995. Extrapolating trends from long-term experiments to farmers' fields: The case of irrigated rice systems in Asia. In *Agricultural Sustainability in Economic, Environmental, and Statistical Terms*, V. Barnett, R. Payne and R. Steiner (Eds.). John Wiley & Sons, London, pp. 63–84.

Chand, R., S. Garg, and L. Pandey. 2009. Regional Variations in Agricultural Productivity: A District Level Study. Discussion Paper No. NPP 01/2009. National Centre for Agricultural Economics and Policy Research, New Delhi.

Chandna, P., M.L. Khurana, J.K. Ladha, M. Punia, R.S. Mehla, and R. Gupta. 2010. Spatial and seasonal distribution of nitrate-N in groundwater beneath the rice–wheat cropping system of India: A geospatial analysis. *Environ. Monit. Assess.* 178:545–562.

Chastain, T.G., K.J. Ward, and D.J. Wysocki. 1995. Stand establishment responses of soft white winter wheat to seedbed residue and seed size. *Crop Sci.* 35:213–218.

Conway, G.R., and J. Waage. 2010. *Science and Innovation for Development*. UKCDS, London.

Derpsch, R., and T. Friedrich. 2009. Global overview of conservation agriculture adoption. In *Proc. 4th World Cong. Conservation Agric.*, February 4–7, New Delhi, India, pp. 429–438.

Dhillon, S.S., P.R. Hobbs, and J.S. Samara. 2000. Investigations on bed planting system as an alternative tillage and crop establishment practice for improving wheat yields sustainably. In *Proceedings of the 15th Conference of the International Soil Tillage Research Organisation*, July 2–7, Fort Worth, TX (CD-ROM).

Dingkuhn, M., B.B. Singh, B. Clerget, J. Chantereau, and B. Sultan. 2006. Past, present and future criteria to breed crops for water-limited environments in West Africa. *Agric. Water Manage.* 80: 241–261.

Duiker, S.W., J.F. Haldeman Jr., and D.H. Johnson. 2006. Tillage × maize hybrid interactions. *Agron. J.* 98:436–442.

Duvick, D.N. 1990. Genetic enhancement and plant breeding. In *Advances in New Crops*, J. Janick and J.E. Simon (Eds.). Timber Press, Portland, OR, pp. 90–96.

Ellis, M. H., G.J. Rebetzke, W. Spielmeyer, R.A. Richards, D.G. Bonnett. 2004. Gibberellin responsiveness and early growth of altérnate dwarfing genes in wheat (*Triticum aestivum* L.). *Funct. Plant Biol.* 31: 583–589.

Ferguson, H., and G.O. Boatwright. 1968. Effects of environmental factors on the development of the crown node adventitious roots of winter wheat (*Triticum aestivum*). *Agron. J.* 60:258–264.

Fick, G.N., and C.O. Qualset. 1976. Seedling emergence, coleoptiles length, and plant height relationships in crosses of dwarf and standard-height wheats. *Euphytica* 25:679–684.

Forcella, F., D.D. Buhler, and M.E. McGiffen. 1994. Pest management in crop residues. In *Advances in Soil Science: Crop Residue Management*, J.L. Hatfield and B.A. Stewart (Eds.). Lewis Publishers, London, pp. 173–189.

Funnermark, W.L., and A.R. Hallauer. 1985. Cold tolerance in corn hybrids as affected by tillage systems and planting dates. *Proc. Iowa Acad. Sci.* 92:58–62.

Gathala, M.K., V. Kumar, P.C. Sharma et al. 2013. Optimizing intensive cereal-based cropping systems addressing current and future. *Agric. Ecosyst. Environ.* 177:85–97.

Godfray, C., J.R. Beddington, I.R. Crute et al. 2010. Food security: The challenge of feeding 9 billion people. *Science* 327:812–818.

Goncharov, N.P. 2003. Genetics of growth habit (spring vs winter) in common wheat: Confirmation of the existence of dominant gene Vrn4. *Theor. Appl. Genet.* 107:768–772.

González, F.G., G.A. Slafer, and D.J. Miralles. 2003. Grain and floret number in response to photoperiod during stem elongation in fully and slightly vernalised wheats. *Field Crop. Res.* 81:17–27.

Gupta, R. 2010. Management of natural resources for enhanced wheat production—New innovations. In *The Brain Storming Workshop on "Prospects of Producing 100 MT of Wheat by 2015."* December 18, TAAS, New Delhi.

Gupta, R.K., and I.P. Abrol. 2000. Salinity build-up and changes in the rice–wheat system of the Indo-Gangetic plains. *Exp. Agric.* 36:273–284.

Gupta, R.K., R.K. Naresh, P.R. Hobbs, J. Jiaguo, and J.K. Ladha. 2003. Sustainability of post-green revolution agriculture: The rice–wheat cropping systems of the Indo-Gangetic plains and China, Chapter 1. In *Improving the Productivity and Sustainability of Rice–Wheat Systems: Issues and Impact*, J.K. Ladha, J. Hill, R.K. Gupta, J. Duxbury and R.J. Buresh (Eds.). ASA Special Publication 65. ASA, Madison, WI, pp. 1–25.

Gupta, R., and K. Sayre. 2007. Conservation agriculture in South Asia. *J. Agric. Sci.* 145:207–214.

Herndl, M., J.W. White, L.A. Hunt, S. Graeff, and W. Claupein. 2008. Field-based evaluation of vernalization requirement, photoperiod response and earliness *per se* in bread wheat (*Triticum aestivum* L.). *Field Crop. Res.* 105(3):193–201.

Hersterman, O.B., F.J. Pierce, and E.C. Rossman. 1988. Performance of commercial corn hybrids under conventional and no tillage systems. *J. Prod. Agric.* 1:202–206.

Hughes, K.A., and W.J.P. Mitchel. 1987. The relationship of coleoptile length and plant height with establishment of cereals under zero tillage. *Proc. Annual Conf. New Zealand Agron. Soc.* 17:67–70.

Iqbal, M., A. Shahzad, and I. Ahmed. 2011. Allelic variation at the Vrn-A1, Vrn-B1, Vrn-D1, Vrn-B3 and Ppd-D1a loci of Pakistani spring wheat cultivars. *Electron. J. Biotechnol.* 14(1). Available at http://dx.doi.org/10.2225/vol14-issue1-fulltext-6.

Jat, M.L., M.K. Gathala, Y.S. Saharawat, J.P. Tetarwal, R. Gupta, and Y. Singh. 2013. Double no-till and permanent raised beds in maize-wheat rotation of north-western Indo-Gangetic plains of India: Effects on crop yields, water productivity, profitability and soil physical properties. *Field Crop. Res.* 149:291–299.

Jat, M.L., R. Gupta, Y.S. Saharawat, and R. Khosla. 2011. Layering precision land leveling and furrow irrigated raised bed planting: Productivity and input use efficiency of irrigated bread wheat in Indo-Gangetic plains. *Am. J. Plant Sci.* 2:578–588.

Joshi, A.K., R. Chand, B. Arun, R.P. Singh, and R. Ortiz. 2007. Breeding crops for reduced tillage management in the intensive, rice–wheat systems of South Asia. *Euphytica* 153:135–151.

Kaspar, T.C., T.M. Crosbie, R.M. Cruse, D.C. Erbach, D.R. Timmons, and K.N. Potter. 1987. Growth and productivity of four corn hybrids as affected by Tillage. *Agron. J.* 79(3):477–481.

Kato, K., M. Yamashita, K. Ishimoto, H. Yoshino, and M. Fujita. 2003. Genetic analysis of two genes for vernalisation response, the former *Vrn2* and *Vrn4*, by using PCR based molecular markers. In *Proceedings of 10th International Wheat Genetics Symposium*, Vol. 3, N.E. Pogna, M. Romano, E.A. Pogna and G. Galterio (Eds.). Instituto Sperimentale per la Cerealicoltura, Paestum, pp. 971–973.

Kirkegaard, J., M. Conyers, J. Hunt, C. Kirkby, M. Watt, and G. Rebetzke. 2011. Sense and nonsense in conservation agriculture: Principles, pragmatism and productivity in Australian mixed farming systems. Resilient Food System for a Changing World. In *Proceeding of the 5th World Congress of Conservation Agriculture Incorporating 3rd Farming Systems Design Conference*, September 25–29, Brisbane, Australia, pp. 3–8.

Kuhlman, D.E., and K.L. Steffey. 1982. Insect control in no till corn. *Res. Conf. Proc. Ann. Corn Sorghum Res. Conf.* 37:118–147.

Kumar, V., Y.S. Saharawat, M.K. Gathala et al. 2013. Effect of different tillage and seeding methods on energy use efficiency and productivity of wheat in the Indo-Gangetic Plains. *Field Crop. Res.* 142:1–8.

Ladha, J.K., D. Dawe, H. Pathak et al. 2003. How extensive are yield declines in long-term rice–wheat experiments in Asia? *Field Crop. Res.* 81:159–180.

Matson, P.A., W.J. Parton, A.G. Power, and M. Swift. 1997. Agricultural intensification and ecosystem properties. *Science* 277:504–509.

Matsui, T., S. Inanaga, Y. Sugimoto, and N. Nakata. 1998. Chromosomal location of genes controlling coleoptiles length in wheat using chromosome substitution lines. *Wheat Inf. Serv.* 87:22–26.

McIntosh, R.A., Y. Yamazaki, K.M. Devos, J. Dubcovsky, W.J. Rogers, and R. Appels. 2003. Catalogue of gene symbols for wheat. In *Proceedings of the 10th International Wheat Genetics Symposium*, Vol. 4, N.E. Pogna, M. Romano, E.A. Pogna and G. Galterio (Eds.). Instituto Sperimentale per la Cerealicoltura, Rome, pp. 1–34.

Mehla, R.S., J.K. Verma, R.K. Gupta, and P.R. Hobbs. 2000. Stagnation in the Productivity of Wheat in the Indo-Gangetic Plains: Zero-till–Seed-Cum-Fertilizer Drill as an Integrated Solution. Rice–Wheat Systems of the Indo-Gangetic Plains. RWC Paper Series No. 8, New Delhi, India, 9 pp.

Miralles, D.F., R.A. Richards, and G.A. Slafer. 2000. Duration of the stem elongation period influences the number of fertile florets in wheat and barley. *Aust. J. Plant Physiol.* 27:931–940.

Mohanty, M., and D.K. Painuli. 2004. Modeling rice seedling emergence and growth under tillage and residue management in a rice–wheat system on a Vertisol in Central India. *Soil Till. Res.* 76:167–174.

Nazareno, de N.R.X., P.E. Lipps, and L.V. Madden. 1993. Effect of levels of corn residue on the epidemiology of gray leaf spot of corn in Ohio. *Plant Dis.* 77:67–70.

Newhouse, K.E., and T.M. Crosbie. 1986. Interactions of maize hybrids with tillage systems. *Agron. J.* 78(6):951–954.

Nyvall, R.F. 1982. Disease control. In *Tillage and Soil Environment*, R.M. Cruse and R. Horton (Eds.). Coop. Ext. Service Bull. CE 1755, Iowa.

Ortiz, R., K.D. Sayre, B. Govaerts et al. 2008. Climate change: Can wheat beat the heat? *Agric. Ecosyst. Environ.* 126:46–58.

Ortiz-Monasterio, J.I., S.S. Dhillon, and R.A. Fischer. 1994. Date of sowing effects on grain and yield components of irrigated spring wheat cultivars and relationship with radiation and temperature in Ludhiana, India. *Field Crop. Res.* 37:169–184.

Passioura, J.B. 1996. Drought and drought tolerance. *Plant Growth Regul.* 20:79–83.

Passioura, J.B. 2002. Environmental biology and crop improvement. *Funct. Plant Biol.* 29:537–546.

Pathak, H., J.K. Ladha, P.K. Aggarwal et al. 2003. Trends of climatic potential and on farm yield of rice and wheat in the Indo-Gangetic Plains. *Field Crop. Res.* 80:223–234.

Pimbert, M. 2008. The role of local organisations in sustaining local food systems, livelihoods and the environment, Chapter 4. In *Towards Food Sovereignty, Reclaiming Autonomous Food Systems*. Available at http://www.iied.org/pubs/pdfs/G02374.pdf:IIED.

Poster, S.L. 1998. Water for food production: Will there be enough in 2025? *BioScience* 48:629–637.

Pretty, J. 2008. Agricultural sustainability: Concepts, principles and evidence. *Philos. Trans. R. Soc. Lond.* 363:447–466.

Pretty, J., C. Toulmin, and S. Williams. 2011. Sustainable intensification in African agriculture. *Int. J. Agric. Sustain.* 9:5–24.

Pugsley, A.T. 1971. A genetic analysis of the spring-winter habit of growth in wheat. *Aust. J. Agric. Res.* 22:21–31.

Rajaram, S., K.D. Syre, J. Diekmann, R. Gupta, and W. Erskine. 2007. Sustainability considerations in wheat improvement and production. *J. Crop Improve.* 19(1/2):105–123.

Rebetzke, G.J., R. Appels, A.D. Morrison et al. 2001. Quantitative trait loci on chromosome 4B for coleoptile length and early vigour in wheat (Triticum aestivum L.). *Aust. J. Agr. Res.* 52(11–12):1221–1234.

Rebetzke, G.J., S.E. Bruce, and J.A. Kirkegaard. 2005. Longer coleoptiles improve emergence through crop residues to increase seedling number and biomass in wheat (*Triticum aestivum L.*). *Plant Soil* 272:87–100.

Rebetzke, G.J., R.A. Richards, V.M. Fisher and B.J. Mickelson. 1999. Breeding long coleoptile, reduced height wheats. *Euphytica* 106(2):159–168.

Regmi, A.P., and J.K. Ladha. 2005. Enhancing productivity of rice–wheat system through integrated crop management in the eastern-Gangetic Plains of South Asia. *J. Crop Improv.* 15:29.

Richards, R.A. 1996. Increasing yield potential in wheat—Source and sink limitations. In *Increasing Yield Potential in Wheat: Breaking the Barriers*, M.P. Reynolds, S. Rajaram and A. McNab (Eds.). CIMMYT, México, DF, pp. 134–149.

Rodell, M., I. Velicogna, and J.S. Famiglietti. 2010. Satellite-based estimates of groundwater depletion in India. doi:10.1038/nature08238. Available at www.nature.com/nature.

Royal Society. 2009. *Reaping the Benefits: Science and the Sustainable Intensification of Global Agriculture*. Royal Society, London.

Saharawat, Y. 2009. *Technical Progress Report, Cereal System Initiative for South Asia, 2009–10*. CIMMYT, New Delhi, 153 pp.

Sayre, K.D., and O.H. Moreno Ramos. 1997. Application of Raised Bed Planting System to Wheat. Wheat Program Special Report No. 31. CIMMYT, Mexico, DF.

Schillinger, W.F., E. Donaldson, R.E. Allan, and S.S. Jones. 1998. Winter wheat seedling emergence from deep sowing depths. *Agron. J.* 90:582–586.

Scopel, E., B. Triomphe, M.F.S. Ribeiro, L. Seguy, J.E. Denardin, and R.A. Kochann. 2004. Direct seeding mulch-based cropping systems (DMC) in Latin America. In *New Directions for a Diverse Planet: Proceedings for the 4th International Crop Science Congress*, T. Fischer, N. Turner, J. Angus, L. McIntyre, M. Robertsen, A. Borrell and D. Llyod (Eds.), September 26–October 1. Brisbane, Australia. Available at http://www.cropscience.org.au.

Shindo, C., and T. Sasakuma. 2002. Genes responding to vernalization in hexaploid wheat. *TAG Theor. Appl. Genet.* 104(6–7):1003–1010.

Singh, S.R., U.S. Gautam, U. Kumar, A. Rehman, and S.K. Sinha. 2001. *Groundwater Development to Enhance Surface and Rainwater Utilization and Agricultural Productivity in Southern Bihar*. ICAR-IWMI Ground Water Policy Initiative, Policy Dialogue on Ground Water Management, CSSRI, Karnal.

Slafer, G.A., J.L. Araus, C. Royo, and L.F. García Del Moral. 2005. Promising eco-physiological traits for genetic improvement of cereal yields in Mediterranean environments. *Ann. Appl. Biol.* 146:61–70.

Slafer, G.A., and E. Whitechurch. 2001. Manipulating wheat development to improve adaptation and to search for alternative opportunities to increase yield potential. In *Application of Physiology to Wheat Breeding*, M.P. Reynolds, J.I. Ortiz-Monasterio and A. McNab (Eds.). CIMMYT, Mexico DF, pp. 160–170.

Stelmakh, A.F. 1998. Genetic systems regulating flowering response in wheat. *Euphytica* 100(1–3):359–369.

Swaminathan, M.S. 2000. An evergreen revolution. *Biologist* 47:85–89.

Swan, J.B., T.C. Kaspar, and D.C. Erbach. 1996. Seed-row residue management for corn establishment in the northern US Corn Belt. *Soil Till. Res.* 40:55–72.

The 2030 Water Resources Group. 2011. Executive Summary. Charting Our Water Future: Economic frameworks to inform decision-making, 30p. Available at http://www.mckinsey.com/client_service/sustainability/latest_thinking/charting_our_water_future.

Tilman, D., K.G. Cassman, P.A. Matson, R. Naylor, and S. Polasky. 2002. Agricultural sustainability and intensive production practices. *Nature* 418:671–677.

Timsina, J., D. Godwin, E. Humphreys et al. 2008. Evaluation of options for increasing yield and water productivity of wheat in Punjab, India using the DSSAT-CSM-CERES-Wheat model. *Agric. Water Manage.* 95:1099–1110.

Tirol-Padre, A., and J.K. Ladha. 2006. Integrating rice and wheat productivity trends using the SAS mixed procedure and meta-analysis. *Field Crop. Res.* 95:75–88.

Tollenaar, M., L.M. Dwyer, and D.E. McCullough. 1994. Physiological basis of genetic improvement of corn. In *Genetic Improvement of Field Crops*, G.A. Slafer (Ed.). Marcel Dekker, New York, pp. 183–236.

Trethowan, R.M., T. Mahmood, Z. Ali, K. Oldach, and A. Gutierrez Garcia. 2012. Breeding wheat cultivars better adapted to conservation agriculture. *Field Crop. Res.* 132:76–83.

Trethowan, R.M., M.P. Reynolds, K.D. Sayre, and I. Ortiz-Monasterio. 2005. Adapting wheat cultivars to resource conserving farming practices and human nutritional needs. *Ann. Appl. Biol.* 146:404–413.

Triplett, G.B., and D.M. Van Doren. 1985. An overview of the Ohio conservation tillage research. In *A Systems Approach to Conservation Tillage*, F.M. D'Intri (Ed.). Lewis, Chelsea, MI, pp. 59–68.

Verhulst, N., A. Carrillo-García, C. Moeller, R. Trethowan, K.D. Sayre, and B. Govaerts. 2011. Conservation agriculture for wheat-based gravity irrigation systems: Increasing resilience though improved soil quality. *Plant Soil* 340:467–479.

Villegas, D., N. Aparicio, M.M. Nachit, J.L. Araus, and C.A. Royo. 2000. Photosynthetic and developmental traits associated with genotypic differences in durum wheat yield across the Mediterranean basin. *Aust. J. Agric. Res.* 51:891–901.

Waddington, S.R., X. Li, J. Dixon, G. Hyman, and M.C. Vicente. 2010. Getting the focus right: Production constraints for six major food crops in Asian and African farming systems. *Food Sec.* 2:27–48. doi:10.1007/s12571-010-0053-8.

Watt, M., J.A. Kirkegaard, and G.J. Rebetzke. 2005. A wheat genotype developed for rapid leaf growth copes well with the physical and biological constraints of unploughed soil. *Funct. Plant. Biol.* 32:695–706.

Watt, M., J.A. Kirkegaard, and J.B. Passioura. 2006. Rhizosphere biology and crop productivity Australian Journal of Soil Research 44, 299–317.

Wuest, S.B., S.L. Albrecht, and K.W. Skirvin. 2000. Crop residue position and interference with wheat seedling development. *Soil Till. Res.* 55:175–182.

Yadav, R., L. Kumar, M.L. Jat, and R. Gupta. 2012. Tailoring wheat genotypes for conservation agriculture in different cropping systems: An innovative and much needed breeding paradigm. In *Resilient Food System for a Changing World. Proceeding of the 5th World Congress of Conservation Agriculture Incorporating 3rd Farming Systems Design Conference*, September 25–29, Brisbane, Australia, pp. 444–446.

Yan, L., D. Fu, C. Li et al. 2006. The wheat and barley vernalization gene VRN3 is an orthologue of FT. *Proc. Natl. Acad. Sci. U.S.A.* 103:19581–19586.

6 Soil Management for Sustainable Agricultural Intensification in the Himalayan Region

Roshan M. Bajracharya, Dil Prasad Sherchan, Bed Mani Dahal, and Nani Raut

CONTENTS

6.1 INTRODUCTION

Soils have been and still are the primary medium for large-scale agricultural production upon which human civilizations have depended for sustenance and expansion. Over millennia, since the first human communities began settled agriculture, people have learned to manage and successfully cultivate their lands to produce the food required for nurturing modern societies and the development of their regions. Previously, communities were small and the demand on land and soil resources was low; hence, traditional agricultural practices could meet the needs of the population (Braidwood 1960; Piggot 1961). Following the industrial revolution in the 1800s, the world's human population has grown tremendously, currently surpassing

7 billion, with growth still rapid in the less developed countries of Asia, Africa, and the Middle East. Thus, while arable land has essentially reached the limits of expansion, pressures on the land resource base continue to increase with ever-greater demands for food production. The need for producing higher amounts of food on the same amount of land has fuelled agricultural intensification. The Green Revolution of the post–World War II era enabled many advanced nations to dramatically increase crop yields and food production through the use of inorganic fertilizers and improved crop varieties. However, by the end of the 20th century, the impacts of intensified production systems, which relied heavily on man-made chemical fertilizers and pesticides, on the environment, as well as soil biophysical quality degradation, erosion, and fertility decline, could no longer be ignored (Thomas et al. 1956). As the nutritive requirements of a growing world population will only increase into the foreseeable future, sustainable agricultural intensification practices and technologies are indispensable. The development and adoption of such options will require concerted efforts on the parts of scientists, governments, and local farming communities merged through contextual knowledge systems. The application of integrated nutrient management has shown that optimum combinations of farmyard manure and inorganic fertilizers increased yields by 16%–35% in the mid-hills of Nepal. Similarly, a combination of compost and fertilizer gave significantly higher yields of wheat and potato crops over conventional farming. Combinations of different crops, such as pigeon pea and groundnut, have been shown to result in yield increases of maize of >100%, while leading to improved soil properties. Inoculation of seeds with *Rhizobium* strains have also been found to lead to yield increases of 40%–67% compared with noninoculated seeds for a variety of crops, such as soybean, lentils, black gram, groundnuts, and broad beans. Apart from crop and seed improvements, soil and water management options, such as conservation tillage, mulching, and microirrigation, are equally important in ensuring sustainable intensive production. Studies in the mid-hill region of the Himalaya have demonstrated that mulching and reduced tillage could lead to a reduction in soil and nutrient losses from upland farms by 46% to >100%, while increasing water retention and providing comparable crop yields. Thus, successful implementation of sustainable agricultural intensification to meet the growing global food demand will involve simultaneously addressing soil and land management, water use efficiency, crop and agrobiodiversity, and adequate policy and institutional support.

Soil and land resources have been the backbone of human civilization ever since prehistoric communities of people established permanent settlements and began settled agriculture some 10 thousand or more years ago (Darlington 1969). Past civilizations, such as the ancient settlements of the Tigris–Euphrates and Nile River valleys, flourished as a direct consequence of having access to fertile soils, and likewise, declined as a result of degradation and loss of fertility of agricultural lands (Hillel 1992). Through the course of history, previous human communities learned to manage their soils and successfully cultivate their lands, even under rather harsh climates and terrains such as the arid region of Egypt and the mountainous regions of South Asia (Hillel 2007).

Over the centuries, and particularly since the beginning of the Industrial Revolution in the 1800s, the human population has seen a tremendous increase in growth rate. The 20th century saw the population doubling rates drop to <40 years, leading to an exponential growth of the world's population, which has now exceeded 7 billion. The ever-growing population continues to exert pressures on the land and soil resources of the world, which are finite. Although modern technologies, such as fertilizers, synthetic pesticides, and mechanization of agriculture, have boosted the production, the global food supplies are stretched to meet the ever-increasing demand. Clearly, with arable land being limited, we are left with no option but to cultivate the existing land more intensively in order to feed the masses. Thus, the past few decades have witnessed agricultural intensification in various forms (Carswell 1997; Dahal et al. 2008).

Agricultural intensification can be regarded as any change in the cropping or livestock-rearing practices that make use of a fixed area of land more frequently or intensely than previous traditional or conventional practices did. Thus, an increase in the number of crops grown per annual cropping cycle, increase in the stocking rates of livestock grazed on a parcel of land, or change in types of crops grown and the sequence in which they are grown (e.g., intercropped or relayed) are all forms of agricultural intensification (Boserup 1965; Carswell 1997; Dahal et al. 2008). Agricultural intensification can have both beneficial and adverse consequences on the environment and human societies. While intensified production systems provide higher yields, and therefore returns, it is often achieved through the use of chemical fertilizers and synthetic pesticides, which have far-reaching and long-term consequences for ecological balance and human health. However, it is not necessary that intensified cropping will inevitably lead to land and environmental degradation as it generally requires a greater degree of care and meticulous planning, which may, in fact, result in an overall improvement of land and soil quality. Thus, while it is evident that the demand for food and pressures on the land resource base will continue to increase into the foreseeable future, fueling the need for agricultural intensification, farmers can choose to adopt sustainable and environmentally friendly approaches to intensified production to avoid the vicious cycle of degradation (McCalla 2001; Dahal et al. 2008; Bajracharya and Dahal 2012). The possible outcomes of different pathways chosen to increase global production on limited agricultural land are schematically illustrated in Figure 6.1.

6.2 IMPACTS OF AGRICULTURAL INTENSIFICATION ON THE ENVIRONMENT

Traditional agricultural practices of the past were done under circumstances of low demand on the land resource base and with minimal technological inputs, and thus, had small impacts on the environment as well as low output. However, with the onset of the Green Revolution and the development of chemical fertilizers along with a wide variety of pesticides, modern agriculture experienced a major shift in production practices. Coupled with large-scale mechanization of farming, the ever-increasing use of agricultural inputs (agrochemicals and improved hybrid varieties of crops) led to a dramatic increase in outputs; however, impacts on the environment also began to magnify in scale and intensity. In South Asia, the shift in agricultural pattern to high input, intensified cropping was somewhat delayed and progressed at

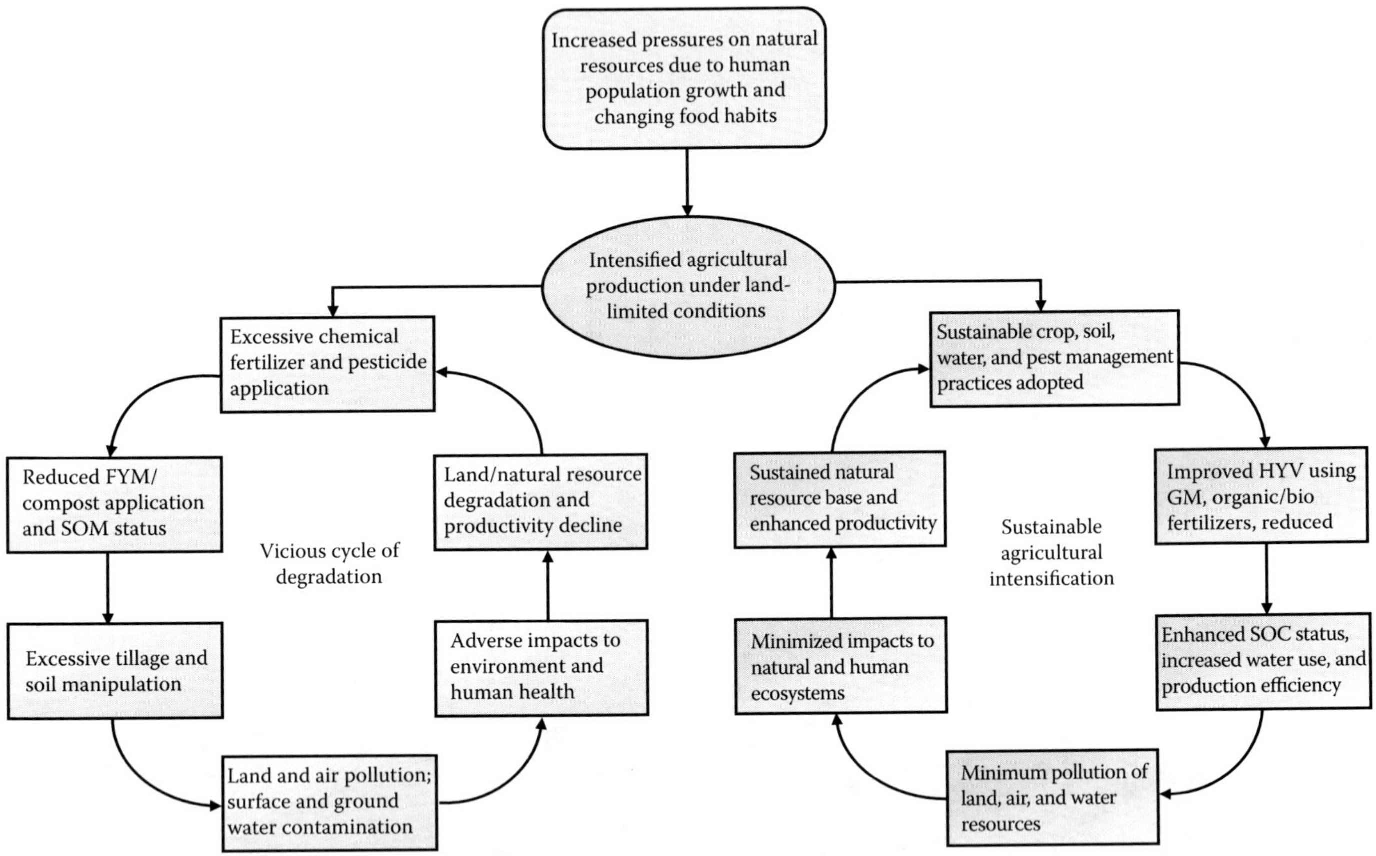

FIGURE 6.1 Potential pathways of intensified production practices in response to increased food demands of a growing world population.

different paces in different countries, many of which are as yet suboptimal (Pagiola 1995; Brown and Shrestha 2000; Murgai et al. 2001; Fan et al. 2012). Numerous factors determine farmers' decisions to intensify production, including market access, infrastructure, input availability (Brookfield 2001; Dahal et al. 2009), caste or ethnicity, farm labor availability, extension services, education level, land management training (Paudel and Thapa 2004), as well as inducing factors such as extreme land shortage, land holding size, environmental and technological constraints (Turner and Shajaat Ali 1996), and poverty, leading to degradation of common property resources (Pinstrup-Andersen and Pandya-Lorch 1994; Lopez 1998).

One of the main ways in which intensification of agriculture influences the environment is through its effects on biodiversity. Most intensified production systems lead to a reduction in crop diversity, as well as the diversity of soil organisms and reduced soil organic matter, thus leading to imbalance and degradation of the soil (Matson et al. 1997; Lal 2007, 2011). This is in part due to large areas cropped in monoculture systems and in part to reduced livestock rearing as farmers become specialized; hence, less farmyard manure (FYM) is applied to the soil (Ali 1996; Thorne and Tanner 2002; Lal 2011). Fragoso et al. (1997) observed from studies in India, Mexico, and Peru that earthworm communities exhibited lower species richness, reduced numbers of ecological groups, and predominance of endogenic groups under intensive farming compared with undisturbed ecosystems. Likewise, Black and Okwakol (1997) noted that land clearing, preparation for cropping, and other agricultural management practices significantly affected termite diversity and activity. Although they are regarded by most farmers as pests, termites likely play a major beneficial role as ecosystem engineers as well as promote essential ecosystem processes, such as decomposition of lignin, in semiarid regions. Kahindi et al. (1997) on the other hand, found that because of a high degree of host specificity, loss of a single rhizobial species from an agroecosystem could result in loss of nitrogen fixation by a particular legume. However, because of the inoculants industry enabling compatible rhizobial genotypes to establish effective nitrogen fixation, a high biodiversity of N-fixing bacteria may not be required to ensure function in intensively cropped soils. In the mid-hills of Nepal, Begum et al. (2011) observed that the diversity and population densities of some soil mesofauna, such as collembola and mites, were significantly higher in forested areas than in agricultural fields.

In Nepalese hill farming, tillage and various land preparation practices have been integral to subsistence farming systems on often marginal and steep sloping lands (Bajracharya 2001). However, past farming practices were typically of low intensity (one or two crops per cropping cycle) and integrated crop–livestock systems where farmers exclusively used FYM and compost. With the increasing availability of modern agrochemicals, changes in food habits, irrigation facilities, and growing demands, farmers have progressively intensified production with multiple and high-value crops now being grown in peri-urban areas (Brown and Shrestha 2000; Dahal et al. 2009). This has been accompanied by the expansion of cultivated land areas and a reduction in organic inputs and manure due to decreasing numbers of livestock reared per farm household (Bajracharya 2001; Karki 2006; Acharya et al. 2007). The disruption in the balance among extraction of nutrients from the soil by crops and return of residues and organic matter has been, therefore, leading to deterioration in soil quality and hence land productivity (Schreier et al. 2001; Sherchan and Karki 2006; Paltridge et al. 2011).

Numerous studies have indicated that cropping intensification, with greater reliance on agrochemicals and decreasing soil organic matter, in the hill regions of Nepal, is unsustainable and leads to land degradation due to soil erosion and nutrient losses during the high rainfall monsoon season (Pilbeam et al. 2002; Gardner and Gerrard 2003; Acharya et al. 2007; Dahal et al. 2009). Moreover, preferential application of nitrogenous and phosphorous fertilizers by farmers has apparently led to micronutrient imbalances in the cultivated soils (Andersen 2001; Paltridge et al. 2011). Recent studies conducted by Raut et al. (2012) revealed that intensive rice crop production in Nepal leads to soil acidification, which increases the rate of denitrification and release of the potent greenhouse gas N_2O into the atmosphere, hence enhancing global climate change (Table 6.1). Soil acidification due to excessive nitrogen fertilization and the consequent enrichment of soil, water, and atmospheric nitrogen through ammonia volatilization and nitrate leaching have also been reported in eastern China and the North China Plain (Ju et al. 2009; Guo et al. 2010). In India, intensified agriculture using extensive irrigation was found to increase the regional moisture flux, which modified the convective available potential energy of the region, leading to a reduction in the surface temperature and modified circulation patterns causing meso-scale changes in monsoon precipitation (Douglas et al. 2009). Lal (2004, 2007, 2011) points out that forest and soil degradation has major implications for environmental quality, climate change, and food security in the South Asia region. Thus, clearly, intensive farming can have local-, regional-, as well as global-scale impacts on the environment and climate.

TABLE 6.1
Effects of Intensified Agriculture on Soil pH and $N_2O/(N_2 + N_2O)$ Product Ratios in Lowland and Upland Farming of the Nepal Mid-Hills

	Soil pH (Mean ± SD)		$N_2O/(N_2 + N_2O)$	
Land Use Type	**Traditional Farm**	**Intensive Farm**	**Traditional Farm**	**Intensive Farm**
Lowland C1	5.42 ± 0.11	4.93 ± 0.11	0.59	0.91
Lowland C2	5.38 ± 0.11	4.60 ± 0.11	0.48	0.66
Lowland C3	4.94 ± 0.05	4.27 ± 0.05	0.81	0.99
Upland C1	5.64 ± 0.06	4.78 ± 0.06	0.74	0.90
Upland C2	5.27 ± 0.04	4.62 ± 0.04	0.62	0.62
Upland C3	4.48 ± 0.06	4.02 ± 0.06	0.82	0.75
Upland C4	5.05 ± 0.05	4.21 ± 0.05	0.50	0.63
Means	5.17	4.49	0.65	0.78

Source: Adapted from Raut, N., P. Dorsch, B.K. Sitaula, and L.R. Bakken, *Soil Biology and Biochemistry* 55, 104, 2012.

6.3 SOCIAL AND ECONOMIC ASPECTS OF INTENSIVE AGRICULTURE

Although less developed countries have seen marked intensification of agriculture in the 20th century, the pace and motivations of intensive farming have varied considerably across nations and regions. For instance, shifting cultivation, a practice feasible only under low population pressure and low demands on the land resource, has been gradually replaced by sedentary agriculture in parts of Asia such as Malaysia, Thailand, and Indonesia (Rasul and Thapa 2003). However, in other countries, such as India, Laos, Bangladesh, and Nepal, it still continues despite increasing population pressures and demands on the land resource base. Rasul and Thapa (2003) point out that a change in agricultural practices does not occur automatically as a result of population pressure (as postulated by Boserup 1965), but requires reinforcement by factors such as land ownership, infrastructure development, government policies, and support systems and facilities.

In Bangladesh, a country facing extreme population pressure and land shortage, farmers adopt both subsistence and market-oriented production strategies depending on the particular set of circumstances they face involving land holding size, environmental constraints, and technology available (Turner and Shajaat Ali 1996). Although from 1950 to 1986, the proportion of landless farmers increased, on average, smallholder farms were able to keep pace with demand. Agricultural changes occurred in a step fashion with periods of intensification giving way to involution or stagnation. Nonetheless, the average response to increase in agricultural demands was to increase land productivity through increased cropping frequency, higher yields, or higher volumes of crops. Practically all farmers adopted some level of market production to meet their own as well as societal needs (Turner and Shajaat Ali 1996).

Nepal also saw a progression toward market-driven production systems as a result of increased access to markets owing to the development of roads and rural infrastructure, as well as expansion of peri-urban areas (Brown and Shrestha 2000). Multiple cropping incorporating high-value crops and vegetables where irrigation is available and expansion of agriculture to steep upland areas have raised concerns about land degradation. Management and conservation options, such as improved composting, nitrogen-fixing plants, liming, and increasing crop water use efficiency, implemented through adequate extension and technical support, are believed to be key to preventing resource degradation (Brown and Shrestha 2000; Dahal et al. 2009). Indeed, Thapa and Paudel (2002) found that, of two western watersheds in Nepal undergoing degradation due to soil erosion, fertility depletion, and landslides, the one having external interventions showed less severe land degradation. Here, watershed management projects had promoted soil conservation practices to counter anthropogenic causes of land degradation. In another study, Paudel and Thapa (2004) reported that the adoption of structural and biological measures to control land degradation depended on factors such as the ethnicity of farmers, farm labor, soil type on land holding, education level, and land management training. However, they noted that the most influential factor for the adoption of technology was the availability of extension services.

Murgai et al. (2001) point out that the post–Green Revolution agricultural productivity growth of the two Punjab states in India and Pakistan exhibited wide spatial and temporal variation. While outputs and crop yields were generally much higher in India, productivity growth was only marginally higher than the Pakistan Punjab. Growth in inputs accounted for most of the growth in outputs for both Punjabs, and intensification in wheat–rice systems resulted in resource degradation in both states as well. The authors concluded that policies were needed to promote agricultural productivity and sustainability through public investments in education, roads, research and extension, while eliminating subsidies that encourage intensification of inputs.

6.4 SUSTAINABLE AGRICULTURAL INTENSIFICATION

To ensure that we can meet the nutritive requirements of human populations well into the future, it is essential that farmers are provided with technological options and knowledge to adopt sustainable intensified cropping and livestock-rearing practices. While no single formula can exist for sustainable agricultural intensification globally, numerous practices and technologies have been locally tested and reported to have long-term conservation and production benefits. Various practices and conservation measures have been developed or adapted locally to suit site-specific conditions. An attempt is made here to identify and present the salient cultural practices as well as technologies most suitable for enhancing productivity while concomitantly protecting the environment and ensuring sustainability of hill agriculture in the Himalayan region of South Asia.

6.4.1 Integrated Nutrient Management

Traditionally, under low population pressures and demand, farmers relied solely on compost or FYM made from animal manure, forest leaf litter, and crop residues (straw, grain husk, etc., after use as livestock bedding). Applied to fields with low-intensity farming, typically two crops per annual cropping cycle, this traditional organic fertilizer had maintained the soil's productive capacity for generations. In recent decades, however, increasing population pressures, changing diet patterns, and reduced livestock numbers per farm household, as well as shortage of farm labor, have necessitated alternative approaches to enhance and maintain the soil's fertility and productivity (Ali 1996; Bajracharya 2001; Acharya et al. 2007).

Numerous studies reveal that modern practices that incorporate substantial amounts of organic matter and manure or cattle urine, along with judicious, moderate amounts of inorganic fertilizers, could increase the yields and sustainability of agricultural systems as well as enhance soil organic carbon (SOC) accumulation (Sherchan and Gurung 1998; Sherchan et al. 1999; Tripathi and Tuladhar 2000; Bajracharya and Atreya 2007; Dahal and Bajracharya 2012). As shown in Table 6.2, a combination of FYM and inorganic fertilizers increased yields by 35% on average in maize–millet and by 16% on average in rice–wheat cropping systems. Similarly,

TABLE 6.2
Average Crop Yields (Mg ha^{-1}) under Different Fertilizer Treatments and Yield Difference Compared with Farmyard Manure and Inorganic Fertilizer Combination

		Yield Difference (± S.E.) Comparing	
Crop Type	**Avg. Yield ± S.E.**	**FYM + IF to FYM Only**	**FYM + IF to IF Only**
Maize	2.63 ± 0.57	+0.563 ± 0.072	0.869 ± 0.128
Millet	1.76 ± 0.35	−0.073 ± 0.024	0.631 ± 0.078
Rice	3.56 ± 0.46	+1.185 ± 0.115	0.590 ± 0.230
Wheat	1.98 ± 0.26	−0.012 ± 0.004	0.312 ± 0.066

Source: Adapted from Sherchan, D.P., C.J. Pilbeam, and P.J. Gregory, *Experimental Agriculture*, 35, 1, 1999.

Note: FYM, farmyard manure; IF, inorganic fertilizer; S.E., standard error.

Sherchan and Karki (2006) reported significantly higher yields with the addition of combinations of compost and chemical fertilizer, as well as FYM and fertilizer for wheat, maize, and rice (Figure 6.2). In Laos, Linquist et al. (2007) also reported that combined applications of crop residues and commercial fertilizers enhanced soil nutrient balance, particularly during years of irregular and fluctuating soil moisture levels. Studies in China gave similar results of increases in SOC, nitrogen, microbial biomass carbon, and nitrogen upon application of pig manure with reduced rates of commercial fertilizers (Xu et al. 2007).

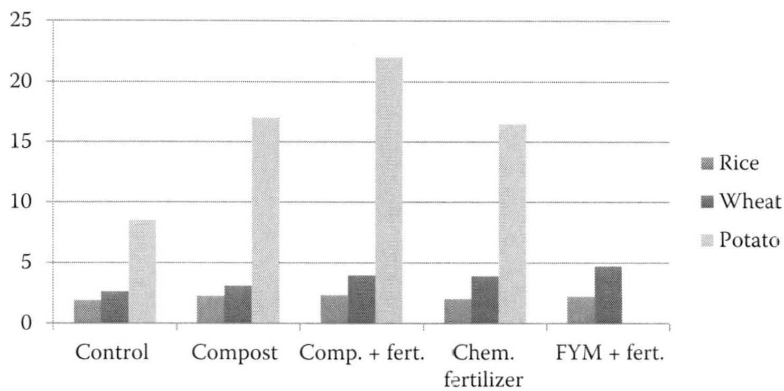

FIGURE 6.2 Wheat, rice, and potato yields under different FYM and chemical fertilizer treatments. (Modified from Sherchan, D.P., and K.B. Karki. 2006. Plant nutrient management for improving crop productivity in Nepal. Improving plant nutrient management for better farmer livelihoods, food security and environmental sustainability. In: Proceedings of a Regional Workshop, Beijing, December 12–16, 2005. FAO RAP Publ. 2006/27, pp. 41–57.)

6.4.2 Crop Rotations and Cropping Patterns

Cropping patterns that include legumes with cereal crops or combinations of crops (intercropping) offer potential for improved soil quality, reduced production risks, and better return to farmers. Sherchan et al. (1997) reported that planting pigeon pea with groundnuts had a beneficial effect on soil physical properties along with increase in maize yields (Table 6.3; 2.42–2.53 Mg ha^{-1} vs. 1.124 Mg ha^{-1}) over the traditional maize–millet cropping pattern. Correspondingly, soil properties such as SOM, bulk density, and available water capacity were more favorable in pigeon pea and groundnut and sole groundnut cropping patterns (Table 6.3). Moreover, Sherchan and Karki (2006) point out that because pulses and beans are an important source of dietary protein for rural communities in the region, development of improved varieties of legumes with higher N-fixing capacity would help enhance the sustainability of cropping systems. Sharma and Sharma (2004) in India also concluded that inclusion of a legume in the crop rotation, such as rice–potato–mung bean, along with a combination of chemical fertilizers and FYM, was highly effective in building up SOC, nitrogen, phosphorus, and potassium levels in soil.

6.4.3 Biofertilization and Seed Inoculation

Another promising technique for enhancing sustainability through efficient uptake of crop nutrients is the use of biological agents such as algae, bacteria, and non-crop plants. Quyen and Sharma (2003) noted that a combination of green manure (*Sesbania* sp.), blue-green algae, and *Psuedomonas striata* inoculations, with FYM application, gave the highest rice yields. The yield (5.2 Mg ha^{-1}) of this organically grown rice was greater than that obtained using high chemical fertilizer doses.

The use of rhizobial inoculants for seeds of leguminous crops such as soybean, black gram, chickpea, cowpea, mung bean, and groundnuts (Table 6.4), as well as other plants like *Desmodium*, *Stylosanthes*, *Medicago*, and vetch, have been shown to be effective

TABLE 6.3
Soil Properties and Yield of Maize Test Crop under Different Cropping Patterns

Cropping Pattern	SOM (%)	Bulk Density (Mg cm^{-3})	AWC (%)	Maize Yield (Mg ha^{-1})
Maize–millet	0.77	1.65	11.47	1.12
Groundnut	0.91	1.54	12.99	2.42
Pigeon pea + groundnut	1.01	1.56	12.84	2.53

Source: Modified from Sherchan, D.P., G.B. Gurung, and B.D. Gurung. 1997. Productivity assessment of red clay soils with legume and cereal combinations on dry, rainfed hill environments in Nepal. PAC Technical Paper No. 180. Pakhribas Agricultural Centre, Dhankuta, Nepal.

Note: AWC, available water capacity; SOM, soil organic matter.

TABLE 6.4
Increase in Yield of Various Crops by *Rhizobium* Inoculation (RI) of Seeds

	(Mg ha^{-1})		
Crop	**Typical Yields**	**Yields with RI**	**Yield Increase (%)**
Soybean	2.04	3.35	65
Lentils	0.86	1.20	40
Black gram	0.25	0.37	49
Groundnut	0.98	1.44	47
Broad beans	0.36	0.60	67

Source: Modified from Sherchan, D.P. and K.B. Karki. 2006. Plant nutrient management for improving crop productivity in Nepal. Improving plant nutrient management for better farmer livelihoods, food security and environmental sustainability. In: Proceedings of a Regional Workshop, Beijing, December 12–16, 2005. FAO RAP Publ. 2006/27, pp. 41–57.

in enhancing yields and survivability in low-fertility soils (Sherchan and Karki 2006). Moreover, owing to the development and large-scale production of compatible bacterial strains, it has been pointed out that high biodiversity in intensively cropped soils may not be necessary to ensure effective nitrogen fixation (Kahindi et al. 1997).

6.4.4 Minimum and Zero Tillage

Changing tillage practices from frequent and soil-inverting cultivation to reduced or minimum tillage systems has been demonstrated to improve soil organic matter status, biophysical properties, and yields in the long term. The Nepal Agricultural Research Council, with support from the Rice–Wheat Consortium, has been conducting trials on no-till, direct seeding of wheat following rainy season rice crop in the hill and plains region. This technique has the advantage of allowing early sowing and reduced labor costs, and hence has proven to be economical and resource conserving (Giri 1998). Furthermore, use of a no-till seed drill offers cost savings of nearly 50% and higher yields by 40%–70% compared with traditional farmer methods (Bajracharya 2001; Sherchan and Karki 2006). Tripathi et al. (2005) confirmed the cost-effectiveness and yield benefits of the zero-till seeding technology in on-farm trials in the Terai region. No-till seed drills allow for timely seeding of wheat, and with the direct seeding rice technology there is considerable cost saving with rice yield either at par or increased compared with transplanted rice.

Atreya et al. (2006) found reduced tillage systems (elimination of one or more cultivation operations) in upland maize-based cropping in central Nepal to be effective against soil and nutrient losses, while maintaining yields at par with conventional practices. Erosion and loss of nutrients such as soil organic matter, total nitrogen, available phosphorus, and exchangeable potassium were all significantly reduced under mulching and reduced tillage practices in the central mid-hills of Nepal as shown in Table 6.5. Similarly, Tiwari et al. (2009) reported considerable reductions in soil and nutrient losses compared with traditional farming practices and commercial vegetable cultivation in a mid-hill watershed of central Nepal (Table 6.6). Although yields were lower under reduced tillage, as can be expected in the early years, the overall income was not much lower owing to reduced labor inputs compared with traditional farming.

TABLE 6.5
Cropping System Effects on Runoff, Soil, and Nutrient Losses from Upland Agriculture in the Nepal Mid-Hills

Cropping System	Runoff (mm)	Soil Loss	SOM Loss	TN Loss	AP Loss	EK Loss
		(Mg ha^{-1})			(kg ha^{-1})	
CT, maize	129	15.8[a]	0.34[a]	18.8[a]	1.9[a]	4.0[a]
CT, maize + SB	117	17.3[a]	0.31[a]	18.8[a]	1.9[a]	4.4[a]
Mulch, maize	123	9.3[bc]	0.17[b]	9.6[b]	1.0[bc]	2.3[b]
Mulch, maize + SB	110	8.3[b]	0.15[b]	9.1[b]	0.8[b]	2.2[b]
RT, maize	128	10.8[bc]	0.23[c]	11.9[b]	0.9[bc]	2.6[b]
RT, maize + SB	134	11.4[c]	0.21[c]	11.7[b]	1.1[c]	2.7[b]

Source: Adapted from Atreya, K., S., Sharma, R.M. Bajracharya, and N.P. Rajbhandari, *Journal of Environmental Management*, 88, 547, 2008.

Note: AP, available phosphorus; EK, exchangeable potassium; SOM, soil organic matter; TN, total nitrogen. Means with the same superscript letter (a, b, or c) are not significantly different.

TABLE 6.6
Annual Soil and Nutrient Losses and Crop Yields (Means ± SE) as Influenced by Farming Systems in the Central Nepal Mid-Hills

Farming System	Soil Loss (Mg ha^{-1} y^{-1})	SOC Loss	TN Loss	Annual Average Yield (Mg ha^{-1} y^{-1})	Gross Income (NRs. ha^{-1} y^{-1})
		(kg ha^{-1} y^{-1})			
Traditional	1.22 ± 0.46	25.5 ± 12.3	5.3 ± 2.1	4.07 ± 0.22[a]	21,860 ± 2393[a]
Reduced till	1.02 ± 0.36	17.5 ± 6.8	3.9 ± 1.5	2.07 ± 0.25[b]	20,353 ± 2469[a]
Commercial veg.	1.27 ± 0.43	24.5 ± 11.4	5.6 ± 2.1	3.64 ± 0.15[a]	34,721 ± 4621[b]

Source: Modified from Tiwari, K.R., B.K. Sitaula, R.M. Bajracharya, and T. Borresen, *Nutrient Cycling in Agroecosystems*, 86, 241–253, 2009.

Note: NRs., Nepali rupees (USD 1 = NRs. 78 approx.); SOM, soil organic matter; TN, total nitrogen. Means with the same superscript letter (a or b) are not significantly different.

In Pakistan, Iqbal et al. (2007) observed that minimum tillage, in conjunction with FYM amendments, led to an increase in soil nutrient (N, P, and K) and organic matter status, as well as maize yields. Another study in China showed that the conservation tillage system involving ridges and furrows with seasonal no-till led to restoration of physical, chemical, and biological properties of a degraded southern hill region soil (Zhu et al. 2002). Such permanent bed or ridge and furrow planting systems have been shown to have potential in the hill regions of Nepal, particularly for high-value crops such as potato and other vegetables (Bajracharya 2001; Sherchan and Karki 2006).

6.4.5 Biogas and Biochar

Biogas slurry applied as a fertilizer and soil amendment has numerous beneficial effects on the soil, as well as enhances yields. As >120,000 biogas plants have been installed in rural areas across Nepal, it could serve as an important component of sustainable farming systems while simultaneously helping conserve forests and reduce carbon dioxide and methane emissions to the atmosphere. Biogas slurry application was observed to increase the yields of maize, rice, wheat, and cabbage by 30%, 23%, 16%, and 25%, respectively, over the conventional farmer practice in the Nepal mid-hills (Karki 2004).

Biochar is a subject of recent scientific investigation as a potential means of enhancing the carbon storage capacity and longevity in soils, while simultaneously increasing the soils' fertility and productive capacity. Biochar, a pyrolysis product of biomass, has been used by ancient civilizations in the Amazon, northwest Europe, and the Andes (Sombroek et al. 1993; Sandor and Eash 1995; Downie et al. 2011). It has recently gained scientific attention as a simple yet potentially powerful tool for climate change mitigation while contributing to sustainable agricultural production (Downie et al. 2011; International Biochar Initiative 2012). The benefits of biochar reportedly result from its high stability, porosity, and resistance to microbial breakdown, thereby acting as sites for increased water and nutrient retention (Sohi 2012).

6.4.6 Microirrigation and Water Management

Water management will no doubt be a subject requiring considerable attention and effort with climate change, erratic precipitation, and water scarcity becoming ever more acute in the years to come. Yet, to meet the increasing demands and livelihood security, farmers have opted to grow high-value and out-of-season vegetables such as tomatoes (*Lycopersicon esculentum* Mill.), cucumber (*Cucumis sativus* L.), squash (*Cucurbita pepo* L.), and cauliflower (*Brassica oleracea* L.). Such crops, however, are sensitive to water stress and require a regular, reliable supply of water (Wan and Kang 2006; Enciso et al. 2007). Under these conditions, drip or microirrigation offers a viable and water-efficient option for irrigating vegetable crops during periods of water scarcity.

Drip or trickle irrigation is a type of microirrigation method intended for both water conservation and efficient use of water by the crop. Microirrigation systems deliver water in the form of drops, tiny streams, or miniature sprays directly to the plant root zone. Such systems may be placed on the soil surface or at varying depths in the crop bed (Schwab et al. 1993; Hla and Scherer 2003; Ensico et al. 2007). Microirrigation systems have the advantage of enabling economic crop yields with minimal use of water (Randhawa and Abrol 1990), as well as the control of nutrient levels, which can be supplied along with the irrigation water (Schwab et al. 1993; Hla and Scherer 2003). Such microirrigation techniques, together with water harvesting, will have to be exploited to the fullest extent if we are to meet agricultural water requirements under intensive crop production in the future.

6.4.7 Other Holistic Management Approaches

Increasing the efficiency and productivity of agricultural systems while ensuring sustainability in the face of ever-increasing demands, will, undoubtedly require holistic, integrated approaches that incorporate environmentally sound land, water, crop, livestock, and pest management (Figure 6.3). Latif et al. (2005) compared three rice cultivation methods in the Comilla District of Bangladesh and found that the existing best management practice of the Bangladesh Rice Research Institute gave significantly better performance, higher yields, higher profit, and lower cost than the System of Rice Intensification or conventional farmer practice. These intensive practices and improved technologies may be site specific and have to be adjusted to local conditions. Other studies in China also indicated that integrated crop–soil management practices hold potential to increase crop yields without posing adverse environmental risks. Chen et al. (2011) used a hybrid–maize simulation model to determine the optimum combination of planting date, crop density, and plant variety to apply at a particular site based on long-term weather data. Nearly double the yield (13 Mg ha^{-1}) of current farmer practice was obtained without increasing the nitrogen fertilizer dose. Thus, substantially

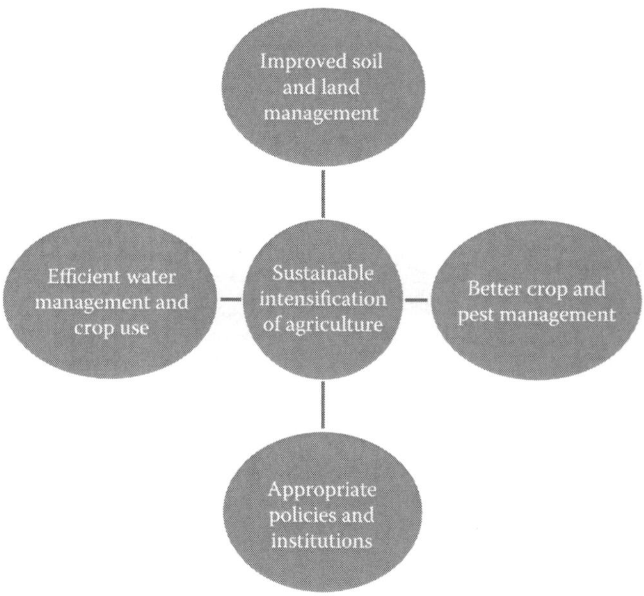

FIGURE 6.3 Simultaneous measures needed to achieve sustainable agricultural intensification in the Himalayan region.

improved yields, through soil and crop management, may be achieved by the right combination of genetically improved crop varieties coupled with better agronomic practices, in the face of deteriorating soil quality, increasing water shortages, and other uncertainties brought about by climate change (Fan et al. 2012).

The Sloping Agricultural Land Technology proposed by Tacio (1993) uses an integrated and holistic approach through a combination of terracing, contour operations, fodder hedge rows, agroforestry, and mixed cropping. The approach, tailored to local conditions, is likely to be compatible with sustainable hill farming in the Himalayan region. Indeed a "whole landscape" approach, using place-specific strategies combining crop diversification and improved soil, crop, and livestock management, along with agroforestry, are likely needed to achieve sustainable land use through targeted policy and management interventions (DeFries and Rosenzweig 2010).

6.5 FUTURE PRIORITIES FOR FOOD SECURITY AND ENVIRONMENTAL PROTECTION

It is amply evident that demands on land and water resources will continue to increase and that further intensification of agriculture will be required to meet global food demands in the future. Under the present trends of increased agricultural intensification in the richer nations and increased land clearing in poorer nations, it has been estimated that about 1 billion hectares of land would be cleared globally by 2050 with greenhouse gas emissions reaching about 3 Pg y^{-1} of CO_2 equivalent and nitrogen use reaching approximately 250 Mt y^{-1} (Tilman et al. 2011). Local land use decisions can have global implications for terrestrial carbon stocks and global climate change in view of the food demands of a growing population, changing diets, and biofuel production. Tropical areas have been calculated to lose nearly twice as much carbon (about 3 Mg C per million grams of annual crop yield), while producing less than half the annual crop yields compared with temperate regions (West et al. 2010). Thus, special emphasis is needed on increasing yields from existing croplands in the tropics rather than clearing new lands.

Agricultural intensification has been typically characterized by a synergy between agronomic innovations and plant breeding (Evans 2003). Globally, agricultural intensification has been able to keep pace with population growth through continuous evolution; it is estimated that about 2 million hectares of land are affected by soil erosion, 1.5 million hectares by salinization and toxification, and even greater amounts of land are permanently lost due to urban expansion onto arable land (Evans 2003). Therefore, more innovative, integrated, and sustainable production systems will be needed to keep pace with demands in the future. Genetically improved seeds and crops are already in use worldwide and likely to be an important part of agricultural sustainability. Hence, the use of high-yielding varieties enhanced through genetic modification is expected to be a crucial part of the solution to raise global yields without further degrading the environment (Ronald 2011). However, studies suggest that the expected future food demand cannot be met by increasing genetic yield potential alone as the gap between average farm yields and genetic potential is narrowing (Cassman 1999). Therefore, a combination of soil quality improvement along with greater precision in agricultural management choices (crop, nutrient, water, pest, etc.) will be needed, which will require breakthroughs in plant physiology, eco-physiology, agroecology, and soil science (Matson et al. 1997; Cassman 1999; Chen et al. 2011; Fan et al. 2012).

The pressures on natural resources in the Himalayan region will most certainly be extreme due not only to the vast population but also to the delicate balance in which ecosystems of this region exist. Further exacerbating the uncertainty of future changes in productive capacity of the land are the escalating impacts of climate change. Thus, a four-pronged strategy, optimally integrating (i) improved soil and land management, (ii) increased water use efficiency and water resource management, (iii) crop and agrobiodiversity management, and (iv) appropriate policy and institutional support, will be necessary to meet the daunting challenge of sustainable agricultural intensification (Figure 6.3). The possible measures and technologies to be applied toward formulating the strategy are described in Table 6.7.

TABLE 6.7
Measures and Technologies Required to Achieve Sustainable Intensification of Agriculture in the Himalayan Region

Resource Category	Measures and Technologies
Soil/land resource—improved soil management	• Enhanced soil fertility and quality through improved composting, use of adequate farmyard manure, and urine application • Increased soil organic matter and soil carbon sequestration • Use of biofertilizers and optimization of rhizosphere microbial activity • Use of biochar and zeolite amendments to improve soil biophysical properties and water retention • Adoption of reduced or minimum tillage, crop residue management, and other conservation practices
Water resource management and use efficiency	• Increased water use efficiency through microirrigation and timing of application; use of laser level to improve irrigation/water productivity • Water harvesting and groundwater recharge • Improved water retention in soil through the application of biochar, zeolites, and mulching • Water recycling, wastewater reclamation and reuse, desalinization
Crop/agrobiodiversity resource management	• More efficient and risk-averting crop production systems, such as agroforestry with high-value crops; mixed cropping, relay cropping, and intercropping • Improved crop varieties with drought, cold, and pest resistance; high-yielding varieties enhanced through genetic modification • Forage species/crop rotations and planting of fruit and fuel wood tree species on private land • Integrated and natural pest control approaches
Policy and institutional initiatives	• Adequate investment in agricultural research • Technical support to farmers through extension and outreach services • Institutional support and strengthening of capacity • Policies and incentives to encourage conservation and sustainable production

However, it is clear that scientists and farmers must work collaboratively, in a mutually reinforcing manner, to find the optimal strategies. Fruitful collaborations will indeed depend on the ability to build contextual knowledge systems enabling understanding of the complexities of the natural systems, farming systems, as well as the sociocultural fabric of the many diverse localities across the Himalayan region (Andersen 2005). This can be achieved by merging scientific knowledge systems with local knowledge systems and the appropriate incorporation of indigenous technical knowledge. It is only through such contextual knowledge that effective and self-perpetuating changes in agricultural and natural resource management will likely be adopted by the local land users (Bajracharya and Sherchan 2009).

6.6 CONCLUSIONS

Ensuring adequate food production in the future will, without a doubt, involve further intensification of agriculture. Among the greatest challenges of modern times is the necessity to meet the ever-growing food demands with limited arable land resources and without causing irreversible damage to the environment in the face of increasing unpredictability due to global climate change (The Royal Society 2009; Lambin et al. 2011). Although agriculture has been generally regarded as a source of greenhouse gases, with intensification, the net emissions on a yield basis has avoided an estimated 161 Gt C despite increased emissions from fertilizer use (Burney et al. 2010). Therefore, it makes sense to invest in yield improvement through intensified cropping as a cost-effective option for avoiding greenhouse gas emissions that would occur owing to the production of the same amount of crops at previous lower yield rates. However, the transition to sustainability will clearly require a stable population and stable levels of material consumption (Ruttan 1999), and sound policies and innovations that harness globalization to increase land use efficiency rather than causing uncontrolled land use expansion (Lambin et al. 2011). This will require land systems to be modeled as open systems with large flows of goods, people, and capital connecting local land use with global-scale factors. Hence, the global-scale cascading effects of local land use decision could possibly be avoided through new forms of global governance linking trade with environmental protection (Lambin and Meyfoiydt 2011). Certainly then, maintenance of the capacity for institutional and technological innovations will be more critical than environmental constraints, which will require increased investments in agricultural research and alleviating poverty, food insecurity, and malnutrition in low-production-potential areas most at risk of environmental degradation (Pinstrup-Andersen and Pandya-Lorch 1994; Ruttan 1999; The Royal Society 2009).

In the Himalayan region, attaining sustainable, intensified production will require an integrated and holistic approach encompassing improved soil management, increased water use efficiency, higher yielding varieties coupled with diversified cropping, and integrated plant nutrient and pest management backed by appropriate policy and institutional support. The combination of measures and technologies, such as reduced/minimum tillage, choice of cropping patterns and crop varieties, biofertilization, natural and biological pest control, microirrigation, and integrated crop–livestock agroforestry systems, will no doubt need to be matched to local conditions with increased precision and site specificity. Success in achieving such intensive and sustainable production systems will clearly depend on farmers, scientists, and policy makers working together in a synergistic manner for the benefit of society at present and in the future.

REFERENCES

Acharya, G.P., M.A. McDonald, B.P. Tripathi, R.M. Gardner, and K.J. Mawdesley. 2007. Nutrient losses from rain-fed bench terraced cultivation systems in high rainfall areas of the mid-hills of Nepal. *Land Degradation and Development* 18:486–499.

Ali, M. 1996. Quantifying the socio-economic determinants of sustainable crop production: An application to wheat cultivation in the Tarai of Nepal. *Agricultural Economics* 14:45–60.

Andersen, P. 2001. Beyond local knowledge and institutional reach: Micronutrient disorders in hill agriculture. Paper presented at the Regional Conference on Physical Mobility and Development in the Mountains, March 15–17, 2000, Kathmandu, Nepal, 7 pp.

Andersen, P. 2005. Building contextual knowledge: The interface between local and scientific knowledge. In: P. Andersen, J.K. Tuladhar, K.B. Karki, and S.L. Maskey (eds.), *Micronutrients in South and Southeast Asia*. Proceedings of the International Workshop held in September 8–11, 2004, Kathmandu, Nepal, pp. 87–95.

Atreya, K., S. Sharma, R.M. Bajracharya, and N.P. Rajbhandari. 2006. Applications of reduced tillage in hills of central Nepal. *Soil and Tillage Research* 88:16–29.

Atreya, K., S. Sharma, R.M. Bajracharya, and N.P. Rajbhandari. 2008. Developing a sustainable agro-system for central Nepal using reduced tillage and straw mulching. *Journal of Environmental Management* 88:547–555.

Bajracharya, R.M. 2001. Land preparation: An integral part of farming systems in the mid-hills of Nepal. *Nepal Journal of Science and Technology* 3:15–24.

Bajracharya, R.M., and K. Atreya. 2007. Carbon sequestration in upland farming systems of the Nepal midhills. Paper presented at the National Conference on Environment, Tribhuvan University, June 22–24, 2007, Kathmandu, Nepal.

Bajracharya, R.M., and B.M. Dahal. 2012. Agricultural intensification. In: R.K. Craig, J.C. Nagle, B. Pardy, O.J. Schmitz, and W.K. Smith (eds.), *The Berkshire Encyclopedia of Sustainability, Volume 5, Ecosystem Management and Sustainability*. Berkshire Publishing Group, New York, pp. 7–10.

Bajracharya, R.M., and D.P. Sherchan. 2009. Fertility status and dynamics of soils in the Nepal Himalaya: A review and analysis. In: D.P. Lucero, and J.E. Boggs (eds.), *Soil Fertility*. Nova Science Publishers Inc., New York, pp. 111–135.

Begum, F., R.M. Bajracharya, S. Sharma, and B.K. Sitaula. 2011. Assessment of soil quality using microarthropod communities under different land system: A case study in the mid-hills of central Nepal. *Journal of the Life Sciences* 5(1):1–8.

Black, H.I.J., and M.J.N. Okwakol. 1997. Agricultural intensification, soil biodiversity and agroecosystem function in the tropics: The role of termites. *Applied Soil Ecology* 6(1): 37–53.

Boserup, E. 1965. The conditions of agricultural growth: The economics of agrarian change under population pressure. London: George Allen & Unwin Ltd., 108 pp.

Braidwood, R.J. 1960. The agricultural revolution. *Scientific American* 203(3):131–148.

Brookfield, H. 2001. Intensification and alternative approaches to agricultural change. *Asia-Pacific Viewpoint* 42(2/3):181–192.

Brown, S., and B. Shrestha. 2000. Market-driven land use dynamics in the Middle Mountains of Nepal. *Journal of Environmental Management* 59:217–225.

Burney, J.A., S.J. Davis, and D.B. Lobell. 2010. Greenhouse gas mitigation by agricultural intensification. *Proceedings of the National Academy of Sciences of the United States of America* 107(26):12052–12057. Published online, http://www.pnas.org/cgi/doi/10.1073/pnas.0914216107.

Carswell, G. 1997. Agricultural intensification and rural sustainable livelihoods: A "think piece." IDS Working Paper No. 64.

Cassman, K.G. 1999. Ecological intensification of cereal production systems: Yield potential, soil quality, and precision agriculture. *Proceedings of the National Academy of Sciences of the United States of America* 96:5952–5959.

Chen, X.-P., Z.-L. Cui, P.M. Vitousek et al. 2011. Integrated soil–crop system management for food security. *Proceedings of the National Academy of Sciences of the United States of America* 108(16):6399–6404. Published online, http://www.pnas.org/cgi/doi/10.1073/pnas.1101419108.

Dahal, B.M., I. Nyborg, B.K. Sitaula, and R.M. Bajracharya. 2009. Agricultural intensification: Food insecurity to income security in a mid-hill watershed of Nepal. *International Journal of Agricultural Sustainability* 7(4):249–260.

Dahal, B.M., B.K. Sitaula, and R.M. Bajracharya. 2008. Sustainable agricultural intensification for livelihood and food security in Nepal. *Asian Journal of Water, Environment and Pollution* 5(2):1–12.

Dahal, N., and R.M. Bajracharya. 2012. Effects of sustainable soil management practices on distribution of soil organic carbon in upland agricultural soils of mid-hills of Nepal. *Nepal Journal of Science and Technology* 13(1):133–141.

Darlington, C.D. 1969. *The Evolution of Man and Society*. Simon & Schuster Publ., New York, pp. 69–70.

DeFries, R., and C. Rosenzweig. 2010. Toward a whole-landscape approach for sustainable land use in the tropics. *Proceedings of the National Academy of Sciences of the United States of America* 107(46):19627–19632. Published online, http://www.pnas.org/cgi/doi/10.1073/pnas.1011163107.

Douglas, E.M., A. Beltrán-Przekuratb, D. Niyogi, R.A. Pielke Sr., and C.J. Vörösmarty. 2009. The impact of agricultural intensification and irrigation on land–atmosphere interactions and Indian monsoon precipitation—A mesoscale modeling perspective. *Global and Planetary Change* 67:117–128.

Downie, A.E., L. Van Zwieten, R.J. Smernik, S. Morris, and R.R. Munroe. 2011. *Terra Preta Australis*: Reassessing the carbon storage capacity of temperate soils. *Agriculture, Ecosystems and Environment* 140:137–147.

Enciso, J., J. Jifon, and B. Wiedenfeld. 2007. Subsurface drip irrigation of onions: Effect of drip tape emitter spacing on yield and quality. *Agricultural Water Management* 92:126–130.

Evans, L.T. 2003. Agricultural intensification and sustainability. *Outlook on Agriculture* 32(2):83–89.

Fan, M., J. Shen, L. Yuan et al. 2012. Improving crop productivity and resource use efficiency to ensure food security and environmental quality in China. *Journal of Experimental Botany* 63(1):13–24.

Fragoso, C., G.G. Browna, J.C. Patróna et al. 1997. Agricultural intensification, soil biodiversity and agroecosystem function in the tropics: The role of earthworms. *Applied Soil Ecology* 6(1):17–35.

Gardner, R.A.M., and A.J. Gerrard. 2003. Runoff and soil erosion on cultivated rainfed terraces in the Middle Hills of Nepal. *Applied Geography* 23:23–45.

Giri, G.S. 1998. No-till Wheat Farming Technology. National Wheat Improvement Programme. Nepal Agricultural Research Council, Bhairahwa, Rupandehi, Nepal, 14 pp.

Guo, J.H., X.J. Liu, Y. Zhang et al. 2010. Significant acidification in major Chinese croplands. *Science* 327:1008–1010. Published online, http://www.sciencemag.org/content/327/5968/1008.full.html.

Hillel, D. 1992. *Out of the Earth: Civilization and the Life of the Soil*. University of California Press, Berkeley, CA, 321 pp.

Hillel, D. 2007. *Soil in the Environment: Crucible of Terrestrial Life*. Academic Press, MA, 320 pp.

Hla, A.K., and T.F. Scherer. 2003. *Introduction to Micro-Irrigation*. North Dakota State University and U.S. Department of Agriculture Publication, 7 pp. Published online, http://www.ag.ndsu.edu/pubs/ageng/irrigate/ae1243w.htm, accessed September 13, 2007.

International Biochar Initiative. 2012. Biochar. Published online, http://www.biochar-international.org/.

Iqbal, M., A. Hassan, and R. Lal. 2007. Nutrient content of maize and soil organic matter status under various tillage methods and farmyard manure levels. *Acta Agriculturae Scandinavica* 57:349–356.

Ju, X.-T., G.-X. Xing, X.-P. Chen et al. 2009. Reducing environmental risk by improving N management in intensive Chinese agricultural systems. *Proceedings of the National Academy of Sciences of the United States of America* 106(19):8071–8078. Published online, http://www.pnas.org/cgi/doi/10.1071/pnas.0902655106.

Kahindi, J.H.P., P. Woomerb, T. Georgec, F.M. de Souza Moreirad, N.K. Karanjaa, and K.E. Giller. 1997. Agricultural intensification, soil biodiversity and ecosystem function in the tropics: The role of nitrogen-fixing bacteria. *Applied Soil Ecology* 6(1):55–76.

Karki, K.B. 2004. Improvement of animal dung, biogas, environment and soil fertility. Proc. Fifth Nat'l. Animal Science Convention, Livestock Development for Socio-economic Empowerment, Nepal Animal Science Association, pp. 241–246.

Karki, K.B. 2006. Impact of cropping intensification on nutritional balance in Nepalese soils. Proc. Int'l. Seminar on Environmental and Social Impacts of Agricultural Intensification in Himalayan Watersheds held in October 15–17, 2006, Kathmandu, Nepal, pp. 27–33.

Lal, R. 2004. Soil and forest degradation in South Asia. *Forestry, Journal of the Institute of Forestry* 12:8–17. Pokhara, Nepal.

Lal, R. 2007. Soil degradation and environmental quality in South Asia. *International Journal of Ecology & Environment* 33(2–3):91–103. International Science Publishers, New Delhi, India.

Lal, R. 2011. Soil degradation and food security in South Asia. In: R. Lal, M.V.K. Sivakumar, S.M.A. Faiz, A.H.M. Mustafizur Rahman, and K.R. Islam (eds.), *Climate Change and Food Security in South Asia*. Springer, New York, pp. 137–152.

Lambin, E.F., and P. Meyfroidt. 2011. Global land use change, economic globalization, and the looming land scarcity. *PNAS* 108(9):3465–3472.

Latif, M.A., M.R. Islam, M.Y. Ali, and M.A. Saleque. 2005. Validation of the system of rice intensification (SRI) in Bangladesh. *Field Crops Research* 93(2–3):281–292.

Linquist, B.A., V. Phengsouvanna, and P. Sengxue. 2007. Benefits of organic residues and chemical fertilizer to productivity of rain-fed lowland rice and to soil nutrient balances. *Nutrient Cycling in Agroecosystems* 79:59–72.

Lopez, R. 1998. Agricultural intensification, common property resources and the farm household. *Environmental and Resource Economics* 11(3–4):443–458.

Matson, P.A., W.J. Parton, A.G. Power, and M.J. Swift. 1997. Agricultural intensification and ecosystem properties. *Science* 277:504–509.

McCalla, A.F. 2001. Challenges to world agriculture in the 21st century. *Agricultural and Resource Economics Update* 4(3). Published online. University of California, Davis, CA.

Murgai, R., M. Ali, and D. Byerlee. 2001. Productivity growth and sustainability in post–green revolution agriculture: The case of the Indian and Pakistan Punjabs. *World Bank Research Observer* 16(2):199–218. Oxford University Press, UK.

Pagiola, S. 1995. Environment and Natural Resource Degradation in Intensive Agriculture in Bangladesh. World Bank Environment Department Paper No. 15, Environmental Economics Series. Land, Water and Natural Habitats Division, World Bank, 32 pp.

Paltridge, N.G., S.P.P. Grover, L. Gouyi et al. 2011. Crop nutrient status and dynamics on small-holder farms in central Tibet, China. *Plant and Soil* 348:219–229.

Paudel, G.S., and G.B. Thapa. 2004. Impact of social, institutional and ecological factors on land management practices in mountain watersheds of Nepal. *Applied Geography* 24:35–55.

Piggot, S. (eds.). 1961. *The Dawn of Civilisation*. Thames and Hudson Publ., London, UK.

Pilbeam, C.J., P.J. Gregory, B.P. Tripathi, and R.C. Munankarmy. 2002. Fate of nitrogen-15-labelled fertilizer applied to maize–millet cropping systems in the hills of Nepal. *Biology and Fertility of Soils* 35:27–34.

Pinstrup-Andersen, P., and R. Pandya-Lorch. 1994. Poverty, agricultural intensification, and the environment. *The Pakistan Development Review* 33(4):463–496.

Quyen, N.V., and S.N. Sharma. 2003. Relative effect of organic and conventional farming on growth, yield and grain quality of scented rice and soil fertility. *Archives of Agronomy and Soil Science* 49:623–629.

Randhawa, N.S., and I.P. Abrol. 1990. Sustaining agriculture: The Indian scene. In: C.A. Edwards, R. Lal, P. Madden, R.H. Miller, and G. House (eds.), *Sustainable Agriculture Systems*. Soil and Water Conservation Society, Ankeny, IA, pp. 438–450.

Rasul, G., and G.B. Thapa. 2003. Shifting cultivation in the mountains of south and south-east Asia: Regional patterns and factors influencing the change. *Land Degradation and Development* 14(5):495–508.

Raut, N., P. Dorsch, B.K. Sitaula, and L.R. Bakken. 2012. Soil acidification by intensified crop production in south Asia results in higher $N_2O/(N_2 + N_2O)$ product ratios of denitrification. *Soil Biology and Biochemistry* 55:104–112.

Ronald, P. 2011. Plant genetics, sustainable agriculture and global food security. *Genetics* 188:11–20.

Ruttan, V.W. 1999. The transition to agricultural sustainability. *Proceedings of the National Academy of Sciences of the United States of America* 96:5960–5967.

Sandor, J.A., and N.S. Eash. 1995. Ancient agricultural soils in the Andes of southern Peru. *Soil Science Society of America Journal* 59:170–179.

Schreier, H., S. Brown, P.B. Shah, B. Shrestha, G. Nakarmi, and R. Allen. 2001. Human interactions in soil and geomorphic processes in Nepal: The role of soil fertility in degradation and rehabilitation processes. *International Journal of Applied Earth Observation and Geoinformation* 3(1):93–98.

Schwab, G.O., D.D. Fangmeier, W.J. Elliot, and R.K. Frevert. 1993. *Soil and Water Conservation Engineering* (4th Edition). John Wiley & Sons Inc., New York, pp. 452–468.

Sharma, S.N., and S.K. Sharma. 2004. Role of crop diversification and integrated nutrient management in resilience of soil fertility under rice–wheat cropping system. *Archives of Agronomy and Soil Science* 50:345–352.

Sherchan, D.P., and G.B. Gurung. 1998. Effect of Long Term Application of Chemical Fertilizer and Manure on Crop Yields and Soil Chemical Properties Under Rice–Wheat Cropping Pattern (The Results of 8 Cropping Cycles). PAC Technical Paper No. 183. Pakhribas Agricultural Centre, Dhankuta, Nepal, 20 pp.

Sherchan, D.P., G.B. Gurung, and B.D. Gurung. 1997. Productivity Assessment of Red Clay Soils with Legume and Cereal Combinations on Dry, Rainfed Hill Environments in Nepal. PAC Technical Paper No. 180. Pakhribas Agricultural Centre, Dhankuta, Nepal.

Sherchan, D.P., and K.B. Karki. 2006. Plant nutrient management for improving crop productivity in Nepal. Improving plant nutrient management for better farmer livelihoods, food security and environmental sustainability. Proceedings of a Regional Workshop, Beijing, December 12–16, 2005, FAO RAP Publ. 2006/27, pp. 41–57.

Sherchan, D.P., C.J. Pilbeam, and P.J. Gregory. 1999. Response of wheat–rice and maize/millet systems to fertilizer and manure applications in the mid-hills of Nepal. *Experimental Agriculture* 35:1–13.

Sohi, S.P. 2012. Carbon storage with benefits. *Science* 338:1034–1035. Published online by AAAS, http://www.sciencemag.org/cgi/collection/ecology, accessed January 10, 2013.

Sombroek, W.G., F.O. Nachtergaele, and A. Hebel. 1993. Amounts, dynamics and sequestering of carbon in tropical and subtropical soils. *Ambio* 22:417–426.

Tacio, H.D. 1993. Sloping agricultural land technology: A sustainable agroforestry scheme for the uplands. *Agroforestry Systems* 22(2):145–152.

Thapa, G.B., and G.S. Paudel. 2002. Farmland degradation in the mountains of Nepal: A study of watershed 'with' and 'without' external interventions. *Land Degradation and Development* 13:479–493.

The Royal Society. 2009. *Reaping the Benefits: Science and the Sustainable Intensification of Global Agriculture*. Royal Society, London, 72 pp.

Thomas, W.L., C.O. Sauer, M. Bates, and L. Mumford (eds.). 1956. *Man's Role in Changing the Face of the Earth*. Chicago University Press, Chicago, IL.

Thorne, P.J., and J.C. Tanner. 2002. Livestock and nutrient cycling in crop–animal systems in Asia. *Agricultural Systems* 71:111–126.

Tilman, D., C. Balze, J. Hill, and B.L. Befort. 2011. Global food demand and the sustainable intensification of agriculture. *PNAS* 108(50):20260–20264.

Tiwari, K.R., B.K. Sitaula, R.M. Bajracharya, and T. Borresen. 2009. Effects of soil and crop management practices on yields, income and nutrient losses from upland farming systems in the Middle Mountains region of Nepal. *Nutrient Cycling in Agroecosystems* 86:241–253. Published online, http://link.springer.com/article/10.1007/s10705-009-9289-0.

Tripathi, B.P., and J.K. Tuladhar. 2000. Sustaining wheat productivity and maintaining soil fertility in maize–wheat system. *Nepal Agriculture Research Journal* 4:31–36.

Tripathi, J., A.P. Regmi, M.R. Bhatta, U. Adhikari, and N. Shakya. 2005. Zero Tillage: A Resource Conserving Technology. NARDF 233/2004/05. Nepal Agric. Res. Council, National Wheat Crop Research Program. Bhairahwa, Rupandei, Nepal, 26 pp.

Turner, B.L., and A.M. Shajaat Ali. 1996. Induced intensification: Agricultural change in Bangladesh with implications for Malthus and Boserup. *Proceedings of the National Academy of Sciences of the United States of America* 96:14984–14991.

Wan, S., and Y. Kang. 2006. Influence of drip-irrigation frequency on radish (*Raphanus sativus* L.) growth and water use. *Irrigation Science* 24:161–174.

West, P.C., H.K. Gibbs, C. Monfreda et al. 2010. Trading carbon for food: Global comparison of carbon stocks vs. crop yields on agricultural land. *PNAS* 107(46):19645–19648.

Xu Y., W. Chen, and Q. Shen. 2007. Soil organic carbon and nitrogen pools impacted by long-term tillage and fertilization practices. *Communications in Soil Science and Plant Analysis* 38:347–357.

Zhu, B., M. Gao, G. Liu, J. Zhang, S. Chen, and X. Zhang. 2002. Soil degradation and soil productivity restoration and maintenance in hilly land of southern China. *Archives of Agronomy and Soil Science* 48:311–318.

7 Long-Term Effects of Different Fertilizer Management Practices on Soil Organic Carbon Pool in Smallholder Farms of the Huang–Huai–Hai Plains, China

Xiangbin Kong, Rattan Lal, Baoguo Li, Lei Han, Hongjun Lei, Kejiang Li, and Youlu Bai

CONTENTS

7.1 INTRODUCTION

The Huang–Huai–Hai plains (HHH) is the principal wheat (*Triticum aestivum*)–maize (*Zea mays*) production area in China and is characterized by high chemical fertilizer inputs since the 1980s. Data from eight long-term experiments for monitoring soil organic carbon (SOC) changes under different fertilizer managements over time were collated to study the trends in SOC for the following six treatments: (i) control without inorganic or organic fertilizers (CK); (ii) chemical fertilizers used as N, P, and K without organic manure (UF); (iii) combined application of NPK without organic manure (CF); (iv) retention of wheat and maize straw along with the use of animal manure (O); (v) combined use of chemical fertilizers and organic manure (FO); and (vi) combined application of chemical fertilizers N, P, or K separately and organic fertilizers (UFO). The data indicated the following: (i) The mean SOC pool at 0–20 cm depth was 18.9 ± 1.8, 19.7 ± 1.7, 20.8 ± 2.1, 20.3 ± 2.5, 23.4 ± 7.2, and 24.3 ± 6.6 Mg ha^{-1} for CK, UF, CF, O, UFO, and CFO managements, respectively, and was in the order of CFO > UFO > O > CF > UF > CK. (ii) In comparison with CK, the SOC pool increased with all fertilizer management treatments, and the mean absolute SOC pool change (ΔSOC over this period) was −0.5, 0.8, 1.9, 1.6, 4.8, and 5.4 Mg ha^{-1} for CK, UF, CF, O, UFO, and CFO treatments, respectively. (iii) The highest rate of increase in SOC pool was measured in CFO, and the maximum rate (MR) of increase ranged from 224.4 to 1969.7 kg ha^{-1} year^{-1}, and the 23 years average rate (AR) ranged from 74.0 to 1529.6 kg ha^{-1} year^{-1}. (iv) The trend line for UFO and CFO within all the experimental sites indicated an increasing trend, which maintained a constant/stable SOC pool for CK, UF, CF, and O fertilizer managements. (v) There was a threshold for CFO and UFO fertilizer managements; the maximum SOC pools for the XinjiA, XinjiB, and Xuzhou sites were 42.1, 34.2, and 26.4 Mg ha^{-1}, and the experimental durations to achieve these rates were 21.0, 41.9, and 15.7 years, respectively. (vi) The maximum SOC pools under UFO for the XinjiA, XinjiB, and Xuzhou sites were 42.8, 26.4, and 34.2 Mg ha^{-1}, and the experimental durations to achieve those SOC pools were 23.0, 20.0, and 15.8 years, respectively. The SOC concentration/pool in the root zone were strongly affected by the fertilizers and management.

Soils are important to the moderation of numerous ecosystem services such as net primary productivity, food and fiber production, water storage and quality, and biodiversity (Lal 2010a). SOC is an important determinant of soil quality and sustainable agriculture, and its quality and quantity influence soil fertility and crop production. The SOC pool in the root zone is influenced by soil fertility management. Most soils in managed ecosystems contain a lower SOC concentration/pool than those under natural environments because of the higher rates of mineralization accelerated by changes in soil temperature and moisture regimes, lower input of biomass C, and higher losses caused by accelerated erosion and leaching (Lal 2004). The SOC pool loss, attributed to historic land misuse and soil mismanagement, can be restored by conversion to a restorative fertilizer and adoption of recommended management practices (RMPs) (Dao 1998; Allmaras et al. 2000; Dao et al. 2002; Lal 2004). Increase in SOC pool by the adoption of RMPs has also been documented in China's croplands at national (Pan et al. 2010; Wang et al. 2010), provincial (Cao et al. 2003), and county scales (Kong et al. 2009), as is evidenced by data from long-term agro-experimental sites (Huang et al. 2006). Adoption of RMPs is important in enhancing the SOC pool, increasing crop productivity, and mitigating climate change. Thus, estimating the SOC sequestration potential under different fertilizer managements is a researchable priority, especially in the HHH agroecosystems of China.

The HHH is the primary wheat (*T. aestivum*)–maize (*Z. mays*) growing area in China. The region comprises ~16% of China's cropland (Lei et al. 2005). The HHH produces 60%–80% of China's wheat and 35%–40% of its maize (Kong et al. 2014), and is the most intensively cultivated region in the country. The average yield (kg ha^{-1}) increase between 1985 and 2009 was from 1582 to 5860 for wheat and 4492 to 5610 for maize. The present yields are 1.25 and 1.07 times the national average yield of wheat and maize in China. The dramatic increase in crop yields and production in the HHH is driven by intensive irrigation and the high rates of N, P, and K use since 1960, and averaging 1.66, 4.43, and 1.46 times the national average rates for China (Kong et al. 2014).

The increase in SOC since the 1980s was reported by Piao et al. (2009) and Huang and Sun (2006) in parts of China. However, the changing trends of SOC and the mechanism of increase across the entire HHH region are not completely understood. Thus, data from eight long-term experiments for monitoring SOC change under different soil fertility management treatments over time were collected and collated to determine the temporal trends in SOC pool. Therefore, the objective of this chapter is to evaluate the long-term effects of different fertilizer and management practices on the SOC pool of representative soils and biomes across the entire HHH region. These sites are representative of the smallholder farms widely distributed in the region.

7.2 MATERIALS AND METHODS

7.2.1 Study Area and SOC Analysis

The HHH plains, located in northern China, are formed by alluvial sediments deposited by three rivers (i.e., the Huang River or Yellow River, Huai River, and Hai River) (Figure 7.1). These are the largest plains and constitute an important agricultural region in China, covering 320,000 km^2, with 18.67 million hectares (Mha) of farmland and a population of 200 million (Liu et al. 2010). The region is characterized by the intensive use of irrigation and chemical fertilizers, and the predominant cropping system in the region is double cropping of winter wheat and summer maize.

The SOC concentration was measured from eight long-term experimental sites in which the plots were managed by the Ministry of Agriculture, People's Republic of China (Figure 7.1). The climate and soil properties of the study sites are shown in Table 7.1. The annual rainfall ranges from 461.9 to 837.3 mm, the annual cumulative temperature from 4874.0 to 5368.2 degree days, and the annual average temperature from 12.8°C to 14.6°C.

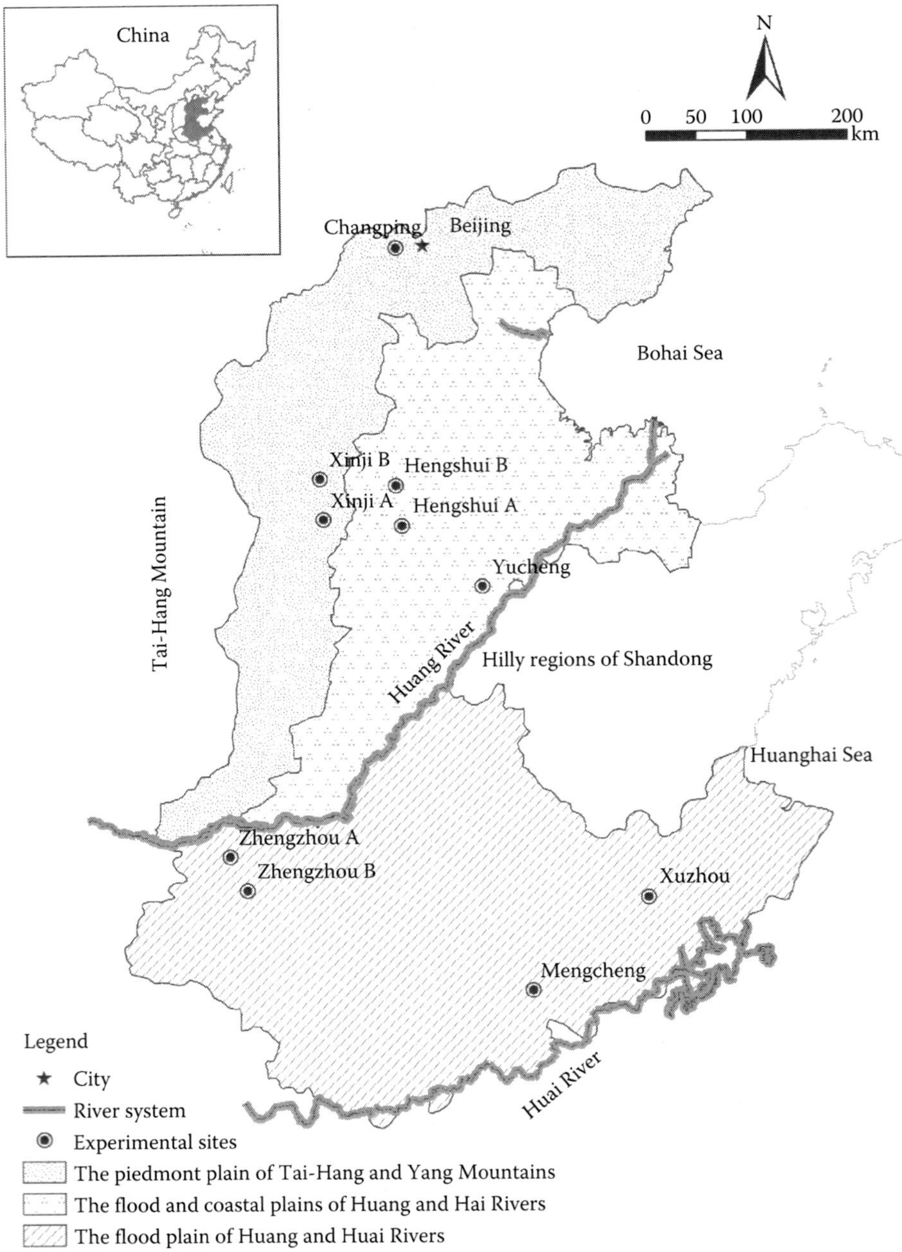

FIGURE 7.1 Location of eight long-term experimental sites.

TABLE 7.1
Soil and Climate Environments of Eight Long-Term Experiments across the HHH

Location/County	Latitude		Annual Rainfall (mm year^{-1})	Annual Accumulative Temperature (Degree Days)	Annual Average Temperature (°C)	Experiment Duration (Year)	Soil Properties		
	Latitude	Longitude					pH	Soil Texture	Bulk Density (Mg m^{-3})
Changping	40°02′	116°10′	574.1	4874.0	12.8	1984–1997	8.6	Silty clay loam	1.5
HenshuiA	37°42′	115°42′	478.1	4996.3	13.2	1979–2002	8.3	Loam	1.4
HenshuiB	37°43′	115°43′	478.1	4996.3	13.2	1979–2002	8.2	Loam	1.5
XinjiA	37°54′	115°13′	461.9	5015.7	13.2	1979–1999	8.3	Silt loam	1.4
XinjiB	37°55′	115°14′	461.9	5015.7	13.2	1979–1999	8.2	Silt loam	1.4
ZhengzhouA	34°46′	113°40′	623.2	5334.0	14.4	1980–2000	8.1	Loam	1.5
ZhengzhouB	34°47′	113°41′	623.2	5334.0	14.4	1990–1999	8.3	Loam	1.5
Xuzhou	33°54′	117°57′	837.3	5368.2	14.6	1980–1987	8.3	Sand loam	1.4

Three soil samples for assessing the SOC concentration from one treatment were obtained for 0–20 cm depth from all sites before the harvest of wheat every year. The SOC concentration was determined by the wet combustion method (Tiessen and Moir 1993), the bulk density by the core method (Soil Survey Staff 1999), and the texture by the hydrometer method. Soil texture ranged from silty clay loam, loam, silt loam, to loam (Soil Survey Staff 1999). Crop rotation involved winter wheat and summer maize in XinjiA, XinjiB, ZhengzhouA, ZhengzhouB, HengshuiA, HengshuiB, and Xuzhou, compared with winter wheat from 1985 to 1986, winter wheat–summer maize and spring maize after 1987 in Changping and Beijing.

7.2.2 Experimental Design and Management

The eight long-term experiments had different fertilizer managements, and the rates of application of chemical fertilizer (N, P, K) and organic manure in these treatments (Table 7.2) were designed according to the local household (small landholder) fertilizer managements and the soil nutrients varying across different soil types in the HHH. Experimental plots varied from 1 to 3 for every treatment across the long-term experiments; however, the plot area varied from 40 to 100 m^2 (Guo et al. 1998; Zhang et al. 2002; Zhang et al. 2005; Xia et al. 2008; Lin et al. 2009).

All farm operations were similar except for fertilizer management for different treatments described above. Winter wheat was irrigated two to three times, and maize was irrigated one to two times depending on the precipitation distribution during the growing season. The volume of water used for each irrigation was 900 (9 cm) to 12,00 (12 cm) m^3 ha^{-1}. Herbicides and pesticides were applied to control weeds and reduce insect pressure, respectively. Organic and chemical fertilizers, including P and K, were applied as basal dose; one-third to two-third of N was applied as basal

TABLE 7.2
Application of Chemical Fertilizer and Organic Manures under Different Fertilizer Management Treatments

Fertilizer Managements	C (kg ha^{-1})	N (kg ha^{-1})	P_2O_5 (kg ha^{-1})	K_2O (kg ha^{-1})
CK	0	0	0	0
UF	0	135–360	0	0
	0	0	60–150	0
	0	0	0	60–150
CF	0	135–360	60–150	60–150
O	500–4050	12–120	12–170	30–200
UFO	500–4050	147–480		
	500–4050		72–320	
	500–4050			90–350
CFO	500–4050	147–480	72–320	90–350

Note: The application amount of fertilization under different treatments was calculated and synthesized on the basis of the experimental design at the eight experimental sites.

dose and the other part was top dressed for wheat; and all chemical fertilizers were top dressed for maize (Guo et al. 1998; Zhang et al. 2002; Zhang et al. 2005; Xia et al. 2008; Lin et al. 2009).

7.2.3 Data Analyses and Calculations

For the purpose of macro-level studies, all treatments are grouped into six fertilizer managements based on >76 different treatments at eight experimental sites across the HHH (Table 7.2): (i) no chemical fertilizer or organic manure application as the control treatment (CK); (ii) chemical nitrogen (N), phosphorus (P), and potassium (K) fertilizers applied separately and without organic manure (UF); (iii) combined application of chemical fertilizer N, P, and K without organic manure (CF); (iv) wheat and maize straw retention or use of manures including that from soybean (*Glycine max*) cake, chicken, horse, and cow dung or manures only (O); (v) combined application of chemical fertilizer N, P, and K and organic fertilizers (CFO); and (vi) combined application of chemical fertilizer N, P, or K separately and organic fertilizers (UFO).

The different fertilizers used were grouped into chemical fertilizers and organic manure; however, the amounts of chemical fertilizer and organic manure applied were not the same for the six categories described above. The application rates ranged from 90 to 360 kg ha^{-1} for N, from 60 to 150 kg ha^{-1} for P, and from 82.5 to 250 kg ha^{-1} for K. The amount of organic fertilizers used also varied among the eight experimental sites. Organic manure fertilizers were converted into C, N, P, and K, according to the ratio of C/N and C/P for different organic manures (Lei et al. 2005). Nutrient application rate ranged from 12 to 120 kg ha^{-1} of N, from 12–170 kg ha^{-1} of P, and from 30 to 200 kg ha^{-1} of K as organic fertilizers among all the experimental sites.

The data on SOC concentration (Table 7.3) were normalized and converted to SOC pool (Li et al. 2007) by using Equation 7.1:

$$\text{SOC pool (Mg ha}^{-1}) = \text{SOC (g kg}^{-1}) \times 10^4\ \text{m}^2\,\text{ha}^{-1} \times 0.2\ \text{m} \times \text{SBD (Mg m}^{-3}) \times 10^{-3} \qquad (7.1)$$

TABLE 7.3
Descriptive Statistics of SOC in Different Fertilizer Management Treatments

		SOC (Mg ha^{-1})			
Fertilizer Treatments	**N**	**Mean**	**Standard Deviation**	**Minimum**	**Maximum**
CF	204	7.2^{b}	0.6	5.5	8.6
O	107	7.2^{b}	0.8	5.5	10.0
UF	172	6.9^{ab}	0.5	5.7	8.1
CFO	282	8.5^{c}	2.4	6.0	20.6
UFO	238	8.3^{c}	2.6	5.5	19.4
CK	110	6.6^{a}	0.6	5.2	8.0
Total	1113	7.7	1.9	5.2	20.6

Note: Means followed by different letters differ from each other at the 5% level of probability.

where the SOC concentration is in g kg^{-1} and soil bulk density (SBD) is in Mg m^{-3}. The change in SOC pool in treatments was computed with reference to the control (CK) by using Equation 7.2:

$$\Delta SOC\ (Mg\ ha^{-1}) = Tr\ (Mg\ ha^{-1}) - CK\ (Mg\ ha^{-1}) \tag{7.2}$$

where ΔSOC is the change in SOC pool under a specific treatment (Tr) in comparison with that under CK. The maximum rate (MR) and average rate (AR) of SOC change were calculated using Equations 7.3 and 7.4:

$$AR = (kg\ ha^{-1}\ year^{-1}) = (SOC_f - SOC_i)/t \tag{7.3}$$

$$MR = (kg\ ha^{-1}\ year^{-1}) = (SOC_m - SOC_i)/t \tag{7.4}$$

where SOC_i is SOC pool at the beginning of the experiment and SOC_f is the final SOC pool after t years of experiment, and SOC_m is the maximum SOC pool during the experimental period.

7.2.4 Data Processing and Statistics

Data were organized into CK, O, CF, UF, CFO, and UFO fertilizers and management treatments among the eight different sites. The SOC pools for different treatments were computed into mean and standard deviation ($\overline{x} \pm \overline{sd}$), and the SOC pool trend line for different treatments was computed using Microsoft Excel 2007. The significance of differences at $P < 0.05$ and $P < 0.01$ between fertilizers and management in different long-term experiment sites were tested using SPSS (version 13.0).

7.3 RESULTS AND DISCUSSIONS

7.3.1 SOC Pool Change under Different Fertilizer Managements across HHH

The mean SOC pool in 0–20 cm depth was 18.9 ± 1.8, 19.7 ± 1.7, 20.8 ± 2.1, 20.3 ± 2.5, 23.4 ± 7.2, and 24.3 ± 6.6 Mg ha^{-1} for CK, UF, CF, O, UFO, and CFO fertilizer management treatments, respectively (Table 7.4). The numbers of observations (n) for different treatments were 204 for CF, 107 for O, 172 for UF, 282 for CFO, 238 for UFO, and 110 for CK. The SOC pool for all fertilizer and managements differed significantly from that of CK, and that for CF, UF, and O from those of CFO and UFO fertilizer and managements. The maximum SOC pool was observed for the CFO, and the least for the CK. The SOC pool was in the order of CFO > UFO > O > CF > UF > CK.

The changes in SOC pool for diverse fertilizer managements can be organized into three groups. First is the baseline SOC pool for CK in arable land without any fertilizer input. Second is the high SOC pool for CFO and UFO treatments with combined use of organic manure and chemical fertilizer. Third is the middle range of SOC pool for UF, CF, and O fertilizer managements. The SOC pool in fertilizer managements with a middle range can be increased to those of the CFO and UFO levels with the adoption of improved management.

TABLE 7.4
Statistical Analysis of SOC Pool and Its Increase for Eight Long-Term Experimental Sites

		SOC Pool (Mg ha^{-1})				ΔSOC Pool (Mg ha^{-1})		
Fertilizer Management	**N**	**Mean**	**Standard Deviation**	**Minimum**	**Maximum**	**Mean**	**Minimum**	**Maximum**
CF	204	20.8[b]	2.1	16	25.1	1.8[b]	−3.4	13.1
O	107	20.3[b]	2.5	15.3	28.1	1.6[b]	−2.3	6.5
UF	172	19.7[ab]	1.7	15.8	23.4	0.8[b]	−3.6	4.2
CFO	282	24.3[c]	6.6	16.8	57.1	5.4[c]	−1	37.3
UFO	238	23.4[c]	7.2	15.3	53.9	4.8[c]	−3.1	33.4
CK	110	18.9[a]	1.8	14.6	23.0	(−)0.5[a]	−3.8	3.8
Total	1113	21.8	5.3	14.6	57.1	3.0	−3.8	37.3

Note: Means followed by different letters differ from each other at the 5% level of probability.

7.3.2 SOC Pool Changes under Different Fertilizer Managements Compared with CK

The data in Table 7.4 show changes in SOC pool for different fertilizer and management treatments compared with CK, and that of CK compared with the antecedent SOC pool of arable land. In general, the mean SOC pool increased by 3.0 Mg ha^{-1}, with the maximum and minimum change from 37.3 to –3.8 Mg ha^{-1}, respectively. Of the total of 1113 observations, 1013 (90%) indicated an increase and 110 (10%) a decrease in the SOC pool.

The magnitude of change in different fertilizer managements vis-à-vis CK differed widely. All fertilizer managements contained significantly more SOC pool compared with that of CK. Furthermore, significant differences in the magnitude of increase in SOC pool were observed for UFO, CFO, and also for O, UF, and CF fertilizer management treatments. However, no significant difference in SOC pool occurred between CFO and UFO, and among O, UF, and CF treatments. The mean ΔSOC were −0.5, 0.8, 1.8, 1.6, 4.8, and 5.4 Mg ha^{-1} for CK, UF, CF, O, UFO, and CFO, respectively. The highest SOC sequestration was observed for CFO, and its magnitude was in the order of CFO > UFO > CF > O > UF > CK.

The trend line was computed to estimate the SOC sequestration potential for different fertilizer managements in the HHH region between 1978 and 2003. The rates of SOC sequestration estimated in this region differ from those reported by Pan et al. (2010) and Wu and Cai (2007), and are discussed below for each treatment.

7.3.2.1 CK

The change in SOC pool with reference to the initial level was −0.5 Mg ha^{-1}. Of the 110 measurements in CK, the maximum and minimum change in SOC pool was 3.8 and –3.8 Mg ha^{-1}, respectively. These results of ΔSOC pool in CK are similar to those reported by Wu and Cai (2007). The baseline of SOC pool in arable land was 18.9 ± 1.8 Mg ha^{-1} in the HHH. Of all the experimental sites, the SOC pool increased only in the XinjiA and ZhengzouB sites, and decreased in all other experimental sites. Therefore, the SOC pool without fertilizer inputs in the HHH remained at a low level between 1978 and 2003. These trends in the SOC pool in cropland soils in the HHH ecosystem are similar to those of cultivated soils of South Asia and sub-Saharan Africa where the attendant decline in soil quality is among the important causes of low and declining agronomic production (Lal 2010b). The trend of SOC pool for CK fertilizer management is suboptimal and good neither for achieving food security nor mitigating the climate change or improving the environment.

7.3.2.2 O

The mean increase in SOC pool is 1.6 Mg ha^{-1}. This result is similar to an increase of 2.1–2.9 g kg^{-1} reported by Ma et al. (2010). The mean range of increase in SOC pool was lower than those for CFO and UFO. The magnitude of increase in SOC pool with the use of organic manure was less than that for the combined use of inorganic fertilizer and organic manure. The maximum and minimum increases were 6.5 and 2.3 Mg ha^{-1}, respectively. The maximum increase of 2.4 Mg ha^{-1} was observed in HenshuiB. The results show that the application of organic manures has similar effects on improving the SOC pool at different experimental sites. The use of organic manures has been widely practiced across China for millennia (Bi et al. 2009). This traditional method of improving soil fertility without chemical fertilizer application has supported the large population in China and maintained soil quality (especially the SOC pool). The present data show that the beneficial impact of using organic manure can be enhanced when used in combination with chemical fertilizers.

7.3.2.3 UF

The impact of using N, P, and K fertilizers separately changed the mean SOC pool by 0.8 Mg ha^{-1}. The maximum and minimum changes in SOC pool were 4.2 and −3.6 Mg ha^{-1}, respectively. While the SOC pool increased in comparison with CK, the magnitude of change was lower than that for O, CF, UFO, and CFO. The impact of chemical fertilizer separately on the SOC pool in the HHH is a debatable issue. Zhang et al. (2002) reported that the use of chemical fertilizer merely maintained the SOC pool. In contrast, the results of the present study show that the impact of using chemical fertilizer separately on the SOC pool was not as strong as that of using combined chemical fertilizer N, P, and K in combination with organics.

7.3.2.4 CF

The average increase in SOC pool through the use of combination of chemical fertilizer was 1.8 Mg ha^{-1}. The maximum and minimum changes in SOC pool were 13.1 and −3.4 Mg ha^{-1}, respectively. In comparison with CK as the baseline, the difference in ΔSOC pool between CF and UF was about 1 Mg ha^{-1}. Thus, use of combined fertilizer is preferred.

7.3.2.5 CFO

The average increase in SOC pool was 5.4 Mg ha^{-1}. The maximum and minimum changes were 37.3 and −1.02 Mg ha^{-1}, respectively. The CFO treatment registered the maximum increase in SOC pool compared with other treatments. Use of inorganic fertilizers in combination with organic amendments is the best strategy to increase the SOC pool for the principal crop rotation of winter wheat and summer maize in the HHH. The straw/grain ratio is 0.8 and 1.0 for summer maize and winter wheat, respectively (Lei et al. 2005). The HHH has a high agronomic production. Thus, recycling of crop residues and a judicious use of fertilizers are the effective techniques to increase the SOC pool. Indeed, application of fertilizer has increased the SOC pool in the HHH (Piao et al. 2009).

7.3.2.6 UFO

The average increase in SOC pool in the UFO treatment was 4.8 Mg ha^{-1}. The maximum and minimum changes were 33.4 and −3.1 Mg ha^{-1}, respectively. The magnitude of increase in SOC pool in UFO was more than those for UF, CF, and O. These results are in accordance with the survey of SOC change (1980–2006) conducted in Daxing County (Kong et al. 2009).

The technical potential of SOC sequestration in the UFO, CF, O, UF, and CK fertilizer managements was 0.91, 3.46, 3.92, 4.6, and 5.4 Mg ha^{-1}, respectively. On the basis of fertilizer use in arable land (Deng et al. 2010), the total technical potential of SOC sequestration with improved fertilizer management for UFO, CF, O, UF, and CK in the HHH was 22.3, 87.4, 89, 116, and 136 Tg C during 23 years.

Thus, there is a strong correlation between the SOC concentration/pool in the root zone and the application of fertilizers and of organic manures. The combined use of inorganic fertilizers and organic manures is the best strategy to maintain/enhance the SOC pool in the HHH.

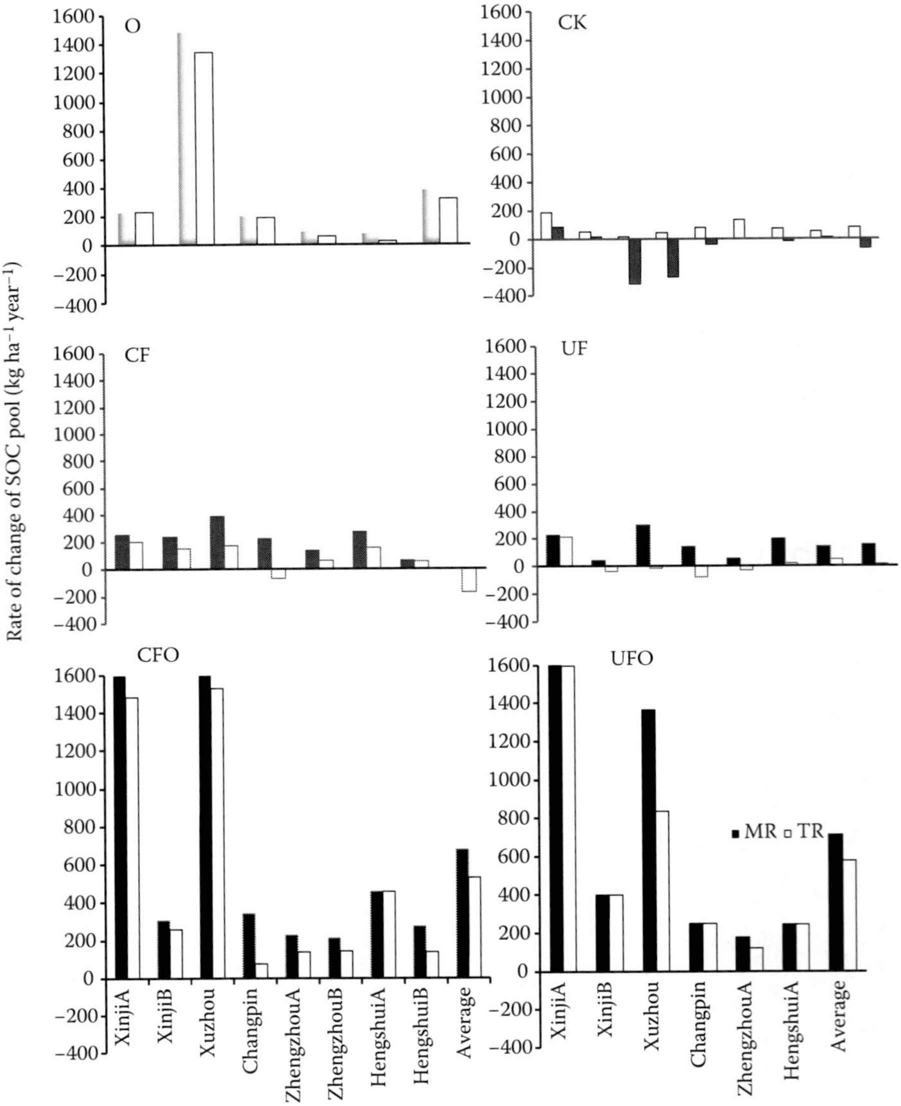

FIGURE 7.2 Mean rate (MR) and average rate (AR) of change of SOC pool at different sites and different fertilizer managements.

7.3.3 Rate of Change of SOC Pool under Different Fertilizer Management Treatments

There are no definite trends in the rate of change in SOC pool, which had both maximum and minimum points. Thus, both the maximum rate (MR) and the average rate (AR) of change in SOC pool are used to establish any trends among different treatments (Figure 7.2). Both MR and AR were in the order of CFO > UFO > O > CF > UF > CK. These results are in accordance with those reported by Pan et al. (2010) and Wu and Cai (2007).

The range of change is 52.0 to 188.0 kg ha^{-1} $year^{-1}$ for MR, and −315.5 to 84.1 kg ha^{-1} $year^{-1}$ for AR in comparison with the CK treatment. The range of increase in SOC pool (kg ha^{-1} $year^{-1}$) for MR and AR, respectively, is 57.4 to 299.4 and −83.1 to 214.0 for UF, 1.9 to 389.7 and −166.3 to 203.2 for CF, 76.3 to 1480.9 and 24.0 to 1345.6 for O, 176.3 to 1823.3 and 119.2 to 1594.3 for UFO, and 224.4 to 1969.7 and 74.0 to 1529.6 for CFO. The rate of change of SOC pool showed an increasing trend over time for all the experimental sites. The analysis of MR and AR for the SOC pool indicated a nonlinear trend exhibiting both the minima and maxima. Thus, the AR of the change of SOC pool is not a good indicator of the instantaneous rate of change.

With CK as the baseline, the change in the SOC pool between 1978 and 2003 was strong for UFO and CFO, with a gradual increase for O but no definite trend for CF. These data indicate that the SOC pool in cropland cannot be increased continuously by the application of chemical fertilizers alone. Maintenance and increase of SOC pool in the HHH depend on the combined use of organic manures and fertilizers.

7.3.4 Change in SOC Pool under Different Fertilizer Managements over Long-Term Experiments

Regression equations indicating the change in SOC pool over time under different fertilizers and management treatments during the experimental period are shown in Figure 7.3. The trend of change in SOC pool can be organized into three groups. Group 1 consists of CK, which maintained its SOC pool at almost 20.0 Mg C ha^{-1} between 1978 and 2003. This is the minimum stable baseline for the SOC pool in cropland in the HHH. Group 2 consists of CF, UF, and O, which exhibited a slight change in the SOC pool compared with that of CK: those of O and CF increased and that of UF decreased only slightly. Group 3 consists of CFO and UFO in which the SOC pool increased from 20.0 to 50.0 Mg ha^{-1} (2.5 times or 150% increase) during the experimental period. Therefore, regular application of organic and inorganic fertilizers can substantially increase the SOC pool and also sustain it at a high level.

Regression equations of trend lines for different fertilizer management treatments are shown in Table 7.5. A linear trend of change in SOC pool over time is observed in CK, UF, CF, and O, compared with a polynomial function for CFO and UFO (Table 7.4).

There was a threshold value of SOC pool for CFO and UFO treatments. The maximum SOC pools for XinjiA, XinjiB, and Xuzhou were 42.1, 34.2, and 26.4 Mg ha^{-1} along with the durations to attain the maximum of 21.0, 41.9, and 15.8 years, respectively. The maximum SOC pools under UFO for the XinjiA, XinjiB,

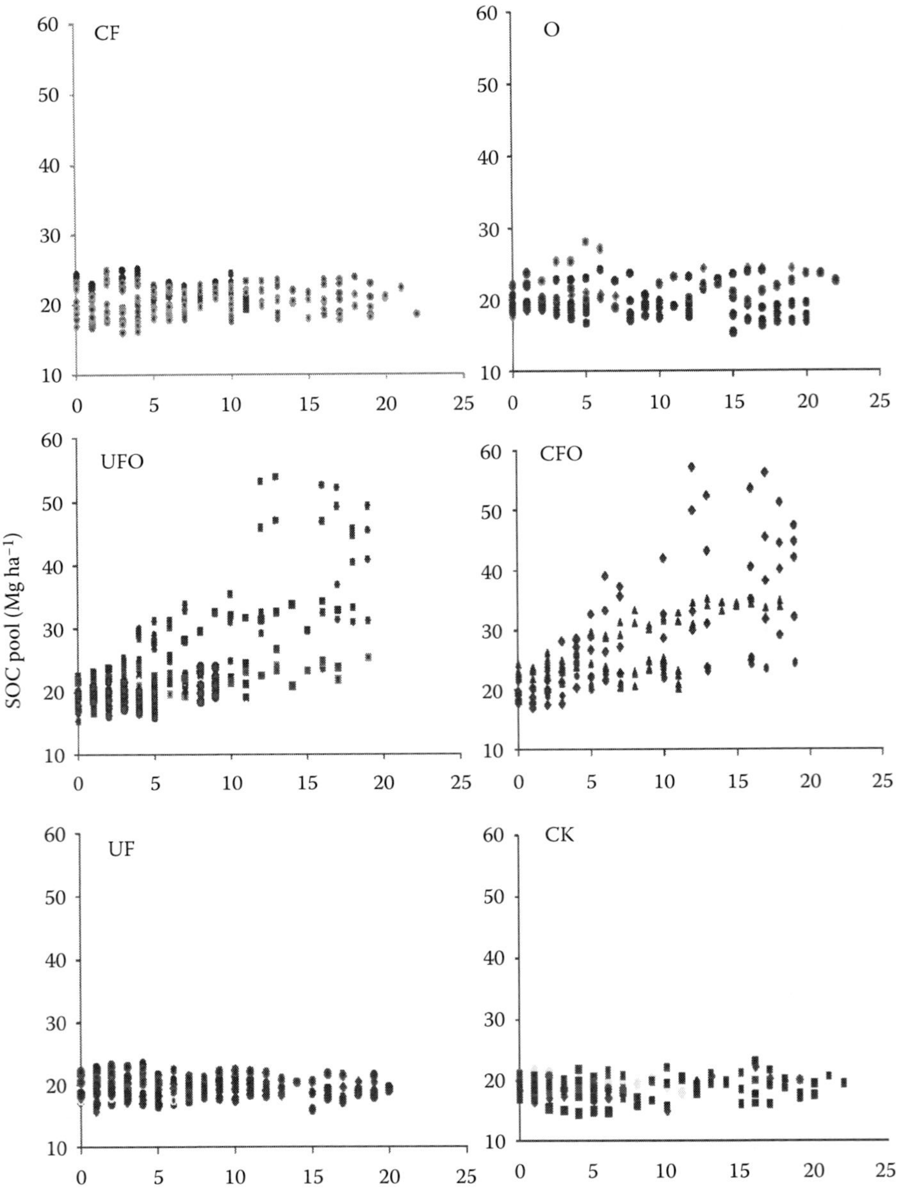

FIGURE 7.3 Temporal change in SOC pool in different fertilizer managements.

and Xuzhou sites were 42.8, 26.4, and 34.2 Mg ha^{-1} along with the durations to achieve the maximum of 23.0, 20.0, and 15.8 years, respectively. The data show that the duration of amendment use was also important for SOC sequestration, and the continuous application of fertilizers and organic manures is the best strategy to increase the SOC pool. Similar results were reported by Wang et al. (2010).

TABLE 7.5
Trend Line Regression Equations of Temporal Changes in SOC Pool for Different Fertilizer Managements

Treatments	Region	Trend Model	R^2	Threshold	
				Years	SOC Pool (Mg ha^{-1})
CF	XinjiA	$y = 0.22x + 17.44$	$R^2 = 0.56$***	20.00	21.80
	XinjiB	$y = 0.08x + 18.54$	$R^2 = 0.12$	20.00	20.20
	Xuzhou	$y = -0.09x + 18.33$	$R^2 = 0.04$	0.00	18.33
	Changping	$y = -0.21x + 23.46$	$R^2 = 0.31$***	0.00	23.46
	ZhengzhouA	$y = -0.01x + 22.48$	$R^2 = 0.00$	0.00	22.48
	ZhengzhouB	$y = 0.04x + 20.20$	$R^2 = 0.00$	8.00	20.53
	HenshuiA	$y = 0.07x + 18.23$	$R^2 = 0.17$	21.00	19.70
	HenshuiB	$y = -0.02x + 20.58$	$R^2 = 0.02$	0.00	20.58
O	XinjiB	$y = 0.30x + 18.84$	$R^2 = 0.82$***	20.00	24.94
	Xuzhou	$y = -0.25x^2 + 3.01x + 18.26$	$R^2 = 0.94$**	6.08	27.41
	ZhengzhouA	$y = 0.10x + 21.37$	$R^2 = 0.10$	18.00	23.11
	HenshuiA	$y = -0.06x + 18.89$	$R^2 = 0.12$	0.00	18.89
	HenshuiB	$y = 0.07x + 20.75$	$R^2 = 0.12$	23.00	22.38
UF	XinjiA	$y = 0.19x + 17.10$	$R^2 = 0.60$***	20.00	20.82
	XinjiB	$y = 0.036x + 17.43$	$R^2 = 0.07$	20.00	18.14
	Xuzhou	$y = -0.17x + 18.64$	$R^2 = 0.21$	0.00	18.64
	Changping	$y = -0.20x + 22.05$	$R^2 = 0.32$**	0.00	22.05
	ZhengzhouA	$y = -0.07x + 22.33$	$R^2 = 0.19$	0.00	22.33
	ZhengzhouB	$y = -0.09x + 19.92$	$R^2 = 0.04$	0.00	19.92
	HenshuiA	$y = 0.02x + 18.35$	$R^2 = 0.02$	21.00	18.77
CFO	XinjiA	$y = -0.06x^2 + 2.42x + 16.72$	$R^2 = 0.59$**	21.00	42.09
	XinjiB	$y = -0.0037x^2 + 0.31x + 19.94$	$R^2 = 0.60$**	41.88	26.42
	Xuzhou	$y = -0.06x^2 + 1.74x + 20.63$	$R^2 = 0.92$***	15.68	34.24
	Changping	$y = -0.01x^2 + 0.05x + 23.15$	$R^2 = 0.02$	2.90	23.23
	ZhengzhouA	$y = 0.09x + 22.24$	$R^2 = 0.18$*	18.00	23.91
	ZhengzhouB	$y = -0.03x^2 + 0.48x + 21.75$	$R^2 = 0.29$*	8.64	23.82
	HenshuiA	$y = 0.24x + 18.45$	$R^2 = 0.42$***	21.00	23.57
	HenshuiB	$y = 0.10x + 20.75$	$R^2 = 0.19$*	23.00	22.98
UFO	XinjiA	$y = -0.05x^2 + 2.32x + 16.18$	$R^2 = 0.64$**	22.95	42.83
	XinjiB	$y = 0.36x + 17.98$	$R^2 = 0.82$***	20.00	25.15
	Xuzhou	$y = -0.05x^2 + 1.51x + 21.00$	$R^2 = 0.88$***	15.75	32.91
	Changping	$y = -0.06x + 22.23$	$R^2 = 0.04$	0.00	22.28
	ZhengzhouA	$y = 0.08x + 21.94$	$R^2 = 0.13$	18.00	23.39
	HenshuiA	$y = 0.20x + 18.50$	$R^2 = 0.14$***	21.00	22.75

(continued)

TABLE 7.5 (Continued)
Trend Line Regression Equations of Temporal Changes in SOC Pool for Different Fertilizer Managements

Treatments	Region	Trend Model	R^2	Threshold Years	Threshold SOC Pool ($Mg\ ha^{-1}$)
CK	XinjiA	$y = 0.21x + 16.51$	$R^2 = 0.51$**	20.00	20.76
	XinjiB	$y = 0.09x + 18.51$	$R^2 = 0.41$*	20.00	20.35
	Xuzhou	$y = -0.42x + 16.83$	$R^2 = 0.73$*	0.00	16.83
	Changping	$y = -0.23x + 21.21$	$R^2 = 0.54$*	0.00	21.21
	ZhengzhouA	$y = 0.03x + 20.37$	$R^2 = 0.00$	18.00	20.89
	ZhengzhouB	$y = -0.05x + 18.91$	$R^2 = 0.00$	0.00	18.09
	HenshuiA	$y = -0.05x + 17.83$	$R^2 = 0.10$	0.00	17.83
	HenshuiB	$y = 0.01x + 19.05$	$R^2 = 0.01$	23.00	19.24

* Significance of the SOC pool trend at $P < 0.05$.
** Significance of the SOC pool trend at $P < 0.01$.
*** Significance of the SOC pool trend at $P < 0.001$.

7.4 CONCLUSIONS

The data presented support the following conclusions:

1. The SOC pool for CK was maintained at the stable baseline level between 1978 and 2003 at 18.9 ± 1.8 Mg ha^{-1}.
2. The SOC pool in other fertilizer and management treatments generally increased in comparison with that of CK for all experimental sites in the HHH. The average change in SOC pool was 0.8, 1.9, 1.6, 4.8, and 5.4 Mg ha^{-1} for UF, CF, O, UFO, and CFO, respectively. The increase in SOC pool was much higher in treatments involving the combined use of chemical fertilizers and organic manures compared with chemical fertilizer alone.
3. The rate of change of SOC pool varied over time; the MR of increase ranged from 224.4 to 1969.7 kg ha^{-1} $year^{-1}$ for CFO, and the AR from 74.0 to 1529.6 kg ha^{-1} $year^{-1}$.
4. The trend line of SOC pool over time for UFO and CFO for all experimental sites indicated a rapid increase throughout the experimental duration, while it maintained a steady-state (stable) level for CK, UF, CF, and O treatment.
5. There were threshold levels of SOC pool for CFO and UFO, and the maximum values for XinjiA, XinjiB, and Xuzhou sites were 42.1, 34.2, and 26.4 Mg ha^{-1} with the durations to attain these maximum of 21.0, 41.9, and 15.7 years, respectively. The maximum SOC pools under UFO for the XinjiA, XinjiB, and Xuzhou sites were 42.8, 26.4, and 34.2 Mg ha^{-1} with the corresponding experimental durations of 23.0, 20.0, and 15.8 years, respectively.

These data indicate that the increase in SOC concentrations and pools were strongly influenced by different fertilizers and management treatments across the whole HHH. The increase in the order of SOC was CFO > UFO > CF > O > UF > CK over a long-term duration, and SOC in most of the fertilizer managements increased over time. The results from eight long-term experiments at the field experimental level show that the increase in chemical fertilizer inputs strongly enhanced the SOC pool across HHH.

ACKNOWLEDGMENTS

This work was supported by the Special Fund for National Land Resource–Scientific Research from the National Ministry of Land and Resources, China (201011006-3), the Ministry of Science and Technology of China (grant no. 2008FY110600), and the Basic Research Development Program of China (973 program, no. 2009CB118607).

REFERENCES

Allmaras, R.R., Schomberg, H.H., Douglas Jr., C.L., Dao, T.H. 2000. Soil organic carbon sequestration potential of adopting conservation tillage in US croplands. *Soil Water Conserv. Soc. J.* 55:365–373.

Bi, L.D., Zhang, B., Liu, G.R. et al. 2009. Long-term effects of organic amendments on the rice yields for double rice cropping systems in subtropical China. *Agric. Ecosyst. Environ. J.* 129:534–541.

Cao, C.F., Kong, L.C., Wang, Z.S., Chen, A.P., Zhang, C.L., Chen, L.T. 2003. Effect of long-term trial on raising fertility of Shajiang Black Soil (in Chinese). *Chin. J. Soil Sci.* 34:102–105.

Dao, T.H. 1998. Tillage and crop residue effects on carbon dioxide evolution and carbon storage in a paleustoll. *Soil Sci. Soc. Am. J.* 62:250–256.

Dao, T.H., Stiegler, J.H., Banks, J.C., Boggle-Boerngen, L., Adams, B. 2002. Effects of post-contract fertilizations on soil carbon and nitrogen of conservation reserve grasslands. *Argon. J.* 94:146–152.

Deng, X.Z., Jian, G.Q., Lin, Y.Z., Han, J.Z. 2010. Simulation of the changes of soil organic carbon stock of cropland in China (in Chinese). *Geogr. Res. J.* 29(11):94–100.

Guo, L.P., Zhang, F.S., Wang, X.R., Xu, S.W. 1998. Effect of different fertilizer management on mineral nitrogen in Fluvo-aquic soil (in Chinese). *J. Chin. Agric. Univ.* 3(a):54–61.

Huang, B., Wang, J.G., Jin, H.Y., Xu, S.C. 2006. Effects of long-term application fertilizer on carbon storage in Calcareous Meadow Soil (in Chinese). *J. Agro-Environ. Sci.* 25:161–164.

Huang, Y., Sun, W.J. 2006. The soil C distribution change from 1980 to 2000 in China. *Sci. Press J.* 51:750–752.

Kong, X.B., Dao, T.H., Qin, J., Qin, H.Y., Li, C.Z., Zhang, F.R. 2009. Effects of soil texture and fertilization interactions on organic carbon in soils in North China cities' urban fringe. *Geoderma J.* 154:86–92.

Kong, X.B., Lal, R., Li, B.G. et al. 2014. Fertilizer intensification and its impacts in China's HHH plains. *Adv. Agron. J.* 145:135–169.

Lal, R. 2004. Soil carbon sequestration impacts on global climate change and food security. *Sci. J.* 304:1623–1627.

Lal, R. 2010a. Beyond Copenhagen: Mitigating climate change and achieving food security through soil carbon sequestration. *Food Sec. J.* 2:169–177.

Lal, R. 2010b. Enhancing eco-efficiency in agro-ecosystems through soil carbon sequestration. *Crop Sci. J.* 50:120–121.

Lei, H.J., Li, B.G., Bai, Y.L., Huang, Y.F., Li, K.J. 2005. Estimation of soil organic matter mineralization in the plough layer and CO_2 emission under different farming practices in China's Huang–Huai–Hai Plain. *J. Hebei Agric. Sci.* 11(1):60–63.

Li, Z.P., Han, F.X., Sub, Y. et al. 2007. Assessment of soil organic and carbonate carbon storage in China. *Geoderma J.* 138:119–126.

Lin, Z.A., Zhao, B.Q., Yuan, L., Hwat, B.S. 2009. Effects of organic manure and fertilizers long-term located application on soil fertility and crop yield (in Chinese). *Sci. Agric. Sin.* 42(8):2809–2819.

Liu, S.X., Mo, X.G., Lin, Z.H. et al. 2010. Crop yield responses to climate change in the Huang–Huai–Hai Plain of China. *Agric. Water Manage.* 97:1195–1209.

Ma, J.Y., Cao, C.Y., Zhang, C., Li, K.J., Ren, T.S. 2010. Effect of long-term application of chemical fertilizers and organic manure on soil organic carbon and bulk density (in Chinese). *China Soil Fertil. J.* 6:38–40.

Pan, G.X., Xu, X.W., Smith, P., Pan, W., Lal, R. 2010. An increase in topsoil SOC stock of China's croplands between 1985 and 2006 revealed by soil monitoring. *Agric. Ecosyst. Environ. J.* 136:133–138.

Piao, S., Fang, J.Y., Ciais, P., Peylin, P., Huang, Y. 2009. The carbon balance of terrestrial ecosystems in China. *Nat. J.* 458:1009–1010.

Soil Survey Staff. 1999. *Soil Taxonomy, A Basic System of Soil Classification for Making and Interpreting Soil Surveys*, 2nd ed.: USDA Natural Resources Conservation Service Agricultural Handbook, vol. 436. US Gov. Printing Office, Washington, DC.

Tiessen, H., Moir, J.O. 1993. Total and organic carbon. In: Carter, M.R. (Ed.), *Soil Sampling and Methods of Analysis*. Lewis Publishers, Boca Raton, FL, pp. 187–199.

Wang, C.J., Pan, G.X., Tian, Y.G., Li, L.Q., Zhang, X.H. 2010. Changes in cropland topsoil organic carbon with different fertilizations under long-term agro-ecosystem experiments across mainland China. *Sci. China J.* 53:858–867.

Wu, L.Z., Cai, Z.C. 2007. Estimation of the change of topsoil organic carbon of croplands in China based on long-term experimental data (in Chinese). *Ecol. Environ. J.* 16(6):1768–1774.

Xia, W., Liang, G.Q., Zhou, W., Wang, H., Wang, X.B., Sun, J.W. 2008. Adsorption and desorption characteristics of soil phosphorus in calcareous in fluvo-aquic soil under long-term fertilization (in Chinese). *Plant Nutr. Fertil. Sci.* 14(3):431–438.

Zhang, A.J., Zhang, M.P. 2002. Study on regularity of growth and decline of soil organic matter under long-term fertilization for yellow fluvo-aquic soil (in Chinese). *J. Anhui Agric. Univ. J.* 29(1):60–63.

Zhang, H., Chen, B.D., Jun, H. et al. 2002. Study of fertilization on variations of soil nutrients and environmental quality with long-term location field experiment in Chao soil—Effect of fertilization on dynamics changes of NO_3-N in soil profile (in Chinese). *Chin. J. Soil Sci.* 33(1):28–29.

Zhang, Y., Liu, H., Li, Z., Lin, B., Zhang, F. 2005. Study of nitrate leaching potential from agricultural land in Northern China under long-term fertilization conditions (in Chinese). *Plant Nutr. Fertil. Sci.* 11(6):711–716.

8 Productivity of Small Landholders of South Asia and Scarcity of Water Resources

Rattan Lal

CONTENTS

8.1 INTRODUCTION

The Anthropocene is influencing the climate change, water resources, and environments of the densely populated South Asia in general and of India in particular. Changing climate and availability of renewable freshwater have affected South Asia since the extinction of the ancient hydric Indus Valley civilization of Harappa and Mohenjo-Daro. Despite an endowment of abundant water resources, an increasing gap between demand and supply is widening because of a large uncertainty, increasing frequency of extreme climatic events, and high risks of soil degradation. The rapid melting of the Himalayan glaciers, source of the 10 largest rivers in Asia, may also exacerbate the water scarcity and conflicts over the shared water resources of the Indus and the Ganges–Brahmaputra–Meghna (GBM) river systems. Conflicts may also emerge among states such as the Indian states of Karnataka and Tamil

Nadu over access to water in the Cauvery River. Energy subsidies for irrigation and undervaluing of the scarce resources are the causes of groundwater overdraft, depleted aquifers, salinized soils, and other cases of the tragedy of commons. There exists a strong link between food security, soil security, water security, and climate security. Following the Cornucopian hypothesis, adverse effects of climate change on water scarcity may enhance adaptation and increase in water productivity (WP). The strategy is to enhance synergisms between science, culture, and religions to resolve conflicts and promote the use of water-saving adaptive technologies. Water occupies a central place in the practice and beliefs of most religions that encourage stewardship and forbid overuse and pollution. While building on the scientific knowledge, diverse cultures and religions offer a distinct perspective related to moral ethics and stewardship.

Global- and continental-scale climate change has been caused by natural processes throughout the history of Earth beginning ~4.5 billion years ago. Over and above the effects of the motion of tectonic plates, volcanism and seismic activities, the earth's climate has also been influenced by the orbital/astronomic cycles (Gunatilaka 2009): (i) the 100,000-year eccentric cycle, (ii) 40,000-year tilt, and (iii) 19,000–21,000-year processional cycles. With the onset of the industrial revolution since ~1750, however, anthropogenic perturbations have become a major force similar to or bigger than even some geologic forces. These anthropogenic forces are strongly influencing the climate, water quality, atmospheric chemistry, biodiversity and species extinction, and other factors that moderate the environment and climate. Thus, the era since 1750 has been appropriately termed "the Anthropocene" (Crutzen and Stoermer 2000; Crutzen 2002). Whereas fossil fuel combustion is a major factor influencing atmospheric chemistry and the radiative forcing related to enrichment of greenhouse gases (GHGs) (i.e., CO_2, CH_4, and N_2O) (Intergovernmental Panel on Climate Change 2013), atmospheric brown cloud, generated by the combustion of traditional fuels, is strongly influencing the climate of South Asia (Ramamathan et al. 2005). Evidences of the influence of the Anthropocene on climate change are also reported in South Asia (Gunatilaka 2009). In this regard, the impact of China and India, the emerging economic giants of the world (Bawa et al. 2010), on regional and global environments cannot be overstated. In addition to rising demands of energy and water by the growing and ever-affluent societies, there are multiple pressures on biodiversity, especially those of subalpine and alpine regions of the Himalayan ranges and of the mangrove ecosystems of coastal and deltaic zones such as the Sundarban of the Gangetic delta. In this regard, the development of roads and other infrastructures also strongly influence the hydrologic processes (Cuo et al. 2008) and the environment. Agricultural expansion, increases in irrigated agriculture, and heavy use of fertilizers and pesticides have strongly influenced soil quality by accelerated erosion, water quality by nonpoint source pollution, and atmospheric quality by the emission of soot and particulate matter (dust) along with trace gases from soil (e.g., CO_2, CH_4, and N_2O).

This chapter describes the scarcity of water resources in India and South Asia, and discusses the impact of the projected climate change. Water, being one of the two principal natural resources (the other is soil) affected by climate change, is used as an example to describe policy, social, and religious approaches for its sustainable use and management.

8.2 CULTURAL AND RELIGIOUS BELIEFS IN ENVIRONMENTAL ISSUES

Climate and the environment are strongly integrated into cultures, religions, and ethnic values of the people in South Asia and elsewhere. The rise and fall of the Indus Valley civilization is strongly linked to the change in climate. The Harappa and Mohenjo-Daro flourished between 2800 and 2600 BC and vanished in 1900 BC, probably because of drastic climate change resulting in aridization of the region (Singh 1971; Lamb 1982). The Indus Valley civilization, larger in area than the Nile Valley and Mesopotamian civilizations combined, flourished when the rainfall of the region was 500–600 mm/year compared with 100–300 mm since its extinction (Lamb 1982). The region, under stronger development of the summer monsoonal rainfall, had dense vegetation and was a habitat for elephants (*Elephas maximus*), rhinoceros (*Rhinoceros unicornis*), water buffaloes (*Bubalus arnee*), and other large animals. The hydric civilization perished with the onset of drought in ~1500 BC, which also coincided with the migration of the Aryans into the Indus Valley. The central theme of the post-Harappan Aryan civilization (the Vedic culture) was based on nature worship (Enzel et al. 1955; Wassan et al. 1984). The nature worship by Indo Aryans is exemplified in the Prasna Upanishad, which states that the human body is made of five elements, "kshiti (soil), jal (water), pawak (energy), gagan (sky/space), sameera (air); panch (five) tatva (elements) yah (from) adham (made) sharira (body)." Thus, these five elements are objects of worship and reverence among several South Asian cultures. Three of these five elements (soil, air, and water) constitute nature or the environment. Buddhism also promotes nature worship, and the principal objects of worship include forests, rivers, and mountains, and the belief in the close interdependence between humans, animals, water, Earth, Sun, Moon, and stars (Singh and Asce 2008). Similar to Hindu and Buddhist beliefs, the importance of water is also highlighted in Islamic teachings and scriptures. Prophet Mohammed stated that, "Do not overuse water even if you are on a running river." The Holy Quran state, "... Heavens and earth were joined together... we made from water every living things" (Holy Quran, Soorah Al-Anbea, verse 30) (Al-Senafy and Akber 2002; Singh and Asce 2008).

South Asian countries have experienced thousands of years of hydric civilization (Gunnell and Krishnamurthy 2003; Gunnell et al. 2007). Some have argued that the Anthropocene may have commenced ~10,000 years ago with the dawn of settled agriculture that prompted deforestation, biomass burning, soil cultivation, and emission of GHGs into the atmosphere (Ruddiman 2003). However, climate change and other environmental issues of the Anthropocene are major challenges now than ever before, and must be effectively addressed. Thus, it is important to understand the impacts of climate change on water resources and the environment.

8.3 CLIMATE CHANGE AND WATER RESOURCES IN SOUTH ASIA

Most countries in South Asia are endowed with abundant water resources (Table 8.1). Yet there is an increasing gap between demand and supply of water throughout South Asia. The scarcity of water, also manifested in regional and international conflicts, is

TABLE 8.1
Renewable Freshwater Resources of South Asian Countries

	Estimates of Renewable Freshwater (km^3/year)	
Country	**UN (2010)**	**Engelman and LeRoy (1993)**
Afghanistan	65	50
Bangladesh	1210	2357
Bhutan	73	95
Nepal	210	170
India	1869	2085
Pakistan	234	468
Sri Lanka	50	43

Source: United Nations. 2011. Environment statistics—Country snapshots: India. United Nations Statistics Division. http://unstats.un.org/unsd/environment/Questionnaires/country_snapshots.htm (accessed April 20, 2012); Engelman, R., and P. LeRoy. 1993. *Sustaining Water*. 56. Washington, DC: Population Action International.

exacerbated by the climate change. The projected warmer climate implies acceleration of the hydrologic cycle with severe impacts on groundwater depletion and reduction in stream flow because of the general reduction in runoff (Saleth 2011) partly caused by changes in rainfall amount and distribution (Mall et al. 2006). Increase in evaporation may accentuate water scarcity by reducing the net recharge of the groundwater and aquifer. Vulnerability to the impact of climate change on renewable groundwater resources may be a serious constraint to achieving food security and maintaining economic growth (Döll 2009). There has also been a significant increasing trend in the frequency and magnitude of extreme rain events corresponding with a decreasing trend in the frequency of moderate events (Goswami et al. 2006). Thus, variation in water availability is a major source of risk for agricultural productivity decline and food insecurity in South Asia (Li et al. 2011). Whereas rainfed cropping systems by small landholders are the most vulnerable to drought stress, reduction in runoff and depletion of groundwater also adversely influence irrigated agriculture. Rapid melting of the Himalayan glaciers may increase water scarcity in the Indo-Gangetic plains.

8.4 CLIMATE CHANGE AND THE HIMALAYAN GLACIERS

There are marked trends of increasing temperature since early 1980 in India, but with significant variations in these trends during different seasons and over different regions (Dash and Hunt 2007). Trends of increased temperatures in Pakistan correspond to an increase in growing degree days and a decrease in growing season lengths (Hussain and Mudasser 2007). The projected climate change may reduce the total and per capita availability of renewable freshwater resources in South Asia (Table 8.2). The Greater Himalayas, also called the Third Pole, hold the largest mass of ice outside of the polar regions. The Himalayan glaciers are the

TABLE 8.2
Estimates of Fresh Groundwater Resources of South Asia under Two Emission Scenarios

	2010		2100	
Emission Scenarios	**Fresh Groundwater (m^3/km^2)**	**Per Capita (m^3/capita/year)**	**Fresh Groundwater (m^3/km^2)**	**Per Capita (m^3/capita/year)**
High	117,000	468	52,000	115.6
Low	110,000	550	35,000	116.7

Source: Adapted from Ranjan P., S. Kazama, and M. Sawemoto, *Global Environ Change*, 16, 388, 2006.

source of the 10 largest rivers in Asia (Xu et al. 2009). Thus, the rate of melting of the Himalayan glaciers has also become a popular topic in the media (Inman 2010; Fischetti 2011; Goldenberg 2011; Nelson and Alleyne 2011; Raloff 2011). Whether the spatially heterogeneous rate of melting (Fujita and Nuimura 2011) is normal or climate induced remains to be seen. However, the potential of an already tense situation between India and Pakistan could become worse as effects of climate change result in shrinkage of water and scramble for control over the dwindling water resources (see Section 8.7, Figure 8.1). Despite the controversy and debates, the risks of the melting of Himalayas on water, biodiversity, and livelihoods are high (Xu et al. 2009;

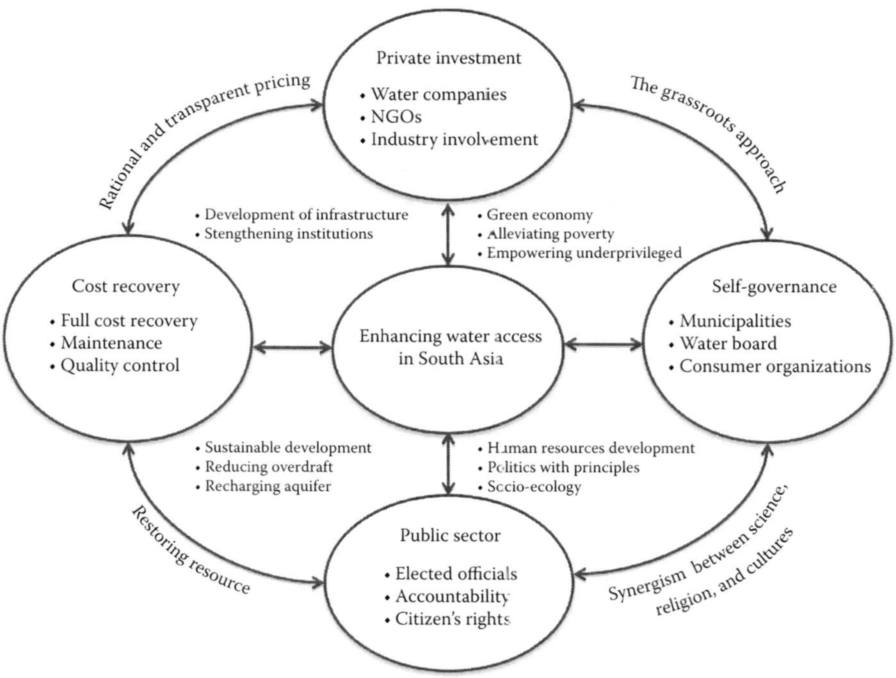

FIGURE 8.1 Options of improving water governance and reducing conflicts.

Ice Age Now 2011). In addition to the impact on glaciers, climate change may also thaw the cryosphere and melt the permafrost of the Qinghai–Tibet Plateau (Cheng and Wu 2007). Even with increasing risks of water scarcity due to glacial melting, the water demand in South Asia is increasing very rapidly.

8.5 WATER RESOURCES AND THE RISING DEMAND

Water scarcity is a problem as old as civilization (Yeston et al. 2006). Globally, only 10% of the maximum available water and 30% of the green water (plant available) is being used (Oki and Kanae 2006). The problem is the high variability (spatial and temporal) and the uncertainties. The projected climate change exacerbates risks and uncertainties (Sophocleous 2004) related to the severity of the water scarcity that already exists. The renewable freshwater resources in arid and semiarid regions, such as those in India and other South Asian countries, are likely to become more unstable and prone to severe drought stress (Shen and Chen 2010). The problem of drought stress on small landholder farms may be exacerbated by high risks of soil degradation because of the widespread use of extractive farming practices and the projected climate change.

Despite the abundance of renewable freshwater resources, the existing water scarcity on a regional basis in India is also likely to be exacerbated with the projected climate change. India receives an annual precipitation of about 4000 km^3; however, it has very high spatial and temporal variability (Kumar et al. 2005). For example, Mousinram near Cherrapunji, in the northwestern region, receives the world's highest rainfall (~1250 cm/year). However, it also frequently suffers from a shortage of water during the nonrainy season, every year (Kumar et al. 2005). The country as a whole faces the problem of flood and drought syndrome, which may be accentuated by the projected change in precipitation and temperature, overexploitation of groundwater (see Section 8.6), water logging and salinity because of the excessive use of canal water, salt water intrusion into aquifers of the coastal areas, and water pollution/eutrophication and contamination.

South Asian rivers carry an annual water flux of ~2100 km^3 (~6% of global runoff) and an annual sediment flux of ~1 billion Mg (~10% of global flux) (Subramanian 2008). High concentration of nitrogen (N) in waters of rivers in South Asia is increasing because of intensive use of fertilizers in agroecosystems. The discharge-weighted average of NO_3-N in Indian rivers is 2 mg/L, and the average sediment-bound N (mostly organic) is 0.2% (Subramanian 2008). In comparison, the reported global average for the uncontaminated river system is 0.028 mg/L of NO_3-N. The concentration of NO_3-N varies among rivers and locations in a river.

There is also contamination of groundwater with fertilizers, pesticides, and other agricultural and industrial pollutants. In addition, arsenic (As) contamination in groundwater in the Southeast Asia region is alarming, and >100 million people are at risk (Rahman et al. 2009). The upper level of As concentration (μg/L) in the groundwater has been reported at 4730 in Bangladesh, 1610 in Cambodia, 4440 in China, 3880 in West Bengal, 3191 in the Uttar Pradesh state of India, 3590 in Taiwan, 3050 in Vietnam, 2620 in Nepal, and 906 in Pakistan (Rahman et al. 2009). The acceptable level for human consumption is 50 μg/L. Furthermore, the As-polluted water used for irrigation is contaminating soils (and rice) and is accentuating the human health risks (Brammer and Ravenscroft 2009).

The emphasis on biofuel is another confounding factor because of the energy and water trade-offs (Mushtaq et al. 2009). Therefore, biofuel policies must focus on short-duration, multipurpose, and proven drought-tolerant crops (Rajagopal 2008) that can be useful to restore degraded lands and create another income stream for the rural poor.

8.6 POLITICS OF WATER

The great rivers of South Asia (e.g., Ganges and Brahmaputra) have contributed, through flood and drought syndrome, to the miseries and impoverishments of the lives of hundreds of millions. Through political consensus and goodwill among the nations involved, however, perils can be transformed into prosperity (Crow and Singh 2000). There is no justification for why the population of resource-rich river systems must be perpetually subject to poverty through recurring cycles of flood and drought. Thus, politicians must work together to harness the riches of the great rivers, conserve soil and water, and transform a desperate and destitute situation into food security and economic growth. It is important to understand the role of politics in sustainable management of water. Some policy issues may be addressed by (i) learning from and using political science to implement water reforms, (ii) identifying the most critical issues, and (iii) developing an action plan (World Water Council 2004). The goal is to enhance water security.

Water security involves the sustainable use and protection of water systems, the protection against water-related hazards (floods and droughts), the sustainable development of water resources, and the safeguarding of (access to) water functions and services for humans and the environment (Schultz and Uhlenbrook 2007). In accordance with the definition of food security, water security also exists when all people at all times have access to sufficient, safe, and clean water to maintain a healthy and active life.

Addressing the water scarcity, the cause of water conflict over the shared water resources can improve relations among neighborly states. The international water conflicts and cooperation is also influenced by domestic water events and vice versa (Giordano et al. 2002). India has two major international watersheds: (i) the GBM and (ii) the Indus. The GBM involves an area of 1.7×10^6 km^2 and covers 30% of India's land area, involving 15 states and one union territory. The water resources are shared with Bangladesh (which lies almost entirely within the GBM watershed), Nepal, China, Bhutan, and Myanmar. The watershed as a whole covers 1% of the earth's total land surface but is home to 10% of the world's population and contains the largest concentration of the world's poor on Earth (Ahmad et al. 2001; Rangachari and Verghese 2001; Shah 2001). Most farmers are small landholders and resource poor.

The Indus River also originates in the Tibetan Plateau but traverses through drier climatic regions of northwestern India and eastern Pakistan. India occupies one-third of the total area of the watershed (~1.1×10^6 km^2), and Pakistan the remaining two-third (Giordano et al. 2002). The Indus Valley was also the home to one of the world's ancient civilizations of Harappa–Mohenjo-Daro (see Section 8.2).

Transborder water woes are affecting South Asian riparian countries (Salman and Uprety 1999; Lamballe 2000; Chatterjee 2008). In addition to international conflicts, such as between India and Pakistan over territorial issues, disputes also exist among individual Indian states because water resources are shared between the national and the state government. When water resources are shared among several states, the national government retains the overall management authority albeit with an active local involvement in the management. The Indian states of Karnataka and Tamil Nadu are in conflict over access to water in the Cauvery River. India controls the headwater of Indus, and the conflict between India and Pakistan may escalate because of the shared water. Yet, the Indus Water Treaty of 1960 has remained remarkably stable (Sahni 2006; Giordano et al. 2002). Indeed, India and Pakistan have thus far demonstrated willingness to keep apart the border issues from the water issues. India's relations with Nepal and Bangladesh have also been overall favorable. The Farakka Barrage, constructed a few kilometers upstream from the Bangladesh border, has been an issue of concern (Salman 1998; Giordano et al. 2002), and a long-term treaty was concluded in 1996 (Nakayama 1997). A similar bilateral treaty (the Mahakali Treaty) was finalized in 1996 between India and Nepal (Shah 2001). Most of these bilateral treaties, especially those with Pakistan, have demonstrated resiliency despite the border-related hostilities. However, the effectiveness of these treaties can be enhanced by also involving the following entities (Crow and Singh 2000): (i) looking beyond diplomacy and involving private economy; (ii) encouraging third parties such as local governments, nongovernmental organizations, and corporations; (iii) sharing the costs and benefits emerging from environmental improvements (e.g., payments for ecosystem services); and (iv) promoting multilateralism to bring in new resources to address the problem.

8.7 WATER GOVERNANCE

Water governance is defined by "the political, social, economic, and administrative systems that are in place, and which directly or indirectly affect the water use, development, and management" (Anand and Easwaran 2010). The strategic options of harnessing the benefits of South Asia's rivers necessitate an effective governance (Tyler 2004). Water governance must adopt a problem-solving approach to look beyond the polarizing debates, create polycentric coalitions, promote community involvement or comanagement, encourage private sector and institutional diversity, and create water markets (Bruns 2010). In addition to being ill governed, the scarce resource is colossally underpriced (Economist 2003; Anand and Easwaran 2010), which leads to overuse of water, wastage, and adverse environmental impacts.

The resource stress caused by water scarcity can lead to three possible scenarios (Warner 2004): (i) Malthusians: scarcity leading to war, honey pot argument, or scramble of gold diggers; (ii) Cornucopians: scarcity leading to adaptation; and (iii) Rousseauian antiglobalists: severe scarcity as a consequence of the structural violence of an inequitable global system. Adaptation to the scarcity of the public good is evidently the most desirable scenario. As a public good, it is difficult to price it using economic principles. Furthermore, water policy also interacts with health, education, and environment, especially the climate. Anand and Easwaran (2010) proposed that water governance be based on four pillars: (i) policy: specifically focused on water problem, must be transparent, unambiguous, and based on credible information; (ii) people: with passion for water, local and scientific knowledge, and commitment to good governance; (iii) procedure: that is consistent and pragmatic in implementing the policy; and (iv) pricing: based on the just value of the scarce resource.

On the basis of these concepts, the basic principles of water governance are listed in Figure 8.2. The water security and sustainability rests on three R's (reduce waste, reuse what is available, and recycle all), three C's (commitment of the people involved, continuity in the program, and coherence in policy implementation), and three I's (inputs, investments, and incentive). There are several models of governance (Mathur 2004). Some of the governance options are outlined in Figure 8.1. These options are based on private investment, self-governance, public sector, and cost recovery. The cost recovery through a rational pricing system is essential to

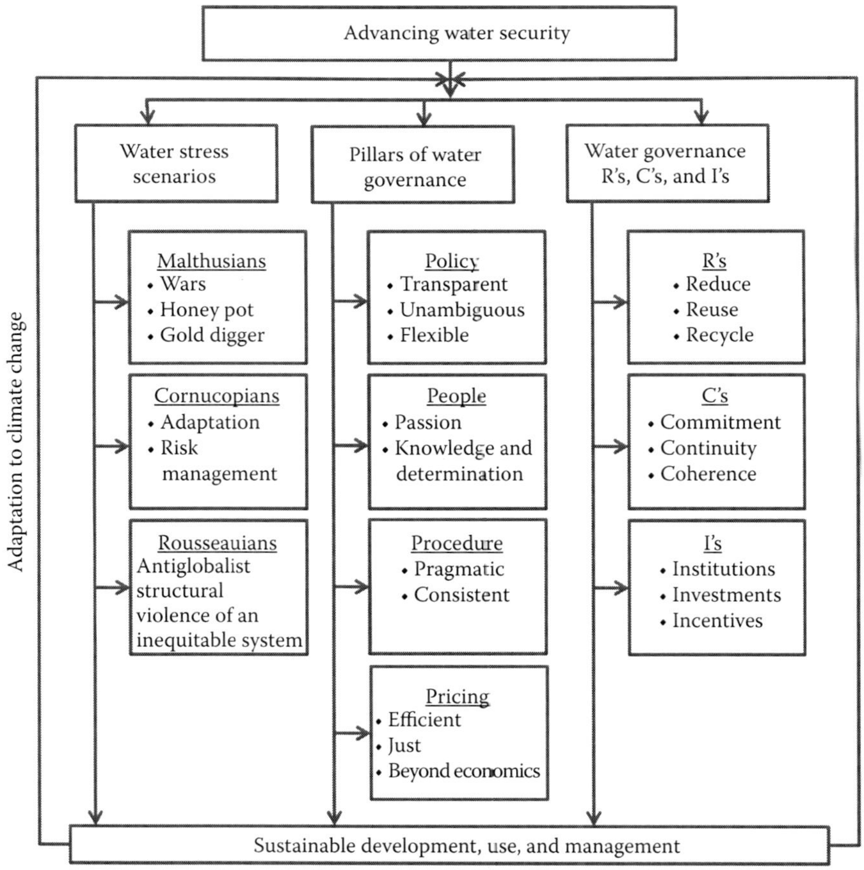

FIGURE 8.2 Options for effective water governance to enhance access to water. (Based on concepts by Warner, J. 2004. Water, wine, vinegar, blood: On politics, participations, violence and conflict over hydrosocial contract. In: *Proceedings of the Workshop on Water and Politics*, Marseille. 7–18. Marseille: World Water Council.)

sustainability. Thus, the viable governance options outlined in Figure 8.3 must carefully consider the following (Tyler 2004): (i) water pricing, (ii) whether water is a social or an economic resource, (iii) and virtual water. Lack of access to safe and clean water is symptomatic of poverty (Mathur 2004), which must be abolished through economic development. Energy subsidies for irrigation (Shah et al. 2003), such as in Punjab, Andhra Pradesh, and Gujarat states of India, are the cause of the groundwater overdraft, depleted aquifers, salinized soils, and bankrupt electricity utility companies (40%–60%). Indeed, the low efficiency of water use in agriculture in India can be improved by rational pricing of water (Singh 2007; Kondepati 2011). While farmers are sensitive to an increase in irrigation price change, water sustainability and security cannot be achieved unless the administered prices are increased to the level that would prevail in the free market.

A significant reliance on participation by local communities is an option that needs careful appraisal (Figure 8.1). Excessive use can be avoided if water is regarded as a precious resource (Alauddin and Quiggan 2008), without falling into the trap of "the tragedy of commons" (Hardin 1968). Adopting concepts of socioecology (Mukherji and Shah 2005) may be relevant to sustainable groundwater development (Limaye 2004). Therefore, an effective community-based approach would call for the following (Alauddin and Quiggin 2008; Gadgil and Rao 1995): (i) greater local control on water (and soil/land), (ii) enhanced capacity for value addition, and (iii) specific financial rewards. Financial rewards can be implemented through payments for ecosystem services. Some of the policy options to resolve the conflicts include the following:

1. *Water privatization*: It involves transferring of water control and/or water management services to private companies (Sampath et al. 2003). Water privatization [at different levels such as (i) service contracts; (ii) design, build, operate, own, and transfer; and (iii) divestiture] is recommended by the national water policy of India (Article 13). Rational price recovery from the consumer is the only option to address inefficiency and excessive use.
2. *Public participation*: A grassroots or a bottom–up approach to conflict resolution involves public participation in water management. Building on traditional knowledge and indigenous water structures and management systems may be cost-effective (Tankha and Fuller 2010). The proposed river-linking scheme, with an estimated budget of $120 billion aimed to transfer India's fierce monsoons into a friend (Bagla 2006; Misra et al. 2007), is also based on the privatization model.
3. *Scarcity value*: Ghosh and Bandyopadhyay (2009) hypothesized a scarcity value–based explanation of transboundary water disputes in the Cauvery River basin. It is the value that could have been generated if the limit on water availability could be relaxed by one unit. It measures the degree of deprivation and creates the basis for disputes. Ghosh and Bandyopadhyay observed that water disputes between the states of Karnataka and Tamil Nadu are not based on the physical scarcity of water but are a temporal coincidence of demand based on scarcity value, and imply that enhanced supply would not necessarily resolve the conflict.

4. *Participatory water management and capacity building*: Tankha and Fuller (2010) emphasized the role of creating institutions to facilitate stakeholder participation, administrative reforms, and capacity building in water resources management. Rational price recovery from the consumer is the only option to address inefficiency and excessive use.
5. *Public participation*: A grassroots or a bottom–up approach to conflict resolution involves participation in water management. Building on traditional knowledge and indigenous water structures and management systems may be cost-effective (Tankha and Fuller 2010).
6. *Payments for ecosystem services*: Farmers in the upstream region of a watershed must be rewarded for conserving water and reducing nonpoint source pollution through payments for the "green water credits."

8.8 LINKING SECURITY OF FOOD, WATER, SOIL, AND CLIMATE IN SOUTH ASIA

There is a strong link between water security, food security, soil security, and climate security. Sustainable management of these three resources is essential to human well-being and ecosystem functions (Figure 8.3). The link is especially important in intensively managed agroecosystems (Alagh 2001; Alauddin and Quiggan 2008; Douglas et al. 2009). The goal is to produce more food with

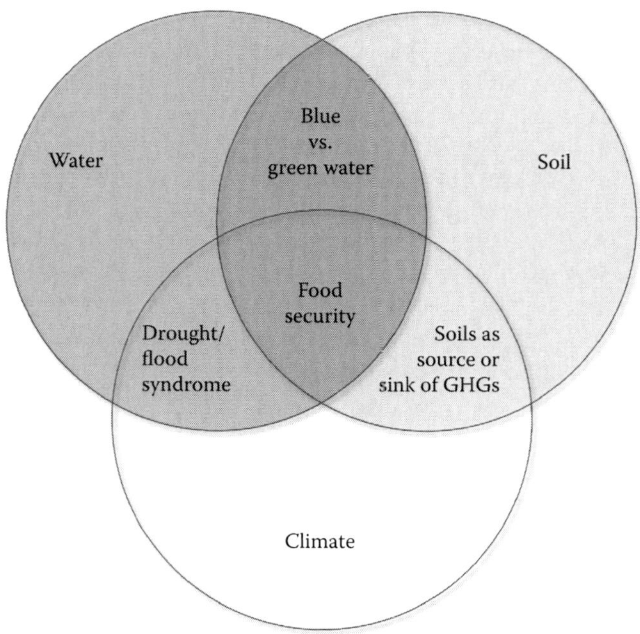

FIGURE 8.3 Dependence of food security on the interactive effects of climate, water, and soil resources.

less water by enhancing WP or WP (Cai et al. 2011), and with less CO_2 emission through the efficient and economic use of water (Wolff and Stein 1999) and C sequestration in soils of agroecosystems (Lal 2004a,b). The question, however, remains about which crop should be chosen and which WP must be enhanced (Bessembinder et al. 2005). Several technological options of increasing WP and adapting agroecosystems to changing climate include deficit irrigation, sequencing of water deficit, surge irrigation management especially in Vertisols, no-till/mulch farming, water harvesting and recycling, modern microirrigation systems, crops and cropping systems, and integrated nutrient management (INM) based on the liberal use of organic amendments (Shah et al. 2006; Rockström et al. 2007; Ali and Talukder 2008). With reference to INM, reducing nutrient input (from a maximum to an optimum level) may maintain agronomic production while reducing the risks of environmental pollution (Good et al. 2004) and reducing emissions of N_2O. Although a debatable issue, selective use of genetically modified crops may benefit the environment while improving WP (Brookes and Barfoot 2011). Mixing saline water with good water (Sharma and Minhas 2005) and using wastewater are also important options. A widespread use of such technological innovations is essential to improve and sustain Asia's groundwater boom (Shah et al. 2003), in an era of warming Earth characterized by extreme climatic events.

8.9 ASSESSING WATER RESOURCES

In accordance with the dictum "what cannot be credibly measured cannot be adequately managed," sustainable management of water necessitates credible assessment (available reserves and the demand) of the scarce but precious resource. The reliable data on water supply are critical to developing programs for sustainable management. The warning about severe depletion of the groundwater resources (Kerr 2009; Rodell et al. 2009) has to be based on credible assessment. In this regard, multispectral remote sensing images can be appropriately used to enhance resource conservation and improve agronomic productivity (Chandna et al. 2012). Similarly, Lobell et al. (2010) used satellite technology for assessing yield growth opportunities in northwest India. Gumma et al. (2011) used MODIS 250 m time-series data to map irrigated areas in the Krishna River basin. The scheme of interlinkage among major rivers can only be implemented if there exist reliable data on river flow on time scales encompassing days, seasons, and years (Jian et al. 2009).

8.10 USING SCIENCE, CULTURE, AND SPIRITUALISM TO RESOLVE SOCIAL AND POLITICAL CONFLICTS IN MANAGING WATER RESOURCES

Resource scarcity and other issues create social and political conflicts, which must be amicably resolved to maintain peace and harmony. India and other countries in South Asia face water controversies as village-versus-state and community-versus-state conflicts (Mollinga 2010). The fast economic development in India since its liberalization in the 1990s has created more diverse interest groups around water and, to some extent, accentuated these conflicts. The irrigation economy of South Asia has its own complexities (Shah et al. 2003). Farmers committing suicide in India have been on the rise, and estimated at 150,000 between 1997 and 2005, because of the disparity between income and expenditure (Agroamoorthy et al. 2009). This complex issue may be more effectively addressed through a synergism between modern science and religions. Albert Einstein opined that "science without religion is lame; religion without science is blind." The strategy is to understand the reasons of controversies and conflicts and address them objectively through use of a science–religion–culture mix. Historical roots of ecological crisis must be identified (White 1974).

8.11 CULTURAL AND SPIRITUAL APPROACH TO CLIMATE CHANGE AND WATER SCARCITY

Soil and water resources, being finite and easily degraded/polluted, must be managed in an eco-friendly and sustainable manner. Presently, ~0.5 billion people face water shortage in 29 countries, 33% of the world population is living under moderate to severe water stress, 1.2 billion people lack access to adequate supply of safe water, 2.5 billion do not have access to proper sanitation, 7 million die annually from waterborne diseases, and half of the world's rivers and lakes are seriously polluted (Singh and Asce 2008). By 2025, about 5.5 billion people will experience some water stress of which >1 billion will be prone to severe and socially disruptive stress. How can we solve this crisis?

Being the basis of all life through its ability to cleanse and wash away impurities and pollutants, water occupies a central place in the practices and beliefs of most religions (Table 8.3). Major religions have addressed the sustainable use of water and other resources (United Nations [UN] 1975; Dwivedi 1989; Abrams 2000; Faruqui 2003; UN/WWWAP 2003; Sutcliffe et al. 2011). The mysticism, originating from its power to destroy as well as create, is central to the beliefs of creation of life in Hinduism, Christianity, Judaism, and other religions. Because of its power to cleanse and purify, water is an object of worship in most religions. Believers of Zoroastrianism think that it was the curse of the evil spirit (Angra Mainyu) that made the water salty. Thus, pollution of water is considered evil. The powers of the gods and their incarnation are outlined in Hinduism and Judaism scriptures. The waters of River Yamuna receded after touching the feet of baby Krishna carried across the river by his father who was being chased by the army of Kansa. Similarly, the water of the Red Sea was parted by Moses so that Israelites being chased by the Egyptian army could escape (Table 8.3). Thus, daily bathing of the Buddha, similar to baptizing in Christianity, is symbolic of spiritual rebirth. As a symbol of purification, water is used in all Hindu religious ceremonies.

The need to combine spiritual/cultural beliefs with scientific principles is greater now than ever, to effectively address the problems of water scarcity, pollution, and exploitation being exacerbated by climate change and demands of the Anthropecene. Yet, the human consumption of water is increasing rapidly because of the burgeoning demands (UN/WWWAP 2003). It is the time to reconsider humanity's roots and linkages with nature.

TABLE 8.3
Significance of Water as Stated in Different Religious Scriptures

Buddhism	"As the rains fill the rivers and overflows into the ocean, so what is given here may reach the departed." Buddhists seek "nirvana" in natural settings amid rivers, forests, and mountains.
Christianity	John 4:1–42, Jesus offers "living water" to a Samaritan woman so that she will never thirst again.
Hinduism	The sacred rivers of Hindus are "Ganga, Sindhu, Srasvati, Yamuna, Godavari, Narmada, Kaveri, Saryu, Kachipra, Vetravati, Mahasurandi, Khyata, Gaya, Gandak."
	During the Great Flood (Pralaya), Manu is rescued by a fish; the fish towed a large boat taking into it seeds and animals, and anchored it on the highest of the Himalayas.
	Baby Krishna was carried to safety by his father (Vasudev) across the river Yamuna when the army of Kansa was chasing them. The water of the Yamuna parted when it touched the feet of Krishna.
Islam	"Heavens and Earth were joined together—we made from water every living thing—."
Judaism	The divine flood washed away all the sins of the world but only Noah survived. Parting of the Red Sea by Moses enabled the Israelites to escape from the Egyptian army.
Shinto	Waterfalls are sacred, used in "suigyo," and standing under can purify the believers.
Zoroastrianism	The evil spirit Angra Mainyu attacked the earth and made pure water salty. Thus, pollution is evil, and pure water is sacred. The dead are not cremated, buried, or immersed in water because fire, earth, and water must be kept pure.

Source: Dwivedi, O.P. 1989. *World Religions and the Environment*. 462. New Delhi, India: Gitanjali; Abrams, P. 2000. The water page. *Water in religion*. Water Policy International Ltd. http://www.africanwater.org/religion.htm (accessed January 2, 2014); Holy Quran-Surah Al Aneba, verse 30; Kalidasanskritam Manglashatkam (2056 Vikram calendar).

In Latin, the term *nature* means "to be born" (natal, prenatal, etc.). In English, the term has successfully evolved into three meanings (Proctor 2004): (i) the quality or character (13th century), (ii) the inherent force directing the world (14th century), and (iii) the physical world as a whole (17th century). It is in the context of its use to denote "the physical world as a whole" that there is a need to review the relevance of theology and spiritualism because its very nature is being jeopardized by human greed and delusion. There also exists a strong connection between "sustainability" and theology (Jenkins and Chapple 2011) because the proper use of spiritual beliefs can affect the environmental behavior and involve the stewardship response of human society. Multiple social and ecological problems that make up the challenge of sustainability can be addressed through religious, spiritual, and philosophical ideals. Indeed, unsustainable use of natural resources (i.e., soil, water, forests, energy, minerals, wildlife) is attributed to humanity's insatiable desires. Consequently, the earth and all its creatures are commodified into a pool of resources to be exploited to satisfy these never-ending desires. The unsustainable use of natural resources is driven by the religion of market:

"the endless hunger ... are we happy yet" (Loy 1997). These endless needs of humanity are the roots of the ecological crisis (White 1976). Becker (1992) stated it directly but pertinently, "we may recycle newspapers and glass, and we may take proper satisfaction in doing so, but we remain caught in a web of spiritual assumptions about success and consumption process and waste, that effectively undermine and trivialize our efforts to escape." Yet, properly interpreted and appropriately used in conjunction with modern science, most religions advocate protection, appropriate use, and stewardship of the finite natural resources (Table 8.3). The Indo–Aryan Vedic philosophy states that God is nature, "Oh King, the river are the veins of the Cosmic Person and the trees are the hairs of his body. The air is his breath, the ocean is his waist, the hills and mountains are the stacks of his bones and the passing ages are his movements" (Srimad Bhagavatam 2.1.32–33). The modern Gaia hypothesis (Lovelock 1979) proposed a similar concept. Similar to the Gaia (Mother Earth), the Vedic scriptures also personify Mother Earth as the goddess Bhumi or Prithvi and pray, "O Mother, with your oceans, rivers and other bodies of water, you give us land to grow grains, on which our survival depends" (Atharva Veda 12.1). Thus, stewardship of Mother Earth is ingrained in similar philosophies among all religious beliefs and cultures. It is also important to realize that "you should take only what is really necessary for yourself, which is set aside for you, and you should not take anything else, because you know to whom it belongs" (Isa Upanishad).

While judicious use of scientific knowledge is a given, diverse cultures and religions offer a distinct perspective related to moral ethics and stewardship. The concept of "dharma" (duty) and karma (so shall you reap) are ingrained in all religions (Table 8.3), albeit under different settings and varying linguistic descriptions. The command by Prophet Mohammad, "Do not overuse water even if you are on a running river," says it all and is the most relevant to solving the global water crisis.

8.12 CONCLUSION

Water scarcity in South Asia/India, likely to be exacerbated by environmental and climate change, may either increase regional conflicts or promote adaptation depending on the adoption of strategies toward sustainable management of the finite resource. The flood and drought syndrome of the great rivers of South Asia can be transformed into prosperity and improved environment through judicious water governance that promotes adoption of conservation-effective technologies and enhances synergism between science, religion, and culture; minimizes conflicts among nations/states; encourages community participation; and replaces subsidies with national pricing policies. The strong link between food security, water security, soil security, and climate security must be enhanced through constructive policies and adaptive strategies for alleviating poverty, eliminating social and gender inequities, and improving the environment. Harnessing knowledge of science through the exploration of synergism with religion(s) and culture(s) is needed to promote community participation and encourage peace and harmony. This is the time to build on the experiences of thousands of years of hydric civilizations to use the diversity (climatic, geologic, physiographic, and ecologic) and polycultural heritage (religions, ethnic, and historic) for developing green economy, restoring the environment, and adapting to and mitigating the climate change.

REFERENCES

Abrams, P. 2000. The water page. *Water in Religion*. Water Policy International Ltd. Available at http://www.africanwater.org/religion.htm (accessed January 2, 2014).

Agroamoorthy, G., S. Chaudhary, and M. J. Hsu. 2009. Sustainable development using small dams: An approach to avert social conflict and relive poverty in India's semi-arid regions. *Asia Pacific J Social Work Dev* 19(2):52–69.

Ahmad, Q. K., A. K. Biswas, R. Rangachari, and M. M. Sainjus. 2001. A framework for sustainable development of the GBM region. In *Ganges–Brahmaputra–Meghna Region: A Framework for Sustainable Development*, eds. Q. K. Ahmad, A. K. Biswas, R. Rangachar, and M. M. Sainjus, 1–29. Bangladesh: The University Press.

Alagh, Y. K. 2001. Water and food security in south Asia. *Water Resour Dev* 17(1):23–36.

Alauddin, A., and J. Quiggan. 2008. Agricultural intensification, irrigation and the environment in south Asia: Issues and policy options. *Ecol Econ* 65:111–124.

Ali, M. H., and M. S. U. Talukder. 2008. Increasing water productivity in crop-production—A synthesis. *Agric Water Manage* 95:1201–1213.

Al-Senafy, M., and A. Akber. 2002. Water management in Islam. In *Water Resources Development and Management*, eds. M. Al-Rashed, V. P. Singh, and M. M. Sherif, 501–508. The Netherlands: Swets and Zeitlinger, Lissie.

Anand, P., and S. K. Easwaran. 2010. New thinking on water governance—Cleaning the clouds. *Proceedings of New Thinking in Water Governance*, 57–82. Singapore: National Univ. of Singapore.

Bagla, P. 2006. Controversial rivers project aims to turn India's fierce monsoon into a friend. *Science* 313:1036–1037.

Bawa, K. S., L. P. Koh, T. M. Lee et al. 2010. China, India, and the environment. *Science* 327:1457.

Becker, W. H. 1992. Ecological sin. *Theol Today* 49(2):152–164.

Bessembinder, J. J. E., P. A. Leffelaar, A. S. Dhindwal, and T. C. Ponsioen. 2005. Which crop and which drop, and the scope for improvement of water productivity. *Agric Water Manage* 73:113–130.

Brammer, H., and P. Ravenscroft. 2009. Arsenic in groundwater: A threat to sustainable agriculture in south and south-east Asia. *Environ Int* 35:647–654.

Brookes, G., and P. Barfoot. 2011. *GM Crops: Global Socio-Economic and Environmental Impacts 1996–2009*, 191. Dorchester, UK: PG Economics Ltd.

Bruns, B. 2010. Water governance as problem solving. *Proceedings of New Thinking in Water Governance*, 33–56. Singapore: National Univ. of Singapore.

Cai, X., D. Molden, M. Mainuddin, B. Sharma, M. D. Ahmad, and P. Karimi. 2011. Producing more food with less water in changing world: Assessment of water productivity in 10 major river basins. *Water Int* 36(1):42–62.

Chandna, P., J. K. Ladha, U. P. Singh, M. Punia, R. Gupta, and B. S. Sidhu. 2012. Using remote sensing technologies to enhance resource conservation and agricultural productivity in underutilized lands of south Asia. *App Geol* 32:757–765.

Chatterjee, A. 2008. Impacts of trans-border water woes in South Asian riparian countries—Assessment and analysis. In: *Water and Urban Development Paradigms—Towards an Integration of Engineering, Design and Management Approaches*, eds. J. Feyen, K. Shannon, and M. Neville, 675–679. The Netherlands: Taylor & Francis.

Cheng, G., and T. Wu. 2007. Responses of permafrost to climate change and their environmental significance, Qinghai–Tibet Plateau. *J Geophys Res* 112:2003–2012. doi:10.1029/2006JF000631.

Crow, B., and N. Singh. 2000. Impediments and innovation in international rivers: The waters of south Asia. *World Dev* 28(11):1907–1925.

Crutzen, P. J. 2002. The geology of mankind. *Nature* 415(6867):23.

Crutzen, P. J., and E. F. Stoermer. 2000. The Anthropocene. *Global Change Newsl* 41:17–18.

Cuo, L., W. Giambelluca, A. D. Ziegler, and M. A. Nullet. 2008. The roles of roads and agricultural land use in altering hydrological processes in Nam Rim watershed, northern Thailand. *Hydrol Process* 22:4339–4354.

Dash, S. K., and J. C. R. Hunt. 2007. Variability of climate change in India. *Curr Sci* 93:782–788.

Döll, P. 2009. Vulnerability to the impact of climate change on renewable groundwater resources: A global-scale assessment. *Environ Res Lett* 4:035006.

Douglas, E. M., A. Beltrán-Przekurat, D. Niyogi, R. A. Pielke Sr., and C. J. Vörösmarty. 2009. The impact of agricultural intensification and irrigation on land–atmosphere interactions and Indian monsoon precipitation—A mesoscale modeling perspective. *Global Planet Change* 67:117–128.

Dwivedi, O. P. 1989. *World Religions and the Environment*, 462. New Delhi, India: Gitanjali.

Economist. 2003. The promise of blue revolution, August 7, 2003. Available at http://www.economist.com/printedition/2003-0809.

Engelman, R., and P. LeRoy. 1993. *Sustaining Water*, 56. Washington, DC: Population Action International.

Enzel, V., L. L. Ely, S. Mishra et al. 1955. High resolutions Holocene environmental changes in the Thai Desert, northwestern India. *Science* 284:125–128.

Faruqui, N. 2003. Balancing between the eternal yesterday and the eternal tomorrow: Economic management innovative approaches to contemporary issues. In *Rethinking Water Management: Innovative Approaches to Contemporary Issues*, eds. C. Figueres, J. Rockstrom, and C. Tortajad, 61. London: Earthscan Publishers Ltd.

Fischetti, M. 2011. Caught on video: A Himalayan glacier deflates. *Scientific American*. Available at http://www.scientificamerican.com/article.cfm?id=caught-on-video-a-himalayan-glacier (accessed January 2, 2014).

Fujita, K., and T. Nuimura. 2011. Spatially heterogeneous wastage of Himalayan glaciers. *Proc Natl Acad Sci U S A* 108(34):14011–14014.

Gadgil, M., and P. R. S. Rao. 1995. Designing incentives to conserve India's bio-diversity. In *Property Rights in a Social and Ecological Context: Case Studies and Applications*, eds. S. Hanna, and M. Munasngha, 53–62. Washington, DC: Beijer International Institute of Ecological Economics and The World Bank.

Ghosh, N., and J. Bandyopadhyay. 2009. A scarcity value based explanation of trans-boundary water disputes: The case of the Cauvery River Basin in India. *Water Policy* 11:141–167.

Giordano, M., M. Giordano, and A. Wolf. 2002. The geography of water conflict and cooperation: Internal pressures and international manifestations. *Geogr J* 168(4):293–312.

Goldenberg, S. 2011. Everest's ice is retreating as climate change grips the Himalayas. *The Guardian*. Available at http://www.theguardian.com/world/2011/sep/25/climate-change-himalayas-glaciers-melting (accessed January 2, 2014).

Good, A. G., A. K. Shrawat, and D. G. Muench. 2004. Can less yield more? Is reducing nutrient input into the environment compatible with maintaining crop production? *Trends Plant Sci* 9(12):597–605.

Goswami, B. N., V. Venugopal, D. Sengupta, M. S. Madhusoodanan, and P. K. Xavier. 2006. Increasing trend of extreme rain events over India in a warming environment. *Science* 314:1442–1445.

Gumma, M. K., P. S. Thenkabail, and A. Nelson. 2011. Mapping irrigated areas using MODIS 250 meter time-series data: A study on Krishna River Basin (India). *Water* 3:113–131.

Gunatilaka, A. 2009. The Anthropocene—A 200 year record of human driven geological impacts: Prelude to global climate changes and implication for south Asia. *J Natl Sci Found Sri Lanka* 37(1):3–11.

Gunnell, Y., and A. Krishnamurthy. 2003. Past and Present status of runoff harvesting systems in dryland peninsular India: A critical review. *Ambio* 32(4):320–324.

Gunnell, Y., A. Krishnamurthy, and B. Sultan. 2007. Response of the south Indian runoff-harvesting civilization to northeast monsoon rainfall variability during the last 2000 years: Instrumental records and indirect evidence. *Holocene* 17(2):207–215.

Hardin, G. 1968. The tragedy of the commons. *Science* 162(3859):1243–1248.

Hussain, S. S., and M. Mudasser. 2007. Prospects for wheat production under changing climate in mountain areas of Pakistan—An econometric analysis. *Agric Syst* 94:494–501.

Ice Age Now. 2011. Glaciers are growing around the world, including the United States. *Ice Age Now*. Available at http://www.iceagenow.com/Growing_Glaciers.htm (accessed January 2, 2014).

Inman, M. 2010. Settling the science on Himalayan glaciers. *Nature*. Available at http://www.nature.com/climate/2010/1003/full/climate.2010.19.html (accessed January 2, 2014).

IPCC. 2013. Climate change 2013, the physical science basis, fifth assessment report, WMO, Geneva, Switzerland.

Jenkins, W., and C. K. Chapple. 2011. Religion and environment. *Annu Rev Environ Res* 36:441–463.

Jian, J., P. J. Webster, and C. D. Hoyos. 2009. Large-scale controls on Ganges and Brahmaputra river discharge on intraseasonal and seasonal time-scales. *Q J Roy Meteorol Soc* 135:353–370.

Kerr, R. A. 2009. Northern India's groundwater is going, going, going. *Science* 325:798.

Kondepati, R. 2011. Agricultural groundwater management in Andhra Pradesh, India: A focus on free electricity policy and its reform. *Int J Water Resour Dev* 27(2):375–386.

Kumar, R., R. D. Singh, and K. D. Sharma. 2005. Water resources of India. *Curr Sci India* 89(5):794–811.

Lal, R. 2004a. Off-setting China's CO_2 emissions by soil carbon sequestration. *Clim Change* 63:263–275.

Lal, R. 2004b. Soil carbon sequestration in India. *Clim Change* 65:277–296.

Lamb, H. H. 1982. *Climate History and the Modern World*, 387. London: Methuen.

Lamballe, A. 2000. Water, a source of conflict in South Asia. *Guerr Mond Conflits Contemp* 195:115–167.

Li, X., S. R. Waddington, J. Dixon, A. K. Joshi, and M. C. de Vicente. 2011. The relative importance of drought and other water-related constraints for major food crops in South Asian farming systems. *Food Sec* 3:19–33.

Limaye, S. D. 2004. Some aspects of sustainable ground water development in South Asia. In *Research Basins and Hydrological Planning*, eds. R. Z. Xi, W. Z. Gu, and K. P. Seiler, 35–45. The Netherlands: Taylor & Francis.

Lobell, D. B., J. I. Ortiz-Manasterio, and A. S. Lee. 2010. Satellite evidence for yield growth opportunities in northwest India. *Field Crop Res* 118:13–20.

Lovelock, J. 1979. *Gaia: A New Look at Life on Earth*, 148. Oxford, UK: Oxford Univ. Press.

Loy, D. R. 1997. The religion of the market. *J Am Acad Relig* 65(2):257–290.

Mall, R. K., A. Gupta, R. Singh, R. S. Singh, and L. S. Rathore. 2006. Water resources and climatic change: An Indian perspective. *Curr Sci India* 90(12):1610–1626.

Mathur, N. 2004. Unbundling 'lack of access' to water in rural India. *Proceedings of the Workshop on Water and Politics*, 249–256. Marseille: World Water Council.

Misra, A. K., A. Saxena, M. Yaduvanshi, A. Mishra, Y. Bhadauriya, and A. Thakur. 2007. Proposed river-linking project of India: A boon or bane to nature. *Environ Geol* 51:1361–1376.

Mollinga, P. P. 2010. The material conditions of a polarized discourse: Clamours and silences in critical analysis of agricultural water use in India. *J Agrar Change* 10(3):414–436.

Mukherji, A., and T. Shah. 2005. Socio-ecology of groundwater irrigation in South Asia: An overview of issues and evidence. *Int Assoc Hydrogeol (Groundwater Intensive Use)* 7:53–77.

Mushtaq, S., T. K. Maraseni, J. Maroulis, and M. Hafeez. 2009. Energy and water tradeoffs in enhancing food security: A selective international assessment. *Energy Policy* 37:3635–3644.

Nakayama, M. 1997. Successes and failures of international organizations in dealing with international waters. *Water Resour Dev J* 13:367–382.
Nelson, D., and R. Alleyne. 2011. Some Himalayan glaciers are advancing rather than melting, study finds. *The Telegraph*. Available at http://www.telegraph.co.uk/earth/environment/climatechange/8284223/Some-Himalayan-glaciers-are-advancing-rather-than-melting-study-finds.html (accessed January 3, 2014).
Oki, T., and S. Kanae. 2006. Global hydrological cycles and world water resources. *Science* 313:1068:1072.
Proctor, J. 2004. Resolving multiple visions of nature, science and religion. *Zygon* 39(3):637–658.
Rahman, M. M., R. Naidu, and P. Bhattacharya. 2009. Arsenic contamination in groundwater in the Southeast Asia region. *Environ Geochem Health* 31:9–21.
Rajagopal, D. 2008. Implications of India's biofuel policies for food, water and the poor. *Water Policy* 10(1):95–106.
Raloff, J. 2011. Glaciers largely stable in one range of Himalayas. *Science News*. Available at https://www.sciencenews.org/article/glaciers-largely-stable-one-range-himalayas (accessed January 3, 2014).
Ramamathan, V., C. Chung, D. Kim et al. 2005. Atmospheric brown clouds: Impacts on south Asian climate and hydrological cycles. *Proc Natl Acad Sci U S A* 102:5326–5333.
Rangachari, R., and B. G. Verghese. 2001. Making water works to translate poverty into prosperity: The Ganga–Brahmaputra–Barak region. In *Ganges–Brahmaputra–Meghna Region: A Framework for Sustainable Development*, eds. Q. K. Ahmad, A. K. Biswas, R. Rangachari, and M. M. Sainjus, 81–142. Bangladesh: The University Press.
Ranjan, P., S. Kazama, and M. Sawemoto. 2006. Effects of climate change on coastal fresh groundwater resources. *Global Environ Change* 16:388–399.
Rockström, J., M. Lannerstad, and M. Falkenmark. 2007. Assessing the water challenge of a new green revolution in developing countries. *Proc Natl Acad Sci U S A* 104(15):6253–6260.
Rodell, M., I. Velicogna, and J. S. Famiglietti. 2009. Satellite-based estimates of groundwater depletion in India. *Nature* 460:999–1003.
Ruddiman, W. F. 2003. The anthropogenic greenhouse era began thousands of years ago. *Clim Change* 61:261–293.
Sahni, H. 2006. The politics of water in south Asia: The case of the Indus waters treaty. *SAIS Rev* 26(2):153–165.
Saleth, R. M. 2011. Water scarcity and climatic change in India: The needs for water demand and supply management. *Hydrol Sci J* 56(4):671–686.
Salman, S. M. A. 1998. Sharing the Ganges waters between India and Bangladesh: An analysis of the 1996 treaty. In *International Watercourses: Enhancing Cooperation and Managing Conflict*, World Bank Technical paper no. 414, eds. S. M. A. Salman, and L. B. de Chazournes, 127–153. Washington, DC: The World Bank.
Salman, S. M. A., and K. Uprety. 1999. Hydro-politics in South Asia: A comparative analysis of the Mahakali and the Ganges treaties. *Nat Resour J* 39(2):295–343.
Sampath, A., B. Kedarnarth, C. Ramanujam et al. 2003. *Water Privatization and Implications in India*, 1–19. Austin, TX: Association for India's Development. Available at http://www.doccentre.org/docsweb/water/water_privatization.pdf (accessed January 3, 2014).
Schultz, B., and S. Uhlenbrook. 2007. *Water Security: What Does it Mean, What May it Imply? UNESCO–Institute for Water Education*. The Netherlands: Deft. Available at http://www.iahr.org/uploadedfiles/userfiles/files/3_paper%20water%20Secur%20Box%20draft.pdf (accessed January 3, 2014).
Shah, R. B. 2001. Ganges–Brahmaputra: The outlook for the twenty first century. In *Sustainable Development of the Ganges–Brahmaputra–Meghna Basins*, eds. A. K. Biswas, and J. I. Uitto, 17–45. Tokyo: United Nations University Press.
Shah, T., A. Deb Roy, A. S. Qureshim, and J. Wang. 2003. Sustaining Asia's groundwater boom: An overview of issues and evidence. *Nat Resour Forum* 27:130–141.

Shah, T., O. P. Singh, and A. Mukherji. 2006. Some aspects of south Asia's groundwater irrigations economy: Analysis from a survey in India, Pakistan, Nepal Terai and Bangladesh. *Hydrogeol J* 14:286–309.

Sharma, B. R., and P. S. Minhas. 2005. Strategies for managing saline/alkali waters for sustainable agricultural production in south Asia. *Agr Water Manage* 78:136–151.

Shen, Y., and Y. Chen. 2010. Global perspective on hydrology, water balance, and water resources management in arid basins. *Hydrol Process* 24:129–135.

Singh, G. 1971. The Indus Valley culture seen in the context of post-glacial climatic and ecologic studies in northwest India. *Rev Paleobot Paleoentol* 64:385–388.

Singh, K. 2007. Rational pricing of water as an instrument of improving water use efficiency in the agricultural sector: A case study in Gujarat, India. *Water Resour Dev* 23(4):679–690.

Singh, V. P., and F. Asce. 2008. Water, environment, engineering, religion, and society. *J Hydrol Eng* 3:118–123.

Sophocleous, M. 2004. Global and regional water availability and demand: Prospects for the future. *Nat Resour Res* 13(2):61–75.

Subramanian, V. 2008. Nitrogen transport by rivers of south Asia. *Curr Sci India* 94(11):1413–1418.

Sutcliffe, J., J. Shaw, and E. Brown. 2011. Historical water resources in South Asia: The hydrological background. *Hydrol Sci J* 56:775–788.

Tankha, A., and B. Fuller. 2010. Getting things done: Bureaucratic and entrepreneurial approaches to the practice of participatory water management reforms in Brazil and India. *Water Policy* 12(1):84–103.

Tyler, S. 2004. Governance and water management in Asia: What do we need to learn? *Proceedings of the Workshop on Water and Politics*, 115–124. Marseille: World Water Council.

United Nations. 1975. Management of international water resources: Institutional and legal aspects. *Natural Resources Water Series*, 271. New York: United Nations, Department of Economic and Social Affairs.

United Nations. 2011. Environment statistics—Country snapshots: India. United Nations Statistics Division. Available at http://unstats.un.org/unsd/environment/Questionnaires/country_snapshots.htm (accessed April 20, 2012).

United Nations/WWWAP. 2003. UN World Water and Development Report: Water for People, Water for Life, August 25, 2003. Available at http://www.unesco.org/water/wwap/wwdr/index.shmtl (accessed April 20, 2012).

Warner, J. 2004. Water, wine, vinegar, blood: On politics, participations, violence and conflict over hydrosocial contract. *Proceedings of the Workshop on Water and Politics*, 7–18. Marseille: World Water Council.

Wassan, R. J., G. I. Smith, and D. P. Agarwal. 1984. Late quaternary sediments, minerals, and infrared geochemical history of Didwane Lake, Thai Desert, India. *Paleogrogr Paleoclimatol Paleoecol* 46:345–372.

White, L. Jr. 1974. The historical roots of our ecologic crisis. In *Ecology and Religion in History*, eds. D. Spring, and E. Spring, 12. New York: Harper Torchbook.

White, L. Jr. 1976. The historical roots of our ecological crisis. *Science*. 155(3767):1203–1207.

Wolff, P., and T. M. Stein. 1999. Efficient and economic use of water in agriculture—Possibilities and limits. *Nat Resour Dev* 49/50:151–159.

World Water Council. 2004. Understanding the role of politics in water management. *Proceedings of the Workshop on Water and Politics*, February 26–27, Marseille. Available at http://www.oslj.org/Foundation/Docs/UN/WorldWaterCouncil-WaterAndPolitics.pdf (accessed January 2, 2014).

Xu, J., R. E. Grumbine, A. Shrestha et al. 2009. The melting Himalayas: Cascading effects of climate change on water, biodiversity, and livelihoods. *Conserv Biol* 23(3):520–530.

Yeston, J., R. Coontz, J. Smith, and C. Ash. 2006. Introduction: A thirsty world. *Science* 313:1067.

9 Conservation Agriculture on Chernozems in the Republic of Moldova

Boris Boincean and Rattan Lal

CONTENTS

9.1 INTRODUCTION

The presently used Moldovan farming systems are not sustainable. The unsustainability of the present farming systems are indicated by the following: (i) the leveling off or declining yields of most crops, both on average for the Republic of Moldova and in the long-term field experiments conducted by Selectia Research Institute of Field Crops (RIFC) in Balti (northern part of Moldova), and (ii) the experimentally monitored decreases in the stocks of soil organic matter (SOM) as result of uncompensated annual losses (e.g., mineralization, leaching, and erosion).

For example, the average increase in the yield of winter wheat (*Triticum aestivum*) for the Republic of Moldova for the period 1972–1981 was 1.75 Mg/ha or 94.6% relative to 1962–1971. However, the increase in yield for the same period in the long-term field experiments with different crop rotations was 1.86 Mg/ha or 56.7%. Since then, the yields of winter wheat have stabilized or even decreased. In general, the yields in the long-term field experiment have been higher than national average by 42.8%–77.3% for the first 30 years (1962–1991) and by 106.1%–119.1% for the last 17 years (1994–2011). Similar trends have been observed for sugar beet (*Beta vulgaris*) and corn for grain (*Zea mays*).

The introduction of new, more intensive varieties of winter wheat have increased yields with crop rotation by 12.3%–12.6% relative to older, less productive varieties, in both fertilized and unfertilized plots. It means that new varieties of winter wheat have higher requirements of soil fertility but not of higher rates of mineral fertilizers.

Mineral fertilizers have decreased the positive impact of crop rotation relative to permanent cropping for winter wheat, sugar beet, and corn for grain by 60.7%, 127.7%, and 136.6%, respectively. Nonetheless, the impact of crop rotation on yield remains to be substantial. Excessive use of mineral fertilizers and pesticides cannot compensate for the lack of crop rotation.

Use of mineral and organo-mineral fertilizers and irrigation increases the yields of winter wheat and sugar beet by 19.6%–24.8% and 38.7%–43.9%, respectively. However, these inputs are reducing the stocks of SOM both in the 0–20 cm and especially in the 0–100 cm soil depth.

Thus, uncompensated losses of SOM in crop rotation on plots fertilized with mineral and organo-mineral fertilizers are estimated at 0.26 and 0.09–0.13 Mg/ha/year in the 0–20 cm soil layer, but 0.45–0.58 Mg/ha/year with irrigation on both fertilized and unfertilized plots.

There is a strong need for a paradigm shift from traditional toward sustainable intensification on the basis of more intensive recycling of nutrients and energy through use of site-specific renewable sources of energy. It is only through integration of crop rotations with perennial legumes and grasses, in conjunction with minimum tillage and organic fertilizers (farmyard manure [FYM]), that crop productivity and soil fertility can be enhanced and sustained.

Indiscriminate agricultural intensification, primarily focused on maximum yields and profitability without taking into account adverse impacts on soil quality, environment, and people's health (Boincean 2013a; Cassman 1999; Gliessman 2000; International Assessment of Agricultural Science and Technology for Development 2009), is not sustainable. These negative consequences are externalized; however, the consequences to soil and biodiversity can be irreversible (Weil and Magdoff 2000). The weakness of the technological reductionist approach, still dominating agriculture in Moldova, is that the main factor of intensification (e.g., lack of crop rotations or long periods of continuous monoculture; mechanization; intensive tillage; mineral fertilizers) have been applied separately and not integrated within the farming system. The new paradigm for agricultural intensification is based on the premise of decreasing the dependence on nonrenewable sources of energy with simultaneous reduction in the risks of pollution and degradation of the environment (Gliessman 2000; Kassam 2011).

The Republic of Moldova is a unique country. Almost 80% of the territory is covered by Chernozems. A typical Chernozem from Balti steppe (the northern part of the Republic of Moldova) was the foundation of soil science in the 19th century. The father of soil science, the Russian scientist V.V. Docuceaev, described "Russian Chernozem" in his book published in 1877. A typical Chernozem is deep, fertile, and rich in SOM on humus and has a good structure. Soil monoliths from Balti have been displayed by V.V. Docuceaev in Chicago and Paris as the best soil in the world. Chernozems are developed under the influence of the predominant climate of these regions (alternating drought and humid conditions); steppe vegetation (mainly from perennial grasses [*Stipa*]) and the characteristic fauna, including animals; local landscape (relief); and granulometric and mineralogical characteristics of parent materials. Anthropogenic activities did not influence soil formation under natural conditions.

Degradation of these soils, similar to most Chernozems from East European regions, began with the deforestration of the territory and plowing of the steppes.

Globally, Chernozems cover a wide region of some 240 million hectares (Mha) of the middle-latitude Eurasian steppes and North American prairies. In the former Union of Soviet Socialist Republics, the total area under Chernozems consisted of 190 Mha on 79% of the world total. The area under typical Chernozems in the Republic of Moldova is 817,000 ha.

Sustainable development of agriculture of Chernozems can be achieved by modeling agroecosystems similar to the attributes of natural ecosystems. Indeed, the industrial model of agricultural intensification has strongly influenced Chernozems, especially during the last 50 years since the 1960s. Krupenikov (2008) mentions >40 types of Chernozem degradation. One of the integrated indices of soil fertility is SOM concentration. The SOM stock in Chernozems has decreased strongly since Docuceaev's earlier estimates of 100–150 Mg/ha.

The awareness regarding the state of Chernozems has been enhanced by the classics of agronomy (Docuceaev 1952; Kosticev 1885, 1940; Villiams 1950–1952). Kovda (1983), in the book *Russian Chernozem: 100 Years After Docuceaev*, has pointed out that Chernozems have been changed significantly since the 1960s.

The long grass fallow at the turn of the 19th century has been replaced by intensive row cropping, and by taking soil for granted as a natural resource according to a typical consumerist attitude. The globally successful intensive row crop system developed after the Second World War by the industrial model of intensification is popularly called the "Green Revolution."

However, the initial increase in yield caused by the heavy inputs (mineral fertilizers, pesticides, etc.) masked the adverse effects of decline in SOM and soil health. The problem cannot be solved without adopting a holistic approach to the farming system.

Long-term field experiments at Selectia RIFC have provided useful data. These experiments include crop rotations, continuous monocultures, different systems of soil tillage, irrigation, and fertilizer use. The data indicate that separate application of modern factors of agricultural intensification cannot compensate for the annual losses of SOM (Boaghii and Bulat 2013; Boincean 2013a; Boincean et al. 2013a,b) and the attendant decline in soil quality. Only a combination of the three main factors of each farming system (e.g., crop rotation, system of soil tillage, and nutrient management for the specific crop rotation) with regular input of fresh organic matter (OM) can restore soil quality. It is only through the adoption of a holistic farming system approach that would make it possible to drastically reduce soil tillage, and decrease the amount of nitrogen (N) input from mineral fertilizers by a more efficient use of biological nitrogen fixation from leguminous crops and FYM. The holistic approach can also reduce the use of pesticides by increasing the diversity of crops within every field and landscape (Boincean 2013a; Cassman 1999; Krupenikov et al.

2010; Lal 2013; Powlson et al. 2013), and strengthen soil resilience and its disease-suppressive attributes.

Therefore, the objectives of this chapter are to discuss the results of a long-term soil management experiment on soil quality and agronomic productivity of a Chernozem in Moldova. In addition, the results are discussed in the context of sustainable management of these soils under similar ecological characteristics elsewhere in Eastern Europe.

9.2 METHODS AND CONDITIONS

The long-term field experiments on a typical Chernozem (heavy clay) on the Balti steppe in northern Moldova have been established for ~40–50 years. The surface 0–20 cm layer of these soils is characterized by an SOM concentration of 4.8%–5.0% (by the Tiurin method); pH_{water} 7.3; pH_{CaCl2} 6.2; and concentration of total N, P, and K of 0.20%–0.25%, 0.09%–0.11%, and 1.22%–1.28%, respectively. Details of soil properties and physiographic characteristics have been reported elsewhere (Dent 2013; Krupenikov et al. 2011).

The long-term field experiment with crop rotation, established in 1962, includes eight crop rotations with different intensities (40%–70%) of row crops. Winter wheat (*T. aestivum*) is grown in three fields of each crop rotation (30%), but after different preceding crops (early harvested—vetch [*Vicia sativa*] and oats [*Avena sativa*] for green biomass; alfalfa [*Medicago sativa*] on the third year after first cutting, etc.; corn for silage and corn for grains). Fertilization with organic and mineral fertilizers is done according the recommended rates for the Republic of Moldova. Simultaneously, several crops are grown as permanent crops. The size of the experimental plots is 283 m^2 in crop rotations with three replicates. The size of the experimental plots for permanent crops is 450 m^2, but without any replications.

The long-term field experiment with four systems of fertilization includes

1. Unfertilized (absolute control)
2. Mineral fertilizer at three rates (NPK_{75}, NPK_{130}, and NPK_{175} kg a.i. per hectare of crop rotation)
3. Organo-mineral fertilizer with two rates of FYM (10 and 15 Mg/ha) and three rates of mineral fertilizers per hectare of crop rotation
4. Organic fertilizers (15 Mg of FYM) per 1 ha of crop rotation

The following mineral fertilizers are used in the experiment:

1. Ammonium nitrate (31.1%–35.0% a.i. of nitrogen)
2. Double superphosphate (45%–50% a.i. of P_2O_5)
3. Muriate of potassium (40% a.i. K_2O)
4. Ammonium phosphate (11%–12% a.i. N and 46%–60% a.i. P_2O_5)

The composted farmyard cow manure used in the experiment contains C—13.2%, N—0.51%, P—0.23%, and K—0.60% on a dry-weight basis.

The size of the experimental plots is 242 m^2. The experiment with four replications was laid out according to the randomized block design. More details of the experiment are available in earlier reports (Boincean and Nica 2007; Boincean et al. 2013a,b; Boincean 2013b,c).

The long-term field experiment with irrigation, established in 1970, includes six treatments (alfalfa–alfalfa–alfalfa–winter wheat–sugar beet–corn for grain). The size of each experimental field is 3 ha. The three systems of fertilization used in the experiment are without fertilization, composted manure, and composted manure together with mineral fertilizers. Mineral fertilizers are used for winter wheat (N_{60} P_{90} K_{40} kg a.i./ha) and sugar beet (N_{70} P_{90} K_{50} kg a.i./ha). For sugar beet, composted manure (80 Mg/ha) is applied each year in crop rotation. The experiment has four replications. The total size of the experimental plots is 20–400 m^2, and 50–100 m^2 for measuring crop yields. All plots are irrigated by a sprinkler system at the rate of water application of 350–400 m^3/ha. The irrigation water is applied when soil water field capacity is <75%–80%.

For winter wheat, irrigation is used before and after sowing, and during the vegetation period, according the design of the experiment.

For sugar beet, the rates of irrigation are from 300–500 to 800–900 m^3/ha. The irrigation is used during three main periods of vegetative growth, taking into consideration the status of soil water field capacity.

Herbicides for weed control have been used only in crop rotations with different systems of fertilization and irrigation: for winter wheat, Esteron (0.8–1.0 l/ha); for corn for grain, Trophy 90 EC (2.0–2.5 l/ha); and for sugar beet, Frontier Optime (1.2–1.4 l/ha). Insecticides used consisted of DecisProfi 250 WG for sugar beets and Actara 25 WG for winter wheat.

The long-term multifactorial field experiment established in 1996 includes two crop rotations (with and without a mixture of perennial grasses and leguminous crops), two systems of soil tillage in crop rotation (combination of moldboard plow with nonreversible soil tillage and nonreversible system of soil tillage), and three systems of fertilization (unfertilized, manure, and manure + NPK). Crop rotations vary in space and time, and have three replications. The experimental layout is randomized split blocks. No pesticides or herbicides are used in the experiment. The size of each experimental plot is 264 m^2. More details about the experiment are available in Boincean et al. (2013a) and Boincean (2013b,c).

The 50-year mean annual precipitation is 544.2 mm. However, the distribution of precipitation is not uniform, especially during the vegetative period. Torrential rains are received during a short period of May–June in Moldova.

9.3 RESULTS AND DISCUSSION

The industrial model of intensification substantially increased the yields of most crops in Moldova during the 1970s and 1980s (Table 9.1).

However, yields leveled off or decreased across the country, and also in the Selectia long-term field experiments, despite the introduction of new varieties and hybrids of field crops. The yield gap between farmers' yields and experimental station yields has increased over the last 20 years because of the collapse of the economy. Indeed,

TABLE 9.1
Temporal Changes in Average Crop Yields for the Republic of Moldova and in the Selectia Long-Term Field Experiments between 1962 and 2011

	National Average for Republic of Moldova		Selectia Long-Term Field Experiments		Yield Increase in Long-Term Field Experiments Relative to National Average	
Years	Mg/ha	± Relative to Initial	Mg/ha	± Relative to Initial	± Mg/ha	%
			Winter Wheat (*Triticum aestivum*)			
1962–1971	1.85	–	3.28	–	+1.43	77.3
1972–1981	3.60	+1.75	5.14	+1.86	+1.54	42.8
1982–1991	3.55	−0.05	5.43	+0.29	+1.88	53.0
1994–2003	2.35	−1.20	5.15	−0.28	+2.80	119.1
2004–2011	2.44	+0.09	5.03	−0.12	+2.59	106.1
			Sugar Beet (*Beta vulgaris*)			
1962–1971	23.65	–	40.06	–	+16.41	69.4
1972–1981	26.58	+3.20	45.79	+5.73	+18.94	70.5
1982–1991	27.58	+0.73	44.96	−0.83	+17.38	63.0
1994–2003	19.81	−7.77	44.04	−0.92	+24.23	122.3
2004–2011	26.97	+7.16	40.90	−3.14	+13.93	51.6
			Maize (*Zea mays*) Grain Yield			
1962–1971	3.17	–	5.60	–	+2.43	76.7
1972–1981	3.55	+0.38	6.78	+1.18	+3.23	91.0
1982–1991	3.96	+0.41	6.86	+0.08	+2.90	73.2
1994–2003	2.79	−1.17	5.84	−1.02	+3.05	109.3
2004–2011	2.94	+0.15	5.87	+0.03	+2.93	99.7

experimental yields are 60%–100% higher than the national average yields. The stagnation or decline in crop yields are attributed to soil degradation, in particular to continuing, uncompensated annual losses of SOM as a result of intensive tillage, insufficient return of fresh OM, and accelerated soil erosion.

The simplistic approach to agricultural intensification has led to numerous adverse consequences. Thus, there is a strong need for a paradigm shift toward a lesser dependence on nonrenewable sources of energy and their derivates. The data from the Selectia long-term field experiments are summarized in the following in terms of the influence of each factor of agricultural intensification on crop yields and soil quality.

9.3.1 Crop Rotation and Continuous Cultivation

The highest yields were obtained from crop rotation as opposed to continuous monocropping, and from fertilized relative to unfertilized treatments. The responses of

TABLE 9.2
Yields of Continuous Winter Wheat, Sugar Beet, and Maize and Crops Grown in Rotation (1994–2011)

Cropping Systems	Fertilizer Use		Increase with Fertilizer		LSD_{05}	Yield Increase in Crop Rotation Relative to Continuous Cultures, Mg/ha and %	
	Unfertilized	Fertilized	Mg/ha	%		Unfertilized	Fertilized
Winter Wheat							
Continuous crop	1.95	2.80	0.85	43.6		—	—
Crop rotation	4.73	5.10	0.37	7.8	0.20	2.78 (142.6)	2.30 (82.1)
Sugar Beet							
Continuous crop	8.54	16.83	8.29	97.1		—	—
Crop rotation	32.55	42.65	10.1	31.0	2.1	24.01 (281.1)	25.82 (153.4)
Maize Grain Yield							
Continuous crop	3.81	5.47	1.66	43.6		—	—
Crop rotation	5.43	5.85	0.42	7.7	0.27	1.62 (42.5)	0.38 (6.9)

individual crops differ (Table 9.2) but the yield increase that can be attributed to fertilizers is higher with continuous monocultures. Furthermore, agronomic yield under crop rotation relative to continuous monoculture is significantly more than through the benefits of fertilization, except for the grain yield of maize.

Despite the apparent increase in yield, fertilizer use decreased the positive effects of crop rotation. Probably, the visible effects of fertilizers masked the benefits of crop rotation. In the beginning of industrial intensification, fertilizers and the use of agrochemicals to control weeds, pests, and diseases appeared to compensate for the simplification of crop rotations during the period of extraordinary specialization and concentration of agricultural production. Low prices for nonrenewable sources of energy and their derivates (mineral fertilizers and agrochemicals) made the new approach attractive. However, in reality, fertilizers are no substitute for crop rotation. Alas, cheap fuel, and fertilizers are now a history.

9.3.2 High-Yielding Varieties and Hybrids

The other misconception has been the perception that agronomic yields can simply be increased by introducing new varieties of field crops along with higher rates of fertilizer. However, experimental data for winter wheat in the Selectia long-term experiments do not support this perception (Table 9.3).

TABLE 9.3
Yield Increase of Winter Wheat under the Influence of New, High-Yielding Varieties Grown in Crop Rotation and as Continuous Wheat, Average for 1994–2011

Crop Rotation, Continuous Wheat	Unfertilized		Fertilized		
	Mg/ha	%	Mg/ha	%	LSD_{05}
Crop rotation	0.53	12.6	0.56	12.3	0.20
Continuous wheat	0.13	7.1	0.06	2.2	—

Any additional yield from new varieties was similar in fertilized and unfertilized plots. Thus, the additional yield is produced from inherent soil fertility because new varieties require additional plant nutrients compared with traditional varieties. Moreover, the yield increase from new varieties (12.3%–12.6%) is significantly lower than the beneficial influence of crop rotation, and such an increase was achieved only in crop rotation. The yield increase from new varieties under continuous wheat was 2.2% and 7.1% on fertilized and unfertilized plots, respectively.

The experience gained by Selectia RIFC during the last 70 years (and farmers' experience) demonstrates the value of using crop varieties that are well adapted to local conditions. Use of appropriate crop rotation also reduces dependence on imported seed, including genetically modified seeds. There has been a lot of discussion about adaptation to global warming using varieties and hybrids with higher tolerance to droughts. The increased risks of drought under a changing climate cannot be ignored. Thus, it is also pertinent to take into account soil conditions and the strategy to enhance soil water storage in the root zone for reducing vulnerability to drought and decreasing expenses to install irrigation facilities.

The goal of increasing yields and profitability has been implemented since the 1960s through a package of practices: new varieties and hybrids of field crops; intensive, mechanized tillage; monoculture or short-duration specialized crop rotations; irrigation; and fertilizers and chemicals for pest, weed, and disease control. Such packages minimized the importance of inherent soil fertility, specifically that of SOM. Indeed, data from the Selectia long-term field experiment with different crop rotations and continuous cropping for >50 years show losses of SOM under crop rotations with different intensity or frequency of row crops and under continuous winter wheat and maize (Table 9.4).

Even crop rotation with 30% of alfalfa and 4 Mg/ha of FYM over the crop rotation could not compensate for the annual loss of SOM (Boincean 2013a). The same is true for the long-term field experiments with different systems of soil tillage, fertilization, and irrigation in crop rotations.

In general, long-term field experiments in many countries have been aimed at studying one or two factors, very often in large gradations, without their integration into the framework of a farming system. The original Selectia long-term field experiments were no exception. Indeed, the industrial intensification of agriculture was stimulated by a reductionist approach to farming systems research. A more holistic

TABLE 9.4
Changes in Stocks of Soil Organic Carbon during 47 Years in Long-Term Field Experiments with Crop Rotations and Continuous Monocultures on Fertilized Plots, 0–20 cm Soil Layer, Field No. 6

		Continuous Cereal		Crop Rotation			
				Share of Row Crops, %			
				40	50	60	
Years	Indices	Winter Wheat	Maize, Grain	4 Mg/ha Manure + 30% Alfalfa (*Medicago sativa*)	4 Mg/ha Manure + 10% Black Fallow	8 Mg/ha Manure	LSD_{05}
1962	Initial SOC stocks, 78.7 Mg/ha						
2009	SOC stocks	73.4	67.4	64.8	59.5	62.2	2.9
	Stock reduction	5.3	11.3	13.9	19.2	16.5	
	Percentage from initial stocks	6.7	14.4	17.7	24.4	21.0	
	Rate of SOC depletion, Mg/ha/year	0.11	0.24	0.30	0.41	0.35	

approach was initiated in 1995–1996 with a multifactorial experiment including three main components of each farming system: alternation of crops, fertilization, and soil tillage in crop rotation. The results from two full rotations are presented below.

9.3.3 Soil Tillage

Another long-term experiment conducted by the Department of Sustainable Farming Systems at Selectia RIFC included eight different systems of soil tillage in a seven-field crop rotation with 57% of row crops. Mineral fertilizers were applied at the recommended rates for different crops along with an average of 5.7 Mg/ha at FYM over the crop rotation. Table 9.5 shows the yield data for winter wheat, sugar beet, and maize for two variants: annual moldboard plowing for all crops in the rotation and noninversion tillage for all crops except sugar beet where the moldboard plow was used only for one rotation. The primary noninversion tillage was performed by blade working tools at the depth of 25–27 cm for corn for grain and 32–35 cm for sugar beet. For winter wheat, disks were used for tillage to the depth of 10–12 cm. Crop yields did not respond to different systems of tillage; however, the stocks of SOM were higher under annual noninverting soil tillage than under the plow tillage (Table 9.6). The total losses of SOM during the 14 years for the 0–20 cm layer were 0.8 and 1.3 Mg/ha, respectively. Reduction in soil tillage decreased the total and annual losses of SOM, but cannot mitigate them completely on arable lands.

TABLE 9.5
Yields in Seven-Field Crop Rotation with Different Systems of Primary Soil Tillage for 1977–1990

	Field Crops, Mg/ha					
	Winter Wheat		Sugar Beet[a]		Maize for Grain	
	Moldboard Plow	Noninversion Tillage	Moldboard Plow	Noninversion Tillage	Moldboard Plow	Noninversion Tillage
	4.56	4.70	46.0	45.3[a]	7.3	7.3
LSD_{05}	0.73		3.33		0.79	

Source: Adapted from Boaghii, IV, and Bulat, L. 2013. Primary soil tillage in the crop-rotation for main field crops in Moldova. In: Dent, DL (editor). *Soil as World Heritage.* Springer, Dordrecht, pp. 273–282.

[a] 1984–1990.

TABLE 9.6
Stocks of Soil Organic Matter (or Carbon) under Different Methods of Primary Soil Tillage in Crop Rotation: Average of Seven-Field Crop Rotation, for 0–20 cm Soil Layer

Stocks of Soil Organic Matter, Mg/ha				Moldboard Plow		Noninversion Tillage	
Moldboard Plow		Noninversion Tillage					
1977	1990	1977	1990	Total Losses	Annual Losses	Total Losses	Annual Losses
62.0	60.7	64.3	63.5	1.3	0.093	0.8	0.057

Source: With kind permission from Springer Science+Business Media: *Soil as World Heritage*, Dent, DL (editor), Primary soil tillage in the crop-rotation for main field crops in Moldova, 2013, pp. 273–282, Boaghii, IV, and Bulat, L.

As in the other long-term experiments with crop rotations and continuous cultivation, high crop yields are masking the true state of soil fertility. Perhaps the increased yields of most crops during the initial stages of the industrial agricultural intensification also masked the state of soil degradation.

9.3.4 Fertilization

In the other long-term field experiments at Selectia RIFC, 12 systems of fertilization are studied using six-field crop rotation in which row crops comprise half of the cropping. More details about this experiment are given in a report by Boincean et al. (2013b). The data in Table 9.7 show mean yields for winter wheat, sugar beet, and maize for 1991–2011 under four levels of fertilization.

TABLE 9.7
Impacts of Different Systems of Fertilization from 1991 to 2011 on Yield of Winter Wheat, Sugar Beet, and Maize (Mg/ha)

Crops	Unfertilized, Mg/ha	Fertilized		Increase by Fertilization		LSD_{05}
		System of Fertilization	Mg/ha	Mg/ha	%	
Winter wheat	4.27	NPK 130 kg a.i./ha	5.22	0.95	22.2	0.41
		NPK 130 kg/ha + 10 Mg/ha manure	5.33	1.06	24.8	
		NPK 130 kg/ha + 15 Mg/ha manure	5.25	0.98	23.0	
		15 Mg/ha manure	5.25	0.98	23.0	
Sugar beet	30.3	NPK 130 kg a.i./ha	41.1	10.8	36.5	4.28
		NPK 130 kg/ha + 10 Mg/ha manure	43.6	13.3	43.9	
		NPK 130 kg/ha + 15 Mg/ha manure	45.2	14.9	49.2	
		15 Mg/ha manure	43.1	12.9	42.2	
Maize for grain	6.24	NPK 130 kg a.i./ha	7.17	0.93	14.9	0.85
		NPK 130 kg/ha + 10 Mg/ha manure	6.92	0.68	10.9	
		NPK 130 kg/ha + 15 Mg/ha manure	7.05	0.81	13.0	
		15 Mg/ha manure	7.00	0.76	12.2	

All systems of fertilization increased the yields of all crops in the rotation, but especially those of sugar beet and winter wheat. There are no differences in the yield response to different systems of fertilization. However, the effects on soil fertility are drastic (Table 9.8). Organo-mineral and organic systems of fertilization are reducing the stocks of SOM to a lesser extent relative to the mineral system of fertilization.

The maximum loss of SOC at the rate of 0.33 Mg/ha/year occurred in the unfertilized control. In comparison, the rate of SOC loss was lower (0.26 Mg/ha/year) with the use of mineral fertilizers. Yet, even optimal rates of mineral fertilizers combined with 15 Mg of FYM per hectare for each crop rotation did not mitigate the annual losses of SOC. Furthermore, the losses are severe while considering the entire soil profile. While mineral, organo-mineral, and organic systems of

TABLE 9.8
Annual Losses of Soil Organic Carbon during 42 Years in Long-Term Field Experiment with Different Systems of Fertilization in Crop Rotation (Field No. 2)

Nr. or No	System of Fertilization	Rate of Loss of Total Carbon, Mg/ha/year, in 0–20 cm Layer
1	Control (without fertilization)	0.33
2	NPK 130 kg a.i./ha	0.26
3	NPK 130 kg a.i./ha + 10 Mg/ha manure	0.13
4	NPK 130 kg a.i./ha + 15 Mg/ha manure	0.09
5	15 Mg/ha manure	0.12

fertilization influenced crop yields similarly, only systems involving liberal applications of organic manure could sustain the SOM level. However, the liberal use of organic manures is not a practical reality in Moldova.

9.3.5 Irrigation

In the long-term experiment with irrigation in crop rotation, the yield of winter wheat responded more than that of sugar beet (Table 9.9). With a deep root system, sugar beet can absorb water from deeper soil layers during periods of peak demand. In comparison, winter wheat has a shallow root system and responds strongly to supplementary irrigation in the autumn, before or after sowing. Thus, changing the structure of land area and seeding wheat after early harvesting of the preceding crop offers an opportunity to reduce expenditures on irrigation.

Surprisingly, even in a six-field rotation with 50% of alfalfa, 13.3 Mg/ha/year manure over the rotation together with recommended rates of mineral fertilizers, there are losses of SOM from the 0–20 cm layer amounting to 0.09–0.18 Mg C/ha/year (Table 9.10), and losses are much greater for the whole soil profile (0–100 cm), especially on irrigated, fertilized plots (0.45 Mg C/ha/year) relative to rainfed, fertilized plots.

9.3.6 Integration of Crop Rotation, Tillage, and Fertilization into a Farming System

A multifactorial long-term field experiment was initiated in 1996 to study the action and interaction between crops, systems of soil tillage, and fertilization within the framework of crop rotation (Boincean et al. 2013b). The experiment includes two crop rotations (with and without perennial legumes and grasses), two systems of soil tillage (minimum tillage and combination of minimum tillage and plowing), and three fertilization systems (without fertilization, manure, and manure + NPK). The data show residual effects of the alfalfa and ryegrass on the yields of the following crops in the rotation: winter wheat, sugar beet, and maize (Table 9.11). The data support the conclusion that including alfalfa and ryegrass in the third year after the first cut is a good preceding crop for winter wheat, compared with maize silage in the other crop rotation.

Soil tillage hardly changed crop yields; at least, minimum tillage showed no improvement compared with the combination of minimum tillage and plowing. The open question is whether the saving of fuel from minimum tillage compensates for the extra expenditure on chemicals for weed control. Mineral fertilization is not efficient for winter wheat after alfalfa and grasses harvested as fodder (for feeding animals), which offers real opportunities to reduce mineral nitrogen use. Nonetheless, fertilizer is very effective on winter wheat after maize silage. The extra yield of wheat from application of organo-mineral fertilizers is equivalent to 1.31 Mg/ha on a plot with a combination of minimum soil tillage with selective moldboard plowing and 1.67 Mg/ha on a plot with minimum tillage alone. The extra yield from fertilization is more for sugar beet sown after winter wheat following maize silage compared with winter wheat sown after alfalfa and grasses in the third year after the first cut: −11.2–8.9 Mg/ha and 5.2–6.2 Mg/ha, respectively. The same trend appears for maize for grain, in particular on a plot with a combination of minimum tillage and moldboard plowing.

TABLE 9.9
Yields of Winter Wheat and Sugar Beet in Long-Term Field Experiment with Fertilization and Irrigation, Averages of 1970–2010 for Winter Wheat and 1968–2009 for Sugar Beet (Mg/ha)

	Rainfed		Irrigated		Increase by Fertilization		Increase by Irrigation		
Index	Unfertilized	Fertilized	Unfertilized	Fertilized	Rainfed	Irrigated	Unfertilized	Fertilized	LSD_{05}, t/ha
				Winter Wheat					
Mg/ha	4.14	4.95	5.45	6.58	0.81	1.13	1.31	1.63	0.21
%					19.6	20.7	31.6	32.9	
				Sugar Beet					
Mg/ha	38.0	52.7	42.0	58.4	14.7	16.4	4.0	5.7	3.3
%					38.7	39.1	10.5	10.8	

TABLE 9.10
Changes in Stocks of Soil Organic Carbon in 0–20 cm Layer during 42 Years in Long-Term Field Experiment with Irrigation in Crop Rotation (Field No. 1)

		Rainfed		With Irrigation		
Years	Indices	Unfertilized	Fertilized	Unfertilized	Fertilized	LSD_{05}, t/ha
1968	Initial stocks of carbon, 70.1 Mg/ha					
2010	Carbon stocks	62.6	64.1	62.6	66.2	2.8
	Stock reduction	7.5	6.0	7.5	3.9	
	Percentage of initial stocks	10.7	8.6	10.7	5.6	
	Rate of SOC loss, Mg C/ha/year	0.18	0.14	0.18	0.09	
Rate of SOC loss for 1-m depth since 1968, Mg C/ha/year		0.59	0.28	0.58	0.45	

Although the different systems of soil tillage in crop rotations had little effect on crop yields, they had a larger effect on the retention of SOM (Table 9.12).

The greatest increase in stocks of SOM were observed in the crop rotation with a mixture of alfalfa and perennial grasses for green biomass used with manure + NPK: 0.96 Mg/ha on the plots with minimum tillage and selective moldboard plowing, and 0.760 Mg/ha on the minimum-tillage plots.

The experimental data indicate synergy between three main components of the farming system: use of perennial legumes and grasses in crop rotation, combination of minimum tillage with selective plowing, and integrated nutrient management combining FYM with NPK. The legumes and grasses alone cannot maintain the stocks of SOM under crop rotations on unfertilized plots. However, combination of selective plowing and minimum tillage can reduce the losses of SOM under arable land use systems. On the other hand, SOC stocks can be maintained or increased if enough OM is added as FYM. A somewhat lesser benefit from the combination of moldboard plowing and minimum tillage on fertilized plots with manure + NPK on stocks of SOM was observed in crop rotation without perennial legumes and grasses. This trend may be attributed to the quality of SOM under the influence of different rotations, systems of soil tillage, and fertilization.

In general, plowing accelerates the oxidation of SOM. However, an optimal combination of the moldboard plow and minimum tillage in crop rotation can be very effective on heavy soils infested with perennial weeds. The danger of using moldboard plowing in crop rotation comes mainly from the insufficiency of energy-rich crop residues and manure, and in aggravating the risks of soil erosion.

The data from the multifactorial experiment were used to appraise the production expenditures for winter wheat in crop rotation after maize silage and after alfalfa and perennial grasses under both minimum tillage and selective use of moldboard plowing with minimum tillage. The economic analyses were done by using a technological map

TABLE 9.11
Yield (Mg/ha) of Winter Wheat, Sugar Beet, and Maize, Grains in Selectia Long-Term Multifactorial Field Experiment, Average for Two Full Rotations (1996–2009)

	Crop Rotation with Alfalfa and Perennial Grasses			Crop Rotation without Legume and Perennial Grasses			
Systems of Fertilization	Minimum Till + Plowing	Minimum Tillage	Change	Minimum Till + Plowing	Minimum Tillage	Change	LSD_{05}, t/ha
Winter Wheat							
Without fertilization	4.21	4.22	+0.01	2.41	2.30	−0.11	
Manure + NPK	4.31	4.38	+0.07	3.72	3.97	+0.25	
±	+0.10	+0.16		+1.31	+1.67		0.32
Sugar Beet							
Without fertilization	35.2	32.0	−3.2	28.9	27.4	−1.5	
Manure + NPK	40.4	38.2	−2.2	40.1	36.3	−3.8	
±	+5.2	+6.2		+11.2	+8.9		2.59
Maize Grains							
Without fertilization	5.54	5.39	−0.15	4.90	5.08	+0.18	
Manure + NPK	5.77	5.57	−0.20	5.51	5.32	−0.19	
±	+0.23	+0.18		+0.61	+0.24		0.34

TABLE 9.12
Changes in Stocks of Soil Organic Matter in 0–20 cm Layer for Crop Rotation with Different Systems of Fertilization and Soil Tillage

	Rotation with Alfalfa and Perennial Grasses				Rotation without Alfalfa and Perennial Grasses				
	Minimum Tillage + Plowing		Minimum Tillage		Minimum Tillage + Plowing		Minimum Tillage		
Year	Unfertilized	Manure + NPK	Unfertilized	Manure + NPK	Unfertilized	Manure + NPK	Unfertilized	Manure + NPK	LSD_{05}, t/ha
1999	101.8	98.6	94.8	98.2	94.8	97.2	95.0	100.8	2.9
2009	95.3	108.2	99.1	105.8	94.6	100.6	97.7	105.1	3.1
Difference	−6.5	+9.6	+4.3	+7.6	−0.2	+3.4	+2.7	+4.3	
Rate of increase of SOM, Mg/ha/year	−0.650	+0.960	+0.960	+0.760	−0.02	+0.340	+0.270	+0.430	

that includes all technological operations for growing winter wheat. The prices for fuel fertilizers and pesticides were in accordance with the real prices for these products in the Republic of Moldova. The yield of wheat was the same for all treatments (4.2 Mg/ha); however, the costs were different. The same wheat yield can be achieved at less cost following the alfalfa–ryegrass mixture compared with winter wheat following maize silage. By using minimum tillage, it was possible to save $US 26.1/ha on fuel after both preceding crops. When sown after silage maize, additional expenditures for weed, pest, and disease control amounted to $39.6/ha and those for mineral N fertilizers to $76/ha. Thus, the total saving by sowing winter wheat after alfalfa–ryegrass mixtures in the third year after the first cut amounted to $115.6/ha. These savings can be realized only by adopting crop rotation; otherwise, the money saved from minimum tillage ($26.1/ha) is not enough to pay for chemical weed, pest, and disease control ($39.6/ha). Conservation farming can be effective by using an optimal combination of minimum tillage and a good crop rotation.

9.4 CONCLUSIONS

1. The leveling off or declining yields of most crops, both on average for the Republic of Moldova and in the long-term field experiments, necessitates reappraisal of the existing farming systems. Thus, there is a strong need for a paradigm shift toward sustainable intensification based on more intensive recycling of nutrients and energy within each farming system, and using predominantly local, renewable sources of energy. This strategy would be more environmentally friendly, and achieve healthy soils and healthy people dependent on it.
2. The neglect of crop rotations and soil fertility during the specialization and concentration of agriculture are attributed to the higher efficiency of chemicals in continuous monocultures relative to more diverse crop rotations, stimulated by cheap fuel and its derivates, as well as ignorance of the damage to the environment and public health.
3. High-yielding varieties of winter wheat are more efficient in crop rotation than in continuous wheat. These varieties require additional nutrients, which are derived not from high rates of mineral fertilizers but by depletion of the inherent soil fertility.
4. Minimum soil tillage does not reduce the yield of field crops relative to moldboard plowing; however, it minimizes risks of soil erosion, reduces the rate of mineralization, and decreases losses of SOM.
5. Mineral, organo-mineral, and organic systems of fertilization all increase yields of field crops; however, their influence on soil fertility vary. Organic and organo-mineral systems of fertilization improve inherent soil fertility *vis-à-vis* the use of mineral fertilizers.
6. Irrigation, especially in combination with fertilizers, increases crop yields but accelerates the loss of SOM from the entire soil profile.
7. No single factor of intensification used alone can compensate for the annual losses of SOM from arable land. However, integration of crop rotation with perennial legumes and grasses, fertilization with liberal use of FYM, and

conversion to minimum tillage can maintain crop productivity and improve soil fertility. The combination of the moldboard plow and minimum tillage can be efficient in crop rotation with perennial legumes and grasses and with sufficient application of FYM or other supplementary sources of organic fertilizers.

8. Integration of optimal systems of soil tillage and fertilization in the framework of crop rotation with perennial legumes and grasses allows significant reduction of production costs by cutting the use of nitrogen from mineral fertilizers; irrigation; and chemicals for weed, pest, and disease control.

ACKNOWLEDGMENTS

The authors would like to thank the research and technical staff of the Department of Sustainable Farming Systems at Selectia RIFC for their field and laboratory activities during the years of fulfilling the research program.

REFERENCES

Boaghii, IV and Bulat, L. 2013. Primary soil tillage in the crop-rotation for main field crops in Moldova. In: *Soil as World Heritage*, Dent, DL (editor). Springer, Dordrecht, pp. 273–282.

Boincean, BP. 2013a. *Alternative Technologies for Growing Winter Wheat in the Republic of Moldova (Guide)*. Balti, p. 67 (in Romanian).

Boincean, BP. 2013b. 50 years of the long-term field experiment with crop rotation and permanent crops at the Selectia Research Institute of Field Crops. In: *Soil as World Heritage*, Dent, DL (editor). Springer, Dordrecht, pp. 175–199.

Boincean, BP. 2013c. Soil tillage for sustainable farming systems. *Pro-Environment Journal of Documentation, Research and Professional Training* (Cluj-Napoca, Romania) 14(6), 14–199 (in Romanian).

Boincean, BP, Martea, MP, Ungureanu, AI and Hropotinschi, PM. 2013a. Results from the long-term field experiment with irrigation on chernozem soils in the Balti steppe of the Republic of Moldova. In: *Soil as World Heritage*, Dent, DL (editor). Springer, Dordrecht, pp. 233–250.

Boincean, BP and Nica, LT. 2007. Productivity, fertilization and fertility of chernozem soil in the steppe zone of Moldova. In: *Mineral versus Organic Fertilization. Conflict or Synergis? Proceeding of the 16th International Symposium of the International Scientific Center for Fertilizers (CIEC)*, Gent, Belgium, September 16–19, pp. 102–109.

Boincean, BP, Nica, LT and Stadnic, SS. 2013b. Long-term changes in soil fertility and productivity on typical chernozem under different systems of fertilization in crop rotation. In: *Soil as World Heritage*, Dent, DL (editor). Springer, Dordrecht, pp. 209–230.

Cassman, KG. 1999. Ecological intensification of cereal production systems: Yield potential soil quality and precision agriculture. *Proceedings of the National Academy of Sciences U S A* 96, 5952–5959.

Dent, D. 2013. Business case for green water credits. In: Dent, D. (editor), *Soil as World Heritage*. Springer, Dordrecht, Germany, pp. 459–470.

Docuceaev, VV. 1952. *Russian Cernoziom,* Second edition. State Publication of Agricultural Literature, Moscow, 634 pp. (in Russian).

Gliessman, SR. 2000. *Agroecology. Ecological Processes in Sustainable Agriculture*. Lewis, Boca Raton, FL.

International Assessment of Agricultural Science and Technology for Development. 2009. *Agriculture at a Crossroads*. IAASTD, Synthesis Report, Washington, DC.
Kassam, A. 2011. *The Future of Farming: What Needs to Change? A Personal View*. Hugh Bunting Memorial Lecture, University of Reading, June 13.
Kosticev, PA. 1885. *Soils from Chernozem Regions of Russia*. SPG, Moscow, Russia (in Russian).
Kosticev, PA. 1940. *Soil Science*, vols. I–III. Selihozgiz, Moscow, Russia (in Russian).
Kovda, VA. 1983. The post and the future of chernozem. In: *Russian Chernozem. 100 Year After Docuceaev*, Moscow, Science, pp. 253–281 (in Russian).
Krupenikov, I. 2008. *Chernozems*. Pontos, Chisinau, p. 285 (in Russian).
Krupenikov, IA, Boincean, BP and Dent, DL. 2010. *The Black Earth. Ecological Principles for Sustainable Agriculture on Chernozem Soils*. Springer, Dordrecht.
Lal, R. 2013. Abating climate change and feeding the world through soil carbon sequestration. In: *Soil as World Heritage*, Dent, DL (editor). Springer, Dordrecht.
Powlson, D, Macdonald, A and Poulton, P. 2013. The continuing value of long-term field experiments as resources to provide insight for achieving food security and environmental protection. In: *Soil as World Heritage*, Dent, DL (editor). Springer, Dordrecht.
Villiams, VR. 1950–1952. *Collected Works*, vols. 5, 6 and 10. Moscow, State Publisher of Agricultural Literature (in Russian).
Weil, R and Magdoff, F. 2000. Significance of soil organic matter to soil quality and health. In: *Soil Organic Matter in Sustainable Agriculture*, Magdoff, F and Weil, R (editors). CRC Press, Boca Raton, FL, pp. 1–45.

10 Experiences on Conservation Agriculture (CA) in Zambia, Malawi, and Ethiopia

Effects on Yield, Soil Properties, Labor Use, Profitability, and Adoption

Jens B. Aune

CONTENTS

10.1 INTRODUCTION

Conservation agriculture (CA) has been actively promoted in many countries in Africa during the last 10 years. The main principles of CA are direct planting of crop seeds, permanent soil cover, and crop rotation (Food and Agriculture Organization of the United Nations [FAO] 2011). This chapter will not use CA in the strict meaning of the word, but rather as a common denominator to describe concepts such as conservation tillage, conservation farming, and reduced tillage. This chapter will summarize experiences on CA from Zambia, Malawi, and Ethiopia CA is also increasingly receiving policy support in Africa and is promoted under Pillar 1 "Land and Water Management" in the Comprehensive Africa Agriculture Development Program (CAADP) of the African Union (CAADP 2011). CA in

Africa has been promoted by actors such as the FAO, the African Tillage Network (ATN), the Department for Development (United Kingdom), the Norwegian Agency for Development Cooperation, and the Conservation Farming Unit (CFU). CA has also become one of the national strategies for promoting agricultural development in Zambia. There are many reasons for this increasing interest in CA in Africa. The success of CA in South America has obviously been an inspiration for also promoting CA on other continents. Yields have been stagnant in sub-Saharan Africa, and key actors like FAO and the ATN see CA as a new approach that can bring about agricultural development. However, it is not possible to transfer the CA methods used in South America directly to Africa as the socioeconomic and environmental conditions are different. CA in South America is mainly a mechanized form of agriculture, making it difficult to transfer this type of CA directly to the small-scale farmers in Africa. CA has also been promoted as an approach in mitigation of greenhouse gases from agriculture and as an adaptation to climate change (Corsi et al. 2012).

The introduction of CA in Africa has been seriously criticized in recent years (Giller et al. 2009) on grounds that it can increase a farmer's labor burden, it gives no yield benefit, it increases weed infestation, it has limited effects on soil organic matter, and there is unavailability of key input for practicing CA.

This chapter will summarize experience with the introduction of CA in Zambia, Malawi, and Ethiopia, and will also draw on results from neighboring countries. The focus is on these countries as they represent contrasting environments with regard to previous tillage systems and ecological conditions, and because of the author's experience in CA research and development in these countries.

10.2 VARIATIONS IN CA METHODS ACROSS AFRICA

CA has not been introduced in a uniform way in Africa, and there are clear differences in approach between neighboring countries such as Zambia and Malawi. However, the agroecological and socioeconomic conditions vary greatly across Africa, and it may therefore make sense to use different approaches.

Tillage by use of the plow and animal traction was introduced in Zambia during the colonial time. Use of the moldboard plow with oxen traction is the dominant form of tillage in southern Zambia as the people in southern Zambia are livestock keepers. Tillage by use of the hoe is more practiced in eastern Zambia. The CFU has been a major actor in promoting conservation agriculture in Zambia. The emphasis has been on establishing basins that are 30 cm long and 15 cm wide and 15–20 cm deep (Umar et al. 2011). The basins are preferably dug with a chaka hoe, which is a hoe that is heavier and has a narrower blade than the traditional hoe. The spacing between the basins is 90 × 70 cm. Basins are also used in CA farming in Zimbabwe. However, the basins here are smaller than those used in Zambia as the size is in the order of 15 × 15 cm with a depth of 15–20 cm (Mazvimavi and Twomlow 2009). The other form of tillage promoted in Zambia is ripping. This is a form of tillage that is gaining popularity mainly because of its labor-saving effect. The soil is ripped to a depth of 20 cm. One reason for introducing both basins and ripping is, according to the CFU, to break the plow pan. However, a soil survey has not confirmed that plow pans are a widespread problem in farms of smallholder farmers in Zambia (Umar et al. 2011). The type of CA practiced in Zambia has only focused on the principle of reduced tillage. The principle of mulching is not practiced in Zambia as livestock grazes freely in the dry season, removing all the residues that are left on the soil surface after harvesting. The dominance of maize in the farming system in Zambia makes it difficult to practice crop rotation. About 60% of the agricultural area is under maize monocropping (Aune et al. 2012). The government of Zambia has been promoting maize production through government subsidies on fertilizer and maize hybrid seeds. For that reason, it has become less appealing for farmers to practice rotation involving grain legumes. There are also other elements added to CA in Zambia such as planting of the tree *Faidherbia albida*.

Traditional tillage in Malawi can be considered as ridge splitting, as new ridges are made by hoe every year by moving the soil from the previous ridge to a new ridge that is created between the previous ridges (Ngwira et al. 2012). This was a tillage system that was introduced during the British colonial time and has since remained the dominant tillage system in Malawi (Ministry of Agriculture, Irrigation and Water 2012). CA is promoted in Malawi by the Ministry of Agriculture, Total Land Care, Sakakawa Global 2000, and the CIMMYT (International Maize and Wheat Improvement Center). Both planting basins and direct sowing have been introduced in Malawi. The planting basins promoted in Malawi have generally been bigger than the planting basins used in Zambia (Ministry of Agriculture, Irrigation and Water 2012). Basins with dimensions of 30 × 30 × 20 cm have frequently been used. Total Land Care has introduced a system of direct sowing. In this new system, a small hole in the soil is opened with a wooden stick and the seeds are thereafter placed in this hole. The crop residues from the previous season are retained as a mulch and there is no movement of soil in this CA system in Malawi. Mulching is feasible in Malawi because the grazing pressure from the animals is low (Ngwira et al. 2012).

Ethiopia has an ancient tradition of tilling the land by using a pair of oxen pulling an ard (the maresha) (McCann 1995) (Figure 10.1). Ethiopians are called "the people

FIGURE 10.1 Oxen tillage in Ethiopia.

FIGURE 10.2 Planting basins as practiced in Zambia.

of the plough" (McCann 1995), and oxen tillage is deeply rooted in the culture of rural Ethiopians. The ard does not turn the soil as the moldboard plow does, but rather breaks up a narrow strip of soil. The soil is tilled up to six times depending on the type of crop that is grown (Aune et al. 2001). This tillage system is under increasing pressure, as it is demanding for the farmers to keep the number of animals necessary for maintaining the system. It has been shown that the farmers constantly need to keep 10 animals to always have access to two oxen (McCann 1995). The animals needed to maintain the system are calves, oxen under training, and cows to ensure reproduction of the system. As farm size in Ethiopia has been shrinking, it is increasingly difficult to support this livestock system. Not all farmers have access to a pair of oxen, and farmers without oxen often have to pay 50% of the harvest to get the land plowed (Aune et al. 2001). The man who plows will, in addition, take the straw. New forms of tillage systems have been introduced in Ethiopia. This includes direct sowing with and without mulch, and different reduced tillage methods.

In the sandy soils of the drylands of West Africa, the seeds are often sown directly. However, this form of tillage is seldom called CA although direct sowing is practiced. Planting basins (called "zai" in Burkina Faso) are also used in West Africa as a method of soil and water conservation and as a method of rehabilitating degraded lands (Sawadogo 2011). These planting basins have similar dimensions as the planting basins in Zambia (Figure 10.2). Organic manure is often added to the planting basins to increase soil fertility.

10.2.1 CA Effects on Yields

When assessing the yield benefits from CA, it is necessary to differentiate between the effects of the different components of CA and the combined effect of the CA component on yield. The effect of CA on yields is variable depending on how it is practiced. Few studies have examined the effect of a full CA package on yield. Often, only one element of CA is studied, such as the effect of tillage or the mulch effect.

A survey of 129 farmers that have been involved in a project on CA in Zambia showed yield levels of 1.8, 5.2, 2.3, and 3.8 tons ha^{-1} for hand hoeing, planting basins, ripping, and plowing, respectively (Umar et al. 2011). On-station experiments by the Golden Valley Agricultural Research Trust in Zambia have shown yield levels of 4.0, 6.3, 5.3, and 5.5 tons ha^{-1} for hoe tillage, basin tillage, ripping, and plowing, respectively (Umar et al. 2011). This shows that the yield level in basins is consistently higher than in other tillage systems in Zambia. The yield-increasing effect of the basins has been particularly apparent in the dry years. The reason why basins have been found to increase yields is probably related to increased water infiltration in the basins and the better nutrient use efficiency, as the fertilizer is also placed adjacent to the seeds in the basins. In years with excessive rainfall, problems with waterlogging in the basins were observed in Zambia (Umar et al. 2011). The reasons why farmers in Zambia find it difficult to adopt CA is related to lack of labor (38%), lack of land (30%), and no timely access to input (9%) (Umar et al. 2012). In Tanzania, no yield difference was found between ripping and basins; however, these CA tillage methods increased yields by 50% as compared with plowing (Rockström et al. 2009). In Zimbabwe, contrasting results have been found on the use of basins. The average yields from 15 regions across Zimbabwe showed an increase from 970 kg ha^{-1} for non-CA farmers to 1546 kg ha^{-1} for CA farmers practicing basins (Marongwe et al. 2011). However, a 4-year study (one site) in Zimbabwe was not able to find any significant difference in maize grain yield between plowing, ripping, and basins (Mupangwa et al. 2013). In Burkina Faso, it has been found that planting basins (zais) can, under certain conditions, double and quadruple yields (Sawadogo 2011). These results show that planting basins will, under most conditions, increase yield.

In Malawi, it has been found that CA can also increase yield as compared with conventional tillage (CT) (Ngwira et al. 2012). The CA system (direct planting, straw mulch, and herbicide application) increased yield compared with CT (ridge splitting, without mulch, and manual weeding) by 41%. The yield benefits of CA in Malawi have been found to be higher in drylands at low altitudes than in higher altitudes with better rainfall (Thierfelder et al. 2013). There was also considerably less risk associated with CA in areas with low rainfall (Ngwira et al. 2013). Multilocational (nine sites) and long-term studies (up to 8 years) in Malawi comparing zero tillage and retention of crop residues (CA) with ridge splitting and burrowing of crop residues (CT) showed that in the initial years, there were no clear yield benefits of the CA system, but after 5 years CA gave clearly higher yields than CT (Thierfelder et al. 2013). Although the yield benefit is not apparent in the initial years, it may still make sense for the farmers to practice CA because of the labor-saving effect. In Zimbabwe, it has been shown, as in Malawi, that CA gives no initial yield benefit over CT; however, the yield in CA increases over time as compared with CT (Thierfelder and Wall 2012).

In Tigray, Ethiopia, no yield difference was found between reduced tillage (one pass with the maresha) and CT (four passes with the maresha) (Habtegebrial et al. 2007). Studies in central Ethiopia on nitosols also confirmed that reduced tillage can produce equal yields as CT (Tulema et al. 2008). However, reduced tillage plots had more weed infestation than plots under CT. Zero tillage for teff was found to give a lower yield and less economic return than conventional and reduced tillage in nitosols and vertisols (Tulema et al. 2008). For wheat, there has been no clear yield difference between CT, reduced tillage, and zero tillage; however, there was a tendency for CT to give the highest yield (Taa et al. 2004). In Ethiopia, considerable research has gone into developing different forms of raised seed beds in vertisols to minimize the problems of waterlogging in this soil type (Nyssen et al. 2011). However, although it has been shown that this is technically feasible, this form of tillage has not been taken up by the farmers (Nyssen et al. 2011). Research on vertisol has been conducted in recent years on more simple permanent raised beds that the farmers can prepare with the maresha. The yield benefits of such permanent beds as compared with CT in vertisols have become visible after 5 years (Araya et al. 2012). The research on tillage methods in Ethiopia shows that there is no clear difference in yield between the different tillage systems. This illustrates that reduced tillage and zero tillage can be interesting alternatives in Ethiopia. The benefits here will not be the increased yield, but rather the reduction in soil erosion and the possibility of changing the livestock system more in the direction of meat- and milk-producing animals, and placing less emphasis on keeping oxen for traction purposes.

A meta-analysis by Rusinamhodzi et al. (2011) across many countries of the world showed that the effect of CA is most prominent in well-drained soil where an N level of >100 kg N ha^{-1} is applied. Mulching as practiced in CA may increase the problems with waterlogging, and this is the reason why the best results with CA are obtained on well-drained soil. Mulching with plant material with a high C/N ratio may cause immobilization of N, and application of N rates >100 kg ha^{-1} can minimize this problem. CA has functioned best in dry years, and waterlogging has been found to be a problem associated with CA in years of high rainfall (Rusinamhodzi et al. 2011; Aune et al. 2012). Rusinamhodzi et al. (2011) found a positive effect of CA in areas where the average rainfall is <600 mm; however, the effect of CA was negative in areas with >1000 mm average rainfall. A study across Ethiopia, Kenya, Tanzania, and Zambia also confirmed that the yield benefits of CA were highest in the driest years (Rockström et al. 2009). This was the case although no mulch was applied in these trials. This illustrates that ripping and basins can have a water-harvesting effect.

Practicing both CA and CT may therefore be considered by the famers as a risk-aversion strategy as the CA will give a positive yield benefit in the dry years while CT may give good results in years with high rainfall.

These results show that CA generally increased yield; however, there were clear differences between countries and the tillage methods practiced. Planting basins give the clearest yield benefits. Direct sowing and mulching as practiced in Malawi also increased yields, but not to the same extent as planting basins in Zambia. The benefits in Malawi were more connected to a reduction in labor.

10.2.2 Labor Use and Economics of CA

Changes in the tillage system also affect the labor demand and the distribution of labor during the cropping season. A study in Zambia showed that the total labor use during the entire season was 124, 145, 61, and 83 person-days ha^{-1} for planting basins, hand hoeing, plowing, and ripping, respectively (Umar et al. 2012). Another study in Zambia showed that labor use was 148, 70, and 63 person-days ha^{-1} for basins, plowing, and ripping (FAO 2011). The gross margin for the four tillage systems was 2212, 34, −270, and 503 Zambian kwachas ha^{-1} for planting basins, hoe tillage, plowing, and ripping, respectively (Umar et al. 2012). The gross margin was highest in the basin systems because of the higher yield. This more than compensated for the higher labor demand of the basin system. The negative gross margin found for the plowing system illustrated that this system cannot pay a regular salary for the work that goes into this tillage method. Studies in Zimbabwe have shown that the gross margin in planting basins and ripping is higher than in plowing (Figure 10.3) (Mazvimavi 2011).

In Malawi, it has been found that a CA system consisting of using a wooden stick (Figure 10.4) and herbicide application has a labor demand of 47 days ha^{-1} as

FIGURE 10.3 Ripping with oxen in Zambia.

FIGURE 10.4 Using the dipple stick to make planting holes in a mulched plot in Malawi.

compared with 65 days ha^{-1} in the CT method (Ngwira et al. 2012). The return labor was double in the CA system as compared with the CT system, and the return per hectare in the CA system was 61% higher than in CT.

CA increases labor demand in relation to weeding, as there is no mechanical disturbance of the weeds. Herbicides have therefore been introduced in Zambia and Malawi as part of the promotion of CA. However, in the 2009–2010 season in Zambia, still only 8% of the farmers in a survey of 429 farmers used herbicides; however, the number is increasing year by year (Nyanga et al. 2011). Glyphosate was the herbicide most frequently used, but many farmers did not use the herbicide correctly (Umar et al. 2011). The concentration of the herbicide was often too low and the water used is often inappropriate (dirty).

Experimental research on CA may not always give a good estimate of the yield effect under real farming conditions. One important factor affecting yield is the sowing time. Sowing is often delayed in Zambia and Ethiopia when plowing is practiced because there are not enough oxen available to plow the land on time (Aune et al. 2001). The average sowing time for the different tillage methods was November 9, 12, 12, and 23 for farmers practicing basins, hand hoeing, ripping, and plowing, respectively (Umar et al. 2012). Changing from plowing to ripping can therefore allow farmers to sow earlier because of the lower labor demand in ripping as compared with plowing. The labor demand of the plowing operation was found to be 3.8 person-days ha^{-1}, while the ripping operation was 0.8 person-days ha^{-1}. The CFU in Zambia has been encouraging farmers to prepare the basins in the dry season to allow them to sow early in the rainy season. However, farmers have generally not taken up this practice as they find it very difficult to till a dry soil and they seem to emphasize other activities in the dry season.

10.2.3 Effect of CA on Soil Properties and Water Availability

The effect of CA on soil properties is dependent on which tillage method is used, additions of organic matter to the systems, and the type of crop rotation practiced.

In Zambia, it was found that CA without recycling of residues did not have any effect on soil chemical properties (Umar et al. 2011). In Burkina Faso and Zimbabwe, it has been found that the soils in the planting basins (zai) have increased soil carbon and nitrogen, and a lowered C/N ratio as compared with the soil between the basins (Sawadogo 2011; Mupangwa et al. 2013). Direct sowing and residue retention have been found to increase soil organic matter content in Zimbabwe (Thierfelder and Wall 2012). The increase in soil organic carbon was higher on a clay soil than on a sandy soil. From Ethiopia, it has been shown that use of permanent ridges and retention of crop residues can significantly increase soil organic matter content and reduce erosion and runoff as compared with a system with CT (Araya et al. 2012).

In Malawi, it was shown in a long-term experiment that the main reason for the yield-increasing effect of zero tillage and mulching is the increase in infiltration by 24%–40% as compared with the CT system in the country (Thierfelder et al. 2013). Results from Zimbabwe showed that zero tillage with residue retention had higher infiltration rates than plowing without residue retention (Thierfelder and Wall 2012).

Another study in Zimbabwe found no difference in the infiltration rate between plowing, ripping, and basins (Mupangwa et al. 2013).

10.2.4 Adoption of CA

True adoption of CA is difficult to assess because the introduction of CA has been accompanied by the distribution of fertilizer, access to planting materials, and extension services. Most of the projects reported here are ongoing, and it is, for this reason, difficult to assess what is true adoption and adoption that is contingent on having access to certain inputs. True adoption can therefore best be studied after the completion of the project, or by determining if there are farmers adopting CA without being involved in CA projects. However, the adoption rates during the project period may still give an indication of true adoption, particularly if the new technologies are used on a big share of the land.

Farmers have embraced CA to various degrees in Zambia. In a sample of 420 households in the CA project areas in Zambia, it was found that 71% had adopted CA (Nyanga 2012). It is estimated that >100,000 farmers are using CA in Zambia. However, this is a very reduced form of CA as only the principle of minimum/reduce tillage is practiced while the farmers are using mulching and crop rotation to only a limited degree. The type of CA practiced in Zambia is therefore more a form of reduced tillage than CA (Aune et al. 2012). The farmers that adopt CA in Zambia also typically practice CA on a minor part of the farm while the rest of the farm is under traditional tillage methods. The average size of land under CA in Zambia for 420 farmers practicing CA (basins and ripping) was 0.78 ha, representing 40% of the land under cropping for these farmers. The rest of the land is under plowing or hoe tillage. Land under ripping has been expanding faster than land under basins. The average size of land under basins is rarely >1 ha. The labor requirement of planting basins is the main reason why planting basins are not expanding more despite the very high return to labor. However, the reason why farmers keep some land under basins is that the maize can be harvested earlier and a higher yield can be expected. The basins here are therefore practiced as a food security strategy. The main reasons for the farmers to practice ripping as compared with hoe tillage and plowing appears to be the labor-saving effects. It is the poorest segments of the population that have accepted CA in Zambia, as the CA farmers have less income and own fewer animals and oxen (Aune et al. 2012). Farmers that have previously practiced plowing do not change to basin tillage; they rather opt for ripping. This allows them to continue using animal traction. Previous hoe farmers find it easier to change to basin tillage.

Adoption of CA is Zambia has also been met with resistance because CA is contrary to how farmers think CA should be practiced (Nyanga et al. 2011). Farmers are used to prepare a clean seed bed and burn the crop residues for that purpose. There is also prestige associated with plowing and cultivating large areas. Farmers also do not accept using animal traction in the dry season because the animals are not well fed during this season, and it is very difficult for the animals to pull tillage equipment through dry soil. The Tonga people in southern Zambia are cattle owners, making it more difficult to introduce basin tillage in this part of Zambia than in the eastern region where hoe tillage is the traditional method for land tillage.

Adoption of CA in Zambia also clearly has a gender dimension. Men are mostly involved with ripping, while women are engaged with digging of basins. The basins are dug with a hoe (chaka hoe) that is considerably heavier than the traditional hoe. Women find it difficult to work with the chaka hoe for long periods, and there is therefore a clear limit to how much land that can be tilled with the chaka hoe. Women in Zambia are often responsible for clearing the land and weeding in the traditional tillage system. CA changes the distribution of labor during the season, as less time is needed to clear the land before sowing; however, the labor demand for weeding increases if herbicides are not applied (Nyanga et al. 2011). There is also clearly a gender issue related to plowing in Ethiopia, as it is socially unacceptable for women to plow in this country.

There is also a considerable adoption of CA in Malawi (Figure 10.5). It is estimated that by 2011, >63,000 farmers had practiced CA (Ministry of Agriculture Irrigation and Water 2012). Both basins and direct sowing are practiced in Malawi.

Planting basins are commonly found on farms in Burkina Faso, and it is estimated that between 30,000 and 60,000 ha are under planting basins (zai) in the country (Sawadogo 2011).

Mulching is the principle of CA most difficult to adhere to (Figure 10.6). Free grazing is practiced in most parts of sub-Saharan Africa, and improved grazing management or zero grazing systems are therefore needed if this principle of CA is to be respected. Livestock management should therefore be an integrated part of CA; however, most development interventions focusing on CA have ignored this issue. It is therefore not surprising that there has not been much progress on this issue. Crop rotation is often a difficult recommendation to follow, as this option is a highly economic question.

Adoption of new technologies by smallholders in Africa is a complicated issue, as farmer behavior is not only based on profit maximization but considerations as drudgery and risk are also factors affecting uptake of new technologies. Other factors constraining the uptake of new technology are access to input, price of input and

FIGURE 10.5 Traditional tillage by using ridge splitting in Malawi.

FIGURE 10.6 Mulched CA maize field in Malawi.

output, and access to credit and assured market. None of these factors are known before the cropping season, making profit maximization inappropriate as the sole farming objective.

10.3 CONCLUSION

CA is not a quick fix to the problem of low food production and environmental degradation in Africa. The effect of CA is variable depending on the socioeconomic and ecological conditions. This chapter shows that the tillage component of CA in Africa is practiced in three main forms: basin tillage, ripping, and zero tillage. The most consistent yield increase has been obtained with basins. Zero tillage and mulching have also been found to increase yields; however, the effect is more variable. The positive effect of zero tillage and basins increases with time. Despite the high yields and profitability of basin tillage, farmers take up this tillage system to a limited degree due to the drudgery associated with this tillage system. Mulching is practiced to a limited degree in most CA systems in Africa; however, Malawi is an exception as the grazing pressure from livestock is less. Direct sowing and mulch as practiced in Malawi gives less initial yield benefits than planting basins, but the advantage of this system is lower labor demand than the tillage system with planting basins and CT system.

It appears that farmers practice both CA and CT. This may reduce risks as CA may give a benefit in dry years and CT can be beneficial in years of high rainfall. Practicing both CA and CT can also lead to a better distribution of labor use. CT has higher labor in relation to land preparation, while CA has higher labor use in weeding if herbicides are not used. CA can also be found to give an earlier harvest because the farmers are less dependent on the availability of traction power. Practicing both CA and CT can therefore be considered as a livelihood diversification and a risk-aversion strategy.

However, although there might not be yield benefits related to CA, it can still be interesting for the farmers to practice CA because of the reduction in labor use in relation to tillage. Changing the tillage system alone is not likely to bring about a lasting increase in agricultural productivity, as plant nutrients are also constraining productivity in these regions. Straw production will be too low without fertilizer, making it impossible for the farmers to retain the straw as a mulch. Less tillage will make the conditions better for weeds, particularly the perennial weeds. Use of herbicides is therefore needed especially on larger farms. It is therefore difficult to promote CA as a low-input system. CA development should therefore be accompanied by measures to increase the availability of fertilizer and herbicides.

CA has too often been promoted as a pure technological approach to agriculture development. CA technologies have been demonstrated to the farmers, and CA equipment has been distributed. None of the development organizations have systematically worked on integrating livestock into CA development. This is a vital issue, as livestock grazing is the primary reason why it is difficult for farmers to practice the principle of mulching in CA. It is only possible to practice mulching if controlled grazing is practiced. Lack of fodder is the primary reason why free grazing is practiced, and there is therefore a need to improve fodder production if mulch-based systems are promoted. This can be in the form of increased straw production, planting of fodder trees, or hay making. Improved grazing will also require that there are local institutions in place that can deal with conflicts that may occur in relation to improved grazing management. Development of CA may also make it possible to change the composition of the livestock from oxen to more productive milk or meat production animals, as there is less need for animal traction in CA. A change from plowing to ripping was found to reduce the need for animal traction by 75% in Zambia. CA development initiatives should therefore, in the future, also take livestock development into consideration and focus on issues related to fodder production, grazing management, and zero grazing of animals.

Agriculture polices have also often frequently worked against the development of CA. A particular issue is the support to maize production in many countries in southern Africa. Hybrid seeds and fertilizer have often been distributed through extension packages, and farmers can sell the maize at a guaranteed minimum price. This has stimulated farmers to practice maize monocropping, making it less interesting for them to practice rotation as recommended in CA.

Papers critical to CA have claimed that CA increases a farmer's labor burden, gives no yield benefit, increases weed infestation, and has limited effects on soil properties (Giller et al. 2009). However, this chapter shows that it is difficult to make general statements on how CA works. Often there is a yield benefit of CA, but the results are variable depending on the methods practiced and the agroecological conditions. More research is clearly needed to identify under which conditions CA works and what the best practice of CA is. This is particularly important in Africa, as CA here is in the early stages of development.

REFERENCES

Araya, T., W.M. Cornelis, J. Nyssen et al. 2012. Medium-term effects of conservation agriculture based cropping systems for sustainable soil and water management and crop productivity in the Ethiopian highlands. *Field Crop Res* 132:53–62.

Aune, J.B., M.T. Bussa, F.G. Asfaw, and A.A. Ayele. 2001. The ox ploughing system in Ethiopia: Can it be sustained? *Outlook Agr* 30:275–280.

Aune, J.B., P. Nyanga, and F.H. Johnsen. 2012. A Monitoring and Evaluation Report of the Conservation Agriculture Project (CAP1) in Zambia. Noragric Report no. 68. Norwegian University of Life Sciences, Aas, Norway, 34 pp. Available at http://www.umb.no/statisk/noragric/publications/reports/2012/noragricrep68.pdf.

Comprehensive Africa Agriculture Development Program. 2011. Pillar 1: Land and water management. Available at http://www.nepad-caadp.net/pillar-1.php.

Corsi, S., T. Friedrich, A. Kassam, M. Pisante, and J. de Moraes Sà 2012. *Soil Organic Carbon Accumulation and Greenhouse Gas Emission Reductions from Conservation Agriculture: A Literature Review*. Integrated Crop Management, Vol. 16, FAO. Rome, 103 pp.

Food and Agriculture Organization. 2011. The main principles of conservation agriculture. Available at http://www.fao.org/ag/ca/1b.html.

Giller, K.E., E. Witter, M. Corbeels, and P. Tittonell. 2009. Conservation agriculture and smallholder farming in Africa: The heretics' view. *Field Crop Res* 114:23–34.

Habtegebrial, K., B.R. Singh, and M. Haile. 2007. Impact of tillage and nitrogen fertilization on yield, nitrogen efficiency of tef (*Eragrostis tef* (Zucc.) Trotter) and soil properties. *Soil Till Res* 94:55–63.

Marongwe, L.S., Z. Kwazira, M. Jenrich, C. Thierfelder, A. Kassam, and T. Friedrich. 2011. An African success: The case of conservation agriculture in Zimbabwe. *Int J Agr Sustain* 9:153–161.

Mazvimavi, K. 2011. Socio-economic Analysis of Conservation Agriculture in Southern Africa. Network Paper 02. FAO, Rome. Available at http://www.fao.org/docrep/013/i2016e/i2016e00.pdf.

Mazvimavi, K., and S. Twomlow. 2009. Socioeconomic an institutional factors influencing adoption of conservation farming by vulnerable households in Zimbabwe. *Agr Syst* 101:20–29.

McCann, J.C. 1995. *People of the Plow. An Agricultural History of Ethiopia, 1800–1990*. University of Wisconsin Press, Madison, WI, 298 pp.

Ministry of Agriculture, Irrigation and Water. 2012. *Approaches to the Implementation of the Conservation Agriculture among Promoters in Malawi*. Ministry of Agriculture, Irrigation and Water, Lilongwe, Malawi. Available at http://alturl.com/znkns.

Mupangwa, W., S. Twomlow, and S. Walker. 2013. Cumulative effects of reduced tillage and mulching on soil properties under semi-arid conditions. *J Arid Environ* 91:45–52.

Ngwira, A., C. Thierfelder, N. Eash, and D.M. Lambert. 2013. Risk and maize-based cropping systems for smallholder Malawi farmers using conservation agriculture technologies. *Exp Agr* 49:483–503.

Ngwira, A.R., J.B. Aune, and S. Mkwinda. 2012. On-farm evaluation of yield and economic benefit of short term maize legume intercropping systems under conservation agriculture. *Field Crop Res* 132:149–157.

Nyanga, P.H. 2012. Factors influencing adoption and area under conservation agriculture: A mixed method approach. *J Agr Res* 1:27–40.

Nyanga, P.H., F.H. Johnsen, and T.H. Kalinda. 2011. Gendered impacts of conservation agriculture and paradox of herbicide use among smallholder farmers. *Int J Tech Dev Stud* 3(1):1–24.

Nyssen, J., B. Govaerts, T. Araya et al. 2011. The use of the maresha ard plough for conservation agriculture in Northern Ethiopia. *Agr Sustain Dev* 31:287–297.

Rockström, J., P. Kaurnbutho, J. Mwalley et al. 2009. Conservation farming strategies in East and Southern Africa: Yields and rain water productivity from on-farm action research. *Soil Till Res* 103:23–32.

Rusinamhodzi, L., M. Corbeels, M.T. Wijk, M.C. Rufino, J. Nyamangara, and K.E. Giller. 2011. A meta-analysis of long-term effects of conservation agriculture on maize grain yield under rain-fed conditions. *Agr Sustain Dev* 31:657–673.

Sawadogo, H. 2011. Using soil and water conservation techniques to rehabilitate degraded lands in northwestern Burkina Faso. *Int J Agr Sustain* 9:120–128.

Taa, A., D. Tanner, T.P. Alan, and A.T.P. Bennie. 2004. Effects of stubble management, tillage and cropping sequence on wheat production in the south-eastern highlands of Ethiopia. *Soil Till Res* 76:69–82.

Thierfelder, C., J.L. Chisui, M.S. Cheesman et al. 2013. Maize-based conservation agriculture systems in Malawi: Long-term trends in productivity. *Field Crop Res* 142:47–57.

Thierfelder, C., and P. Wall. 2012. Effects of conservation agriculture on soil quality and productivity in contrasting agro-ecological environments of Zimbabwe. *Soil Use Manage* 28:209–220.

Tulema, B., J.B. Aune, F. Johnsen, and B. Vanlauwe. 2008. The prospects of reduced tillage in tef (*Eragrostis tef* Zucca) in Gare Arera, West Shawa Zone of Oromiya, Ethiopia. *Soil Till Res* 99:58–65.

Umar, B.B., J.B. Aune, H. Johnsen, and O.I. Lungo. 2012. Are smallholder Zambian farmers economists? A dual-analysis of farmers' expenditure in conservation and conventional agriculture systems. *J Sustain Agr* 36:908–209.

Umar, B.W., J.B. Aune, F.H. Johnsen, and O.I. Lungo. 2011. Options for improving smallholder agriculture in Zambia. *J Agr Sci* 3:50–62.

11 Agroforestry for Small Landholders of Eastern and Southern Africa

Tracy L. Beedy, George Nyamadzawo, Eike Luedeling, Dong-Gill Kim, Frank Place, and Kiros M. Hadgu

CONTENTS

11.1 INTRODUCTION

Soil and water resources worldwide are under stress from the accelerated demands of increasing population. Reduction in soil fertility is driven by increased human population that has reduced land availability and caused a breakdown of traditional fallow systems that smallholder farmers relied on for soil fertility replenishment. Agroforestry is among a suite of sustainable agricultural practices that can rebuild soil fertility and soil organic matter (SOM), and break the cycle of poverty. Many agroforestry species are used for biological fixation of atmospheric nitrogen (N) into available N, root uptake, and recycling of nutrients. Nitrogen that accumulates in the leaves of planted tree fallows and intercrops is released when the biomass decomposes after being incorporated into the soil. Because farming practices of African smallholders tend toward multipurpose mosaics rather than uniform field management, the number of useful combinations of crops with agroforestry is constantly increasing. New tools such as "Useful Tree Species for Africa" facilitate the choice of trees within farming systems. Short-term agroforestry species have increased cereal yields from 10% to 200%, while yield differences under long-term parkland species such as *Faidherbia albida* have ranged from slight decreases to doubling of yields. Parkland systems have long been used by farmers but are now being recognized by the development community. The multiple sources of the yield benefits under parkland management are currently being documented by researchers. While rebuilding soil fertility, agroforestry also increases biomass buildup and carbon (C) sequestration in farming systems. This increase, however, is highly variable throughout eastern and southern Africa, and the residence time of soil organic carbon (SOC) is controversial. All agroforestry systems for which data are available accumulate biomass faster than the natural systems they emulate. The range of C sequestration by smallholder agroforestry in the tropics has been bracketed between 1.5 and 3.5 Mg C ha^{-1} $year^{-1}$. Addition of agroforestry species to farming systems has the potential to either enhance or reduce soil C and greenhouse gas (GHG) emissions. Thus, the study of GHG emissions with agroforestry practices is critical in describing the trade-offs between smallholder and ecosystem benefits from agroforestry.

11.1.1 African Smallholders, Soil Management, and Food Security

Africa is the only region in the world where the per capita food production has been consistently falling (Payne 2010, p. 45). While the number and prevalence of the undernourished has declined worldwide since 1990, in sub-Saharan Africa the number of undernourished has increased from 175 million to 239 million people (Food and Agriculture Organization of the United Nations, World Food Programme, and International Fund for Agricultural Development 2012). Impoverished populations in Africa remain concentrated in rural areas. The United Nations Environment Program (2009) estimated that 80% of the most desperately poor in Africa are subsistence farmers. According to the World Bank (2008), the majority of the poor are expected to remain in rural areas until at least 2040.

Soil and water resources worldwide are under stress from the accelerated demands of increasing population (Lal 2010). These stresses are of great concern in Africa because the rural poor rely most on these soil and water resources. Many smallholder farmers in Africa are located in areas where rainfall is low, erratic, and unreliable. Rainfall variability is a major challenge as many smallholder agricultural systems of eastern and southern Africa are predominantly rainfed and irrigation systems are not well developed (Camberlin et al. 2009). Most smallholder farmers in Africa practice low-input subsistence farming based on fertility-mining and extractive practices in which output exceeds input (Lal 2007). As yields decrease, the pressure to cultivate marginal lands increases. Deforestation and land degradation accompanying agricultural expansion have fragmented ecosystem provisioning, regulating, and supporting services previously provided by woodlands. Sustainable intensification (Figure 11.1) of agricultural systems in sub-Saharan Africa is the key to reducing food insecurity (Garrity et al. 2010; Payne 2010, p. 45). Agroforestry is

FIGURE 11.1 Mr. Mariko Majoni demonstrates his *Gliricidia sepium*–maize intercrop near Chiradzulu, Malawi. (From ICRAF.)

among a suite of sustainable agricultural practices that can rebuild soil fertility and SOM, and break the cycle of poverty (Lal 2004).

11.1.2 Definition and Objectives

Agroforestry is broadly defined as a set of land use practices that deliberately combine trees, shrubs, palms, or bamboo with agricultural crops or animals (Sileshi et al. 2007). This chapter will describe how agroforestry practices increase soil fertility and cereal yields for small landholders, as well as increase biomass and soil C accumulation. Emerging data on the increase in GHG emissions with agroforestry will also be presented, which will inform discussions of the trade-offs between smallholder households and the ecosystem-wide benefits of agroforesty.

The benefits of agroforestry are often classified as ecosystem services, which are conditions and processes through which ecosystems sustain human life (Tallis and Kareiva 2005). Ecosystem services may be divided into four categories: provisioning, regulating, supporting, and cultural services (Sileshi et al. 2007; Carpenter et al. 2009). Provisioning services of agroforestry/crop combinations include provision of food from increased soil fertility and improved soil water balance, and timber and fuelwood from rotational woodlots. Regulating services include erosion control, improved water infiltration, and C sequestration. Supporting services include biomass production and soil fertility improvement. Finally, cultural services include spiritual, cognitive, and aesthetic services.

11.1.3 Benefits and Constraints of Agroforestry

For smallholder farmers, the major benefit from using the soil-fertility-enhancing agroforestry practices described above is the increased yield that can lead to greater food security or additional income. At the same time, few farmers are aware of the possibility of receiving additional payments for the C sequestration service provided by the systems. Only a very small number are actually benefiting from soil C payments (the first ever scheme in western Kenya is very new), and the price paid for sequestered C is very low. Thus, the fact that agroforestry practices can sequester C through tree biomass and soil C buildup is not an important factor in farmers' *ex ante* or *ex post* evaluations of the practices.

What motivates farmers to adopt and manage agroforestry practices for soil fertility depends on the context, which includes how distinct these practices are from traditional farming practices. There are some locations where farmers have long practiced the integration of naturally regenerated trees in their crop fields. The main example of this is the parkland system in the West African Sahel (Boffa 1999); however, establishment of similar systems is possible in many dryland areas throughout Africa. Almost all drylands are important areas for tree regeneration. In typical parkland systems, trees regenerate naturally from roots or seed, and farmers retain those trees that are beneficial to them. From an economic point of view, the integration of trees for yield improvement in these types of systems is attractive for the following reasons: dryland areas have lower labor-to-land ratios than humid and subhumid areas, and thus labor-saving practices such as tree regeneration and management are compatible with resources; there is often poor access to markets in sparsely populated areas so that fertilizer costs are high; and trees offer fodder and other tree products that help households diversify agricultural enterprises and reduce climatic risks.

A recent study from the Sahel (Place and Binam 2013) found that crop yields were significantly higher in fields where mature trees beneficial to soils were found than in treeless plots. This was due to the direct effects of the trees and also an indirect effect by which farmers were more likely to apply manure and fertilizer on fields on which such trees were found. In quantitative terms, yields of millet and sorghum were often found to be 15%–30% higher on fields with an average cover of mature soil-enhancing trees such as *Faidherbia*, controlling for other inputs, rainfall and soil type. A similar finding is reported in the case of mature stands of *Faidherbia* in Malawi, where yields of maize (*Zea mays*) were 15% higher controlling for other factors (Glenn 2012). Given the low costs involved in establishing and managing trees through regeneration, the main constraints are related to abiotic and biotic factors such as aridity, fire, grazing, and the lack of germplasm in the soil. However, one policy constraint is particularly notable in the Sahel. That is the restriction on cutting, pruning, and transport of trees under the forest codes common in West Africa. These are intended to protect valued indigenous parkland trees; however, now that many trees are actually on farms, it has the perverse effect of inducing farmers to remove young seedlings before they mature (Yatich et al. 2013).

There are more opportunities for using agroforestry to improve soil fertility in the subhumid and humid regions. There, trees are commonly planted as seeds or seedlings, and this allows farmers much more control over the densities and species of trees they wish to use. Growing faster in the more humid areas, the trees can also provide quicker impacts on yields, which is desirable to farmers. For example, an improved fallow or dense intercrop practice can already raise yields after 2 years due mainly to the nutrients from leaf fall and incorporation (Sileshi et al. 2008), while soil physical and biological properties are slower to improve from trees. In both of these practices, trees are not allowed to reach mature height as they are either removed after a short time in the case of fallowing or regularly pruned in the case of intercrops.

The economic return from practicing improved fallows has been well studied for southern Africa and to some extent elsewhere. In Zambia, improved fallows performed much better than continuous maize production without fertilizer (Franzel 2004; Ajayi et al. 2007, 2009). In the recommended 5-year cycle of 2 years fallow and 3 years cropping, the net profit from unfertilized maize was only US$130 per hectare against US$269 and US$309 for maize grown under fallows using *Gliricidia* (*Gliricidia sepium*) or *Sesbania* (*Sesbania sesban*) species. The use of mineral fertilizer for 5 years provided higher net profits; however, returns to labor were similar to those of improved fallows. Studies of farmer behavior in eastern Zambia revealed that almost three-quarters of farmers exposed to the practice in the 1990s and early 2000s were still using it by 2009 (Kabwe 2010).

The practice of planting trees for soil fertility is not a traditional farming practice, probably anywhere in the world. Thus, a major constraint to the practice is awareness of its potential and understanding of how to put it into practice. A related constraint is availability of germplasm of trees with soil amelioration benefits. Further work with farmers has found that even if those barriers are removed, other difficulties or constraints emerge. These include lack of land in densely populated areas for fallowing, lack of labor for intensive shrub management, lack of water for nursery operations, and gender considerations when women may manage food production but have fewer rights to plant trees (Place et al. 2005; Ajayi et al. 2009). Thus, although agroforestry has many proven benefits, these benefits may not be well known to all farmers, and adoption is an involved process of working through constraints.

11.1.4 Agroforestry Practices and Smallholder Cropping Systems

The cropping systems developed by subsistence farmers depend on the available natural resource base and the dominant pattern of farm activities (Dixon et al. 2001, p. 11). Suitable agroforestry technologies vary depending on agroecology and prevalent cropping systems.

11.1.4.1 Agroforestry Options

Agroforestry practices may take many forms, and farming practices of African smallholders tend toward multipurpose mosaics rather than uniform field management. Thus, the number of useful combinations of crops with agroforestry is constantly increasing. To facilitate identification of useful tree–crop combinations, researchers at the World Agroforestry Centre (ICRAF) and the University of Copenhagen have developed the "Useful Tree Species for Africa" tool (Figure 11.2) in Google Earth, mapping the prevalence of tree species and linking to a database of their useful properties (Lillesø et al. 2011).

The largest smallholder farming system in eastern and southern Africa is the maize mixed system, although the cereal–root crop mixed system involves a similar number of farmers (Dixon et al. 2001, p. 37; Akinnifesi et al. 2010). Correspondingly,

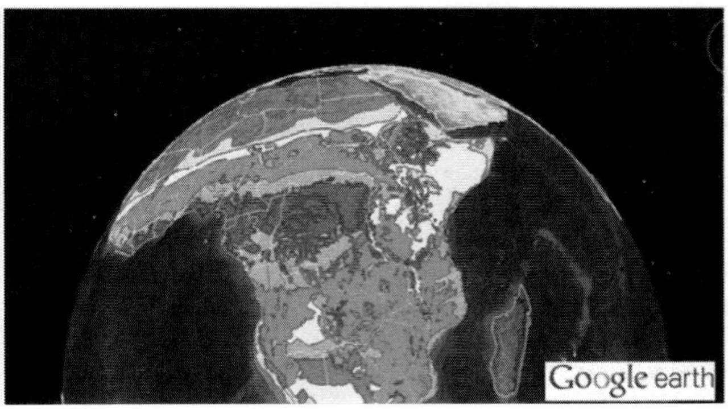

FIGURE 11.2 "Useful Tree Species for Africa"—online tool for tree species selection. (From ICRAF. 2012. Useful Tree Species for Africa. [Online] http://www.worldagroforestry centre.org/our_products/databases/useful-tree-species-africa; Lillesø, J.-P.B., P. van Breugel, R. Kindt, M. Bingham, S. Demissew, C. Dudley, I. Friis, F. Gachathi, J. Kalema, F. Mbago, V. Minani, H.N. Moshi, J. Mulumba, M. Namaganda, H.J. Ndangalasi, C.K. Ruffo, R. Jamnadass, and L. Graudal. 2011. *Potential Natural Vegetation of Eastern Africa, Volume 1: The Atlas.* Forest and Landscape Working Paper 61, University of Copenhagen.)

many critical agroforestry practices for food security in eastern and southern Africa involve strategic use of fertilizer trees to fix N for these annual cropping systems (Akinnifesi et al. 2010). Agroforestry trees that will tolerate coppicing, such as *Gliricidia*, *Leucaena* (*Leucaena leucocephala*), and *Senna* (*Senna siamea*), are planted as intercrops in subhumid areas (Table 11.1, Figure 11.1). The advantages

TABLE 11.1
Examples of Common Agroforestry Practices in Eastern and Southern Africa

Agroforestry Type	Food Crop–Agroforestry Combination	Function	Location (Biome) Rainfall	Reference
Intercrop	*Gliricidia sepium* Maize	Soil fertility Food	Zomba, Malawi (wetter Zambezian miombo woodland) 937 mm	Makumba et al. 2006
Agropastoral parkland	*Faidherbia albida* *Eucalyptus camaldulensis* Barley	Wood Wood Food	Adwa, Tigray, Ethiopia (Afromontane vegetation and east African evergreen bushland)[a] 740–900 mm	Hadgu et al. 2009
Improved fallow	*Tephrosia vogelli* *Sesbania sesban* Maize	Soil fertility Soil fertility Food	Zomba, Malawi (wetter Zambezian miombo woodland) 937 mm	Harawa et al. 2006
Rotational woodlot	*A. crassicarpa* *A. julifera* Maize *A. nilotica* *A. polyacantha* *L. leucocephala* Maize	Wood Wood Food Wood Wood Soil fertility Food	Tabora, Tanzania (drier Zambezian miombo woodland) 928 mm; Shinyanga, Tanzania (Somalia–Masai *Acacia–Commiphora* deciduous bushland) 700 mm	Nyadzi et al. 2003
Fodder bank	*Calliandra calothyrsus* *Pennisetum purpureum* Maize Coffee	Dairy fodder Erosion control Food Cash	Embu, Kenya (Afromontane vegetation, *Acacia* wooded grassland) 1200–1500 mm	Franzel et al. 2003
Multistorey/home garden	Enset Coffee *Millettia ferruginea* *Cordia africana*	Food Cash Shade Shade	Gedeo, Ethiopia (Afromontane vegetation) 800–1200 mm	Negash et al. 2012
Contour hedgerows	*Calliandra calothyrsus* *Pennisetum purpureum* Maize	Erosion barrier Erosion barrier Food	Embu, Kenya (moderately sloping land cleared from Afromontane vegetation) 1200–1500 mm	Angima et al. 2002

[a] Vegetation types from "Useful Tree Species for Africa" (http://www.worldagroforestrycentre.org/our_products/databases/useful-tree-species-africa).

of the intercrop are the supply of nutrient-bearing leaf biomass directly to the soil, with no land displaced from food crops. One disadvantage is the difficulty of tilling the soil near the tree because of woody tree roots. Coppicing species may be planted in hedgerows for alley cropping where precipitation is ample and the hedgerows of trees will not compete with the cereal crop for moisture (Sanchez 1995). Noncoppicing species, such as *Tephrosia* (*Tephrosia vogelii*), *Sesbania*, and *Acacia angustissima* are planted in 2-year improved fallows, to enrich the soil before 2–3 years of maize cropping. Improved fallows also provide N-rich leaf matter directly to the farmed soil. Although cereal grain yields after the improved fallow usually more than repay the time lost to fallow, the most food-insecure farmers may not be able to take the land out of production to take advantage of the benefits.

Fast-growing native and nonnative species have been selected by ICRAF (2013) and by national agricultural research services (Snapp et al. 2002) to provide rapid rehabilitation of soils. *Tephrosia vogelii* has often been used in improved fallows, and is a common shrub in eastern and southern Africa. A more productive cultivar, formerly called *Tephrosia candida*, was selected in India from the original *T. vogelii* germplasm, and has been introduced into eastern and southern Africa. *Gliricidia* was brought as an intercropping species from central America (ICRAF 2013) to eastern and southern Africa. *Tephrosia* and *Gliricidia* are fast-developing species that quickly produce nutrient-rich leaf biomass for soil fertility improvement.

In silvopastoral agroforestry (Figure 11.3), trees are either planted or protected among naturally regenerating populations, and are left scattered in grazing land to increase the carrying capacity of the grazing system (Hadgu et al. 2009). In some areas with high population density and rainfall, agroforestry forage species are planted in fodder banks, and the vegetation is cut and carried to confined livestock (Figure 11.4), especially dairy stock (Chakeredza et al. 2007).

FIGURE 11.3 Silvopastoral agroforestry near Shinyanga, Tanzania. (From Constance Neely, Charlie Pye/ICRAF.)

FIGURE 11.4 Smallholder farmer feeding a mixture of rich tree fodder and crop residue to her dairy animal in Kenya. (From ICRAF.)

Both traditional and introduced fertilizer trees have been used to improve smallholder farm productivity. In parkland agroforestry (Figure 11.5), large-statured trees, such as *Faidherbia*, have been intercropped with cereal crops across Africa for many generations (Wood 1992). *Faidherbia* is a long-lived tree that provides nutrient-rich biomass to cropping and livestock systems, and requires little space and labor, but is a slow-growing species that may take a decade after planting to make an appreciable impact on farming systems.

FIGURE 11.5 *Faidherbia albida* parkland with maize in Tanzania. (From ICRAF.)

FIGURE 11.6 Multistorey (multistrata) coffee production in Ethiopia. (From Dong-Gill Kim, Wondo Genet College of Forestry and Natural Resources, Hawassa University, Ethiopia.)

Rotational woodlots have the advantage of accumulating biomass quickly and providing much-needed wood for building and cooking fuel. However, where population densities are high, rotational woodlots also compete with food crops for space in smallholder farming systems. Multistrata (multistorey) agroforestry protects shade-grown coffee (Figure 11.6) and enset plantings in the highland temperate areas of Ethiopia (Negash et al. 2012) and other highland areas in eastern and southern Africa. The shade-giving species reduce evapotranspiration and erosion and, in some cases, fix N or provide fruit or timber (Sileshi et al. 2013). Home gardens provide fruit (Figure 11.7), medicinal products, timber, fuelwood, shade, and other benefits in dense, multistrata plantings around homesteads in humid areas; however, the shade from the trees prevents planting food crops adjacent to the home, where they are most easily protected.

FIGURE 11.7 Home garden in southern Ethiopia. (From Dong-Gill Kim, Wondo Genet College of Forestry and Natural Resources, Hawassa University, Ethiopia.)

11.1.5 Tree–Crop Interactions

Several traditional smallholder farming systems in eastern and southern Africa utilize tree–crop interactions for sustainable agricultural production (Buresh and Tian 1998; Kidanu et al. 2004). Trees can explore a relatively large space compared with crop plants, and they can have the capacity to capture and use aboveground and belowground resources efficiently (Goldberg and Barton 1990; Garcıa-Barrios and Ong 2004), thereby becoming more resistant to cyclic environmental changes than annual crops (Hiremath et al. 2002). They can increase available nutrients for crops by root exudates and leaf drop (Jung 1970; Radersma and Grierson 2004).

Aboveground and belowground resources are partitioned between trees and crops such that relative interspecific competition is lower than relative intraspecific competition, resulting in niche differentiation (Malézieux et al. 2009). Thus, there can be a total resource increase in the system or increased resource use efficiency (Cannell et al. 1996; Holmgren et al. 1997). System productivity can also be increased by reducing nutrient losses through leaching in deep soil, reducing soil erosion, protection against wind, reduced weed populations (Liebman and Gallandt 1997), or nutrient capture through N fixation and myccrrhizal associations (Young 1989; Giller 2001). Moreover, trees can add considerable amounts of organic matter to the soil, improving soil fertility and physical structure, stabilizing soil structure, and reducing erosion (Young 1997; Roose and Barthes 2001). Thus, trees and crops are complementary since enhanced soil fertility in the presence of trees can increase crop productivity in the vicinity of trees (Verinumbe 1987).

11.1.5.1 Water Relationships

A tree can modify and improve the growth of other trees or crops by changing the biophysical environment (Hunter and Aarssen 1988; Garcıa-Barrios and Ong 2004). Trees may affect soil water content by increasing it (Caldwell and Richards 1989; Dawson 1993; Bayala et al. 2008) or by decreasing it (Odhiambo et al. 2001), and thereby influence nutrient transport to crop roots and root growth (Radersma et al. 2004). Although trees can increase the potential soil water-holding capacity, they can also have negative effects on the actual water volume available in the tree–crop–soil system. Trees can reduce soil evaporation by shading crops and reducing air movement through understories, improve microclimatic conditions by reducing air temperature and wind speed, and reduce water stress in crops (Monteith et al. 1991; Vandenbelt and Williams 1992). Trees reduce exposure to heat stress, which minimizes tissue temperature to optimize the phenology and productivity of understory crops (Monteith et al. 1991; Vandenbelt and Williams 1992) and thus offsets water losses by evaporation from tree canopies (Ong and Swallow 2003). Depending on the slope and soil characteristics, trees can also increase infiltration (Ong and Swallow 2003). Tree roots can use water accumulated deeper in the soil profile, which can benefit crop growth. Besides, they can use residual available water outside the crop growing season (Ong et al. 2002; Garcıa-Barrios and Ong 2004). Integrated tree–crop systems have existed for generations and may enhance crop production (Saka et al. 1994). In general, available water can be used more efficiently in a tree–crop system than a sole crop system owing to favorable microclimate and improved water use efficiency. Trees can reduce the unproductive components of the water balance such as runoff, soil evaporation, and drainage (Ong and Swallow 2003).

11.1.5.2 Weed Suppression

Improved fallows can control *Striga* (*Striga hermonthica*), a parasitic weed that causes large yield losses in many cereal crops (Barrios et al. 1998; Gacheru and Rao 2001). Although the processes are not well understood, it is suspected that the fallow species excrete substances that cause suicidal early germination of *Striga* (Jama et al. 2006). Improved fallows also act as a break crop by smothering weeds, shading weeds, and outcompeting weeds for nutrients. Increased weed suppression increases nutrient availability for plant uptake and thus results in increased crop yields.

11.1.5.3 Soil Nutrient Content

System productivity can also be increased by trees through the reduction of nutrient losses from leaching in deep soil, reduced soil erosion, protection against wind, and reduction of weed populations and aggressiveness (Leibman and Gallandt 1997). Trees have the potential to increase overall system productivity by increasing nutrient availability through N fixation. Trees with perennial mycorrhizal associations may improve the efficiency of phosphorus (P) cycling because of greater absorptive area and of increasing pools of organic P (de Carvalho et al. 2010). Deep-rooting trees may pump nutrients from below the crop rooting zone (Harawa et al. 2006), both those made available by weathering of the bedrock and those leached down from the upper layers (Young 1989).

11.2 SOIL FERTILITY REPLENISHMENT

11.2.1 Soil Fertility Replenishment with Agroforestry

Many smallholder farming areas in southern and eastern Africa have low potential for agriculture because of poor soil fertility and low SOC content, which also causes soils to have a low water-holding capacity. The low soil fertility in these soils is widely recognized as a major factor contributing to low agricultural productivity in southern Africa (Vanlauwe and Giller 2006). Reduction in soil fertility is driven by increased human population that has reduced land availability and caused a breakdown of traditional fallow systems that smallholder farmers relied on for soil fertility replenishment (Jurion and Henry 1969). This has forced people to farm in more fragile lands. In Zambia, unavailability and inability of most smallholder farmers to purchase inorganic fertilizers following the removal of government agricultural subsidies have resulted in a reduction in crop productivity (Howard and Mungoma 1996). A combination of low soil fertility, lack of fertilizer amendments, and poor rainfall has often resulted in decreased crop production and widespread food shortages. This reduction eventually led to a reduction in the ability of most countries in the region to provide a stable food supply for their people.

Agroforestry is one of the many sustainable options that can assist smallholder farmers to replenish their impoverished soils (Ajayi et al. 2009). The use of agroforestry technologies has increased crop yields in smallholder farming areas (Kwesiga et al. 1999; Sileshi et al. 2008). In these systems, soil fertility benefits from fallowing are derived from the use of annual, biannual, or perennial N-fixing trees or "leguminous fertilizer trees," which are either planted in rotation (e.g., improved fallows) or together with crops (e.g., alley or intercropping). Leguminous fallow trees that have been used successfully include coppicing perennials such as *Gliricidia* and *A. angustissima* and others such as *Sesbania* and pigeon pea (*Cajanus cajan*), which do not coppice well and need to be replanted during the fallowing phase of crop–fallow rotations.

11.2.1.1 Nitrogen Fixation

Species such as *Sesbania*, *Gliricidia*, and *Tephrosia* replenish soil fertility through biological fixation of atmospheric N into available N, root uptake, and recycling of nutrients. Nitrogen that accumulates in the aboveground biomass (especially leaves) of planted tree fallows and intercrops is released when the biomass decomposes after being incorporated into the soil, and it is utilized by crops during the cropping phase. Different tree species can fix different amounts of N, and the total amount of N that is released is also dependent on the decomposition rates of the leguminous tree biomass (Mafongoya and Dzowela 1999). For example, Mafongoya and Dzowela (1999) reported that biomass from *S. sesban* decomposed faster than biomass from *A. angustissima* under similar field conditions.

A study in Zimbabwe by Chikowo (2004) showed that total fixed N (estimated using *Hyparrhenia* spp. as reference plant) in *A. angustissima* (nonwoody components + leaf litter) was 122 kg N ha^{-1} during the 2-year fallow period, while pigeon pea, *Sesbania*, and cowpea fixed 97, 84, and 28 kg N ha^{-1}, respectively (Table 11.2).

TABLE 11.2
Total N Fixed by Different Legumes during the Fallowing Phase

Species	Total N (kg ha^{-1})	Location	Reference
	Improved Fallow		
A. angustissima	122	Zimbabwe	Chikowo 2004
S. sesban	84	Zimbabwe	Chikowo 2004
C. cajan	97	Zimbabwe	Chikowo 2004
S. sesban, T. vogelii	150	Zambia	Ajayi et al. 2005
S. sesban (1 year)	60	Malawi	Chirwa and Quinion 2012
S. sesban	128	Kenya	Sanchez and Palm 1996
C. cajan	75–200	Malawi	Kumwenda et al. 1996
	Rotational Woodlot		
A. crassicarpa	78[a]	Tanzania	Kimaro et al. 2008
A. mangium	87[a]	Tanzania	Kimaro et al. 2008
A. polyacantha	104[a]	Tanzania	Kimaro et al. 2008
G. sepium	114[a]	Tanzania	Kimaro et al. 2008

[a] Maize uptake during the following 3 years, rather than supply.

Several other studies have also reported N fixation by legume trees, with reported values ranging from 100 to 200 kg ha^{-1} (Table 11.2). The benefits of the N contribution of fertilizer trees on a regional scale (Zambia, Malawi, Zimbabwe, Tanzania, and Mozambique) as of 2003 were estimated to be from \$6.27 to \$7.13 million per year in savings on the purchase of mineral fertilizer (Ajayi et al. 2005).

Trees may supply crop nutrients through leaf drop, root exudates, and senescence, as well as supply timber and fuelwood. In a study in Tanzania, *Acacia polyacantha* and *Gliricidia* leaves had higher nutrient concentrations and lower C-to-N ratios, reflecting higher leaf quality compared with those of *Acacia crassicarpa* and *Acacia mangium* (Kimaro et al. 2008). Nitrogen uptake of maize follows a similar pattern (Table 11.2), with *A. polyacantha* and *Gliricidia* supplying >100 kg ha^{-1} of N to the maize and the exotic acacias supplying less N for subsequent crops.

11.2.1.2 Other Macro- and Micronutrients

The ability of agroforestry trees to fix N for cereal crops is often emphasized. However, a closer analysis shows that the benefits of agroforestry go beyond N fixation. The use of agroforestry trees has proven to also enhance P availability to subsequent crops (Chikowo 2004; Ayuk and Mafongoya 2002; Ajayi et al. 2005; Jose et al. 2000) Many soils in southern Africa are P deficient, and expensive inorganic P supplements are needed for crop production. Fertilizer trees can economically close this gap as they can improve P availability through the secretion of organic acids and increased mycorrhizal fungi populations in the soil. The fungi that are associated with increased P availability in agricultural soils are arbuscular mycorrhizal (AM) fungi (phylum Glomeromycota) (Harrier and Watson 2003). Bagayoko et al. (2000) reported that N-fixing legumes resulted in better colonization of cereal roots and an increase in AM fungal populations in the soil. Fertilizer trees may act as reservoirs for AM fungi populations in crop root zones (Ingleby et al. 2007). The mixing of crops and trees that occurs in agroforestry systems may also result in a more equitable occurrence of mycorrhizae throughout the root zone because deep-rooted trees will distribute the AM fungi to deeper layers and increase the volume of soil from which nutrients are extracted (Cardoso et al. 2003). Management practices such as agroforestry that allow a buildup of AM fungi in soils would alleviate P deficiency while enhancing N fixation (Houngnandan et al. 2000). AM fungi have been proposed as a partial solution to nutrient deficiencies in tropical soils (Cardoso and Kuyper 2006), and the combination of AM fungi and agroforestry may have a role in the alleviation of micronutrient deficiencies in depleted smallholder soils.

11.2.1.3 Role of Fertilizer Trees as Nutrient Pumps and Safety Nets

In cropping systems located in high-rainfall areas, where net N mineralization exceeds N uptake by crops, infiltrating water carries nutrients to the subsoil, resulting in a buildup of subsoil N that ranges from 70 to 315 kg ha^{-1} (Hartemink et al. 1996). This is supported by Chikowo (2004) who reported that up to 45 kg N ha^{-1} is lost through leaching during the cropping phase of maize fallow rotations. Under conditions where the soils have substantial anion exchange capacity, e.g., Oxisols and oxic Alfisols, leached N and other nutrients are retained in the subsoil beyond the reach of most crops and can only be accessed by tree roots (Mekonnen et al. 1997).

Legume trees act as "safety nets" and nutrient pumps as they help in closing the nutrient cycle by taking nutrients leached into deeper layers up to the surface (Jose et al. 2004; Harawa et al. 2006). In addition, legume trees break up hardened soil layers and help increase infiltration rates since they have deeper rooting systems (Nyamadzawo et al. 2008a). The roots also help by bringing water and nutrients to the surface from deeper soil layers through hydraulic lift (Jose et al. 2004; Bayala et al. 2008).

The soil fertility benefits of using fertilizer trees are a result of complex factors that work together to provide a conducive environment for plant growth. This results in enhanced plant growth and improved maize yields, due to, for instance, improved N fixation and N availability in fallow crop rotations or tree–crop intercropping. Increased SOC, a better safety net from deep-rooted tree fallows, increased P availability from increased soil AM fungi in the soil, and increased weed suppression all result in improved soil fertility in legume tree–crop intercrops or rotations. We concluded that the use of fertilizer trees provides a window of opportunity for smallholder farmers to improve soil fertility and maize yields. The use of fertilizer trees is a sustainable option that can result in improved food security in households that are resource constrained, besides providing other goods and services such as fuelwood.

11.2.2 Maize Yield Increases from Improved Soil Fertility

Studies from the region have shown that fertilizer trees improve maize yields beyond continuous maize production without fertilizers or with natural fallows. Controls used to calculate response ratios in Table 11.3 may be fertilized or unfertilized maize or natural fallow, and are described in the text. The cereal crop is maize unless noted otherwise. These studies are not an exhaustive list, but are given to illustrate the scope of possible yield benefits and to illustrate important issues in management of these combined systems. For a more detailed treatment of maize yield increases with agroforestry, see Sileshi et al. (2008).

11.2.2.1 Improved Fallows

Maize yields obtained after fallowing are highly variable. For example, Nyamadzawo et al. (2012) reported maize yields that ranged between 10% and 250% higher after fallows compared with continuous maize cropping without fertilizers (Table 11.3). They reported yields of 1.8 and 0.7 Mg ha^{-1} for *A. angustissima* and *Sesbania* under conventional tillage (CT), while no tillage had even lower yields (1.3 and 0.8 Mg ha^{-1}) for *A. angustissima* and *Sesbania*, respectively, in the first year of cropping after 2 years of fallowing. However, in the second year of cropping, the maize yields were 1.6 and 0.5 t ha^{-1} for *Sesbania* and *A. angustissima*, respectively, under CT, while the maize yields were 1.5 and 0.3 t ha^{-1} for *Sesbania* and *A. angustissima*, respectively, under no tillage. Maize yields are also affected by additional factors such as pests, diseases, and competition for water. In the first year after fallowing in the same study by Nyamadzawo et al. (2012), maize after *Sesbania* was infested by cutworms and this resulted in lower crop yields, while in the second cropping season, there was a mid-season drought that affected the maize crop in coppicing *A. angustissima*. This resulted in lower crop yields as a result of competition for water and not because of N deficiencies.

TABLE 11.3
Cereal Yield Increases with Agroforestry Compared with Unfertilized or Fertilized Maize or Grass Fallow

Species	Fallow Years	Crop Years	Yield[a] (Mg ha^{-1})	Response[b] Ratio	Location	Reference
			Improved Fallow			
A. angustissima	2	2	3.1[c,d]	2.5	Zimbabwe	Nyamadzawo et al. 2012
S. sesban	2	2	1.5[d]	1.1	Zimbabwe	Nyamadzawo et al. 2012
S. sesban	2	2	5.6[d]	2.8	Zambia	Ayuk and Mafongoya 2002
T. vogelii	2	2	5.5[d]	3.1	Zambia	Ayuk and Mafongoya 2002
T. diversifolia	1	1	5.5	1.2	Kenya	Niang et al. 2002
S. sesban	1	1	7.1	1.6	Kenya	Niang et al. 2002
T. vogelii	1	1	6.2	1.4	Kenya	Niang et al. 2002
			Intercrop			
G. sepium	1	1	0.8	1.6	Malawi	Harawa et al. 2006
S. sesban	1[e]	1	0.7	1.4	Malawi	Harawa et al. 2006
T. vogelii	1[e]	1	0.9	1.8	Malawi	Harawa et al. 2006
G. sepium	1	1	5.2	3.4	Malawi	Makumba et al. 2006
G. sepium	1	1	4.2	4.3	Malawi	Akinnifesi et al. 2007
			Rotational Woodlot			
A. crassicarpa	5	2	4.0[d]	1.3	Tanzania	Kimaro et al. 2008
A. mangium	5	2	4.7[d]	1.6	Tanzania	Kimaro et al. 2008
A. polyacantha	5	2	5.9[d]	2.0	Tanzania	Kimaro et al. 2008
G. sepium	5	2	6.2[d]	2.1	Tanzania	Kimaro et al. 2008
A. crassicarpa	4	1	2.0	3.3	Tanzania	Nyadzi et al. 2003
A. julifera	4	1	1.75	2.9	Tanzania	Nyadzi et al. 2003
A leptocarpa	4	1	1.1	1.8	Tanzania	Nyadzi et al. 2003
S. siamea	4	1	0.75	1.3	Tanzania	Nyadzi et al. 2003
			Parkland			
F. albida	Decades	1	2.2	0.95	Malawi	Saka et al. 1994
F. albida	Decades	1	2.0	1.5	Malawi	Saka et al. 1994
F. albida	Decades	1	1.9	1.76	Ethiopia	Poschen 1986
F. albida	Decades	1	1.6[g]	1.36	Ethiopia	Poschen 1986
F. albida	Decades	1	1.4[f]	1.5	Ethiopia	Hadgu et al. 2009
F. albida	Decades	1	1.4[f]	1.4	Ethiopia	Hadgu et al. 2009
F. albida	Decades	1	3.0	2.0	Ethiopia	Dechasa 2010
F. albida	Decades	1	3.8[g]	1.9	Ethiopia	Dechasa 2010
F. albida	Decades	1	1.0[h]	1.3	Ethiopia	Dechasa 2010
F. albida	Decades	1	0.7[i]	1.3	Ethiopia	Dechasa 2010

[a] Yield of the agroforestry treatment.
[b] Yield of the agroforestry treatment divided by the control.
[c] Yield figures are for maize except where otherwise indicated.
[d] Combined yield over 2 cropping years following the agroforestry rotation.
[e] Relay intercrop.
[f] Barley.
[g] Sorghum.
[h] Wheat.
[i] Tef.

Ayuk and Mafongoya (2002) reported a successive decline in maize yields with the number of cropping years after fallow termination. After fallowing with *Sesbania*, maize yields were 3.6, 2.0, and 1.6 Mg ha^{-1} for the first, second, and third year, respectively, after fallow termination, while after *Tephrosia* maize yields were 3.1, 2.4, and 1.3 Mg ha^{-1} for the first, second, and third year, respectively, after fallow termination. Mafongoya et al. (2003) reported that fallow species had a positive N balance in the first year of cropping after the fallow phase. In the second year of cropping, the N balance in the fallow systems decreased, partly due to large offtake in the large yields and partly due to leaching through the soils and other associated losses of N due to immobilization, volatilization, and denitrification. These observations were supported by Mafongoya and Dzowela (1999), who suggested that the postfallow cropping phase should be restricted to 2 years.

In Kenya, Niang et al. (2002) tested *Tithonia* (*Tithonia diversifolia*), *Tephrosia*, and *Sesbania* in a single year of fallow, followed by two rainy season maize crops in the following year. The fallow species produced 1.2, 1.4, and 1.6 times as much maize as the sole maize control. *Tithonia* was propagated from cuttings, *Tephrosia* was direct-seeded in the maize, and *Sesbania* seedlings were transplanted into the fields.

Improved fallows usually follow a 2-year fallow plus 2-year cropping pattern, so that the yields in the cropping years must be more than double the yields under sole maize for the fallow to be worthwhile for the smallholder. In Table 11.3, all of the 2-year yield increments save one in Zimbabwe are well above two, so that the practice produces worthwhile yield gains for the smallholders, while at the same time improving soil and ecosystem health.

The addition of half the recommended fertilizer dose to tree fallow plots may provide higher yields than fertilizer alone or tree fallows alone. In Zambia, cumulative maize yields were 10.7 Mg in fallows with half of the recommended fertilizer rate, compared with 10.4 Mg for sole fallow over a 3-year period (Ajayi et al. 2005). Similar results were also reported by Chikowo (2004) and Makumba et al. (2006), who reported better maize yields in plots where half the recommended fertilizer rates were applied compared with sole fertilizers or unfertilized fallows. From these observations, it was concluded that at certain levels of fertilizer use, there is some synergy between mineral fertilizers and improved fallow species such as *Sesbania* and *Tephrosia* (Ajayi et al. 2005).

11.2.2.2 Intercrops

Harawa et al. (2006) found yields to be 1.6 times higher with *Gliricidia* intercrop, compared with the sole maize control, across all slope positions. Maize yields were 1.4 and 1.8 times those in the sole maize control with *Sesbania* and *Tephrosia* relay intercrops in the same study. In the relay intercrop, the fallow species are planted annually directly following maize planting. Leaf biomass developed during the dry season is incorporated annually during land preparation.

Each of the agroforestry species fits into specific agroecological niches in cereal-based cropping systems. Harawa et al. (2006) tested *Sesbania*, *Tephrosia*, and *Gliricidia* on upper-slope, mid-slope, and bottom-slope positions near Zomba in southern Malawi. *Sesbania* was found better adapted for the bottom-slope position, while *Gliricidia* was best placed in the well-drained middle slope positions. *Tephrosia* yields were similar across slope positions. However, Brewbaker (1990) has demonstrated that *Sesbania* may successfully establish in various soils and environments.

In Malawi, Makumba et al. (2006) found that an 11-year *Gliricidia* intercrop produced an average of 2.6 times as much grain as sole maize, while a combination of *Gliricidia* plus 48 kg N ha^{-1} produced 3.5 times as much maize as unfertilized sole maize. The performance of sole maize plus 48 kg ha^{-1} N was similar to the *Gliricidia* intercrop with no N added. In two of the study years, Makumba at al. (2006) noted a statistically significant interaction of the *Gliricidia* intercrop with the N fertilizer. Akinnifesi et al. (2007) also found 4.3 times as much maize yield with the *Gliricidia* intercrop.

Yield data reported by Harawa et al. (2006), Makumba et al. (2006), and Akinnifesi et al. (2007) differ considerably. The first study was done in farmer's fields with a combination of farmer and researcher management. The yields reported here were obtained after 1 year of intercropping. This study contrasts species performance at different slope positions, but also illustrates the deficit in SOM and nutrients that is typical in smallholder agriculture in this region. The second and third studies were done at Makoka Agricultural Research Station in southern Malawi. The yields given in these studies were averaged over multiple cropping seasons, during which *Gliricidia* leaf biomass was incorporated each year. The research station, with ample means for inputs, land, and labor, is expected to have a much greater yield over a longer period, while the means of the smallholder to provide inputs are more restricted and variable across years.

11.2.2.3 Rotational Woodlots

Although intercrops and improved fallows provide some fuelwood and poles to the farmstead, smallholders with relatively large landholdings may choose to rotate 0.5–0.8 ha into a woodlot to provide additional wood for the household (Kimaro 2009). These woodlots are of longer duration than improved fallows. Maize is intercropped with the woodlot trees for the first 2–3 years; then the trees are left to develop for 1 or 2 years. The trees are then harvested, and maize is planted to take advantage of the nutrients in the decomposing leaf litter and root mass.

Total maize yields over 2 years after the woodlot rotation were higher (Kimaro et al. 2008) after *A. polyacantha* and *Gliricidia* than after *A. crassicarpa* and *A. mangium* (Table 11.3). Yields after the Australian acacias were only 1.3–1.6 times those after natural fallow, while yields after *A. polyacantha* and *Gliricidia* were more than double the yields compared with the natural fallow. This difference probably reflects higher soil fertility improvement during the fallow period (Kimaro et al. 2008). Yields after *A. polyacantha* and *Gliricidia* were also similar to fully fertilized maize, while unfertilized sole maize yields were similar to those after natural fallow. In contrast, the amount of aboveground biomass produced was much greater in *A. crassicarpa* at 51 Mg ha^{-1} than in the other three species, with *A. mangium* and *A. polyacantha* producing 38 and 36 Mg ha^{-1}, respectively, and *Gliricidia* producing 29 Mg ha^{-1}. Thus, in choosing species for this rotation, the smallholder must consider the trade-offs between timber and food production.

Nyadzi et al. (2003) reported maize yields on a single cropping year after a 4-year fallow. The yields were lower than those in Kimaro et al. (2008), while the response ratios were higher. This may be due to the very low initial SOM (0.4–0.8 g kg^{-1}) in Nyadzi et al. (2003), and to the fact that response ratios tend to be higher in the first year of cropping than in the second year.

Although short- or long-lived species may be used for agroforestry practices, the management and structure of the tree component is very different in parklands compared with improved fallows, intercrops, or rotational woodlots. While the agroforestry component in the short-term systems is coppiced or harvested in 1- to 5-year cycles, parkland trees are allowed to develop into mature large-scale specimens, with crops and livestock benefiting in various ways from the understory location. Parklands are integrated into various crop and livestock systems across Africa from north to south, and from east to west.

11.2.2.4 Parkland Systems in Eastern and Southern Africa

Parkland tree species are usually naturally occurring tree species (Maranz 2009), which are also protected and managed by farmers (farmer managed natural regeneration—FMNR) (Haglund et al. 2011). The main parkland tree species available in eastern and southern Africa include *F. albida*, *Cordia africana*, *Croton macrostachyus*, *Acacia tortilis*, *Moringa stenopetala*, *Terminalia brownii*, *Acacia senegal*, *Acacia seyal*, *Ziziphus mauritiana*, *Balanites aegyptiaca*, *Ficus sur*, and *Millettia ferruginea* (Hailu et al. 2000; Kassa et al. 2010).

Some studies reported that parkland trees do not provide short-term crop yield benefits, which could be attributed to the competition for resources between trees and crops (Bayala et al. 2012). However, parkland trees are protected by farmers for their sustainable and long-term multiple benefits, including direct products such as fodder, fruit, fuelwood, medicinal, or vegetable products. They are also valuable for long-term ecosystem benefits, such as C accumulation, reduction in soil erosion, maintenance of soil structure and fertility, improvement of crop microclimate, reduction of wind incidence, and provision of shade (Dechasa 2010).

11.2.2.5 *Faidherbia albida*: A Successful Parkland Species

Among parkland tree species, the potential of *Faidherbia*, an indigenous African acacia, is well recognized. It is widespread on millions of farmers' fields throughout the eastern, western, and southern regions of the continent. It is highly compatible with food crops because it is usually dormant during the rainy season (see Figure 11.5). Thus, it exerts minimal competition with annual crops, while enhancing crop yields and soil health (Barnes and Fagg 2003). In eastern and southern Africa, most smallholders cannot afford to buy inorganic fertilizers, often because of cash constraints. *Faidherbia* creates a unique opportunity for increasing smallholder productivity by input of high-quality leaf residue for increased soil fertility (Garrity et al. 2010), reducing the need for inorganic N fertilizer. As many smallholders are engaged in both crop and livestock agriculture and their available fodder resources are often inadequate (Giller et al. 2009), *Faidherbia* also increases livestock production through supplying high-quality fodder. It also enhances C storage in farmed landscapes. *Faidherbia* is considered a keystone species for climate-smart (evergreen) agriculture in much of Africa (Garrity et al. 2010).

Studies in Africa have documented increases in maize grain yield (Table 11.3) under *Faidherbia* (Barnes and Fagg 2003). In Malawi, 50% maize yield increases have been recorded under *Faidherbia* trees compared with sole maize (Saka et al. 1994). Thirty percent to 70% yield increases have also been observed under *Faidherbia* on most staples in Ethiopia, including maize, sorghum, and wheat (Poschen 1986; Dechasa 1989; Hadgu et al. 2009). Yield increments of 100%, 70%, 40%, and 10% for sorghum, maize, wheat, and teff, respectively, were also reported (Dechasa 2010).

Improvements have been made to traditional agroforestry practices with *Faidherbia*. This practice has resulted in significant contribution to food security as farmers who have adopted *Faidherbia* produced 1.5 Mg more maize per hectare than conventional practice (Haggblade and Tembo 2003). In this regard, Zambia has taken a lead in systematically utilizing the potential of *Faidherbia*, releasing a national recommendation to plant *Faidherbia* trees at a density of 100 trees ha^{-1} in crop fields as a permanent canopy to increase soil fertility and crop productivity (Garrity et al. 2010). The density may later be reduced to 25–30 trees ha^{-1} as the trees mature. To date, >200,000 families have adopted this practice (Garrity et al. 2010). The government of Ethiopia has recently launched an initiative to plant 100 million *Faidherbia* trees. Alongside promoting natural regeneration of this important tree in farmers' fields, planting *Faidherbia* at such a scale on smallholder farms is expected to have significant economic and environmental benefits and add climate resilience to farming systems, particularly in semiarid areas of Ethiopia.

Faidherbia is a slow-growing tree, often taking 20 years or more to contribute to crop growth (Poschen 1986). Yield increases also differ with location (Saka et al. 1994). Combining fast-growing multipurpose shrubs such as *Gliricidia* and *Sesbania* with parkland trees should speed soil fertility increases and also increase wood and fodder availability. Integrating fast-growing shrubs along with natural regeneration of *Faidherbia* may alleviate the concern of some smallholders regarding the slow establishment of *Faidherbia* seedlings.

Moreover, trees should be spatially and systematically configured (Figure 11.8) to match species to sites and farmer circumstances. Hadgu et al. (2009) noted a pattern of land use intensification in Tigray, northern Ethiopia, which reduced the density of *Faidherbia*, and was accompanied by an increase in timber cultivation and a decline in barley yields.

(Note in Figure 11.8 the loss in ecosystem services associated with cutting *Faidherbia* within a field.) The effect of the trade-off between barley production and timber on food security should be considered at the farm level. More data are also required on other promising parkland tree species such as *Moringa stenopetala, Balanites aegyptiaca*, *Millettia ferruginea*, *Cordia* spp., *Croton* spp., and *Ziziphus* spp., with regard to their effects on crop productivity. Studies similar to those on *Faidherbia* should also be done with such promising tree species.

Agroforestry practices used in appropriate agroecological niches hold great promise for recuperating soil fertility while providing greater food and feed production in agricultural landscapes. Intercrops are more appropriate in areas with high population densities, while improved fallows and rotational woodlots are more appropriate where land can be allocated to fallow and/or wood production. *Faidherbia* parklands have mostly been developed by farmer-managed natural regeneration in river valleys

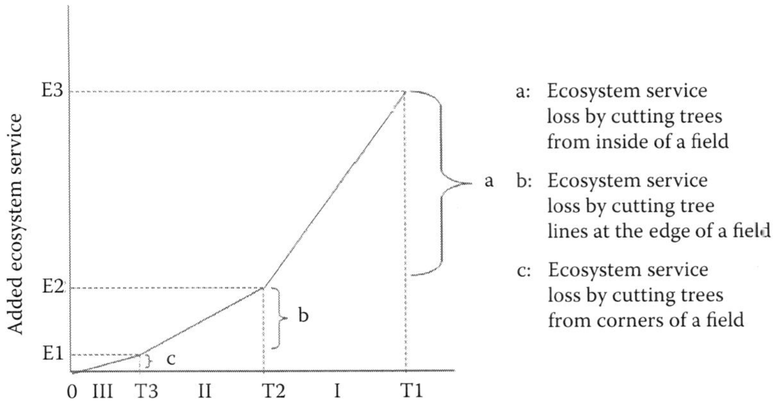

FIGURE 11.8 Theoretical model for added ecosystem services of increasing tree density (e.g., *F. albida*) on barley yield at farm level where E1, E2, and E3 refer to increasing barley yield levels for three spatial density configurations of a tree on the corner, edge, and within agricultural fields. (From Hadgu, K.M., L. Kooistra, W.A. Rossing, and A.H.C. van Bruggen, *Food Sec.*, 1, 337, 2009. With permission.)

and lowland lakeshore areas in eastern and southern Africa, but are increasingly being planted in uplands, especially in Zambia.

Use of different agricultural technologies to improve food security has been a perennial subject in sub-Saharan Africa. However, managing the fate of C in food, energy, and climate systems has become increasingly difficult (Lal 2010). It is increasingly important to manage cropping systems with both food security and C balance in mind.

11.3 ACCUMULATION OF CARBON IN BIOMASS AND SOIL

11.3.1 Biomass Accumulation in Agroforestry in Eastern and Southern Africa

Biomass buildup in agroforestry systems throughout eastern and southern Africa is highly variable. This variation is partly explained by edaphic and climatic site conditions, but it depends at least as strongly on the type of agroforestry that is practiced. Management of the system is also an important factor. Unmanaged natural regeneration of miombo woodlands, a typical vegetation type of eastern and southern Africa, occurs slowly, with annual biomass increments estimated at 1 Mg ha^{-1} in Zambia (Stromgaard 1985) and 0.43 Mg ha^{-1} in Tanzania (Aune et al. 2005).

11.3.1.1 Intercropping Agroforestry Systems

All agroforestry systems for which data are available accumulate biomass faster than the natural systems they emulate. Montagnini and Nair (2004) bracket the range of C sequestration by smallholder agroforestry in the tropics between 1.5 and 3.5 Mg C ha^{-1} year^{-1}. Most published studies of biomass accumulation in simultaneous agroforestry systems, i.e., farming systems where trees and crops are grown simultaneously on the same plots, confirm this range. In a study on intercropping of *Faidherbia* with beans and maize in the Morogoro region of Tanzania, Okorio and Maghembe (1994) reported biomass production in trees ranging between 2.1 and 4.7 Mg ha^{-1} year^{-1} on average during 6 years, depending on tree spacing. Intercropping of *Gliricidia* with maize during 10 years in Malawi produced a total of 20.5 Mg ha^{-1} of tree prunings for incorporation into the soil, corresponding to 2.9 Mg ha^{-1} year^{-1} (Makumba et al. 2007).

Results from a 7-year trial in the same study with slightly different tree management showed substantially lower biomass accumulation in tree prunings of only around 1.1 Mg ha^{-1} year^{-1} (Makumba et al. 2007). Considering that destructive sampling of trees at the end of the trial also revealed 17 Mg ha^{-1} in tree stumps and structural roots, the total biomass accumulation per year should nevertheless have been around 2.7 Mg ha^{-1} year^{-1}, on average, so that this trial also falls within the range estimated by Montagnini and Nair (2004).

Much lower biomass accumulation rates have been reported from agroforestry practices in western Kenya (Henry et al. 2009). In this study, aboveground biomass was measured on 35 farms across two administrative districts. Assuming biomass levels on all farms could be raised to reach the third quartile of the distribution of measured biomass for the respective district, aboveground biomass accumulation potentials were estimated for different land use types. With the exception of windrows and rotational woodlots, all agroforestry systems evaluated had biomass buildup potentials below the range specified by Montagnini and Nair (2004). Converting figures from Henry et al. (2009) from C stocks to biomass stocks (at 45%–50% C in biomass), individual trees in home gardens were expected to raise biomass levels by 0.3 and 0.6 Mg ha^{-1} year^{-1}. Individual trees in food crops had biomass accumulation potential between 0.2 and 0.3 Mg ha^{-1} year^{-1}, individual trees in cash crops between 0.1 and 0.4 Mg ha^{-1} year^{-1}, and individual trees in pastures around 0.1 Mg ha^{-1} year^{-1}. Windrows had higher potential at 2.9–3.5 Mg ha^{-1} year^{-1}, and biomass buildup in rotational woodlots was substantial at 2.3–12.2 Mg ha^{-1} year^{-1}. Two scenarios of intensifying hedgerow biomass had potential of accumulating between 0.2 and 0.5 Mg ha^{-1} year^{-1} of biomass.

11.3.1.2 Improved Fallows

Several studies have estimated biomass buildup in improved fallow systems. Albrecht and Kandji (2003) tabulated data from several studies using a range of tree species, in which aboveground biomass stocks ranged from 7.0 to 21.0 Mg ha^{-1} year^{-1} after 12 months, from 19.8 to 31.0 Mg ha^{-1} year^{-1} after 18 months, and from 27.0 to 43.4 Mg ha^{-1} year^{-1} after 22 months. These values appear on the high side of what is realistic, and are in contrast to modeling results by Walker et al. (2008) for biomass buildup in the IMPALA project, in which some of the figures in the table were generated. They estimated biomass buildup of only 10–32 Mg ha^{-1} after 10 years, corresponding to an average accumulation rate of only 1.0–3.2 Mg ha^{-1} year^{-1} (Walker et al. 2008).

Kaonga and Coleman (2008) measured aboveground C inputs in coppiced fallows in Zambia. These were in the range of 2.6–3.2 Mg C ha^{-1} year^{-1} for multiple species

(*L. leucocephala*, *Gliricidia*, *Calliandra* [*Calliandra calothyrsus*], and *S. siamea*), corresponding to around 5.2–7.1 Mg biomass ha^{-1} $year^{-1}$. For eastern Zambia, Kaonga and Bayliss-Smith (2009) quantified C stocks in improved fallows at 2.9–9.8 Mg ha^{-1}, which corresponds to between 5.8 and 21.8 Mg biomass ha^{-1}.

11.3.1.3 Rotational Woodlots and Tree Plantations

Much higher biomass accumulation rates are possible in the tree phases of rotational woodlots. After 7 years of growth, woodlots in Tanzania had between 26.0 and 57.6 Mg biomass ha^{-1}, corresponding to mean accumulation rates of 3.7–8.2 Mg ha^{-1} $year^{-1}$ (Nyadzi et al. 2003). The best performance was obtained from *Acacia leptocarpa*, followed by *A. crassicarpa*, *Acacia julifera*, *Senna*, and *Leucaena pallida*. These data also formed some of the basis for assumptions by Palm et al. (2010), who estimated that woodlots in Mbola, Tanzania, accumulated 5.3 Mg ha^{-1} $year^{-1}$ during a 5-year rotation.

Carbon accumulation in woodlots in Morogoro, Tanzania, ranged between 2.3 Mg ha^{-1} $year^{-1}$ under *Acacia nilotica* and 5.1 Mg ha^{-1} $year^{-1}$ under *A. crassicarpa* (Kimaro 2009). These rates correspond to between 4.6 and 11.3 Mg ha^{-1} $year^{-1}$ of biomass accumulation. Aune et al. (2005) found that the amount of C sequestered during a 4-year tree phase of rotational woodlots in Uganda was between 4.9 Mg ha^{-1} $year^{-1}$ for *Eucalyptus camaldulensis* and 3.9 Mg ha^{-1} $year^{-1}$ for *Alnus acuminata*, corresponding to biomass increments by 7.8–10.9 Mg biomass ha^{-1} $year^{-1}$. For rotational woodlot systems, biomass accumulation is naturally much higher during the tree phases, during which all the above-mentioned studies were conducted, than during the crop phases, before which essentially all tree biomass is removed.

Even higher biomass accumulation rates than in rotational woodlots are achieved in tree plantations. Ståhl et al. (2002) found that plantations of *Sesbania*, *Calliandra*, eucalyptus (*Eucalyptus saligna*), and grevillea (*Grevillea robusta*) produced 31.5, 24.5, 32.5, and 43.5 Mg aboveground biomass ha^{-1}, respectively, during 22 months in the highlands of eastern Kenya. Palm et al. (2010) used this source to assume a biomass accumulation rate of 12.2 Mg ha^{-1} $year^{-1}$ for 5-year-old woodlots in Sauri, Kenya. A plantation of *Pinus patula* in Tanzania was shown to build up 5.86 Mg C ha^{-1} $year^{-1}$, corresponding to 11.7–13.0 Mg biomass ha^{-1} $year^{-1}$.

Thus, while aboveground biomass buildup in the natural miombo forest in this region ranges from 0.43 to 1 Mg biomass ha^{-1} $year^{-1}$, agroforestry species added to smallholder cropping systems in improved fallows and parklands usually produce 1–5 Mg biomass ha^{-1} $year^{-1}$. Rotational woodlots may add up to 8 Mg biomass ha^{-1} $year^{-1}$ when the figures include the cropping phase. Plantations may produce twice the biomass of rotational woodlots, but require the land to be removed from cropping for long periods. Although building up aboveground biomass in farmed landscapes is an important component of the global biotic C pool, the soil C pool is 4.5 times the size of the biotic pool (Lal 2004), and thus may serve as an important sink for C, in addition to offering advantages in soil fertility and cropping system yield described above. Building up SOC offers important benefits on both the smallholder and ecosystems scales.

11.3.2 SOC in Smallholder Settings

11.3.2.1 SOC and Soil Fertility

Improved fallows, intercrops, and woodlot rotations increase the labile fractions of SOM, which supply nutrients to crops following fallows (Barrios et al. 1997; Beedy et al. 2010; Kimaro et al. 2011), and can also contribute to improving soil structure, buildup of SOM, and C stocks, thus contributing to C sequestration. Buildup of SOM is critical to soil productivity and generally corresponds to nutrient exchange capacity. Release of N from SOM may contribute most of the 40 kg N ha^{-1} taken up by the average maize crop of 1 Mg ha^{-1} (Sanchez and Palm 1996; Makumba et al. 2006). SOC increases the cation exchange capacity (CEC) of the surface soil, which is especially important for nutrient storage in kaolinitic soils and other light-textured soils with low CEC. Increasing SOM can reduce P fixation in soils with high iron and aluminum oxide contents, thus making the P available for plant uptake. High SOC in fallows results in reduced rates of nutrient leaching due to reduced mineralization rates (Nyamadzawo et al. 2009). Agboola (1994) reported that 80% of CEC, and available P, K, Mg, and Ca were highly correlated with SOM levels in some West African Alfisols. Under fallow systems, the microbial biomass has been shown to be higher (Nyamadzawo et al. 2009), the microbial community more diverse, and the rate of plant material decomposition much faster than in nonfallowed systems (Sarmiento and Bottner 2002), thus ensuring nutrient recycling and timely release of N and other nutrients. Leguminous fertilizer trees increase SOC levels of soils and thereby indirectly improve soil fertility. Fernandes et al. (1997) suggested that our greatest opportunity is that SOM is a renewable resource whose level can be replenished by additions of organic inputs.

11.3.2.2 Potential SOC Increases with Agroforestry

Agroforestry land use systems have been reported to sequester more C than other forms of agriculture. The amounts of biomass and SOC additions vary with tree species, soil type, rainfall, and environmental conditions. The extent to which agroforestry practices will build soil C is controlled by the ability of the tree–crop combination to produce biomass residue to be transformed into SOM. Plant biomass residues may be deposited on the soil surface for decomposition, incorporated by tillage, or added to the soil profile at varying depths by root exudates and root decomposition. Carbon sequestration from cropping system residues varies with soil temperature and moisture, litter quality, and root dynamics (Post and Kwon 2000). Albrecht and Kandji (2003) also specify fire, decomposition, leaching, and erosion as four major avenues of SOC loss. Burning of crop residues and tillage oxidize soil C. Erosion transports soil C offsite, sometimes into rivers and lakes. Leaching transports soil C downward in the soil profile and sometimes removes it into streams and rivers via runoff from fields.

The ability of a given soil to retain organic matter varies strongly with soil texture (Albrecht and Kandji 2003). SOC oxidizes more quickly in sandy soils and those with weak aggregation. Organic matter that is not adsorbed to clay colloids or protected within soil aggregates is quickly decomposed by soil microorganisms (Six et al. 2000). In fine-textured soils, SOC is adsorbed to clay colloids, protecting it from decomposition. After an improved fallow of 1–1.5 years with various species in Kenya, Albrecht and Kandji (2003) found an increase in SOC in the top 30 cm by 1.69–2.15 Mg ha^{-1} C in coarse soils, and by 2.58–8.34 Mg ha^{-1} C in fine-textured soils. Because agroforestry increases biomass addition to SOC, it increases soil aggregation, while woody cover and leaf litter from agroforestry protect topsoils from erosion, increasing the SOC that remains in the soil profile.

11.3.2.3 Intercrops, Improved Fallows, and Woodlots

Smallholders in eastern and southern Africa manage agroforestry species as intercrops with food crops such as maize or as improved fallows to replenish soil fertility and/or provide timber and fuelwood (Akinnifesi et al. 2010). Three of the most widely used management patterns for agroforestry in eastern and southern Africa are intercrops, improved fallows, and rotational woodlots.

Intercropping systems are most appropriate for smallholders in relatively population-dense areas with small landholdings, as no land has to be removed from cereal production to include the agroforestry species. Intercrop populations will vary across different cropping systems and ecological conditions. *Gliricidia*, for example, was planted in alternating planting ridges, at 0.9 m distance between trees within a ridge, and 0.75 m distance between ridges in Malawi, with maize planted in the ridges at 44,400 plants ha^{-1} (Makumba et al. 2007). These *Gliricidia* populations may also be used in minimum tillage systems. Makumba et al. (2007) reported 123 and 149 Mg ha^{-1} of soil C (Table 11.4) in 10 and 7-year intercrops in Malawi (Makumba et al. 2007). Soil C in the intercrop was roughly double that in unfertilized sole maize, at 64 and 73 Mg ha^{-1} after 7 and 10 years, respectively. The 7-year study also accumulated 1.2 times as much soil C under the maize–*Gliricidia* intercrop as under an adjoining grass fallow. Unfertilized maize in this study represents the loss of soil C when such soils are cropped continuously to maize with no inputs.

A 10-year intercrop in Zambia with three agroforestry intercrop species ranged from 225 to 245 Mg ha^{-1} of soil C (Kaonga and Bayliss-Smith 2009), similar to the 245 Mg ha^{-1} found in natural fallow, but 1.3–1.6 times as much soil C as found in fertilized maize and miombo treatments. These differences may be, in part, due to a decline in soil C in the fertilized maize treatment. Thus, including agroforestry intercrops in a cropping system can maintain and increase soil C compared with the natural miombo vegetation and to cropping systems with only mineral fertilizer added. In another study from the same location, Kaonga and Coleman (2008) reported an increase from an initial soil C of 26.2 Mg ha^{-1} up to 37.4 Mg ha^{-1} during 10 years in the upper 20 cm of the soils.

Improved fallow agroforestry is appropriate for smallholders in low-population-density areas, where landholdings are larger and land can be spared for improved fallows. The trees are established and left for 1.5–2 years to develop leaf biomass that is incorporated into the soil before the crop planting phase. Kaonga and Coleman (2008) reported that *Tephrosia*, pigeon pea, and *Sesbania* were established at 1 m by 1 m spacing, and maintained for 2 years. During land preparation for the third year, the trees were cut to 10 cm, the stems and branches harvested for fuelwood, and the leaves and twigs incorporated into the soil. Maize then replaced the trees in a 2-year cropping phase.

TABLE 11.4
Development of Soil Organic Carbon with Agroforestry Practices

Woody Species	Depth (cm)	SOC ($Mg\ ha^{-1}$)	Period (Years)	Country	Reference
			Intercrop		
G. sepium	200	123[a]	10	Malawi	Makumba et al. 2007
G. sepium	200	149	10	Malawi	Makumba et al. 2007
G. sepium	200	225	10	Zambia	Kaonga and Bayliss-Smith 2009
L. leucocephala	200	245	10	Zambia	Kaonga and Bayliss-Smith 2009
S. siamea	200	245	10	Zambia	Kaonga and Bayliss-Smith 2009
Various	20	30.5–37.4	10	Zambia	Kaonga and Coleman 2008
			Improved Fallow		
C. cajan	200	149	4	Zambia	Kaonga and Bayliss-Smith 2009
S. sesban	200	150	4	Zambia	Kaonga and Bayliss-Smith 2009
T. vogelii	200	155	4	Zambia	Kaonga and Bayliss-Smith 2009
Various	20	27.3–31.2	4	Zambia	Kaonga and Coleman 2008
A. angustissima	120	26.3	2	Zimbabwe	Nyamadzawo et al. 2008b
S. sesban	120	25.4	2	Zimbabwe	Nyamadzawo et al. 2008b
			Rotational Woodlot		
A. polyacantha	15	21.6	5	Tanzania	Kimaro et al. 2011
G. sepium	15	18.8	5	Tanzania	Kimaro et al. 2011
A. crassicarpa	15	15.8	5	Tanzania	Kimaro et al. 2011
A. mangium	15	25.6	5	Tanzania	Kimaro et al. 2011
A. nilotica	15	22.7	5	Tanzania	Kimaro et al. 2011

Note: SOC, soil organic carbon.
[a] Comparative values from these studies are found in the text.

Between the first and second years of the cropping phase in Kaonga and Coleman (2008), soils were sampled to 20 cm and analyzed for total SOC (Table 11.4). The total soil C after 1 year of cropping was 1.0 and 1.2 times that of the initial soil C. SOC to 200 cm depth (Kaonga and Bayliss-Smith 2009) was comparable to the 150 $Mg\ ha^{-1}$ C found in nearby natural miombo vegetation and more than nine-tenth of the 165 $Mg\ ha^{-1}$ C found in an adjacent natural fallow. Nyamadzawo et al. (2008b) reported that fallows of *A. angustissima* and *S. sesban* accumulated 26.3 and 25.4 $Mg\ ha^{-1}$ in SOC (Table 11.4) after 2 years of fallowing, and fallowing resulted in 3.7–9.1 $Mg\ ha^{-1}$ more SOC compared with continuous maize cropping. After an improved fallow of 1–1.5 years with various species in Kenya, Albrecht and Kandji (2003) found an increase in SOC in the top 30 cm of 1.69–2.15 $Mg\ ha^{-1}$ C in coarse soils, and from 2.58 to 8.34 $Mg\ ha^{-1}$ C in fine-textured soils.

Rotational woodlots differ from improved fallows in that the tree species may be selected for wood production rather than N fixation. However, rotational woodlots also usually increase SOC and promote increased yields in subsequent cereal crops. Tree and crop components are established at the same time, and the intercropping continues for 2–3 years with crop yield declining because of competition from the trees. This is followed by a 2–3-year tree fallow period during which little or no management is required to maintain the trees. After this, the woodlot is cleared to supply wood for household use, such as building poles, firewood, and tobacco curing. Subsequently, crops are grown between tree stumps to benefit from the ameliorated soil conditions (Kimaro 2009).

In Kimaro et al. (2011), two Australian acacias (*A. crassicarpa*, *A. mangium*), two African acacias (*Acacia nilotica*, *A. polyacantha*), and one fertilizer tree (*Gliricidia*) were compared at Morogoro, Tanzania. After 5 years, all of the tree fallows developed more total SOC than the 13.0 Mg ha^{-1} measured under continuous maize, and *A. mangium* had double the SOC of the continuous maize, probably because of declining C in the control. The natural grass fallow comparison had 17.8 Mg ha^{-1} SOC, and the SOC under the tree fallows ranged from 0.89 to 1.4 times the SOC compared with the grass fallow.

11.3.2.4 Connection with Carbon Markets

The studies cited in Table 11.4 fall broadly within the area of southern and eastern Africa dominated by open, dry miombo woodlands, which occupy about 10% of the African land mass (Malmer and Nyberg 2008). According to Williams et al. (2008), soil C stocks in a Mozambican site had a narrower range (21–74 Mg C ha^{-1}) in the top 0.3 m on abandoned land than in the miombo woodland soils (18–140 Mg C ha^{-1}). A growing proportion of miombo woodlands have been cut for fuelwood and converted to smallholder agriculture, which in very few years of extractive cropping reduces SOM and nutrients such that maize production becomes unsustainable. Each of the agroforestry interventions described increases SOC, allowing for production of food crops and fuelwood with less soil degradation. These technologies allow the restoration of some of the ecosystem services formerly provided by the miombo forest, while increasing the provisioning services to provide for the increasing human population resident on the land. Although these practices hold great promise, their use in smallholder settings to generate C credits comes with several constraints. SOM is generally lower and more variable on smallholder land than on larger landholdings or at research sites, decreasing payments and increasing monitoring costs among smallholder households. The residence time of SOC is controversial (Davidson and Janssens 2006; Lal 2004), and studies are especially needed in the area of belowground C cycling and GHG evolution within different cropping systems.

Two ongoing projects operating in Kenya show the difficulties in establishing a price at the farm level. The TIST program (TIST 2011) is paying 1 shilling per surviving tree per year (similar to about \$12 per year per hectare) but hopes to be able to have a larger payment in later years. The Vi Agroforestry soil C program plans to pay \$11 per Mg of C sequestered, which it estimates may be about 2.25 Mg ha^{-1} during a 20-year period (which, therefore, is just over \$1 per hectare per year) (Vi Agroforestry Strategy 2013–2015). The actual price will, however, depend on the actual C sequestered. Given these limitations, the major impetus for the promotion and adoption of agroforestry practices remains the potential to increase food and fuelwood productivity for smallholder households, and reduce land, watershed, and ecosystem degradation.

11.3.3 GHG Emissions

Decomposition of SOM in cropping systems releases nutrients for crop production, but also returns carbon dioxide (CO_2) to the atmosphere. At the same time, some of the C from decomposition of organic matter is retained in soil aggregates and adsorbed to soil colloids, some of which will later be eroded (Lal 2004). Addition of agroforestry species has the potential to either enhance or reduce soil C storage (Kim 2012) and GHG emissions. Thus, the study of GHG emissions is critical to describing the trade-offs between smallholder and ecosystem benefits from agroforestry.

11.3.3.1 Soil Carbon Dioxide, Methane, and Nitrous Oxide Emissions

Carbon dioxide (CO_2) is the dominant pathway of C loss in most terrestrial ecosystems, as well as the most important GHG in the atmosphere (Forster et al. 2007). Soil CO_2 is produced primarily by both heterotrophic (i.e., decomposer organisms) and autotrophic activity (i.e., living roots and mycorrhizae) (Raich and Schlesinger 1992; Schlesinger and Andrews 2000). Soil CO_2 efflux amounts to 75–80 Pg of CO_2 per year globally (Raich and Potter 1995; Raich et al. 2002) and made up 20%–40% of the total annual input of CO_2 into the atmosphere in the 1990s (Raich and Schlesinger 1992; Schimel 1995). Soil temperature, soil moisture, soil C content, litter quality, root dynamics, and plant photosynthesis or growth are known control factors for soil CO_2 flux (e.g., Raich and Schlesinger 1992; Rustad and Fernandez 1998; Davidson et al. 2000; Vargas and Allen 2008).

Methane (CH_4) has the second-largest radiative forcing of the long-lived GHGs after CO_2 (Forster et al. 2007). The net CH_4 flux is the result of the balance between the two offsetting processes of methanogenesis (microbial production under anaerobic conditions) and methanotrophy (microbial consumption) (Dutaur and Verchot 2007). Methanogenesis occurs via the anaerobic degradation of organic matter by methanogenic archaea within the archaeal phylum Euryarchaeota (Thauer 1998). Methanotrophy occurs through methanotrophs metabolizing CH_4 as their source of C and energy (Hanson and Hanson 1996). In anoxic soils, emergent vegetation also influences CH_4 flux to the atmosphere, as plants enable oxygen transport to the rhizosphere, through aerenchymateous tissue, and through the production of labile substrates via root exudation (Joabsson et al. 1999). In general, CH_4 production rates are controlled by the availability of suitable substrates, alternative electron acceptors for competing redox reactions (i.e., sulfate reduction), the nutritional status of the ecosystem (i.e., bog vs. fen), water table position or soil moisture content, temperature, and soil salinity (Hanson and Hanson 1996; Dutaur and Verchot 2007).

Atmospheric nitrous oxide (N_2O) contributes to both the greenhouse effect (Wang et al. 1976) and ozone layer depletion (Crutzen 1970). Nitrous oxide has a relatively high global warming potential (i.e., 298 times greater than CO_2 in a 100-year time horizon; Intergovernmental Panel on Climate Change [IPCC] 2006; Forster et al. 2007) and agricultural soils provide 3.5 Tg N_2O-N year^{-1} of total anthropogenic N_2O emissions (5.7 Tg N_2O-N year^{-1}) (IPCC 2006). Use of N fertilizers and animal manure are the main anthropogenic N_2O sources, which together account for roughly 24% of total annual emissions (Bouwman 1996; Forster et al. 2007). The main processes that produce N_2O in soils are nitrification, the stepwise oxidation of

NH_3 to nitrite (NO_2^-) and then to nitrate (NO_3^-) (Kowalchuk and Stephen 2001), and denitrification, the stepwise reduction of NO_3^- to NO_2^-, NO, N_2O, and ultimately N_2 In denitrification, facultative anaerobic bacteria use NO_3^- as an electron acceptor in the respiration of organic material under low oxygen (O_2) conditions (Knowles 1982). In nitrifier denitrification, which is carried out by autotrophic NH_3-oxidizing bacteria, NH_3 is oxidized to nitrite NO_2^-, followed by the reduction of NO_2^- to nitric oxide NO, N_2O, and molecular nitrogen (N_2) (Wrage et al. 2001).

11.3.3.2 Emissions of GHGs in Agroforestry in Eastern and Southern Africa

Nitrogen-fixing tree and crop intercropping systems can be a sustainable agroforestry practice in eastern and southern Africa (Makumba et al. 2006; Akinnifesi et al. 2010), and they can also contribute to mitigation of climate change through enhanced soil C sequestration. Makumba et al. (2007) reported soil C and soil CO_2 emissions in a 7-year-old *Gliricidia* and maize intercropping system, and a sole maize cropping site in southern Malawi. They estimated that while soil C in the intercropping system was about double that in the sole maize cropping site, soil CO_2 emissions from the intercropping system were up to three times higher. The increased soil CO_2 emissions in the intercropping system could be due to increased SOM and enhanced root respiration in extended root systems from the intercropping system (Makumba et al. 2007). Using the data provided in Makumba et al. (2007), a C loss as soil CO_2 emissions (51.2 ± 0.4 Mg C ha^{-1}) was estimated, amounting to 67.4% of the sequestered soil C (76 ± 8.6 Mg C ha^{-1} in 0–2 m soil depth) for the first 7 years in the intercropping system (Kim 2012). These results suggest the need to account for the C loss as soil CO_2 emissions in assessing the overall impact of the agroforestry system on soil C dynamics.

Maize yields in the *Gliricidia* and maize intercropping systems without additional synthetic N fertilizer input were similar to the yields in the sole maize cropping with 48 kg N ha^{-1} $year^{-1}$ fertilizer applied in southern Malawi (Makumba et al. 2006). These results support the premise that additional N is provided to the crop through N fixation by *Gliricidia* (e.g., Makumba et al. 2006; Akinnifesi et al. 2010). These results also suggest that up to 48 kg N ha^{-1} $year^{-1}$ of fertilizer could be reduced in the intercropping system. Globally, 1% of applied N fertilizer converts to N_2O emission (IPCC 2006), and it was observed that 0.25%–4.1% of applied N fertilizer converts to N_2O emission in sub-Saharan Africa (Kim et al. 2012). Therefore the reduced N fertilizer use through the intercropping system may result in reduced N_2O emissions. In contrast, N_2O emissions in the intercropping system may not be lower than in the conventional cropping system where N fertilizer is applied. Soil collected under N-fixing tree species produced significantly more N_2O than soil collected under non-N-fixing trees and N-fixing crop species in Senegal (Dick et al. 2006). Nitrous oxide emissions from a maize field that previously had a 2-year fallow of *A. angustissima* and *Sesbania* were significantly higher than those from an unfertilized maize field in Zimbabwe (Chikowo 2004). Increased soil organic C and N in the intercropping system can enhance the denitrification process, one of the major processes that produce N_2O gas in soil (e.g., Knowles 1982). It is therefore important to consider N_2O emissions to better understand the contributions of agroforestry to N_2O dynamics.

Soils have been shown to both produce and consume CH_4 (Topp and Pattey 1997; Le Mer and Roger 2001). It is well known that forest soils are the most active sink for CH_4, followed by grass lands and cultivated soils, and that the CH_4 uptake potential of many upland soils is reduced by cultivation and application of ammonium-N fertilizer (e.g., Topp and Pattey 1997; Le Mer and Roger 2001; Dutaur and Verchot 2007). These results suggest that synthetic N fertilizers used in conventional cropping systems may increase CH_4 emissions in sub-Saharan Africa. By contrast, the intercropping system uses less or no synthetic N fertilizer and may have the potential to mitigate CH_4 emissions. Overall, GHG emissions from the agroforestry systems either reduce benefits gained from enhanced soil C sequestration or add new benefits from reduced N_2O and CH_4 emissions. However, there is little data on GHG emissions from agroforestry systems in eastern and southern Africa.

11.3.3.3 Suggested Future Studies

GHG emission in agroforestry has not been well understood, although it is recognized that agroforestry can be a source of GHG emissions or mitigate GHG emissions. First, studies quantifying the source and the mitigation capacity of GHG in various agroforestry systems in eastern and southern Africa are urgently needed. Especially, careful comparison of GHG emissions in agroforestry with monocropping will provide a better understanding of the contribution of agroforestry to mitigating GHG emissions. It is worth noting that field measurements of GHG emissions in eastern and southern Africa should accurately observe peak GHG emissions following rewetting of dry soils (e.g., start/onset of the rainy season), since several reports indicate that peak GHG emissions occur following soil rewetting in the areas (e.g., Dick et al. 2006; Makumba et al. 2007), and these peak emissions may significantly affect annual GHG budgets as has been shown in other areas (e.g., Lee et al. 2004; Goldberg et al. 2010; Kim et al. 2012). These peaks could be measured by using an automated measurement system (e.g., Wolf et al. 2010; Kim et al. 2010a) or by increasing the frequency of manual chamber measurements during these periods (e.g., Beare et al. 2009; Kim et al. 2010b). An area of significant promise involves combining microbial community analyses and/or stable isotope techniques with flux measurements. Models are promising tools for evaluating the importance of GHG emissions in agroforestry systems. Initially, simple linear regressions and empirical models can be developed on the basis of the relationships between environmental factors, including soil moisture and/or soil temperature and soil GHG fluxes. With improved understanding of C and N biogeochemistry and hydrological dynamics in agroforestry systems, process-based models can be developed to more accurately simulate GHG flux. It is critical to enhance the communication between field scientists and the modeling community, as models can be used to generate hypotheses to be tested in the field and laboratory (Kim et al. 2012).

11.4 CONCLUSIONS, CHALLENGES, AND FUTURE NEEDS

Including agroforestry species in smallholder cropping systems has well-documented benefits in reduced land degradation and increased food production. Agroforestry also has the potential to increase carbon storage in soils and aboveground wood biomass.

However, the effects of smallholder soil/agroforestry management decisions on carbon-related agroecosystem processes are not well documented and need further study.

The continuing use of agroforestry practices that are biotically compatible with farming systems depends not only on benefits to food security and the environment but also on compatibility with farming practices already in place, positive policy context, awareness among farmers of the management practices and potential benefits, and constraints to seed supply, land, and labor.

Use of carbon credits to recapitalize soil fertility is developing very slowly in eastern and southern Africa. Further research is needed, especially in the area of soil carbon cycling and the effects on carbon cycling of different management decisions.

REFERENCES

Agboola, A.A. 1994. A recipe for continuous arable crop production in the forest zone of western Nigeria. In: *Alternative to Slash-and-Burn Agriculture*, eds. P.A. Sanchez, H. Van Hauten, pp. 107–120. 15th International Soil Science Congress, Mexico.

Ajayi, O.C., F.K. Akinnifesi, G. Sileshi, and W. Kanjipite. 2009. Labour inputs and financial profitability of conventional and agroforestry-based soil fertility management practices in Zambia. *Agrekon* 48: 276–292.

Ajayi, O.C., F. Place, F. Kwesiga, and P. Mafongoya. 2007. Impacts of improved tree fallow technology in Zambia. In: *International Research on Natural Resource Management: Advances in Impact Assessment*, eds. H. Waibel, D. Zilberman, pp. 147–168. Science Council/CGIAR, CABI Wallingford, UK.

Ajayi, O.C., F.K. Akinnifesi, G. Sileshi, and S. Franzel. 2005. Impact of Fertilizer Tree Fallows in Eastern Zambia. A Study on Impacts of Agroforestry. World Agroforestry Centre 3rd EPMR.

Akinnifesi, F., W. Makumba, G. Sileshi, O. Ajayi, and D. Mweta. 2007. Synergistic effect of inorganic N and P fertilizers and organic inputs from *Gliricidia sepium* on productivity of intercropped maize in Southern Malawi. *Plant Soil* 294: 203–217.

Akinnifesi, F.K., O.C. Ajayi, G. Sileshi, P.W. Chirwa, and J. Chianu. 2010. Fertiliser trees for sustainable food security in the maize-based production systems of East and Southern Africa. A review. *Agron. Sustain. Dev.* 30: 615–629.

Albrecht, A., and S.T. Kandji. 2003. Carbon sequestration in tropical agroforestry systems. *Agric. Ecosyst. Environ.* 99: 15–27.

Angima, S.D., D.E. Stott, M.K. O'Neill, C.K. Ong, and G.A. Weesies. 2002. Use of calliandra–Napier grass contour hedges to control erosion in central Kenya. *Agric. Ecosyst. Environ.* 91: 15–23.

Aune, J.B., A.T. Alemu, and K.P. Gautam. 2005. Carbon sequestration in rural communities Is it worth the effort? *J. Sustain. Forest* 21: 69–79.

Ayuk, E.T., and P.L. Mafongoya. 2002. Risk and returns characteristics of improved fallows in eastern Zambia: An application of stochastic dominance methods. In: *Proceedings of the 14th ICRAF Southern African Regional Review and Planning Workshop*, eds. F. Kwesiga, E. Ayuk and A. Agumya, pp. 28–34. September 3–7, 2001, Harare, Zimbabwe.

Bagayoko, M., A. Buerkert, G. Lung, A. Bationo, and V. Roemheld. 2000. Cereal/legume rotation effects on cereal growth in Sudano–Sahelian West Africa: Soil mineral nitrogen, mycorrhizae and nematodes. *Plant Soil* 218: 103–116.

Barnes, R.D., and C.W. Fagg. 2003. *Faidherbia Albida*: Monograph and Annotated Bibliography. Tropical Forestry Papers 41. Oxford Forestry Institute, UK.

Barrios, E., F. Kwesiga, R.J. Buresh, and J.I. Sprent. 1997. Light fraction soil organic matter and available nitrogen following trees and maize. *Soil Sci. Am. J.* 61: 826–831.

Barrios, E., F. Kwesiga, R.J. Buresh, J.I. Sprent, and R. Coe. 1998. Relating pre-season soil nitrogen to maize yield in tree legume–maize rotations. *Soil Sci. Am. J.* 62: 1604–1609.

Bayala, J., L.K. Heng, M. van Noordwijk, and S.J. Ouedraogo. 2008. Hydraulic redistribution study in two native tree species of agroforestry parklands of West African dry savanna. *Acta Oecol.* 34: 370–378. doi: 10.1016/j.actao.2008.06.010.

Bayala, J., G.W. Sileshi, R. Coe et al. 2012. Cereal yield response to conservation agriculture practices in dry-lands of West Africa: A quantitative synthesis. *J. Arid Environ.* 78: 13–25. doi: 10.1016/j.jaridenv.2011.10.011.

Beare, M.H., E.G. Gregorich, and P. St-Georges. 2009. Compaction effects on CO_2 and N_2O production during drying and rewetting of soil. *Soil Biol. Biochem.* 41: 611–621.

Beedy, T.L., S.S. Snapp, F.K. Akinnifesi, and G.W. Sileshi. 2010. Impact of *Gliricidia sepium* intercropping on soil organic matter fractions in a maize-based cropping system. *Agric. Ecosyst. Environ.* 138: 139–146.

Boffa, J.M. 1999. Agroforestry Parklands in Sub-Sahara Africa. FAO Conservation Guide No. 34. Food and Agriculture Organization of the United Nations, Rome.

Bouwman, A.F. 1996. Direct emission of nitrous oxide from agricultural soils. *Nutr. Cycl. Agroecosyst.* 46: 53–70.

Brewbaker, J.L. 1990. Breeding systems and genetic improvement of perennial sesbanias. In: *Perennial Sesbania Species in Agroforestry Systems*, eds. B. Macklin, D.O. Evans, pp. 39–44. Proceedings of a Workshop, Nairobi, Kenya, NFTA/ICRAF, Hawaii, March 27–31, 1989.

Buresh, R.J., and G. Tian. 1998. Soil improvement in sub-Saharan Africa. *Agroforest. Syst.* 38: 51–76.

Caldwell, M.M., and J.H. Richards. 1989. Hydraulic lift: Water efflux from upper roots improves effectiveness of water uptake from deep roots. *Oecologia* 79: 1–5.

Camberlin, P., V. Moron, R. Okoola, N. Philippon, and W. Gitau. 2009. Components of rainy seasons variability in Equatorial East Africa: Onset, cessation, rainfall frequency and intensity. *Theor. Appl. Climatol.* 98(3–4): 237. doi: 10.1007/s00704-009-0113-1.

Cannell, M.G.R., M. van Noordwijk, and C.K. Ong. 1996. The central agroforestry hypothesis: The trees must acquire resources that the crop would not otherwise acquire. *Agroforest. Syst.* 34: 27–31.

Cardoso, I.M., C. Boddington, B.H. Janssen, O. Oenema, and T.W. Kuyper. 2003. Distribution of mycorrhizal fungal spores in soils under agroforestry and monocultural coffee systems in Brazil. *Agroforest. Syst.* 58: 33–43.

Cardoso, I.M., and T.W. Kuyper. 2006. Mycorhizzas and tropical soil fertility. *Agric. Ecosyst. Environ.* 116: 72–84.

Carpenter, S.R., H.A. Mooney, J. Agard et al. 2009. Science for managing ecosystem services: Beyond the millennium ecosystem assessment. *Proc. Natl. Acad. Sci. U.S.A.* 106: 1305–1312. doi: 10.1073/pnas.0808772106.

Chakeredza, S., L. Hove, F.K. Akinnifesi, S. Franzel, O.C. Ajayi, and G. Sileshi. 2007. Managing fodder trees as a solution to human–livestock food conflicts and their contribution to income generation for smallholder farmers in southern Africa. *Nat. Resour. Forum* 31: 286–296.

Chikowo, R. 2004. Nitrogen cycling in subhumid Zimbabwe: Closing the loop. PhD thesis, University of Wageningen. Wageningen, the Netherlands. ISBN 90-5808-968-X.

Chirwa, P., and A.F. Quinion. 2012. Impact of soil fertility replenishment agroforestry technology adoption on the livelihoods and food security of smallholder farmers in central and southern Malawi. In: *Crop Production Technologies*, ed. P. Sharma, pp. 53–72. InTech, ISBN: 978-953-307-787-1. Available from: http://cdn.intechopen.com/pdfs-wm/25579.pdf.

Crutzen, P. 1970. The influence of nitrogen oxides on the atmospheric ozone content. *Q. J. Roy. Meteorol. Soc.* 96: 320–325.

Davidson, E.A., and I.A. Janssens. 2006. Temperature sensitivity of soil carbon decomposition and feedbacks to climate change. *Nature* 440: 165–173.

Davidson, E.A., L.V. Verchot, J.H. Cattânio, I.L. Ackerman, and J.E.M. Carvalho. 2000. Effects of soil water content on soil respiration in forests and cattle pastures of eastern Amazonia. *Biogeochemistry* 48: 53–69.

Dawson, T.E. 1993. Hydraulic lift and water use by plants, implications for water balance, performance and plant–plant interactions. *Oecologia* 95: 565–574.

de Carvalho, A.M.X., R. de Castro Tavares, I.M. Cardoso, and T.W. Kuyper. 2010. Mycorrhizal associations in agroforestry systems. In: *Soil Biology and Agriculture in the Tropics, Soil Biology*, ed. P. Dion, p. 21. Springer-Verlag, Berlin. doi: 10.1007/978-3-642-05076-3_9.

Dechasa, J. 1989. Evaluation of crop yields under *Acacia albida* shade in semi-arid farms. In: *Proceedings of IAR/ICRAF National Workshop*, pp. 33–38. Held November 14–18, 1988. Awasa Institute of Agricultural Research, Addis Ababa, Ethiopia.

Dechasa, J. 2010. Indiscriminate de-vegetation under improper farming system: A root cause for surface and underground water and food crisis in Ethiopia. *Ecohydrol. Hydrobiol.* 10: 201–209.

Dick, J., U. Skiba, R. Munro, and D. Deans. 2006. Effect of N-fixing and non N-fixing trees and crops on NO and N_2O emissions from Senegalese soils. *J. Biogeogr.* 33: 416–423.

Dixon, J., A. Gulliver, and D. Gibbon. 2001. Farming systems and poverty. In: *Improving Farmers' Livelihoods in a Changing World*, ed. M. Hall. FAO and World Bank, Rome and Washington, DC.

Dutaur, L., and L.V. Verchot. 2007. A global inventory of the soil CH_4 sink. *Global Biogeochem. Cycles* 21(4): GB4013. doi:10.1029/2006GB002734.

Fernandes, E.C.M., P.P. Motavalli, C. Castilla, and L. Mukurumbira. 1997. Management control of soil organic matter dynamics in tropical land-use systems. *Geoderma* 79: 49–67.

Food and Agriculture Organization of the United Nations, World Food Programme, and International Fund for Agricultural Development. 2012. Economic growth is necessary but not sufficient to accelerate reduction of hunger and malnutrition. In: *The State of Food Insecurity in the World 2012*. FAO, Rome. Available at http://www.fao.org/docrep/016/i3027e/i3027e.pdf.

Forster, P., V. Ramaswamy, P. Artaxo et al. 2007. Changes in atmospheric constituents and in radiative forcing climate change 2007: The physical science basis. In: *Contribution of Working Group 1 to the Fourth Assessment Report of the Intergovernmental Panel on Climate Change*, eds. S. Solomon, D. Qin, M. Manning et al., pp. 129–234. Cambridge Univ. Press, New York.

Franzel, S. 2004. Financial analysis of agroforestry practices. In: *Valuing Agroforestry Systems*, eds. J.R.R. Alavalapati, D. Mercer, pp. 9–37. Kluwer Academic Publishers, the Netherlands.

Franzel, S., C. Wambugu, P. Tuwei, and G. Karanja. 2003. The adoption and scaling-up of the use of fodder shrubs in central Kenya. *Trop. Grasslands* 37: 239–250.

Gacheru, E., and M.R. Rao. 2001. Managing striga infestation in maize using organic and inorganic nutrient sources in western Kenya. *Int. J. Pest Manage.* 47: 233–239.

García-Barrios, L., and C.K. Ong. 2004. Ecological interactions, management lessons and design tools in tropical agroforestry systems. In: *Advances in Agroforestry*, Volume 1, ed. P.K.R. Nair, pp. 221–236. Springer, Dordrecht, Netherlands.

Garrity, D.P., F.K. Akinnifesi, O.C. Ajayi et al. 2010. Evergreen agriculture: A robust approach to sustainable food security in Africa. *Food Sec.* 2: 197–214. doi: 10.1007/s12571-010-0070-7.

Giller, K.E. 2001. *Nitrogen Fixation in Tropical Cropping Systems*, 2nd edn. CAB International, Wallingford, UK.

Giller, K.E., E. Witter, M. Corbeels, and P. Tittonell. 2009. Conservation agriculture and smallholder farming in Africa: The heretics' view. *Field Crop. Res.* 114: 23–34.

Glenn, J.V. 2012. Economic assessment of landowner incentives: Analyses in North Carolina and Malawi. MSc thesis, Department of Natural Resources, North Carolina State University, Raleigh, NC.

Goldberg, D.E., and A.M. Barton. 1990. Patterns and consequences of interspecific competition in natural communities: A review of field experiments with plants. *Am. Nat.* 139: 771–801.

Goldberg, S., W. Borken, and G. Gebauer. 2010. N_2O emission in a Norway spruce forest due to soil frost: Concentration and isotope profiles shed a new light on an old story. *Biogeochemistry* 97: 21–30.

Hadgu, K.M., L. Kooistra, W.A. Rossing, and A.H.C. van Bruggen. 2009. Assessing the effect of *Faidherbia albida* based land use systems on barley yield at field and regional scale in the highlands of Tigray, Northern Ethiopia. *Food Sec.* 1: 337–350.

Haggblade, S., and G. Tembo. 2003. Early evidence on conservation farming in Zambia. In: *Workshop on Reconciling Rural Poverty and Resource Conservation: Identifying Relationships and Remedies*, May 2–3, 2003. Cornell University, Ithaca, NY.

Haglund, E., J. Ndjeunga, L. Snook, and D. Pasternak. 2011. Dry land tree management for improved household livelihoods: Farmer managed natural regeneration in Niger. *J. Environ. Manage.* 92(7): 1696–1705.

Hailu, T., L. Negash, and M. Olsson. 2000. *Millettia ferruginea* from southern Ethiopia: Impacts on soil fertility and growth of maize. *Agroforest. Syst.* 48: 9–24.

Hanson, R.S., and T.E. Hanson. 1996. Methanotrophic bacteria. *Microbiol. Mol. Biol. R.* 60: 439–471.

Harawa, R., J. Lehmann, F. Akinnifesi, E. Fernandes, and G. Kanyama-Phiri. 2006. Nitrogen dynamics in maize-based agroforestry systems as affected by landscape position in southern Malawi. *Nutr. Cycl. Agroecosyst.* 75: 271–284.

Harrier, L.A., and C.A. Watson. 2003. The role of arbuscular mycorrhizal fungi in sustainable cropping systems. *Adv. Agron.* 79: 185–229.

Hartemink, A.E., R.J. Buresh, B. Jama, and B.H. Janssen. 1996. Soil nitrate and water dynamics in sesbania fallows, weed fallows and maize. *Soil Sci. Am. J.* 6: 568–574.

Henry, M., P. Tittonell, R.J. Manlay, M. Bernoux, A. Albrecht, and B. Vanlauwe. 2009. Biodiversity, carbon stocks and sequestration potential in aboveground biomass in smallholder farming systems of western Kenya. *Agric. Ecosyst. Environ.* 129: 238–252.

Hiremath, A.J., J.J. Ewel, and T.G. Cole. 2002. Nutrient use efficiency in three fast-growing tropical trees. *Forest. Sci.* 48: 662–672.

Holmgren, M., M. Scheffer, and M.A. Huston. 1997. The interplay of facilitation and competition in plant communities. *Ecology* 78: 1966–1975.

Houngnandan, P., N. Sanginga, P. Woomer, B. Vanlauwe, and O. Van Cleemput. 2000. Response of *Mucuna pruriens* to symbiotic nitrogen fixation by rhizobia following inoculation in farmers' fields in the derived savanna of Benin. *Biol. Fertil. Soils* 30: 558–565.

Howard, J.A., and C. Mungoma. 1996. Zambia's Stop-and-Go Revolution: The Impact of Policies and Organizations on the Development and Spread of Maize Technology. Working Paper 61. Michigan State University International Development, East Lansing, MI.

Hunter, A.F., and L.W. Aarssen. 1988. Plants helping plants. *Bioscience* 38: 34–40.

ICRAF. 2012. Useful tree species for Africa [Online]. Available at http://www.worldagroforestrycentre.org/our_products/databases/useful-tree-species-africa.

ICRAF. 2013. Agroforestree online database [Online]. Available at http://www.worldagroforestry.org/resources/databases/agroforestree.

Ingleby, K., J. Wilson, R.C. Munro, and S. Cavers. 2007. Mycorrhizas in agroforestry: Spread and sharing of arbuscular mycorrhizal fungi between trees and crops: Complementary use of molecular and microscopic approaches. *Plant Soil* 294: 125–136.

Intergovernmental Panel on Climate Change. 2006. Guidelines for national greenhouse gas inventories [Online]. Available at http://www.ipcc-nggip.iges.or.jp/public/2006gl/index.html.

Jama, B., F. Kwesiga, and A. Niang. 2006. Agroforestry innovations for soil fertility management in sub-Saharan Africa: Prospects and challenges ahead. In *World Agroforestry into the Future*, eds. D. Garrity, A. Okono, M. Grayson, S. Parrott, pp. 53–60. World Agroforestry Centre, Nairobi, Kenya.

Joabsson, A., T.R. Christensen, and B. Wallén. 1999. Vascular plant controls on methane emissions from northern peat forming wetlands. *Trends Ecol. Evol.* 14: 385–388.

Jose, S., A.R. Gillespie, and S.G. Pallardy. 2004. Interspecific interactions in temperate agroforestry. *Agroforest. Syst.* 61: 237–255.

Jose, S., A.R. Gillespie, J.R. Seifert, D.B. Mengel, and P.E. Pope. 2000. Defining competition vectors in a temperate alley cropping system in the mid-western USA: Competition for nitrogen and litter decomposition dynamics. *Agroforest. Syst.* 48: 61–77.

Jung, G. 1970. Variations saisonnieres des caracteristiques microbiologiques d'un sol ferrugineux tropicale peu lessive (dior.), soumis ou non l'influence *Acacia albida* (Del.). *Oecologia* 5: 113–135.

Jurion, F., and J. Henry. 1969. *Can Primitive Farming be Modernized?* Publication INEAC, Hors Série, Bruxelles.

Kabwe, G. 2010. Uptake of agroforestry technologies among smallholder farmers in Zambia. PhD thesis, Lincoln University, New Zealand.

Kaonga, M.L., and T.P. Bayliss-Smith. 2009. Carbon pools in tree biomass and the soil in improved fallows in eastern Zambia. *Agroforest. Syst.* 76: 37–51.

Kaonga, M.L., and K. Coleman. 2008. Modelling soil organic carbon turnover in improved fallows in eastern Zambia using the RothC-26.3 model. *Forest Ecol. Manag.* 256: 1160–1166.

Kassa, H., K. Gebrehiwet, and C. Yamoah. 2010. *Balanites aegyptica*, a potential tree for parkland agroforestry systems with sorghum in Northern Ethiopia. *J. Soil Sci. Environ Manage.* 1: 107–114.

Kidanu, S., T. Mamo, and L. Stroosnijder. 2004. Eucalyptus–wheat interaction on Ethiopian Nitosols. *Agric. Syst.* 80: 151–170.

Kim, D.-G. 2012. Estimation of net gain of soil carbon in a nitrogen-fixing tree and crop intercropping system in sub-Saharan Africa: Results from re-examining a study. *Agroforest. Syst.* 86: 175–184.

Kim, D.-G., M. Mishurov, and G. Kiely. 2010a. Effect of increased N use and dry periods on N_2O emission from a fertilized grassland. *Nutr. Cycl. Agroecosyst.* 88: 397–410.

Kim, D.-G., S. Mu, S. Kang, and D. Lee. 2010b. Factors controlling soil CO_2 effluxes and the effects of rewetting on effluxes in adjacent deciduous, coniferous, and mixed forests in Korea. *Soil Biol. Biochem.* 42: 576–585.

Kim, D.-G., R. Vargas, B. Bond-Lamberty, and M. Turetsky. 2012. Effects of soil rewetting and thawing on soil gas fluxes: A review of current literature and suggestions for future research. *Biogeosciences* 9: 2459–2483.

Kimaro, A.A. 2009. Sequential agroforestry systems for improving fuelwood supply and crop yield in semi-arid Tanzania. PhD dissertation, University of Toronto, Toronto, Canada.

Kimaro, A.A., M.E. Isaac, and S.A.O. Chamshama. 2011. Carbon pools in tree biomass and soils under rotational woodlot systems in eastern Tanzania. In: *Carbon Sequestration Potential of Agroforestry Systems: Opportunities and Challenges*, eds. B.M. Kumar, P.K.R. Nair, pp. 129–144. Springer, Dordrecht.

Kimaro, A.A., V.R. Timmer, S.O.A. Chamshama, A.G. Mugasha, and D.A. Kimaro. 2008. Differential response to tree fallows in rotational woodlot systems: Post-fallow maize yield, nutrient uptake, and soil nutrients. *Agric. Ecosyst. Environ.* 125: 73–83.

Knowles, R. 1982. Denitrification. *Microbiol. Rev.* 46: 43–70.

Kowalchuk, G.A., and J.R. Stephen. 2001. Ammonia-oxidizing bacteria: A model for molecular microbial ecology. *Annu. Rev. Microbiol.* 55: 485–529.

Kumwenda, J.D.T., S.R. Waddington, S.S. Snapp, R.B. Jones, and M.J. Blackie. 1996. Soil Fertility Management Research for the Maize Cropping Systems of Smallholders in Southern Africa: A Review. *NRG Paper 96-02*. CIMMYT, Federal District, Mexico.

Kwesiga, F.R., S. Franzel, F. Place, D. Phiri, and C.P. Simwanza. 1999. *Sesbania sesban* improved fallows in eastern Zambia: Their inception, development and farmer enthusiasm. *Agroforest. Syst.* 47: 49–66.

Lal, R. 2004. Soil carbon sequestration impacts on global climate change and food security. *Science* 304: 1623–1627.

Lal, R. 2007. Carbon management in agricultural soils. *Mitigat. Adapt. Strat. Global Change.* 12: 303–322. doi: 10.1007/s11027-006-9036-7.

Lal, R. 2010. Managing soils for a warming earth in a food-insecure and energy-starved world. *J. Plant Nutr. Soil Sci.* 173: 4–15.

Lee, X., H.J. Wu, J. Sigler, C. Oishi, and T. Siccama. 2004. Rapid and transient response of soil respiration to rain. *Global Change Biol.* 10: 1017–1026.

Le Mer, J., and P. Roger. 2001. Production, oxidation, emission and consumption of methane by soils: A review. *Eur. J. Soil Biol.* 37: 25–50.

Liebman, M., and E.R. Gallandt. 1997. Many little hammers: Ecological management of crop–weed interactions. In: *Ecology in Agriculture*, ed. L.E Jackson, pp. 291–343. Academic Press, San Diego, CA.

Lillesø, J.-P.B., P. van Breugel, R. Kindt et al. 2011. *Potential Natural Vegetation of Eastern Africa, Volume 1: The Atlas*. Forest and Landscape Working Paper 61. University of Copenhagen, Frederiksberg, Denmark.

Mafongoya, P.L., R. Chintu, T.S. Chirwa, J. Matibini, and S. Chikale. 2003. *Tephrosia* species and provenances for improved fallows in southern Africa. *Agroforest. Syst.* 59: 279–288.

Mafongoya, P.L., and B.H. Dzowela. 1999. Biomass production of 542 tree fallow and their residual effects on maize in Zimbabwe. *Agroforest. Syst.* 47: 139–151.

Makumba, W., F.K. Akinnifesi, B. Janssen, and O. Oenema. 2007. Long-term impact of a gliricidia–maize intercropping system on carbon sequestration in southern Malawi. *Agric. Ecosyst. Environ.* 118: 237–243.

Makumba, W., B. Janssen, O. Oenema, F.K. Akinnifesi, D. Mweta, and F. Kwesiga. 2006. The long-term effects of a gliricidia–maize intercropping system in Southern Malawi on gliricidia and maize yields and soil properties. *Agric. Ecosyst. Environ.* 116: 85–92.

Malézieux, E., Y. Crozat, C. Dupraz et al. 2009. Mixing plant species in cropping systems: Concepts, tools and models. A review. *Agron. Sustain. Dev.* 29: 43–62.

Malmer, A., and G. Nyberg. 2008. Forest and water relations in miombo woodlands: Need for understanding of complex stand management. Working Papers of the Finnish Forest Research Institute 98: 70–86.

Maranz, S. 2009. Tree mortality in the African Sahel indicates an anthropogenic ecosystem displaced by climate change. *J. Biogeogr.* 36: 1181–93.

Mekonnen, K., R.J. Buresh, and B.A. Jama. 1997. Root and inorganic nitrogen distributions in sesbania fallow, natural fallow and maize fields. *Plant Soil* 188: 319–327.

Montagnini, F., and P.K.R. Nair. 2004. Carbon sequestration: An underexploited environmental benefit of agroforestry systems. *Agroforest. Syst.* 61–62(1–3): 281–295.

Monteith, J.L., C.K. Ong, and J.E. Corlett. 1991. Microclimate interactions in agroforestry systems. *Forest Ecol. Manag.* 45: 31–44.

Negash, M., E. Yirdaw, and O. Luukkanen. 2012. Potential of indigenous multistrata agroforests for maintaining native floristic diversity in the south-eastern Rift Valley escarpment, Ethiopia. *Agroforest. Syst.* 85: 9–28.

Niang, A.I., B.A. Amadalo, J. de Wolf, and S.M. Gathumbi. 2002. Species screening for short-term planted fallows in the highlands of western Kenya. *Agroforest. Syst.* 56: 145–154.

Nyadzi, G.I., R.M. Otsyina, F.M. Banzi et al. 2003. Rotational woodlot technology in northwestern Tanzania: Tree species and crop performance. *Agroforest. Syst.* 59: 253–263.

Nyamadzawo, G., R. Chikowo, P. Nyamugafata, J. Nyamangara, and K.E. Giller. 2008b. Soil organic carbon dynamics of improved fallow–maize rotation systems under conventional and no- tillage in Central Zimbabwe. *Nutr. Cycl. Agroecosyst.* 81: 85–93. doi: 10.1007/s10705-007-9154-y.

Nyamadzawo, G., J. Nyamangara, A. Mzulu, and P. Nyamugafata. 2009. Soil microbial biomass and organic matter pools in fallow systems under conventional and no-tillage practices from central Zimbabwe. *Soil Till. Res.* 102: 151–157. doi: 10.1016/j.still.2008.08.007.

Nyamadzawo, G., J. Nyamangara, M. Wuta, and P. Nyamugafata. 2012. Effects of pruning regimes on soil water and maize yields in coppicing fallow–maize intercropping systems in Central Zimbabwe. *Agroforest. Syst.* 84: 273–286. doi: 10.1007/s10457-011-9453-9.

Nyamadzawo, G., P. Nyamugafata, P. Chikowo, and K.E. Giller. 2008a. Residual effects of fallows on infiltration rates and hydraulic conductivities in a kaolinitic soil subjected to conventional tillage (CT) and no tillage (NT). *Agroforest. Syst.* 72: 161–168. doi: 10.1007/s10457-007-9057-6.

Odhiambo, L.O., R.E. Yoder, D.C. Yoder, and J.W. Hines. 2001. Optimization of fuzzy evapotranspiration model through neural training with input–output examples. *T. Am. Soc. Eng.* 44: 1625–1633.

Okorio, J., and J.A. Maghembe. 1994. The growth and yield of *Acacia albida* intercropped with maize (*Zea mays*) and beans (*Phaseolus vulgaris*) at Morogoro, Tanzania. *Forest Ecol. Manag.* 64: 183–190.

Ong, C.K., and B.M. Swallow. 2003. Water productivity in forestry and agroforestry. In: *Water Productivity in Agriculture: Limits and Opportunities for Improvement*, eds. J.W. Kijne, R. Barker, D. Molden. CAB International, Nairobi, Kenya.

Ong, C.K., J. Wilson, J.D. Deans, J. Mulayta, T. Raussen, and N. Wajja-Musukse. 2002. Tree-crop interactions: Manipulations of water use and root function. *Agr. Water Manage.* 53: 171–186

Palm, C.A., S.M. Smukler, C.C. Sullivan, P.K. Mutuo, G.I. Nyadzi, and M.G. Walsh. 2010 Identifying potential synergies and trade-offs for meeting food security and climate change objectives in sub-Saharan Africa. *Proc. Natl. Acad. Sci.* 107: 19,661–19,666.

Payne, W.A. 2010. Farming systems and food security in sub-Saharan Africa. In: *Food Security and Soil Quality*, eds. R. Lal, B.A. Stewart, pp. 23–56. Taylor & Francis, Boca Raton, FL.

Place, F., M. Adato, P. Hebinck, and M. Omosa. 2005. The Impact of Agroforestry-Based Soil Fertility Replenishment Practices on the Poor in Western Kenya. Research Report 142. International Food Policy Research Institute (IFPRI), Washington, DC.

Place, F., and J. Binam. 2013. Economic Impacts of Farmer-Managed Natural Regeneration in the Sahel. End-of-Project Technical Report for Free University Amsterdam and IFAD. World Agroforestry Centre, Nairobi, Kenya.

Poschen, P. 1986. An evaluation of the *Acacia albida*-based agroforestry practices in the Hararghe highlands of Eastern Ethiopia. *Agroforest. Syst.* 4: 129–143.

Post, W.M., and K.C. Kwon. 2000. Soil carbon sequestration and land-use change: Processes and potential. *Global Change Biol.* 6: 317–328.

Radersma, S., and P.F. Grierson. 2004. Phosphorus mobilization in agroforestry: Organic anions, phosphatase activity, and phosphorus fractions in the rhizosphere. *Plant Soil* 259: 209–219.

Raich, J., and W. Schlesinger. 1992. The global carbon dioxide flux in soil respiration and its relationship to vegetation and climate. *Tellus B* 44: 81–99.

Raich, J.W., and C.S. Potter. 1995. Global patterns of carbon dioxide emissions from soils. *Global Biogeochem. Cycles* 9: 23–36.

Raich, J.W., C.S. Potter, and D. Bhagawati. 2002. Interannual variability in global soil respiration, 1980–94. *Global Change Biol.* 8: 800–812.

Roose, E., and B. Barthes. 2001. Organic matter management for soil conservation and productivity restoration in Africa: A contribution from Francophone research. *Nutr. Cycl. Agroecosyst.* 61: 159–170.

Rustad, L.E., and I.J. Fernandez. 1998. Experimental soil warming effects on CO_2 and CH_4 flux from a low-elevation spruce–fir forest soil in Maine, USA. *Global Change Biol.* 4: 597–605.

Saka, A.R., W.T. Bunderson, O.A. Itimu, H.S.K. Phombeya, and Y. Mbekeani. 1994. The effects of *Acacia albida* on soils and maize grain yields under smallholder farm conditions in Malawi. *Forest Ecol. Manag.* 64: 217–230.

Sanchez, P.A. 1995. Science and agroforestry. In: *Agroforestry: Science, Policy, and Practice*, ed. F.L. Sinclair. Kluwer Academic Publishers, Dordrecht, the Netherlands.

Sanchez, P.A., and C.A. Palm. 1996. Nutrient cycling and agroforestry in Africa. In: *Forest Influences. Unasylva—No. 185—An International Journal of Forestry and Forest Industries*, Volume 47-1996/2. FAO, Food and Agriculture Organization of the United Nations, Rome, Italy.

Sarmiento, L., and P. Bottner. 2002. Carbon and nitrogen dynamics in two soils with different fallow times in the high tropical Andes: Indications for fertility restoration. *Appl. Soil Ecol.* 19: 79–89.

Schimel, D.S. 1995. Terrestrial ecosystems and the carbon cycle. *Global Change Biol.* 1: 77–91.

Schlesinger, W.H., and J.A. Andrews. 2000. Soil respiration and the global carbon cycle. *Biogeochemistry* 48: 7–20.

Sileshi, G.W., F.K. Akinnifesi, O.C. Ajayi, S. Chakeredza, M.L. Kaonga, and P.W. Matakala. 2007. Contributions of agroforestry to ecosystem services in the Miombo eco-region of eastern and southern Africa. *Afr. J. Environ. Sci. Technol.* 1: 68–80.

Sileshi, G.W., F.K. Akinnifesi, O.C. Ajayi, and F. Place. 2008. Meta-analysis of maize yield response to woody and herbaceous legumes in sub-Saharan Africa. *Plant Soil* 307: 1–19.

Sileshi, G.W., P.L. Mafongoya, F.K. Akinnifesi et al. 2013. Chapter 22. Agroforestry: Fertilizer trees. In: *Encyclopedia of Agriculture and Food Systems*, eds. R. Leakey, N. van Alfen. Elsevier Publishers, Oxford, UK.

Six, J., E.T. Elliot, and K. Paustian. 2000. Soil macroaggregate turnover and microaggregate formation: A mechanism for C sequestration under no-tillage agriculture. *Soil Biol. Biochem.* 32: 2099–2103.

Snapp, S.S., G. Kanyama-Phiri, B. Kamanga, R. Gilbert, and K. Wellard. 2002. Farmer and researcher partnerships in Malawi: Developing soil fertility technologies for the near-term and far-term. *Expl. Agric.* 38: 411–431.

Ståhl, L., G. Nyberg, P. Högberg, and R.J. Buresh. 2002. Effects of planted tree fallows on soil nitrogen dynamics, above-ground and root biomass, N_2–fixation and subsequent maize crop productivity in Kenya. *Plant Soil* 243: 103–117.

Stromgaard, P. 1985. Biomass, growth, and burning of woodland in a shifting cultivation area of South Central Africa. *Forest Ecol. Manag.* 12: 163–178.

Tallis, H., and P. Kareiva. 2005. Ecosystem services. *Curr. Biol.* 15: 746–748.

Thauer, R. 1998. Biochemistry of methanogenesis: A tribute to Marjory Stephenson. *Microbiology* 144: 2377–2406.

TIST. 2011. The International Small Group and Tree Planting Programme [Online]. Available at http://www.tist.org/ (verified June 10, 2013).

Topp, E., and E. Pattey. 1997. Soils as sources and sinks for atmospheric methane. *Can. J. Soil Sci.* 77: 167–178.

United Nations Environment Program. 2009. UNEP year book: New science and development in our changing environment. Available at http://www.unep.org/geo/yearbook/yb2009.

Vandenbelt, R.J., and J.H. Williams. 1992. The effect of soil surface temperature on the growth of millet in relation to the effect of *Faidherbia albida* trees. *Agric. For. Meteorol.* 60: 93–100.

Vanlauwe, B., and K.E. Giller. 2006. Popular myths around soil fertility management in sub-Saharan Africa. *Agric. Ecosyst. Environ.* 116: 34–46.

Vargas, R., and M.F. Allen. 2008. Environmental controls and the influence of vegetation type, fine roots and rhizomorphs on diel and seasonal variation in soil respiration. *New Phytol.* 179: 460–471.

Verinumbe, I. 1987. Crop production on soil under some forest plantations in the Sahel. *Agroforest. Syst.* 5: 185–188.

Vi Agroforestry Strategy 2013–2015. Vi-skogen, 105 33 Stockholm, Sweden. Available at http://www.viskogen.se/wp-content/uploads/2012/03/strategiskt-dokument_eng_final.pdf (verified June 10, 2013).

Walker, A.P., M. van Noordwijk, and G. Cadisch. 2008. Modelling of planted legume fallows in Western Kenya. (II) Productivity and sustainability of simulated management strategies. *Agroforest. Syst.* 74: 143–154.

Wang, W.C., Y.L. Yung, A.A. Lacis, T. Mo, and J.E. Hansen. 1976. Greenhouse effects due to man-made perturbations of trace gasses. *Science* 194: 685–690.

Williams, M., C.M. Ryan, R.M. Rees, E. Sambane, J. Fernando, and J. Grace. 2008. Carbon sequestration and biodiversity of re-growing miombo woodlands in Mozambique. *Forest Ecol. Manag.* 254: 145–155.

Wolf, B., X. Zheng, N. Bruggemann et al. 2010. Grazing-induced reduction of natural nitrous oxide release from continental steppe. *Nature* 464: 881–884.

Wood, P.J. 1992. The botany and distribution of *Faidherbia albida*. In: *Faidherbia Albida in the West African Semi-arid Tropics: Proceedings of a Workshop*, April 22–26, 1991, Niamey, Niger, ed. R.J. Vandenbeldt, 212 pp. ICRISAT, Hyderabad, India and ICRAF, Nairobi, Kenya. ISBN 92-9066-220-4.

World Bank. 2008. *World Bank Development Report 2008: Agriculture*. World Bank, Washington, DC.

Wrage, N., G.L. Velthof, M.L. van Beusichem, and O. Oenema. 2001. Role of nitrifier denitrification in the production of nitrous oxide. *Soil Biol. Biochem.* 33: 1723–1732.

Yatich, T., A. Kalinganire, J.C. Weber et al. 2013. How Do Forestry Codes Affect Access, Use and Management of Protected Indigenous Tree Species: Evidence from West African Sahel. Occasional Paper No. 15. World Agroforestry Centre, Nairobi.

Young, A. 1989. *Agroforestry for Soil Conservation*. CAB International, Exeter, UK.

Young, A. 1997. *Agroforestry for Soil Conservation*. CAB International, Wallingford, UK.

12 Adaptation by Smallholders in Eastern Africa to Climate Change through Conservation Agriculture

Didas N. Kimaro, Hildelitha B. Msita, Andrew Kabanza, Stefaan Dondeyne, Jozef A. Deckers, and Jean Poesen

CONTENTS

12.1 INTRODUCTION

Management of agricultural lands in East Africa has remained a matter of concern since the German colonial times. Soil and water conservation in this area has progressed very slowly owing to farmers' poor response to interventions on conservation. Inappropriate land husbandry in the area is a result of pressure on the land resources, which has led to the development of various forms of land degradation. Several studies have been initiated in the region aimed at sustainable land husbandry and appropriate management of agricultural lands. This chapter focuses on land degradation processes, plant nutrient depletion, and adaptation measures based on the principles of conservation agriculture (CA) in the context of climate change. It provides broad experiences and an in-depth knowledge of smallholder farming adaptation to climate change using the principles of CA in East Africa. The highlights from this chapter show that interrill and rill erosion, tillage erosion, and landslides are dominant land degradation processes in the region of East Africa. Very high soil losses are reported particularly from interrill and rill erosion ranging from 91 to 258 Mg/ha/year in mountainous areas. Studies have also demonstrated significant rates of soil flux due to tillage erosion increasing with slope gradient, from 16 kg/m/tillage pass at a slope gradient of 31% to 60 kg/m/tillage pass at a slope gradient of >60%. Landslides dominate on the mountain ridges with slopes >45%, where they have affected >20% of the land with significant crop damage, reducing crop yields by 50%. Low soil fertility is also a form of land degradation contributing to low crop yields. Most of the studied soils have very low organic matter content with organic carbon <1.2%, nitrogen <0.1%, and available phosphorus <5 mg/kg. Inappropriate land uses also have contributed significantly to soil and nutrient losses. In the Lake Victoria Basin settlements, compounds and footpaths produced soil losses of up to 199 Mg/ha/year, while agricultural lands cropped with cotton and cassava produced soil loss of up to 27 Mg/ha/year. CA is among the adaptation strategies that have been practiced in East Africa by the smallholder farming community to reduce land degradation. It comprises three principles that have been applied simultaneously: minimum soil disturbance, permanent soil cover, and complex rotations/crop associations with some land management options, including application of farm yard manure, green manure, composting, mineral fertilizer, weed control, and *in situ* rainwater harvesting. Recent studies have indicated a positive trend in practicing CA in many areas of East Africa owing to the integration of both indigenous and scientific knowledge. In Ethiopia, for example, two local tillage systems, *terwah*+ and *derdero*+, using the traditional *mahresha ard* plough on Vertisol under crop rotation (wheat, grass, pea, wheat) significantly increased soil organic matter to 2.0% when compared with conventional tillage (1.4%) for 0–15 and 15–30 cm soil depths. In this study, the mean yield of wheat for 3 years increased from 2.8 Mg/ha for conventional tillage to 3.7 Mg/ha for *terwah*+ and *derdero*+ tillage systems. CA has been also demonstrated to increase the yield of vegetables in the Uluguru Mountains, Tanzania, from 0.88 to 44.68 Mg/ha. The combination of *miraba* (an indigenous soil erosion control practice), farmyard manure (FYM), and mulching reduced annual soil and nutrient loss from 132 to 0.5 Mg/ha/year, total nitrogen from 342 to 9 kg/ha/year, and available phosphorus from 0.4 to <0.1 kg/ha year, respectively, in the West Usambara Mountains, Tanzania. In this study, grain maize and bean yield increased from 0.7 to 2.6 Mg/ha and 0.2 to 1.3 Mg/ha, respectively. It is obvious from these studies that CA is a promising strategy that can address the complexities and peculiarities of soil quality on smallholder farms. It has proven to help low resource endowed farmers to mitigate problems of poverty, food insecurity, and low income. It enhances the resilience of soil productive capacity in the context of climate change (Bationo et al. 2003). Furthermore, CA technologies together with the application of organic and inorganic fertilizers, maize stover combined with inorganic fertilizers, and crop rotations and intercropping have resulted in yield gains over the farmers' practice in most farming systems in the region. However, experiences show that although the CA technologies discussed in this chapter have shown promising results, most of them remained limited to participating farmers within the small project sites. Therefore, there is a need for further research aimed at assessing, improving, and upscaling the potential contribution of CA practices to sustainable smallholder agriculture, particularly in semiarid areas of East Africa in the context of climate change, soil restoration, gender equity, and agricultural productivity. It is apparent from this chapter that CA has increased the yield of most crops particularly in the fragile ecosystems. It can be concluded that CA for crop production is the best model that can be used to promote small-scale farming on fragile ecosystems in East Africa.

12.1.1 Background

East African countries depend largely on agriculture. However, land degradation and soil fertility depletion remain major biophysical constraints to agricultural productivity, partly attributable to limited investment in soil-improving technologies, lack of appropriate information, and low adoption of available technologies owing to inadequate incentives. Nutrient mining, which is rampant in many smallholder farms of East Africa, is exacerbated by continuous cropping; inadequate nutrient replenishment in relation to plant demand; and high rates of soil erosion, leaching, and removal of crop residues from the fields (Lal 2001; Lynam et al. 1998). As a result, soil fertility has continued to decline to levels that are currently prohibitive to profitable agriculture. According to the Food and Agriculture Organization of the United Nations (FAO) Global Land Degradation Assessment, almost a quarter of the global land area was degraded between 1981 and 2003, with one of the most severely affected areas being Africa south of the equator (FAO 2006a). Major nutrients such as N, P, and K balances for 13 countries in sub-Saharan Africa show negative trends with about 200 million ha of cropland having lost 660 kg N/ha, 75 kg P/ha, and 450 kg K/ha in the last 30 years, and with high to very high depletion rates in East Africa (Cobo et al. 2010). As a result, the originally fertile lands that yielded 2–4 t/ha of cereal grains have been degraded, with cereal crop yields of <1 t/ha becoming common (Sanchez 2002). The new "Vision for African Agricultural Research" developed by Forum for Agricultural Research in Africa (FARA) and its member organizations calls for an annual growth rate of 6% in agricultural productivity by 2020 to achieve sustainable development (FARA 2006). Given the current trends in agricultural growth, achieving this annual growth rate is a major challenge that has been worsened by climate change and variability (World Bank 2011).

12.1.2 Conservation Agriculture

The CA approach involves use of the combined application of organic and mineral resources, resilient germplasm, and nutrient cycling and conservation (Vanlauwe et al. 2010). It is reported to be an overarching approach to restoring and maintaining soil productivity, and results into synergy and improved conservation and synchronization of nutrient release and crop demand, leading to increased fertilizer and nutrient use efficiency and higher yields (Vanlauwe et al. 2002). For example, CA technologies conducted in Kenya (DAP, FYM, and soil and water conservation) demonstrated higher (>10 times) marginal returns from grain amaranth (*Amaranthus hypochondriacus*) compared with maize, with the former being a better climate change/variability adaptation strategy compared with the latter, and in Uganda resulting in >30% increase in groundnut yield with 50–150 kg single superphosphate per hectare (Semalulu et al. 2011a). Furthermore, CA (application of organic and inorganic fertilizers, maize stover combined with inorganic fertilizers, and intercropping maize with *Dolichos lablab*) resulted in yield gains of 55%, 130–270%, and 160%, respectively, over the farmers' practice in the cereal–legume–livestock systems (Matowo et al. 2011). In addition, yield increases of 80%, 130%, and 200% above the farmers' practice were obtained using CA technologies under soybean (*Glycine max* (L.) Merr. and groundnut (*Arachis hypogaea* L.), sorghum (*Sorghum bicolor* (L.) Moench), and for both maize (*Zea mays* L.) and rice (*Oryza sativa* L.), respectively (Mugwe et al. 2011). Other trials in Kenya showed that maize grain yield increased from 1247 to 2678 kg/ha with the addition of 25 kg DAP/ha and 2 Mg of stover (Semalulu et al. 2011b). Related CA work in Siaya county, Kenya, by KEFRI/ICRAF/KARI based on fertilizer tree technology utilizing calliandra (*Calliandra calothyrsus* Meissn) showed that short-duration improved fallows (fertilizer trees) of 6–12 months increased the yield of subsequent maize crops by 1–3 Mg/ha in the first season and had high economic benefits compared with continuous maize cropping or natural weed fallows (Semalulu et al. 2011b). Although these technologies have shown promising results, most of them were limited to participating farmers within the project sites. The main goal of this chapter was to provide further insights into the potential contribution of CA practices to sustainable smallholder agriculture in East Africa in the context of climate change, soil restoration, and agricultural productivity.

12.2 LAND DEGRADATION PROCESSES IN EAST AFRICA

12.2.1 Interrill and Rill Erosion

Soil erosion is a major environmental problem and is widespread worldwide (Xiao et al. 2011). Soil erosion by water is rampant in both semiarid and mountainous areas, leading to low productivity and increased poverty (Kimaro 2003; Liu et al. 2012). The Highlands of East Africa suffer severely from soil erosion since the deforestation of the natural mountain forests and the cultivation of large areas (Kimaro et al. 2008). High rates of soil losses due to severe soil erosion by water are reported from arable lands in the mountainous areas of East Africa (Oldeman et al. 1990). In these areas, past studies show that rates of soil loss by combined processes of interrill and rill erosion are very high, i.e., in excess of 50 Mg/ha/year, which exceeds the tolerable values generally recommended to be 10–12 Mg/ha/year (Milliman and Meade 1983). Soil erosion is a serious problem in mountainous areas of Tanzania (Kimaro 2003; Msita 2013). In the Usambara Mountains, soil erosion by water is estimated to vary from 72 to 120 Mg/ha/year (Lundgren 1980; Pfeiffer 1990), while a soil loss of 28–72

TABLE 12.1
Soil Loss Rates due to Interrill and Rill Erosion on Different Geopedologic Units in Uluguru Mountains, Tanzania

Geopedologic Units	Slope (%)	Soil Loss (Mg/ha/year) Interrill Mean	s.d	Min	Max	Rill Mean	s.d	Min	Max	*N*
Mountain Ridges										
Hyperferrallic Cambisols	>60	115***	35	70	143	258***	15	240	275	5
Leptic Cambisols	50–60	61**	15	45	79	161***	15	141	175	4
Mountain Foothills										
Hyperferrallic Cambisols	>60	98***	12	81	110	235***	11	222	254	6
Endoleptic Cambisols	50–60	44**	6	37	49	111***	9	96	121	5
Chromic Lixisols	40–50	41**	15	21	60	91***	14	68	107	9
Profondic Acrisols	30–40	13[NS]	4	7	24	28*	5	24	35	7

Source: Kimaro, D.N., J. Poesen, B.M. Msanya, and J.A. Deckers, *Catena*, 75, 38, 2008. With permission.

Note: Max, maximum; Min, minimum; *N*, number of observations; NS, not significant; s.d, standard deviation; s.e, standard error.

s.e = 11; $*P < .05$; $**P < .01$; $***P < .001$.

Mg/ha/year was observed in the arable lands on the slopes of Mount Kilimanjaro, Tanzania (Temple 1972). In the Uluguru Mountains, Tanzania, interrill and rill erosion processes are dominant and result in mean soil loss ranging from 91 to 258 Mg/ha/year (Table 12.1). The soil erosion processes demonstrated in this area vary spatially along the landscape both in terms of the type of process and the degree of severity. Such spatial information could form an important guide for understanding the behavior and occurrence of soil erosion in a complex mountainous environment. The information could be taken as a guide for setting up soil and water conservation priorities in East African Highlands.

12.2.1.1 Tillage Erosion

Tillage erosion, also referred to as "arable erosion" (Zachar 1982) or mechanical soil erosion (Kiburys 1995), is the process of soil movement caused by the force applied by agricultural tools and by gravity (Lindstrom et al. 1992; Govers et al. 1994; Dercon 2001). At the scale of farm plot, tillage erosion can be recognized by a less fertile tillage step at the top of the field (due to soil profile truncation) and the development of a soil bank with a high soil fertility status at the bottom of the field if the physical barrier is present (Turkelboom et al. 1999; Nyssen et al. 2000; Poesen et al. 2000; Nyssen 2001). Although tillage erosion became better documented recently (Govers et al. 1994, 1996; Guiresse and Revel 1995), there is still inadequate information on this process particularly in tropical mountainous areas (Nyssen 2001). In view of these observations, a study was conducted in the Uluguru Mountains, Tanzania, on tillage erosion to investigate the effects of slope gradient, tillage depth, and surface ground conditions on tillage erosion due to manual hoeing (Kimaro et al. 2010). In this study, soil flux by manual tillage was measured by means of an on-farm experiment in the farmers' fields during the dry season of the years 2000 and 2001. Twenty-four plots (1.2 m wide and 8 m long) at different slopes (31%, 47%, 51%, 54%, 58%, 61%, 65%, and 67%) were demarcated before seasonal land preparation. Two sets of treatment at each slope site were done: deep tillage and shallow tillage under conditions of bare soil surface and soil surface with residues to both treatments. In the field, local farmers were requested to till the plots in a traditional way. The plots were tilled starting from the bottom of the field and moving up the slope. Farmers used a hoe with a steel blade of about 18.5 cm long and 16.5 cm wide and a wooden handle about 100 cm long. To measure soil translocation, the Gerlach trough method (Gerlach 1967) was used. This method provided quick assessment of soil movement in the field by manual tillage (Poesen et al. 2000).

The results from this study (Tables 12.2 and 12.3) showed that rates of soil flux due to tillage erosion generally increased with increasing slope gradient (Kimaro

TABLE 12.2
Effect of Slope Gradient on Soil Flux Due to Tillage by Manual Cultivation in Uluguru Mountains, Tanzania

Slope (%)	Mean Soil Flux (kg/m/tillage pass)	Standard Deviation (kg/m/tillage pass)	Minimum (kg/m/tillage pass)	Maximum (kg/m/tillage pass)	*N*
31	16[a]	6	7	23	12
47	31[b]	16	8	67	12
51	53[c]	27	21	81	12
54	59[dc]	30	21	92	12
58	59[c]	21	35	95	12
61	67[de]	24	34	101	12
65	75[e]	23	40	100	12
67	70[e]	27	44	111	12

Source: Kimaro, D.N., J. Poesen, J.A. Deckers, and H.B. Msita. 2010. Effects of agroecological conditions and slope gradient on tillage erosion in the northern slopes of the Uluguru Mountains, Tanzania. Chapter 9 in Earl T. Nardali (Ed.). *No-Till Farming: Effects on Soil, Pros and Cons and Potential.* Agriculture Issues and Policies Series. ISBN: 978-1-60741-402-5. Nova Science Publishers Inc., New York. pp. 159–172. With permission.

Note: Means with the same superscript letters are not significantly different ($P < .05$). *N*, number of observations.

TABLE 12.3
Effect of Tillage Depth and Soil Surface Condition on Soil Flux Due to Manual Tillage Cultivation in Uluguru Mountains, Tanzania

Parameter	Mean Soil Flux (kg/m/tillage pass)	Standard Deviation (kg/m/tillage pass)	Minimum (kg/m/tillage pass)	Maximum (kg/m/tillage pass)	*N*
Shallow tillage with residues	31[a]	3	7	47	24
Shallow tillage with bare soil surface	44[b]	4	12	79	24
Deep tillage with residues	62[c]	6	17	89	24
Deep tillage with bare soil surface	78[d]	7	22	111	24

Source: Kimaro, D.N., J. Poesen, J.A. Deckers, and H.B. Msita. 2010. Effects of agroecological conditions and slope gradient on tillage erosion in the northern slopes of the Uluguru Mountains, Tanzania. Chapter 9 in Earl T. Nardali (Ed.). *No-Till Farming: Effects on Soil, Pros and Cons and Potential.* Agriculture Issues and Policies Series. ISBN: 978-1-60741-402-5. Nova Science Publishers Inc., New York. pp. 159–172. With permission.

Note: Means with the same superscript letters are not significantly different ($P < .05$). *N*, number of observations.

et al. 2010). Mean soil flux increased significantly ($P < .05$) from 16 kg/m/tillage pass at a slope gradient of 31% to >60 kg/m/tillage pass at a slope gradient >60%. The rates of soil flux from bare soil surfaces for both shallow and deep tillage were significantly higher by 20% and 30% ($P < .05$), respectively, than soil flux from soil surface with residues. Bare soil surface promoted tillage erosion in contrast to soil surface with residues. Further analysis showed that a combination of shallow tillage and the presence of residues on the soil surface can cut down the rates of soil flux by 60% compared with deep tillage on a bare soil surface. The study concluded, therefore, that farmers should be discouraged from clearing their fields by removing all the residues before cultivation.

12.2.1.2 Landslides

Landslides are a result of gravity affecting the stability and evolution of ridge slopes in the major areas of East Africa, particularly the mountainous areas (Westerberg 1999; Kimaro 2003). The triggering of landslides depends on several complex but interrelated factors, such as prolonged heavy rainfall (Temple and Rapp 1972; Larsson 1989), soil characteristics (Westerberg 1999; Mburu 2001), slope characteristics (Lopez and Zinck 1991) and geomorphology (Ahmad and McCalpin 1999; Westerberg and Christiansson 1999), vegetation, land use, and infrastructure (Larsen and Torres-Sanches 1998). In the Uluguru Mountains, Tanzania, a study was conducted to investigate the extent of cropland damage due to the El Niño rains that occurred in 1997–1998 (Kimaro 2003). The rain-induced landslides dominated on the mountain ridges with a slope gradient of >47% (Table 12.4), where they have affected >20% of the land. Their distribution follows the following trend:

TABLE 12.4
Effects of 1997–1998 El Niño–Induced Landslides on Crop Land in Uluguru Mountains, Tanzania

Land Unit	Land Use	Area of Damaged Cropland (m^2)	Total Damaged Cropland Area (m^2)	Percent of Damaged Crop Land (%)	Volume of Soil Loss (m^3)	Total Volume of Soil Loss (m^3)	Rate of Soil Loss (Mg/ha/year)	*N*
Debris slopes	Beans, maize, pigeon peas, vegetables	250	1240	38	103	557.8	8.3	9
Amphitheaters	Vegetables, bananas	94	414	13	418	618.7	123.7	5
Incisions and V-shaped valleys	Bananas	510	1618	49	532	1286.6	13.8	8

Source: Kimaro, D.N. 2003. Assessment of major forms of soil erosion in the Morningside catchment, Uluguru Mountains, Tanzania. PhD thesis, Sokoine University of Agriculture, Morogoro, Tanzania p. 264. Author's own unpublished PhD thesis.

Note: *N*, number of observations.

on the debris slopes (41%), on the incisions and V-shaped valleys (33%), and on the amphitheaters (23%). The adverse consequences of landslides in the study area include a significant reduction by 50% of crop yield and farmers' income. This study demonstrated the importance of rain-induced landslide, particularly its adverse consequences on cropland and farmers' income. It is an important process of land degradation and should be considered as a major constraint to agricultural development when planning for soil and water conservation measures on tropical mountainous areas.

12.2.1.3 Soil Loss Due to Root and Tuber Harvesting

Previous studies have also demonstrated that significant soil masses are lost from arable land during harvesting of root, tuber, and bulb crops such as carrot (*Daucus carota*), onion (*Allium cepa* L.), round potato (*Solanum tuberosum* L.), and cassava (*Manihot esculenta*) (Ruysschaert et al. 2007). Soil sticking to the harvested crops exported from the field is rarely returned back to the field, and is therefore referred to as soil loss due to crop harvesting (Ruysschaert et al. 2004; Isabirye et al. 2007). In East Africa, the amount of soil lost due to the harvesting of cassava roots and sweet potato tubers under low-input agriculture was assessed on the northern fringe of Lake Victoria in Mayuge District, Uganda (Isabirye et al. 2007). In this study, it was observed that the mean annual soil loss for cassava was 3.4 Mg/ha and for sweet potato was 0.2 Mg/ha (Isabirye et al. 2007). The ammonium acetate lactate extractable soil nutrient losses for cassava were N = 1.71 kg/ha/harvest, P = 0.16 kg/ha/harvest, K = 1.08 kg/ha/harvest, and for sweet potato were N = 0.14 kg/ha/harvest, P = 0.01 kg/ha/harvest, K = 0.15 kg/ha/harvest. Many studies of this nature have been done under highly mechanized agriculture (Poesen et al. 2001), and only a single research (Isabirye et al. 2007) in Uganda that was conducted under low-input agriculture is available. This is an area that calls for further research in the region, particularly with the introduction of root and tuber crops as one of the strategies for adaptation to climate change.

12.2.2 Soil Fertility

The per capita agricultural production and crop yields per unit area of production in East Africa are declining (Sanchez et al. 1997; FAO 1999). The main contributing biophysical factors are nutrient/soil fertility depletion (Vlek 1993) and low soil fertility particularly N and P deficiencies (Bekunda et al. 1997). Loss of nutrients through crop harvests, surface runoff, and soil erosion is on the rise for most of the farming systems. All these have contributed to the negative nutrient balances reported for East Africa (Stoorvogel and Smaling 1990; Wortmann and Kaizzi 1998; Walaga et al. 2000). Most of the studied soils in East Africa have very low organic matter content (%OC <1.2), nitrogen (<1.0%), and available phosphorus (<5 mg/kg) (Tables 12.5 through 12.7). It is clear that the chemical fertility of soils is low, which implies that such soils are often poorly structured and therefore susceptible to erosion (Moberg et al. 1982). The observed low soil fertility in the region could also be attributed to poor land husbandry practices, which include clearing, burning, and continuous cultivation without proper management (Kimaro 2003). Generally, these

TABLE 12.5
Soil Chemical Properties of Some Selected Topsoils (0–30 cm) in Maswa District, Tanzania

						Exchangeable Bases (cmol(+)/kg Soil)			
Village	**pH Water**	**% TN**	**% OC**	**Olsen P (mg/kg Soil)**	**CEC (cmol/kg Soil)**	**Ca**	**Mg**	**K**	**Na**
Isulilo	8.75	0.04	0.63	14.46	24.0	13.97	3.89	0.78	7.27
Isulilo	7.73	0.05	0.61	14.46	17.2	10.47	2.52	0.26	3.63
Njapanda	6.68	0.04	0.63	16.52	16.6	8.11	1.73	0.36	7.79
Njiapanda	7.92	0.03	0.46	15.14	16.0	6.09	2.41	0.14	7.36
Njiapanda	8.03	0.05	0.63	17.16	20.6	14.19	2.41	0.23	4.57
Bakangilija	7.55	0.06	0.82	15.69	18.0	12.65	2.94	0.30	3.41
Bakangilija	7.03	0.03	0.38	14.24	7.0	2.26	0.64	0.11	2.09
Bakangilija	8.92	0.04	0.52	16.60	16.0	12.0	1.68	0.18	6.78
±SE	0.27	0.00	0.05	0.40	1.72	1.48	0.34	0.08	0.78

Source: Kajiru, G.J., J.P. Mrema, F.B. Rwehumbiza, N. Hatibu, and H.F. Mahoo. 2006. Evaluation of the fertility status of the soils of the Ndala River Catchment, Maswa District, Tanzania for rice production under rainwater harvesting systems. In: Msanya, B.M., Kimaro D.N., Kilasara M., Mrema J.P., and Kaaya A.K. (Eds.). *Land Resources Management to Enhance Livelihood of Land Users in East Africa. Reviewed Proceedings of the 22nd Conference*, Held at New Safari Hotel, Arusha, Tanzania, November 29–December 3, 2004.

Note: Ca, calcium; CEC, cation exchange capacity; K, potassium; Mg, magnesium; Na, sodium; OC, organic carbon; Olsen P, available phosphorus; TN, total nitrogen.

practices lead to low biomass productivity and, hence, a low contribution to plant nutrients and organic matter in the soil.

12.2.3 Effect of Land Use on Soil and Nutrient Loss in East Africa

Land use changes that have occurred in East Africa have resulted into a decline in natural vegetation (Kabanza 2013). Land use change coupled with anticipated climate change has led to accelerated land degradation in terms of soil erosion, declining soil fertility, loss in biodiversity, and changes in hydrological cycles (Kabanza et al. 2013b). In the southeastern Tanzania, it was also observed that tied ridges (a local practice in the Makonde and Rondo plateaus) (Figures 12.1 and 12.2) is the best localized form of CA, and effective in storing moisture and organic carbon conservation in Arenic Ferrasols or Acrisols of the area (Kabanza et al. 2013a). Studies conducted in East Africa (Uganda and Tanzania) have demonstrated the linkage between land utilization types, runoff, and soil and nutrient loss (Isabirye et al. 2010; Semalulu et al. 2012; Msita 2013).

Studies conducted in the Lake Victoria shoreline to determine the rate of sediment generated by agricultural and settlement land use types show that settlements

TABLE 12.6
Soil Chemical Characteristics of Selected Topsoils (0–30 cm) in Uluguru Mountains, Tanzania (Mean ± Sed)

Soil Type	pH H_2O	OC (%)	N (%)	Available P (mg/kg)	CEC	Ca^{++}	Mg^{++} (cmol(+)/kg)	K^+	Na^+
Hyperskeleti–Lithic Leptosols	5.6 ± 0.1	0.4 ± 0.1	0.04 ± 0.01	8 ± 1	7.6 ± 0.8	4.1 ± 0.3	2.7 ± 0.1	0.04 ± 0.01	0.07 ± 0.02
Epidystri–Endoskeletic Cambisols	6.1 ± 0.1	1.2 ± 0.2	0.1 ± 0.02	2 ± 1	13.5 ± 1.6	3.3 ± 0.2	2.6 ± 0.1	0.3 ± 0.01	0.1 ± 0.01
Hyperskeleti–Lithic Leptosols	5.5 ± 0.1	0.4 ± 0.1	0.04 ± 0.01	7 ± 1	7.2 ± 1.2	4.1 ± 0.4	2.7 ± 0.2	0.04 ± 0.01	0.07 ± 0.01
Orthieutri–Leptic Cambisols	6.3 ± 0.2	0.9 ± 0.2	0.11 ± 0.01	3 ± 1	8.8 ± 0.9	4.7 ± 0.1	2.9 ± 0.1	0.22 ± 0.02	0.2 ± 0.1
Orthieutri–Leptic Cambisols	6.4 ± 0.2	1.1 ± 0.2	0.12 ± 0.01	3 ± 2	9.1 ± 1	4.7 ± 0.1	2.9 ± 0.1	0.23 ± 0.02	0.1 ± 0.03
Hapli–Profondic Lixisols	5.2 ± 0.1	0.8 ± 0.3	0.08 ± 0.02	13 ± 1	8.7 ± 1.4	4.2 ± 0.3	2.9 ± 0.3	0.17 ± 0.02	0.09 ± 0.01
Hapli–Pachic Phaeozems	6.6 ± 0.2	2.1 ± 0.3	0.05 ± 0.01	6 ± 1	8.9 ± 0.8	3.8 ± 0.1	3.6 ± 0.1	0.47 ± 0.02	0.20 ± 0.03
Hapli–Orthieutric Regosols	5.1 ± 0.2	0.9 ± 0.3	0.08 ± 0.01	12 ± 2	9.0 ± 1.0	4.3 ± 0.2	2.8 ± 0.1	0.18 ± 0.02	0.09 ± 0.01
Hyperskeleti–Lithic Leptosols	5.6 ± 0.2	0.3 ± 0.1	0.03 ± 0.01	8 ± 2	7.4 ± 0.2	4.2 ± 0.1	2.7 ± 0.2	0.04 ± 0.01	0.07 ± 0.01
Episkeleti–Hyperferralic Cambisols	6.5 ± 0.2	1.8 ± 0.2	0.12 ± 0.01	2 ± 1	12.2 ± 1.1	5.3 ± 0.1	4.7 ± 0.2	0.52 ± 0.03	0.16 ± 0.02
Chromi–Endoleptic Cambisols	6.8 ± 0.2	1.9 ± 0.3	0.10 ± 0.01	2 ± 1	14 ± 1	7.7 ± 0.1	3.0 ± 0.2	0.53 ± 0.03	0.18 ± 0.02
Hapli–Chromic Lixisols	6.1 ± 0.2	1.0 ± 0.2	0.19 ± 0.03	3 ± 1	16.9 ± 0.4	8.1 ± 0.3	5.6 ± 0.3	0.77 ± 0.08	0.30 ± 0.02
Chromi–Profondic Acrisols	6.4 ± 0.2	1.2 ± 0.2	0.22 ± 0.01	2 ± 1	14.2 ± 0.5	3.0 ± 1	4.4 ± 0.5	1.06 ± 0.04	0.22 ± 0.01

Source: Kimaro, D.N. 2003. Assessment of major forms of soil erosion in the Morningside catchment, Uluguru Mountains, Tanzania. PhD thesis, Sokoine University of Agriculture, Morogoro, Tanzania pp. 264. Author's own unpublished PhD thesis.

Note: CEC, cation exchange capacity; N, total nitrogen; OC, organic carbon.

TABLE 12.7
Chemical Properties of Cutanic Acrisol (Hyperdystric, Arenic, Rhodic) (Kanhaplic Haplustult) from Makonde Plateau in South Eastern, Tanzania

Horizon	Depth (cm)	pH H_2O	pH KCl	OC (%)	N (%)	Av. P (mg/kg)	BS (%)
Ap	0–15	5.4	4.6	0.51	0.05	1.53	42
AB	15–47	5.5	4.5	0.24	0.06	0.53	36
Bt1	47–108	5.0	4.2	0.03	0.06	0.70	30
Bt2	108–200+	4.9	4.3	0.20	0.06	0.73	23

Source: Kabanza, A. 2013. Insights into soil and water conservation measures for major land utilization types in South Eastern Tanzania. PhD thesis. KU Leuven, Belgium. p. 166. With permission.
Note: % BS, % base saturation; N, total nitrogen; OC, organic carbon.

generate significantly higher sediment yields, i.e., between 17 and 87 Mg/ha/year (Table 12.8) (Isabirye et al. 2010). Other recent studies in the same area have demonstrated further that settlement compounds and footpaths produced soil losses up to 199 Mg/ha/year (De Meyer et al. 2011). Agricultural land use types produced between 0 and 27 Mg/ha/year, with cassava generating the highest rates of 27.3 Mg/ha/year in the Ugandan shoreline (Isabirye et al. 2010). Sediment yield from runoff plots in the gardens on the Tanzania shoreline side were in a range of 1.2–22.7 Mg/ha/year. Cotton, unlike cassava, in Uganda generated medium rates (12–25 Mg/ha/

FIGURE 12.1 Preparation of tied ridges (local CA practice in Makonde and Rondo plateaus) for effective soil moisture storage and organic carbon conservation in Arenic Ferrasols in southeastern Tanzania.

FIGURE 12.2 Maize crop on tied ridges (local CA practice in Makonde and Rondo plateaus) for effective soil moisture storage and organic carbon conservation in Arenic Ferrasols in southeastern Tanzania.

TABLE 12.8
Mean Annual Soil Loss (Mg/ha) Measured on Runoff Plots with Various Land Use Types in Ugandan Shoreline of Iguluibi District

Land Use (2005)	*N*	Mean	Median	SD	Min	Max
Agriculture						
Bananas	3	2.1	0.0	4.3	13.4	0.0
Cassava	3	27.3	8.7	37.9	0.2	138.8
Cassava–maize–sorghum intercrop	3	8.5	7.7	11.0	0.0	41.6
Coffee	3	0.0	0.0	0.0	0.0	0.0
Groundnuts	3	4.0	0.0	12.0	0.0	40.0
Maize	3	2.6	0.2	5.2	0.0	15.6
Millet	3	0.7	0.03	1.7	0.0	6.9
Sugarcane	3	0.0	0.0	0.0	0.0	0.0
Sweet potatoes	3	1.1	0.2	2.6	0.0	10.4
Built Up		Cumulative Yield (t/ha)		Mean Yield (t/ha)		
Settlement 1 (mean compound area = 0.25 ha)	1	87		52		
Settlement 2 (mean compound area = 0.09 ha)	1	17				

Source: Adapted from Isabirye, M., D. Kimaro, and O. Semalulu, *Tropicultura*, 28, 89, 2010. With permission.

Note: *N*, number of plots/path where soil loss was measured during the year 2005.

TABLE 12.9
Mean Annual Soil Loss (Mg/ha) Measured on Runoff Plots within Various Land Use Types in Tanzanian Shoreline of Magu District

Land Use (2004–2005)	*N*	Mean	Median	SD	Min	Max
Rice	3	1.2	1.0	0.4	0.6	2.2
Cassava	3	7.9	6.3	4.8	1.3	19.5
Maize	3	15.6	13.8	5.8	6.7	29.1
Cotton	3	22.7	20.3	9.6	10.5	50.9

Source: Adapted from Isabirye, M., D. Kimaro, and O. Semalulu, *Tropicultura*, 28, 89, 2010. With permission.
Note: *N*, number of sediment plots where soil loss was measured during the years 2004–2005.

year) of sediment (Table 12.9). The size and cropping pattern of the field parcels, and the overall farming techniques in the riparian zone of Lake Victoria have an overall effect of protecting soil structure against raindrop impact, encouraging water infiltration, obstructing runoff, and therefore minimizing sediment yield. The high rates of sediment yield observed under cassava and the settlements in Uganda and cotton in Tanzania call for soil conservation practices, including targeting settlements as well that ensure complete surface cover and encourage good soil structure that enhance infiltration.

Other studies conducted in Bududa District, Uganda (Semalulu et al. 2012), to investigate the effect of different farmer cropping and soil conservation practices on runoff, soil loss, and nutrient loss demonstrated that soil loss was significantly ($P < .05$) higher on annuals than on banana or banana–coffee farming systems in the order of 38.5 > 6.6 > 0.87 Mg/ha/year (Table 12.10). Soil loss values for fields without conservation structures were much higher than those where there were structures. Soil loss values were much higher than the tolerable limit for Uganda, which is 5 t/ha/year, and calls for immediate action to scale up sustainable land management practices.

In West Usambara Mountains, the effectiveness of the *miraba* land utilization type in reducing runoff, soil loss, and nutrient loss was determined for the purpose of establishing insights to support soil and water conservation planning in the area. *Miraba* represents a generic word for an indigenous technology and a specific agricultural landscape type (Figure 12.3). *Miraba* is also referred to as a management unit: one field can be subdivided in different numbers of *Miraba*, each being a square or a rectangular plot fully surrounded by a strip of tall grass of napier grass (*Pennisetum purpureum*), Guatemala grass (*Tripsacum laxum*), or *Hyparrhenia* grass (*Hyparrhenia* spp.) (Msita et al. 2010a). Results from this study have demonstrated that the annual soil loss is different ($P < .05$) among the bare plot, cropped

TABLE 12.10
Effect of Land Use and Conservation Structures on Soil Loss (Mg/ha/year)

Land Use	No Structures	Structures
Annuals	38.5	15.5
Banana	6.6	3.2
Banana–coffee	0.9	0.2
sed	6.1	8.11

Source: Adapted from Semalulu, O., D. Kimaro, V. Kasenge, M. Isabirye, and P. Makhosi, *International Journal of Agricultural Sciences*, 2, 256, 2012. With permission.
Note: sed, standard deviation.

FIGURE 12.3 *Miraba*: An indigenous soil and water conservation technology on the sloping land of Usambara Mountains Tanzania. (Adapted from Msita H.B. et al. 2012. Effectiveness of miraba an indigenous soil and water conservation measures on reducing runoff and soil loss in arable land of Western Usambara Mountains. Paper presented to the EGU General Assembly 2012, in Vienna Austria, 22nd–27th April 2012.)

plot, and *miraba* treatments (Table 12.11). However, the soil loss was not different ($P < .05$) between *miraba* and *miraba* with FYM and mulching (Table 12.11). *Miraba* is effective in controlling soil loss by 99%. There is a significant ($P < .05$) difference among nutrient losses obtained in the bare plot, cropped plot, and *miraba* land utilization types (Table 12.11). *Miraba* is effective in controlling all nutrient losses by >83%. The soil losses are lower in bare and control plots in the short rainy season

TABLE 12.11
Annual Runoff Coefficient and Soil and Nutrient Loss under Different Land Use Types in West Usambara Mountains, Tanzania, Measured on Runoff Plots ($N = 3$)

Treatment	Annual Runoff Coefficient (mm/mm/year)	Annual Soil Loss (Mg/ha/year)	Annual Total N Loss (kg/ha/year)	Annual P Loss (kg/ha/year)	Annual K Loss (kg/ha/year)
Bare	0.15[a]	169.0[a]	464[a]	0.6[a]	10.77[a]
Control	0.10[a]	132.0[b]	342[b]	0.4[b]	6.88[b]
Miraba	0.04[c]	0.7[c]	9[c]	0.1[c]	0.02[c]
Miraba + FYM + mulching	0.03[c]	0.5[c]	9[c]	<0.1[d]	<0.02[c]

Source: Adapted from Msita, H.B. 2013. Insights into indigenous soil and water conservation technologies in Western Usambara Mountains, Tanzania. PhD thesis. KU Leuven, Belgium. p. 194. With permission.

Note: FYM, farmyard manure; Means with the same letter in columns are not significantly different by Duncan multiple range test comparison, $\alpha = 0.05$. *Miraba* = rectangular bound grass strip that does not follow contour (an indigenous soil and water conservation technology on the sloping land of Usambara Mountains Tanzania).

and higher in the long rainy season (Table 12.12). On the other hand, in *miraba* and *miraba* with FYM and mulching, soil losses are higher in the off-season and lower in the short rainy season. The nutrient losses are lower in bare and control plots in the short rainy season and higher in the long rainy season. On the other hand, in *miraba* and *miraba* with FYM and mulching, nutrient losses are higher in the off-season and lower in the short rainy season. These results have demonstrated that *miraba* and *miraba* coupled with FYM and mulching are effective soil and water conservation technologies for the tropical highlands of East Africa.

12.3 CONSERVATION AGRICULTURE IN EAST AFRICA

12.3.1 Conservation Agriculture and Farming Practices

Agriculture is a notable source of the major greenhouse gases: carbon dioxide, methane, and nitrous oxide (Commission of the European Communities Brussels 1996; FAO 2010). Agricultural activities and land use changes contribute about one-third of the total carbon dioxide emissions and are the largest sources of methane (from livestock and flooded rice production) and nitrous oxide (primarily from application of inorganic nitrogenous fertilizer). The Ecological Society of America (2000) provides a list of management techniques that improve agricultural land qualities that consequently increase carbon sequestration. These techniques include the planting of cover crops, mulch farming combined with zero tillage, and agroforestry. Some of these practices would influence agricultural land qualities that increase aboveground

TABLE 12.12
Annual Runoff Coefficient and Soil and Nutrient Loss in Different Seasons in West Usambara Mountains, Tanzania

	Treatment	Long Rainy Season	Off-Season	Short Rainy Season
Annual runoff coefficient (mm/mm/year)	Bare	0.03	0.12	0.004
	Control	0.02	0.8	0.003
	Miraba	0.01	0.03	0.003
	Miraba + FYM+ mulching	0.01	0.02	0.002
Soil loss (Mg/ha/year)	Bare	103	49	17
	Control	80	34	17
	Miraba	0.1	0.6	0.02
	Miraba + FYM+ mulching	0.1	0.5	0.01
Total N loss (kg/ha/year)	Bare	308	155	0.9
	Control	248	94	0.4
	Miraba	0.4	8	0.2
	Miraba + FYM + mulching	<0.4	<8	<0.2
P loss (kg/ha/year)	Bare	0.3	0.3	0.0009
	Control	0.2	0.2	0.0004
	Miraba	0.0004	0.1	0.0001
	Miraba + FYM + mulching	<0.0004	<0.1	<0.0001
K loss (kg/ha/year)	Bare	8.3	2.5	0.02
	Control	4.5	2.4	0.01
	Miraba	0.02	<2.4	0.01
	Miraba + FYM + mulching	<0.02	<2.4	<0.01

Source: Msita, H.B. 2013. Insights into indigenous soil and water conservation technologies in Western Usambara Mountains, Tanzania. PhD thesis. KU Leuven, Belgium. p. 194. With permission.

carbon stock. Such practices constitute what is embedded in the principles of CA. The range of sequestration potential for the different CA practices considered is far from negative for continuous cultivation practices to around 40 Mg/ha with the retention of crop residues and substantial addition of FYM (The Ecological Society of America 2000).

In South America where millions of hectares of farmland have been under CA, several studies reported gains in yields and income (Hobbs and Gupta 2004; IIRR and ACT 2005; Derpsch 2005; Bolliger et al. 2006). The gains were attributed to the improvements in soil-related qualities associated with CA, such as soil moisture availability, soil nutrient availability, and reduction of erosion hazards (Roldan et al.

2003; Fabrizzi et al. 2005). Studies conducted in the volcanic highlands of Central Mexico demonstrated positive effects, including nutrient improvement and increased crop yields during 12 years of zero tillage seeding systems when practiced with proper crop rotation and crop residue management (Govaerts et al. 2006). Maize–wheat rotation combined with full residue retention under zero tillage was observed to increase crop yields significantly. High water infiltration rates and favorable moisture dynamics were reported to explain the observed high crop yields in this area (Govaerts et al. 2007).

In Ethiopia, two local tillage systems, *terwah*+ and *derdero*+, using the traditional *mahresha ard* plough on Vertisol under crop rotation (wheat, grass, pea, wheat) significantly increased soil organic matter to 2.0% when compared with conventional tillage (1.4%) for 0–15 and 15–30 cm soil depths. In this study, the mean yield of wheat for 3 years increased from 2.8 Mg/ha for conventional tillage to 3.7 Mg/ha for the *terwah*+ and *derdero*+ tillage systems (Araya et al. 2012). The effect of contour furrows at 60–70 cm interval without any soil movement in the bed (permanent raised bed system), as well as contour furrows at 1.5–2 m spacing after ploughing (*terwah* cultivation) on runoff and soil loss in comparison with the conventional ploughing system, was evaluated in Tigray, northern Ethiopia (Gebregziabher et al. 2009). In this study, the total soil loss decreased from 19.5 Mg/ha for the conventional ploughing system to 7.6 Mg/ha for *terwah* cultivation and to 4.7 Mg/ha for the permanent raised bed system (Gebregziabher et al. 2009). The study concluded that contour furrows and permanent raised beds can be part of an agricultural intensification process that includes physical soil and water conservation, slope reforestation, irrigation development, and agroforestry in croplands. In addition, the use of permanent raised beds if combined with crop mulching and crop diversification is an important component for the development of sustainable CA practices in the region.

Thus far, in sub-Saharan Africa, particularly in East Africa, CA has been restricted to only large estate farming (IFAD/FAO 2004) and the practices remain poorly implemented among smallholder farmers, particularly in the highland farming communities (Tittonell et al. 2008). CA that has been practiced in East Africa, especially in Kenya and Tanzania, comprised three principles that have been applied simultaneously: minimum soil disturbance, permanent soil cover, and complex rotations/crop associations (Owenya et al. 2011). Recent studies have indicated a positive trend in practicing CA in many areas of East Africa owing to the integration of both indigenous and scientific knowledge (Shetto and Owenya 2007). However, in Tanzania, for example, knowledge on the application and implementation of CA practices toward improvement of soil qualities is still scanty (Tenge et al. 2003; Triomphe et al. 2007), and limited research efforts have been put toward this direction.

Studies conducted on three sites in West Usambara Mountains, Tanzania, identified farmers' CA practices linked to different cropping and management patterns (Table 12.13). The identified CA practices include (i) terrace with animal manure, inorganic fertilizers, crop residues, and crop rotation; (ii) *miraba* with animal manure, crop residues, crop rotation, and agroforestry; and (iii) raised beds with animal manure, inorganic fertilizers, crop residues, and crop rotation. Crop rotation was the most common management practice identified in the studied sites. The most common rotations were maize (*Z. mays*) mixed with beans (*Phaseolus vulgaris*) in

TABLE 12.13
CA Practices Associated to Cropping and Management Patterns as Practiced by Small-Scale Farmers in West Usambara Mountains, Tanzania

	Cropping Patterns in Different Seasons			
CA and Management Practices	**Sept–Nov**	**Dec–Feb**	**June–Sept**	**March–June**
Terrace: Animal manure, inorganic fertilizers, crop residues, crop rotation	Beans, maize	NA	NA	Potatoes
Miraba: Animal manure, crop residues, crop rotation, agroforestry	Beans, maize	NA	NA	Potatoes
Terrace: Animal manure, inorganic fertilizers, crop residues, crop rotation	Beans, maize	NA	NA	Potatoes
Miraba: Animal manure, crop residues, crop rotation, agroforestry	Beans, maize	NA	NA	Potatoes
Terrace: Animal manure, inorganic fertilizers, crop residues, crop rotation	Beans, maize	NA	NA	Potatoes
Miraba: Animal manure, inorganic fertilizers, crop residues, crop rotation, agroforestry	Beans, maize	NA	NA	Potatoes
Raised beds: Animal manure, inorganic fertilizers, crop residues, crop rotation	Cabbage	Tomatoes	Potatoes	NA

Source: Senior author's own field data.
Note: NA, not applicable.

September to November, followed by potatoes (*S. tuberosum*) and cabbage (*Brassica oleracea*) in March to June for rainfed cropping on the sloping lands. For valley bottoms, the most common rotations were cabbage in September to November, tomatoes (*Solanum lycopersicum*) in December to February, and potatoes from June to September. The raised beds are constructed in valley bottoms to provide artificial drainage (June–September). The drainage channels are then used for irrigation by using buckets during the dry spell in September to December. During long rains, in the months of March to June, the valley bottoms are not cultivated because of floods. The drainage channels also act as a drainage system during flush floods. Beans (*P. vulgaris*) were the only grown legume and known by the farmers to enhance nitrogen availability in the soils. Most of the farmers (about 60%) were aware that beans fix nitrogen through a symbiotic association with nitrogen-fixing bacteria. They were also aware that residues from beans had higher nitrogen content, which upon application on the farm as mulch can readily release nitrogen into the soil (Peel 1998; Yusuf et al. 2009). Beans were the sole legume crop commonly used by farmers to improve soil fertility.

The comparison of soil chemical characteristics of fields with CA and non-CA fields show that total nitrogen (TN) observed in terraces and *miraba* exhibits the way CA can influence soil fertility when compared with non-CA fields (Table 12.14). Relatively higher levels of organic carbon (OC) and TN in terraces and *miraba* were

TABLE 12.14
Comparison of Soil Chemical Characteristics with CA and Non-CA Fields in West Usambara Mountains, Tanzania

Site	CA vs. No CA	TN	OC	Olsen. P	CEC	Ca	Mg	K	Na
1	Terraces	0.34	4.39^{bc}	9.18	17.87	6.54	2.27^{cd}	0.62	0.25
	Miraba	0.36	4.24^{bc}	7.87	22.47	10.79	4.18^{a}	0.44	0.25
	No CA	0.24	2.93^{de}	5.38	14.27	7.31	2.86^{bcd}	0.38	0.21
2	Terraces	0.43	5.93^{a}	4.43	20.06	7.71	1.85^{d}	0.26	0.24
	Miraba	0.35	4.30^{bc}	5.09	21.86	7.82	3.87^{ab}	0.15	0.23
	No CA	0.34	4.02^{cd}	3.34	18.26	11.05	2.47^{cd}	0.24	0.22
3	Terraces	0.48^{a}	5.50^{a}	9.18	22.3^{a}	6.21^{c}	3.73^{ab}	0.57	0.22
	Miraba	0.26^{cd}	3.99^{cd}	6.36	11.06^{d}	11.28^{a}	1.19^{d}	0.26	0.23
	No CA	0.17^{d}	2.39^{e}	4.55	12.70^{cd}	8.53^{abc}	2.09^{d}	0.24	0.24
	Mean	0.36	4.44	8.68	18.82	9.19	3.05	0.37	0.26
	CV (%)	21.64	15.76	88.40	18.99	31.38	22.05	63.46	0.10
	F(0.05)	**	***	ns	**	ns	**	ns	ns

Source: Kimaro, D., own field data.

Note: Ca, calcium; CEC, cation exchange capacity; K, potassium; Mg, magnesium; Na, sodium; OC, organic carbon; Olsen P, available phosphorus; TN, total nitrogen; Means with the same superscript letters are not significantly different ($P < .05$); **, $P < .01$; ***, $P < .001$; ns, not significant.

observed compared with non-CA fields, and could be attributed to reduced soil erosion and supportive soil fertility management practiced by farmers on CA fields, which involve application of manure, crop residues, and crop rotation (Aziz et al. 2011). On the other hand, the mean available phosphorus from terraced fields was relatively better followed by *miraba* and non-CA fields. This could be due to comparatively better organic matter content of the terrace and *miraba* fields than non-CA fields that might be, in turn, attributed to the application of manure and crop residues. Glendinning (2000) asserted that in most soils, the amount of organic phosphorus is highly correlated with the amount of organic carbon. The rate of mineralization of organic phosphorus increases as the organic phosphorus content of the soil increases (Kabba and Aulakh 2004).

12.3.2 Effect of Conservation Agriculture on Crop Productivity

In a case study conducted in Karatu District, Tanzania, on CA using farmer field school (FFS), several CA options (Owenya et al. 2011) were tested, including (i) ripped plot, planted with maize intercropped with lablab; (ii) ripped plot, planted with maize intercropped with pigeon pea; (iii) nonripped plot, planted with maize intercropped with lablab, nonripped plot, planted with maize intercropped with pigeon pea; and (iv) farmers' normal practice of ploughing twice and then planting maize intercropped with pigeon pea, beans (*P. vulgaris*), and pumpkin (*Cucurbita maxima*) (Table 12.15). In this case study, farmers realized that crop yield under CA increased with time (Table 12.16), i.e., from 1.2 Mg/ha of maize in 2004 when

TABLE 12.15
CA Treatments Used by Farmer Field School (FFS) Group for Verification of Technologies and Crop Yields in Karatu District, Tanzania

No	CA Technology Tested	Maize Yields (Mg/ha)	Lablab (t/ha)	Pigeon Pea (Mg/ha)
1	Ripped plot, planted with maize intercropped with lablab	3.75	1.63	N/A
2	Ripped plot, planted with maize intercropped with pigeon pea	3.38	N/A	0.75
3	Nonripping plot, planted with maize intercropped with lablab	3.50	1.00	N/A
4	Nonripping plot, planted with maize intercropped with pigeon pea	2.00	N/A	0.75
5	Farmers' normal practice; ploughing twice and then planting maize intercropped with pigeon pea, beans, and pumpkin	1.88	Beans (0.50)	0.63

Source: Adapted from Owenya, M.Z., W.L. Mariki, J. Kienzle, T. Friedrich, and A. Kassam, *International Journal of Agricultural Sustainability*, 9, 145, 2011.

TABLE 12.16
Maize Yield Trend in Karatu District Tanzania during Years of CA Adoption

Year	Average Yield (Mg/ha)	Comments
2004	2.1	CA adopted
2006	2.4	After CA adoption
2009	2.8	After CA adoption

Source: Adapted from Owenya, M.Z., W.L. Mariki, J. Kienzle, T. Friedrich, and A. Kassam, *International Journal of Agricultural Sustainability*, 9, 145, 2011.

the CA was adopted to 3 Mg/ha in 2009 (Owenya et al. 2011). The increase in crop production in the 2009 season, despite that the crop had suffered from drought, is an indication that was brought about by improved soil conservation and water management under CA. Farmers in this area also learned that intercropping of maize with cover crops (pigeon pea [*Cajanus cajan* (L.) Millsp.] and *Dolichos lablab L* [*synonym: Lablab purpureus* (L.)] or *Mucuna* [*Mucuna pruriens*]) provided three harvests per season instead of two.

The CA Involving Farmer Field Schools (CA-FFS) were also established in several districts in Kenya and Tanzania, supporting 101 CA-FFS groups—48 groups in Kenya and 31 in Tanzania (FAO 2006b). It was noted in this project that adoption of

CA has been, characteristically, taken up by households on the basis of some specific elements (options) in varying combinations. Farmers consistently practicing a combination of CA practices that took on board all the principles of CA were still rather few in all the project sites. Depending on the varying local circumstances, communities/households innovated on the most feasible and attractive "entry points" in terms of which practices were tried/adopted first. In Kenya, especially in drier areas, options related to reduced tillage were the most popular, while in Tanzania, the good markets and price for lablab seed provided a good incentive for farmers to "enter" along the soil cover/live cover crop (lablab) option.

Löfstrand (2005) observed in Babati District, Tanzania, that low soil fertility was a limiting factor of crop production in the district, and recycling of plant nutrients to the arable land was low. In this study, soils from three different CA practices, FYM and intercropping with legumes, intercropping with rock phosphate, and intercropping under leguminous *Faidherbia albida* tree, were compared to investigate how different treatments have influenced soil fertility (Löfstrand 2005). FYM and intercropping with legumes improved soil fertility, and application of rock phosphate enhanced the availability of phosphorus and also increased soil pH. When only rock phosphate was applied, crop yields were low and the amount of nutrients either decreased or showed no significant change. Intercropping under leguminous *F. albida* trees produced the highest maize yields and also resulted in the highest level of P and N in the soil. In all these case studies, however, the effectiveness of CA practices to mitigate the adverse impacts of, and to adapt to, climate change were not explicitly taken into account, including assessment of soil and ecosystem carbon pools and carbon/water footprints of production systems.

In Tanzania, conservation tillage is the focus of farmers and researchers who are working in areas prone to soil erosion. This is because in conservation tillage, crops are grown with minimal cultivation of the soil, and most or all crop residues remain on top of the soil rather than being ploughed or disked into the soil. Tillage has long been a traditional agricultural system. In Tanzania, fallowing and using organic matter are practices that farmers traditionally used to maintain and to restore soil fertility. These practices have been used for a long time until the 1950s, when the government agricultural extension programs promoted soil and water conservation practices to control surface runoff, such as stone and earth bunds, ridging, pitting, infiltration or cutoff drains, bench terraces, and contours (Shetto and Owenya 2007). Strategies that are aimed at combating land degradation through mechanical and biological measures—reforestation activities, agroforestry, protection of water catchments, improved land husbandry, and environmental conservation in general—were initiated in Tanzania by the late 1980s (Shetto 1999). As a way to complement the Tanzanian government efforts, a study was conducted in the Uluguru Mountains, Tanzania, to investigate the effect of conservational tillage on vegetable yield in the area (Msita et al. 2010b). The practices that were examined include control conservational tillage (CCT), conservational tillage with manure (CT), conservational tillage with *in situ* composting (CTG), control traditional tillage (CTT), traditional terrace with manure (TT), and traditional tillage with *in situ* composting (TTG). The results obtained demonstrated that all conservational tillage treatments gave higher yield (CCT =

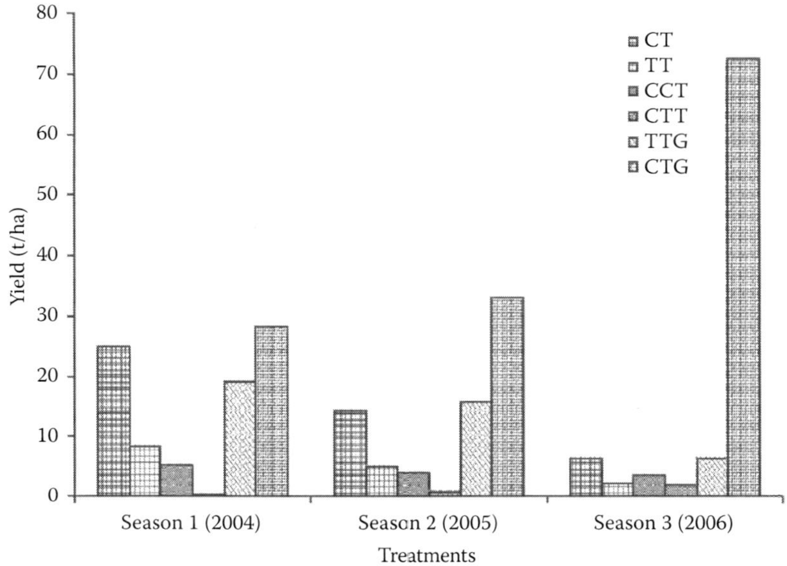

FIGURE 12.4 Fresh yield of vegetable of each season in the Uluguru Mountains, Tanzania, from 2004 to 2006 (CV% season 1 = 37.29, season 2 = 28.07, season 3 = 24.44). CCT, control conservational tillage; CT, conservational tillage; CTG, conservational tillage with *in situ* composting; CCT, control traditional tillage; TT, traditional terrace with manure; TTG, traditional tillage wtih *in situ* composting. (From Msita H.B. et al. 2010b. Effect of conservational tillage on soil loss and plant nutrient status on vegetable yield, northern slopes of Uluguru Mountains, Morogoro, Tanzania. In: Msaky J.J., Kanyama-Phiri G.N., Shongwe G.N. (Eds.), *Enhancing Dissemination of Soil and Water Research Options of SADC Universities. Proceedings of the Workshop on Information Sharing Among Soil and Water Management Experts from SADC Universities*, September 11–13, 2010, Dar es Salaam, Tanzania.)

4.33, CTG = 44.68, and CT = 15.28 Mg/ha) when compared with traditional tillage, which exhibited a lower yield (CTT = 0.88, TTG = 13.85, and TT = 5.23 Mg/ha) (Figure 12.4). From this study, it was apparent that all conservation tillage increased the yield of vegetables in the sloping lands of the Uluguru Mountains, Tanzania. It can be concluded from this study that conservation tillage for vegetable production on sloping land is the best practice that can be promoted for small-scale vegetable farming on fragile ecosystems.

12.4 CONCLUDING REMARKS

The following concluding remarks are made in the light of gaps revealed from the discussion in this chapter so as to provide further insights into optimal soil and water conservation practices based on the principles of CA for enhanced agricultural production in East Africa:

1. CA that has been practiced in East Africa is composed of three principles that have been applied simultaneously: minimum soil disturbance, permanent soil cover, and complex rotations/crop associations with some land management options, including application of FYM, green manure, composting, mineral fertilizer, and *in situ* rainwater harvesting.
2. Positive trends in practicing CA have been noted in many areas of East Africa where integration of both indigenous and scientific knowledge is promoted.
3. CA in East Africa has been demonstrated as a strategy that can address the complexities and peculiarities of soil quality on smallholder farms, and help low resource endowed farmers mitigate problems of poverty, food insecurity, income, and resilience of soil productive capacity in the context of climate change.
4. CA technologies that include application of organic and inorganic fertilizers, crop stover combined with inorganic fertilizers, and crop rotations and intercropping have resulted in yield gains over the farmers' practice in most farming systems in the region.
5. Experiences show that although the CA technologies discussed in this chapter have shown promising results, most of them were limited to participating farmers within the small project sites.
6. There is, therefore, a need for further research aimed at assessing and improving the potential contribution of CA practices to sustainable smallholder agriculture, particularly in semiarid areas of East Africa in the context of climate change, soil restoration, gender equity, and agricultural productivity.
7. It is apparent from this chapter that CA has increased the yield of most crops particularly in the fragile ecosystems. It can be concluded that CA for enhanced crop production is the best model that can be used to promote small-scale farming on fragile ecosystems in East Africa.

There are some considerations of interest in this review.

In fact, many of the indigenous farming systems in East Africa are very close to what "official CA" wants to see promoted, for example, the Kihamba farming system of the Chagga tribe on slopes of Mount Kilimanjaro in Tanzania and the tied ridges systems of the Makonde/Rondo plateau in southeast Tanzania. Alongside these indigenous CA, there is good scope to promote CA as long as it is packaged in a proper way; that is, to adjust the technology to some already preexisting indigenous practices. In Ethiopia, this was done by adjusting the local *Maresha* plough to shape the land into permanent beds and furrows. A big issue coming up after introducing CA on farmers' fields is a lowering of the crop yields during the first 3–4 years. Afterward, as the soil and the system are adjusting to the new situation of reduced till, the yields increase and a win–win situation develops. Surprisingly, in Ethiopia, this yield dip after the introduction of CA was very small if not nonexistent. This observation was not so much the case either in Tanzania. This would be a very important thing to discuss and argue about.

A last thing that comes out from CA, from CIMMYT (International Maize and Wheat Improvement Center) research and also some long-term trials at the

International Institute of Tropical Agriculture in Nigeria, is the stability of the crop yields in extreme climatological years, e.g., poor distribution of rain or poor rains. This aspect of CA makes it particularly suitable for East Africa in view of its erratic rains in most growing seasons, especially in the semiarid parts of the region. This aspect is also an important consideration when it comes to climate change mitigation: under a climate change scenario, the rains are expected to become even more erratic.

ACKNOWLEDGMENTS

This account developed through the years from many discussions with colleagues from Sokoine University of Agriculture (SUA), Tanzania; Katholieke Universiteit Leuven (KU Leuven), Belgium; Ministry of Agriculture, Food Security and Cooperatives (MAFC), Kawanda Agricultural Research Institute (KARI), Uganda Kenya Agricultural Research Institute (KARI), Kenya; and the villagers and farmers of Uluguru and Usambara mountains, Tanzania, where the authors conducted long-term research on indigenous soil and water conservation using principles of conservation agriculture. It is our pleasure to acknowledge these people. Some of the materials reported in this chapter are collected from previous works initially funded by various projects of which the authors would like to acknowledge: Promoting Sustainable Natural Resource Management through Effective Governance and Farmer–Market Linkages Project funded by the European Union through the Association for Strengthening Agricultural Research in Eastern and Central Africa (ASARECA); the VLIR-UOS funded projects including the North South–South Cooperation Programme (Conservation Agriculture for Restored Environment—CARE VLIR NSSCP) project, the RIP-DSS Project on "Enhancing Indigenous Knowledge on Conservation Agriculture for Poverty Alleviation and Sustainable Livelihood, Usambara Mountains, Lushoto, Tanzania"; and the Soil and Water Conservation on the Uluguru Mountains Research Project under SUA-Flemish Inter-University Council (SUA–VLIR Collaboration Programme).

REFERENCES

Ahmad, R., and J.P. McCalpin. 1999. *Landslide Susceptibility Maps for the Kingston Metropolitan Area, Jamaica with Notes on Their Use*. Unit for disaster studies (UDS) and Caribbean Mitigation Projects (CDMP). University of West Indies, Mona, Kingston, Jamaica, UDS Publication No. 5, p. 24.

Araya, T., W. Cornelis, J. Nyssen et al. 2012. Medium-term effects of conservation agriculture based cropping systems for sustainable soil and water management and crop productivity in the Ethiopian highlands. *Field Crops Research* 132(2012): 53–62 (IF: 2.232).

Aziz, I., M. Ashraf, T. Mahmood, and K.R. Islam. 2011. Crop rotation impact on soil quality. *Pakistani Journal of Botany* 43(2): 949–960.

Bationo, A., U. Mokwunye, P.L.G. Vlek, S.B.I. Koala, and Shapiro. 2003. Soil fertility management for sustainable land use in the West African Sudano–Sahelian zone. In: M.P. Gichuri, A. Bationo, M.A. Bekunda, H.C. Goma, P.L. Mafongoya, D.N. Mugendi, H.K. Murwuira, S.M. Nandwa, P. Nyathi, and M.J. Swift (Eds.), *Soil Fertility Management in Africa: A Regional Perspective*. Academy Science Publisher & Tropical Soil Biology and Fertility, Nairobi, Kenya, pp. 253–292.

Bekunda, M.A., A. Bationo, and H. Ssali. 1997. Soil fertility management in Africa. A review of selected research trials. In: R.J. Buresh, P.A. Sanchez, and F. Calhoun (Eds.), *Replenishing Soil Fertility in Africa*. SSSA spec. Publ. 51 SSA, Madison WI, pp. 63–79.

Bolliger, A., J. Magid, A.T.J. Carneiro, N.F. Skorra, A. Calegari, and R. Ralisch. 2006. Taking stock of the Brazilian 'zero-till revolution': A review of landmark research and farmer's practice. *Advances in Agronomy* 91: 48–110.

Cobo, J.G., G. Dercon, and G. Cadisch. 2010. Nutrient balances in African land use systems across different spatial scales: A review of approaches, challenges and progress. *Agriculture, Ecosystems and Environment* 136: 1–15.

Commission of the European Communities Brussels. 1996. Strategy paper for reducing methane emissions. *Communication from the Commission to the Council and to the European Parliament* COM(96): 557 final.

De Meyer, A., J. Poesen, M. Isabirye, J. Deckers, and D. Raes. 2011. Soil erosion rates in tropical villages: A case study from Lake Victoria Basin, Uganda. *Catena* 84(3): 89–98.

Dercon, G. 2001. Tillage erosion assessment in the Austro Ecuatoriano. PhD thesis, Catholic University of Leuven, Belgium, p. 212.

Derpsch, R. (Ed.). 2005. The extent of conservation agriculture adoption worldwide: Implications and impact. *Proceedings of the III World Congress on Conservation Agriculture*, Nairobi, Kenya, October 3–7, 2005, 28 pp.

Ecological Society of America (ESA). 2000. Carbon sequestration in soils. Available at http://www.esa.org/education_diversity/pdfDocs/carbonsequestrationinsoils.pdf. Accessed May 13, 2011.

Fabrizzi, K.P., F.O. Garcia, J.L. Costa, and L.I. Picone. 2005. Soil water dynamics, physical properties and corn and water response to minimum and no-tillage systems in the southern Pampas of Argentina. *Soil and Tillage Research* 81: 57–69.

FAO. 1999. Uganda. Soil Fertility Initiative. Draft Concept Paper. Report No: 99/024 CP-UGA. Food and Agriculture Organization of the United Nations, Rome.

FAO. 2006a. Efficiency of soil and fertilizer phosphorus use: Reconciling changing concepts of soil phosphorus behaviour with agronomic information. *Fertilizer and Plant Nutrition Bulletin* no. 18. Rome, Italy.

FAO. 2006b. Conservation Agriculture for SARD and Food Security in Southern and Eastern Africa (Kenya and Tanzania). AG: GCP/RAF/390/GER (KEN/URT) Terminal Report, Food and Agriculture Organization of the United Nations, Rome, p. 27.

FAO. 2010. *Greenhouse Gas Emissions from the Dairy Sector: A Life Cycle Assessment*. Food and Agriculture Organization of the United Nations Animal Production and Health Division, Rome.

FARA. 2006. *Framework for African Agricultural Productivity (FAAP)*. Forum for Agricultural Research in Africa (FARA), Accra, Ghana, 72 pp.

Gebregziabher, T., J. Nyssen, B. Govaerts et al. 2009. Contour furrows for *in situ* soil and water conservation, Tigray, Northern Ethiopia. *Soils and Tillage Research* 103(2): 257–264 (IP: 1.846).

Gerlach, T. 1967. Hillslope troughs for measuring sediment movement. *Revised Geomorphology Dynamics* 42: 197.

Glendinning, J.S. 2000. *Australian Soil Fertility Manual*. CSIRO Publishing, Collingwood, 49 pp.

Govaerts, B., K.D. Sayre, and J. Deckers. 2006. A minimum data set for soil quality assessment of wheat and maize cropping in the highlands of Mexico. *Soil and Tillage Research* 87: 163–174 (IF: 1.846).

Govaerts, B., K.D. Sayre, K. Lichter, L. Dendooven, and J. Deckers. 2007. Influence of permanent raised bed planting and residue management on physical and chemical soil quality in rain fed maize/wheat systems. *Plant and Soil* 291: 39–54 (IF: 1.821).

Govers, G., T. Quine, P. Desmet, and D. Walling. 1996. The relative contribution of soil tillage and overland flow erosion to soil redistribution on agricultural land. *Earth Surface Processes and Landforms* 21: 929–946.

Govers, G., K. Vandaele, P. Desmet, J. Poesen, and K. Bunte. 1994. The role of tillage in soil redistribution on hillslopes. *European Journal of Soil Science* 45: 469–478.

Guiresse, M., and J.C. Revel. 1995. Erosion due to cultivation of calcareous clay soils on the hillsides of southwest France: II Effect of ploughing down the steepest slope. *Soil and Tillage Research* 35: 157–166.

Hobbs, P.R., and R.K. Gupta. 2004. Problems and challenges of no-till farming for the rice–wheat systems of the Indo-Gangetic Plains in South Asia. In: R. Lal et al. (Eds.), *Sustainable Agriculture and the Rice–Wheat System*. Marcel Dekker Inc., Ohio, pp. 101–119.

IFAD/FAO. 2004. *Conservation Agriculture as a Labour Saving Practice for Vulnerable Households*. FAO, Rome, 73 pp.

IIRR and ACT. 2005. *Conservation Agriculture: A Manual for Farmers and Extension Workers in Africa*. International Institute of Rural Reconstruction, Nairobi, 45 pp.

Isabirye, M., D. Kimaro, and O. Semalulu. 2010. Sediment production from settlements and farmlands within Lake Victoria shoreline zone in Uganda and Tanzania. *Tropicultura* 28(2): 89–95.

Isabirye, M., G. Ruysschaert, L. Van Linden, J. Poesen, M.K. Magunda, and J. Deckers. 2007. Soil losses due to cassava and sweet potato harvesting: A case study from low input traditional agriculture. *Soil and Tillage Research* 92: 96–103.

Kabanza, A. 2013. Insights into soil and water conservation measures for major land utilization types in South Eastern Tanzania. PhD thesis, KU Leuven, Belgium, p. 166.

Kabanza, A., S. Dondeyne, J. Tenga et al. 2013b. More people, more trees in South Eastern Tanzania: Local and global drivers of land-use/cover changes. *African Geographical Review* 32(1): 44–58. doi:10.1080/19376812.2012.746093.

Kabanza, A.K., S. Dondeyne, D.N. Kimaro, E. Kafiriti, J. Poesen, and J. Deckers. 2013a. Effectiveness of soil conservation measures in two contrasting landscape units of South Eastern Tanzania. *Zeischrift für Geomorfologie* 57(3): 269–288. doi:10.1127/0372-8854/2013/0057-0102.

Kabba, B.S., and M.S. Aulakh. 2004. Climatic conditions and crop-residue quality differentially affect N, P and S mineralization in soils with contrasting P status. *Journal of Plant Nutrition and Soil Science* 167(5): 596–601.

Kajiru, G.J., J.P. Mrema, F.B. Rwehumbiza, N. Hatibu, and H.F. Mahoo. 2006. Evaluation of the fertility status of the soils of the Ndala River Catchment, Maswa District, Tanzania for rice production under rainwater harvesting systems. In: B.M. Msanya, D.N. Kimaro, M. Kilasara, J.P. Mrema, and A.K. Kaaya (Eds.), *Land Resources Management to Enhance Livelihood of Land Users in East Africa. Reviewed Proceedings of the 22nd Conference*, held at New Safari Hotel, Arusha, Tanzania, November 29–December 3, 2004.

Kiburys, B. 1995. Mechanical soil-erosion caused by tillage on 5°, 10° and 15° slopes. *Zemes Ukio Mokslai* 4: 10–13.

Kimaro, D.N. 2003. Assessment of major forms of soil erosion in the Morningside catchment, Uluguru Mountains, Tanzania. PhD thesis, Sokoine University of Agriculture, Morogoro, Tanzania, p. 264.

Kimaro, D.N., J. Poesen, J.A. Deckers, and H.B. Msita. 2010. Effects of agroecological conditions and slope gradient on tillage erosion in the northern slopes of the Uluguru Mountains, Tanzania, Chapter 9. In: E.T. Nardali (Ed.), *No-Till Farming: Effects on Soil, Pros and Cons and Potential*. Agriculture Issues and Policies Series. Nova Science Publishers Inc., New York, pp. 159–172. ISBN: 978-1-60741-402-5.

Kimaro, D.N., J. Poesen, B.M. Msanya, and J.A. Deckers. 2008. Magnitude of soil erosion on the northern slope of the Uluguru Mountains, Tanzania: Interrill and rill erosion. *Catena* 75: 38–44.

Lal, R. 2001. Soil degradation by erosion. *Land Degradation and Development* 12: 519–539.

Larsen, M.C., and A.J. Torres-Sanches. 1998. The frequency and distribution of recent landslides in three montane tropical regions in of Puerto Rico. *Geomorphology* 24: 309–331.

Larsson, M. 1989. Landslides in the mountain areas of Kenya: Comparative studies of different slopes within the Nyandarua Range. In: D.B. Thomas, B.K. Biamah, A.M. Kilewe, L. Lundgren, and B.O. Muchoge (Eds.), *Proceedings of the Third National Workshop*. Department of Agricultural Engineering, University of Nairobi and Swedish International Development Authority (SIDA) Nairobi, Kenya, September 16–19, 1986, pp. 123–136.

Lindstrom, M.J., W.W. Nelson, and T.E. Schumachter. 1992. Quantifying tillage erosion rates due to mouldboard plowing. *Soil and Tillage Research* 24: 243–255.

Liu, Y., Y. Tao, K.Y. Wan et al. 2012. Runoff and nutrient losses in citrus orchards on sloping land subjected to different surface mulching practices in the Danjiangkou Reservoir area of China. *Agricultural Water Management* 110: 34–40.

Löfstrand, F. 2005. Conservation agriculture in Babati district, Tanzania: Impacts of Conservation agriculture for small-scale farmers and methods for increasing soil fertility. Master of science thesis (thesis 145:2005), Swedish University of Agricultural Science, Department of Soil Sciences, Section of Plant Nutrition and Soil Fertility, Uppsala, Sweden, p. 83.

Lopez, J.H., and J.A. Zinck. 1991. GIS-assisted modelling of soil-induced mass movement hazards: A case study of the upper Coello river basin, Tolima, Colombia. *ITC Journal* 4: 202–220.

Lundgren, L. 1980. Comparison of surface runoff and soil loss from runoff plots in forest and small-scale agriculture in the Usambara Mts., Tanzania. *Geografiska Annaler: Series A, Physical Geography* 62(3/4): 113–148.

Lynam, J.K.S., S.M. Nandwa, and E.M.A. Smaling. 1998. Editorial. *Agriculture, Ecosystems & Environment* 71: 1–4. doi:10.1016/S0167-8809(98)00127-3.

Matowo, P., G. Ayaga, M. Tenywa et al. 2011. Efficient Use of Crop Residues: Animal Feed versus Conservation Agriculture. Project Technical Report. ASARECA Project No. NRM/09/05. Selian Agricultural Research Institute, Ministry of Agriculture and Food Security, Arusha, Tanzania.

Mburu, D.M. 2001. Mass movement in agricultural land in the eastern slopes of Nyandarua Mountain ranges (Kenya) and the effect on agricultural production. Paper presented at the *19th Annual Conference of the Soil Science Society of East Africa*, Moshi, Tanzania, December 2–7, 2001, p. 15.

Milliman, J.D., and R.H. Meade. 1983. World-wide delivery of river sediment to the oceans. *Journal of Geology* 91: 751–762.

Moberg, J.P., B.M. Msanya, and M. Kilasara. 1982. Pedological and edaphological properties of the soils of two soil associations in the Morogoro area, Tanzania. *Tropical Agriculture (Trinidad)* 59(2): 139–148.

Msita, H.B. 2013. Insights into indigenous soil and water conservation technologies in Western Usambara Mountains, Tanzania. PhD thesis. KU Leuven, Belgium, p. 194.

Msita, H.B., D.N. Kimaro, J. Deckers, and J. Poesen. 2010a. Identification and assessment of indigenous soil erosion control measures in the Usambara Mountains, Tanzania, Chapter 3. In: E.T. Nardali (Ed.), *No-Till Farming: Effects on Soil, Pros and Cons and Potential*. Agriculture Issues and Policies Series. Nova Science Publishers Inc., New York, pp. 49–74. ISBN: 978-1-60741-402-5.

Msita, H.B., P.W. Mtakwa, M. Kilasara et al. 2010b. Effect of conservational tillage on soil loss and plant nutrient status on vegetable yield, northern slopes of Uluguru Mountains, Morogoro, Tanzania. In: J.J. Msaky, G.N. Kanyama-Phiri, and G.N. Shongwe (Eds.),

Enhancing Dissemination of Soil and Water Research Options of SADC Universities. Proceedings of the Workshop on Information Sharing among Soil and Water Management Experts from SADC Universities, Dar es Salaam, Tanzania, September 11–13, 2010.

Msita, H.B., D.N. Kimaro, P.W. Mtakwa et al. 2012. Effectiveness of miraba an indigenous soil and water conservation measures on reducing runoff and soil loss in arable land of western Usambara Mountains. Paper presented to the EGU General Assembly 2012, in Vienna Austria, 22nd–27th April 2012.

Mugwe, J.N., D.N. Mugendi, K. Kaizzi, and S. Ikerra. 2011. Accelerated Uptake and Utilization of Soil Fertility Management Best-Bets Practices in Eastern and Central Africa Sub-Region Project submitted to ASARECA. Final technical report. Kenyatta University, Nairobi, Kenya.

Nyssen, J. 2001. Erosion processes and soil conservation in a tropical mountain catchment under threat of anthropogenic desertification—A case study from northern Ethiopia. PhD thesis, Catholic University of Leuven, Belgium, p. 379.

Nyssen, J., J. Poesen, H. Mitiku, J. Moeyersons, and J. Deckers. 2000. Tillage erosion on slopes with soil conservation structures in the Ethiopian Highlands. *Soil and Tillage Research* 57: 115–127.

Oldeman, L.R., R.T.A. Hakkeling, and W.G. Sombroek. 1990. *World Map of the Status of Human-Induced Soil Degradation. An Explanatory Note.* Global Assessment of Soil Degradation (GLASOD). International Soil Reference and Information Centre, United Nations Environment Programme, in cooperation with Winand Staring Centre, International Society of Soil Science, Food and Agriculture Organization of the United Nations, and International Institute for Aerospace Survey and Earth Sciences. ISRIC, Wageningen, the Netherlands and UNEP, Nairobi, Kenya, pp. 1–27.

Owenya, M.Z., W.L. Mariki, J. Kienzle, T. Friedrich, and A. Kassam. 2011. Conservation agriculture (CA) in Tanzania: The case of the Mwangaza B CA farmer field school (FFS), Rhotia Village, Karatu District, Arusha. *International Journal of Agricultural Sustainability* 9(1): 145–152.

Peel, D.M. 1998. Crop rotations for increased productivity. Available at http://www.ag.ndsu.edu/pubs/plantisci/crops/eb48-.htm. Accessed October 12, 2010.

Pfeiffer, R. 1990. Sustainable agriculture in practice—The production potential and environmental effects of macro-contour lines in west Usambara Mountains, Tanzania. PhD dissertation, University of Stuttgart-Hohenheim, West Germany, p. 195.

Poesen, J., F. Turkelboom, I. Ohler, A.S. Somchai Ongprasert, and K. Vlassak. 2000. Tillage erosion in northern Thailand: Intensities and implications. *Mededelingen Koninklijke Academie Voor Overzeese Wetenschappen* 46(4): 489–512.

Poesen, J.W.A., G. Verstraeten, R. Soenens, and L. Seynaeve. 2001. Soil losses due to harvesting of chicory roots and sugar beet: An underrated geomorphic process? *Catena* 43(1): 35–47.

Roldan, A., F. Caravaca, M.T. Hernandez et al. 2003. No-tillage, crop residue additions, legume cropping effects on soil quality characteristics under maize in Patzcuaro watershed (Mexico). *Soil and Tillage Research* 72: 65–73.

Ruysschaert, G., J. Poesen, G. Verstraeten, and G. Govers 2004. Soil loss due to crop harvesting: Significance and determining factors. *Progress in Physical Geography* 28(4): 467–501.

Ruysschaert G., J. Poesen, G. Verstraeten, and G. Govers. 2007. Soil loss due to harvesting of various crop types in contrasting agro-ecological environments. *Journal of Agriculture, Ecosystems and Environment* 120 (2–4): 153–165.

Sanchez, P.A. 2002. Soil fertility and hunger in Africa. *Science* 295: 2019–2020.

Sanchez, P.A., K.D. Sheperd, M.J. Soule et al. 1997. Soil Fertility replenishment in Africa: An investment in natural resource capital. In: R.J. Buresh, A. Sanchez, and F. Calhoum (Eds.), *Replenishing Soil Fertility in Africa*. SSSA spec. Publ. 51. SSSA, Madison, WI, USA, pp. 1–46.

Semalulu, O., D. Kimaro, G. Ayaga, V. Kasenge, and C. Gumisiriza. 2011a. Promoting Sustainable Natural Resource Management through Effective Governance and Farmer–Market Linkages. Project Final Technical Report. ASARECA.

Semalulu, O., D. Kimaro, G. Ayaga, V. Kasenge, and C. Gumisiriza. 2011b. Promoting sustainable natural resource management through effective governance and farmer–market linkages. Project achievements and lessons learnt. Paper presented at the *ASARECA 1st General Assembly*, Entebbe, Uganda, December 14–16, 2011.

Semalulu, O., D. Kimaro, V. Kasenge, M. Isabirye, and P. Makhosi. 2012. Soil and nutrient losses in banana-based cropping systems of the Mount Elgon hillsides of Uganda: Economic implications. *International Journal of Agricultural Sciences* 2(9): 256–262.

Shetto, R., and M. Owenya (Eds.). 2007. *Conservation Agriculture as Practised in Tanzania: Three Case Studies*. African Conservation Tillage Network, Centre de Coopération Internationale de Recherche Agronomique pour le Développement, Food and Agriculture Organization of the United Nations, Nairobi.

Shetto, R.M. 1999. Indigenous soil conservation tillage systems and risks of animal traction on land degradation in Eastern and Southern Africa. In: P.G. Kaumbutho, and T.E. Simalenga (Eds.), *Conservation Tillage with Animal Traction*. A resource book of Animal Traction Network for Eastern and Southern Africa (ATNESA), Harare, Zimbabwe. French Cooperation, Namibia. Available at http://www.fao.org/ag/ags/agse/agse_s/3ero/namibia1/til_nam.htm#t. Accessed September 23, 2007.

Stoorvogel, J.J., and E.M.A. Smaling. 1990. Assessment of Soil Nutrient Depletion in Sub-Saharan Africa: 1983–2000. Rep. no. 28, Vols. 1–4. Winand Staring Ctr., Wageningen, the Netherlands.

Temple, P.H. 1972. Soil and water conservation policies in the Uluguru Mountains, Tanzania. *Geografiska Annaler* 54A(3–4): 110–123.

Temple, P.H., and A. Rapp. 1972. Landslides in the Mgeta area, Western Uluguru Mountains, Tanzania. *Geografiska Annaler* 54A(3–4): 157–193.

Tenge, A.J., J. de Graaff, and J.P. Hella. 2003. Social and economic factors affecting the adoption of soil and water conservation in West Usambara Highlands, Tanzania. *Land Degradation and Development* 15: 99–114.

Tittonell, P., M. van Wijk, M. Herrero, M.C. Rufino, N. de Ridder, and K.E. Giller. 2008. *Inefficiencies and Resource Constraints—Exploring the Physical Feasibility of Options for the Intensification of Smallholder Crop–Livestock Systems in Vihiga District, Kenya*. Agriculture Systems, Nairobi, 33 pp.

Triomphe, B., J. Kienzle, M. Bwalya, and S. Damgaard-Larsen (Eds.). 2007. Conservation Agriculture in Africa series. In: P. Boahen, B. Addo Dartey, G. Delali Dogbe, E. Asare Boadi, B. Triomphe, S. Daamgard-Larsen, and J. Ashburner (Eds.), *Conservation Agriculture as Practised in Ghana*. African Conservation Tillage Network (ACT), Nairobi, pp. 11–20.

Turkelboom, F., J. Poesen, I. Ohler, and S. Ongprasert. 1999. Reassessment of tillage erosion rates by manual tillage on steep slopes in northern Thailand. *Catena* 29: 29–44.

Vanlauwe, B., J. Chianu, K.E. Giller et al. 2010. Integrated soil fertility management: Operational definition and consequences for implementation and dissemination. *19th World Congress of Soil Science, Soil Solutions for a Changing World*, Brisbane, Australia, August 1–6, 2010. Published on DVD.

Vanlauwe, B., J. Diels, N. Sanginga, and R. Merckx. 2002. *Integrated Plant Nutrient Management in Sub Saharan Africa. From Concept to Practice*. CABI, International, Wallingford, UK, 352 pp.

Vlek, P.L.G. 1993. Strategies for sustaining agriculture in sub-Saharan Africa. In: J. Rogland, and R. Lal (Eds.), *Technologies for Sustaining Agriculture in the Tropics*. ASA Spec. Publ. 56. ASA, CSSA, and SSSA, Madison, WI, pp. 265–277.

Walaga, C., B. Egulu, M. Bekunda, and P. Ebanyat. 2000. Impact of policy change on soil fertility management in Uganda. In: T. Hilhorst, and F.M. Muchena (Eds.), *Nutrients on the Move—Soil Fertility Dynamics in Africa Farming Systems*. International Institute for Environment and Development, London.

Westerberg, L.O. 1999. Mass movements in East Africa highlands: Processes, effects and scar recovery. Doctoral dissertation No. 14, The Department of Physical Geography, Stockholm University, Sweden, p. 147.

Westerberg, L.O., and C. Christiansson. 1999. Highlands in East Africa: Unstable slopes, unstable environments? *Ambio* 28(5): 419–429.

World Bank. 2011. Fertilizer consumption (kilograms per hectare of arable land). Available at http://data.worldbank.org/indicator/AG.CON.FERT.ZS.

Wortmann, C.S., and C.K. Kaizzi. 1998. Nutrient balances and expected effects of alternative practices in farming systems of Uganda. *Agriculture, Ecosystems and Environment* 71: 115–129.

Xiao, P.Q., W.Y. Yao, and C.G. Wang. 2011. Soil erosion process in sloped shrub plots under simulated rainfall. *Advanced Materials Research* 347–353: 2094.

Yusuf, A.A., E.N.O. Iwuafor, R.C. Abaidoo, O.O. Olufajo, and N. Sanginga. 2009. Effect of crop rotation and nitrogen fertilization on yield and nitrogen efficiency in maize in the northern Guinea savanna of Nigeria. *African Journal of Agricultural Research* 4(10): 913–921.

Zachar, D. 1982. *Soil Erosion Development in Soil Science*. Elsevier Scientific Publishing Company, Amsterdam, p. 547.

13 Community, Climate Change, and Sustainable Intensification

Why Gender Is Important

Cornelia B. Flora

CONTENTS

13.1 INTRODUCTION

Adaptation to climate change is greatly enhanced when it occurs across a landscape. Agricultural and herding communities are important loci of adaptation. Women's knowledge and practices can help in determining new community livelihood strategies. Community adaptation is hindered when land grabs limit access to land for crops and animals. New forms of public–private partnerships that are often designated as Foreign Direct Investment further complicate adaptation. Using the Community Capitals Framework, community-based adaptation strategies that take gender into account are identified.

Small producers, their livelihoods, and the land associated with both are threatened by changes in climate. They are among the vulnerable of populations in that their natural and social systems are susceptible to damage from climate change (Intergovernmental Panel on Climate Change [IPCC] 2012). Their adaptive capacity, which is the degree to which adjustments in practices, processes, or structures can moderate or offset the potential for damage or take advantage of opportunities created by a given change in climate (IPCC 2012), is often limited by the political economy that determines the access to and control of the land.

Social science distinguishes two kinds of responses to climate change: adaptation and coping. Adaptation involves anticipating, planning, and acting to decrease vulnerability to climate change. Coping is the use of available skills, resources, and opportunities to address, manage, and overcome adverse conditions with the aim of achieving basic functioning in the short to medium terms. In some cases, coping can increase long-term vulnerability.

While individuals cope or adapt to climate changes, community adaptation is often more effective in preserving soils and the dignity of the small producers. Community adaptation is most effective when women are included, particularly in Africa where women produce subsistence food crops and small animals for local markets (Gladwin 2002).

Structural adjustment in the 1980s and 1990s had major negative impacts on farming communities, particularly on women (Due and Gladwin 1991; Meena 1991; MacKenzie 1993). Governments dismantled and privatized extension systems that distributed agricultural inputs and technical assistance. As government spending on health and education decreased, fees for these services increased dramatically. Wage freezes meant a large reduction in remittances from fathers and sons with urban employment. Farming systems designed for subsidized inputs were no longer affordable (Hassan et al. 2013; Tchingulou et al. 2013). National emphasis on export crop production to generate foreign exchange often led to monocultures and increasing land concentration (Gladwin 1991).

The 21st century brought a new development model, where public–private partnerships brought new sources of foreign exchange to governments in the form of Foreign Direct Investment (FDI). Under the right circumstances, when FDI invests in roads, ports, railways, electricity, and water, for example, it can contribute to international integrations and encourage transfer of technology between countries, and under the right policy environment, FDI can be an important vehicle for development (Organisation for Economic Co-operation and Development [OECD] 2012a). However, land-based investment often is purely extractive. Most FDI studies look at it as a totality, and their analyses combine FDI in industry and infrastructure with FDI in land, finding that it is not extremely negative (Xu and Sheng 2011). Thus, it is difficult to systematically address its impacts in different contexts.

In many parts of the world, the government holds title to communal lands, which is a primary resource for smallholders and pastoralists. When this land is leased to foreign interests (often with a national partner within or close to the national government), the money goes to the government, and the new investors can extract what they please from the land and people. While some countries have used portions of these payments to provide stipends to the poor, such poverty reduction does not produce local development nor is it sustainable once the resource of land, minerals, and/or water are depleted.

De Schutter (2011) is extremely cautious about FDI in land. He points out that giving land away to investors who have better access to capital to "develop" implies huge opportunity costs. As a result, a type of farming is brought in that has much less powerful poverty-reducing impacts than if access to land and water were improved for the local farming communities. This type of FDI farming directs agriculture toward crops for export markets, increasing the vulnerability to price shocks of the target countries. Even where titling schemes seek to protect land users from eviction, individual titling accelerates the development of a market for land rights with potentially destructive effects on the livelihoods of the current land users that will face increased commercial pressure on land and on groups that depend on the commons—grazing and fishing grounds and forests.

Africa is faced with continuing mean temperature increases. Rising temperatures impact the water cycle, with more extreme storms and longer periods between rains. While total precipitation remains the same, local men and women perceive that there is a decline in rainfall. Increases in temperature result in more rapid rates of evapotranspiration, which makes the water seem scarcer despite systematic measurements showing no change in total precipitation (Kassie et al. 2013). Other significant extreme weather events include droughts, floods, freezes, and hail, all of which can damage crops, animals, and soils. They also result in an increased number of bacteria and pests. These changes are particularly felt by smallholders and pastoralists. Moreover, in these groups, the most vulnerable are women and children.

13.2 COMMUNITY CAPITALS AND CLIMATE CHANGE

The Community Capitals Framework (CCF) is a useful tool for analyzing the context and process of social change, and has been used in the context of development in the United States and Latin America (Cepeda Gómez et al. 2008; Gasteyer and Araj 2009; Ashwill et al. 2011; Flora and Bregendahl 2012; Flora and Delaney 2012; Flora and Flora 2013; Pigg et al. 2013; Siles et al. 2013). The CCF defines "capital" as resources invested to create new resources over a long time frame. It was developed to show that many resources are needed to achieve a healthy ecosystem, social inclusion, and economic security for all people (see Figure 13.1).

Capitals in this framework are viewed as collective resources, not just individual property. The capitals are shown in a particular order, with Natural Capital, the natural environment, being the first and the basis for all the others. Next is Cultural Capital, which is the belief system that links people to their environment and each other by connecting the seen and the unseen. Human Capital encompasses the attributes of human beings, including education, abilities, health, and self-esteem. These attributes are determined by structural factors, but are part of each human actor. Social Capital includes the relationships that link people to each other through groups. Social capital can be ties of trust and reciprocity within a group or

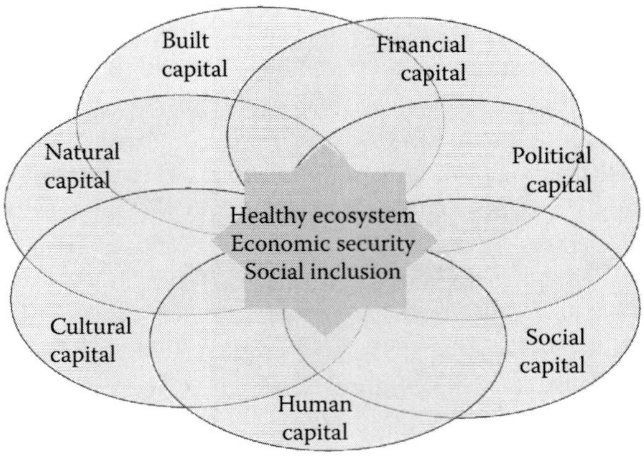

FIGURE 13.1 Community capitals diagram.

community (bonding social capital) and with other groups and communities (bridging social capital). Political Capital is the ability of a group to turn its values and norms into the standards for the society (such as universal education and access to clean water) that are implemented through enforced rules and regulations. It involves voice and power. Financial Capital includes the financial instruments, including but not limited to money, that can be easily traded and monetized. There is a tendency to put all the other capitals in terms of financial capital. While the role of the state (political capital) under capitalism is to make it profitable to do what is for the common good, increasingly anything that is profitable, even if that profit is only for a few, is assumed to be for the common good. Built Capital is the physical infrastructure, buildings, roads, dams, sewer systems, railroads, and electronic technology that can make other capitals more efficient—or can destroy them.

13.2.1 Natural Capital

Natural capital includes air quality and the quality and quantity of water and of soil, biodiversity, and landscape. It can be viewed as a set of resources to be extracted or as a source of life that needs to be tended and cared for, depending on a group's cultural capital.

Soil security is key to food security, and community action is often the best way to increase both through sustainable intensification. For many agriculturalists, the community is where the norms and values, as well as the appropriate cultural and agricultural practices, are embedded. Communities can create healthy ecosystems, economic security, and social inclusion by investing their multiple resources to create new resources, or they can exhaust those resources by consuming them for immediate benefit. When resources are invested to create new resources, they can be referred to as *capitals*. Soil security, and thus food security, depends on how each community invests its natural, cultural, human, social, political, financial, and built capitals. Women are critical in achieving soil security, food security, and sustainable intensification (Gladwin et al. 2002).

Climate change creates new environmental conditions that stress all community resources, although the most obvious is the impact on natural capital. These stresses are felt unequally across communities and within communities, based on access and control of resources. Both gender and ethnicity can limit access to and control of different community capital stocks, making it particularly difficult for women and ethnic minorities to adapt to the climate-induced changes.

Understanding that context and how different actors in the community truly understand the soil as a resource and feel capable of improving it is critically important if sustainable intensification is to occur. This is particularly so when confronted with climate change. A gender-aware perspective that acknowledges the influences of social relations, cultures, beliefs, values, and attitudes on our understanding, experience, and perceptions of the risks of climate change could contribute to a more balanced, nuanced approach to the problem that can consider these multiple understandings (Skinner 2011).

Women in vulnerable communities adapt to changes in the environment caused by climate and other forces. Many need to venture further to get wood and water. They save seed from diverse sources, plant different varieties or species depending on when it rains, have complex and diverse agricultural systems to minimize risk, access a variety of wild foods, closely observe changes in flora and fauna, and adapt their livelihood strategies to them. Access to and control of natural capital is often contested (Allen 2003).

Technologies to improve natural capital often work at the community level. Putting in small dikes and putting on more organic matter to retain moisture may lead to the improvement of soil. Kassie et al. (2013) found that to be the case in Ethiopia. Increasing organic matter in the soil increased its water-holding capacity. This was far easier for local communities to understand than an abstract notion of soil organic matter as an end in itself. As natural capital is extremely heterogeneous, attention to place is critical for adaptation strategies.

Women and men in the same community tend to grow different crops; the specific crop varies by place and gender (Doss 2002). Generally, women grow for home consumption and local markets, while men are tied into longer value chains (Gladwin et al. 2001). As pressure on land has increased, the crops that women grow have become more diverse in southern Cameroon. There is more frequent and more intensive cultivation of the women's crops, which increasingly provide family sustenance. That diversification means changes in land use, including amount of lands cleared for cultivation, planting of different crops, different crop densities, and different fallow periods. Gender and changes in the community shape agro-ecological practices (Guyer 1986).

Climate change has an immediate impact on natural resources. Farm communities and individual farmers have a long record of adapting to changes in rainfall and temperature over time (OECD 2012b). However, current changes and the increasing number of extreme events make the use of natural capital even more problematic than before, influencing farm management practices, land use, what is produced, and where it is produced.

Smallholders around the world are turning to indigenous roots and grains for their basic food supply. Colonial regimes introduced exotic grains into soils that had not coevolved with them. Examples are wheat and soybeans into the Americas and corn and soybeans into Africa. These crops have quickly depleted soils of their organic matter and nutrients and have proved ever-more dependent on external inputs, generally purchased at a highly subsidized rate as the national diet shifted to these new crops. Others are turning to locally produced sources of nutrients such as the *Faidherbia* tree, which improves microclimates, soil organic matter, and nutrients, and provides fodder for livestock at no external cost (Garrity 2013). However, for them to be successfully introduced, the community must agree to protect the new shoots from livestock for the trees to generate properly.

13.2.2 Cultural Capital

Cultural capital determines how communities and groups within communities see the world, how they connect the seen and the unseen, what they take for granted, and what they think is possible to change. Cultural capital is often highly determined by and determines natural capital. Like natural capital, cultural capital is heterogeneous and gendered. Women's cultural capital often includes understanding what to plant under what conditions, and where to take animals as pastures become dry. Community culture gives meaning to animals beyond their market value, and food often has specific ritual meanings. Rituals of respect are important for linking community members to each other and to the land. When one group's cultural capital predominates and devalues other ways of seeing and knowing, it is called hegemony.

Cultural capital determines what women are supposed to do and what men are supposed to do. In much of the world, women are to raise crops and animals for household use, while men produce them for sale. Women search for and haul wood and water for household use, while men undertake the same tasks to sell into a larger market (Nombo et al. 2013).

In much of Africa, farming for food is a critical part of gender identity (Gladwin 2002). At one point, this led to gender complementarity between male and female roles; however, that changed with increasing globalization (Flora 1985). Today, men's tasks are often privileged in the distribution of resources within the household, community, and state.

13.2.3 Human Capital

Human capital represents the skills, abilities, and knowledge that each human being possesses in a community. Gender, in part, determines which skills are taught by grandmothers and mothers to daughters and by grandfathers and fathers to sons. Formal education can be limited by whether a child is a boy or a girl, particularly when the daily work of girls in agricultural production or boys in herding keeps them from school. Norms that stem from cultural capital can define what is appropriate for boys and for girls to change, with many communities seeing the value of girls' education. When males seek work by migrating as part of a household strategy to adapt to climate change, women take over many productive activities traditionally performed by males.

In cases of drought, household food consumption is often cut back as a way of coping. This generally means that women wind up eating less while men and children get a little more to eat, although all remain hungry.

Often, bridging social capital can help women's groups in an agricultural community increase human capital by learning skills from other communities that have adapted successfully or from nongovernmental organizations (NGOs) that build individual capacity in the context of the community.

13.2.4 Social Capital

Social capital consists of interactions among people and groups for mutual support. It involves trust, shared norms, reciprocity, and working together. Social capital has two dimensions: bridging and bonding. Bridging social capital is the linking of local groups or institutions to resources and external partners with similar goals, while bonding social capital is the strengthening of internal organization and capacity to take collective action based on the common backgrounds and experiences of the individuals or groups involved. In traditional societies, interactions are often determined by gender and age, which determine what groups are appropriate for which kinds of people. Often, cultural and political capitals limit both their bonding and bridging social capital. Capacity is the ability of individuals and organizations or organizational units to perform functions effectively, efficiently, and sustainably (United Nations Development Programme 1998).

Women's social capital in many traditional societies is bonding, and there are barriers to women forming bridging social capital. Increasingly, women use bonding social capital to form women's associations and mothers' clubs. Through these community groups, they create bridging social capital with technical assistance providers.

Local bridging and bonding networks are key to community adaptation (Skinner 2011). Since women's networks are frequently informal, they are often ignored when outsiders come in to facilitate adaptation. Many times, external projects are subject to elite capture (taken over by those who have most resources), and these elites are almost always male. Women's groups often create internal mechanisms to protect the assets they generate in their groups for family use.

13.2.5 Political Capital

Political capital refers to the codification of community's norms and values into standards that are supported by rules and regulations that are enforced equally. Policy incentives can hurt or hinder adaptation to climate change. Very often, insurance schemes reward producers for doing what they previously have done (as that makes it easier to determine the value of the loss), rather than encouraging innovative adaptations to changing climatic conditions.

Traditional agricultural communities, particularly women in those communities, are not able to get larger political entities to recognize their norms and values, particularly those related to common land use, so they have little ability to influence the distribution of resources at the family, community, or regional level. Women's organizations are often not officially recognized. Women's strategic interests regarding access and control of land, pasture, water, and fuel are often not recognized in official adaptation strategies. Skinner (2011) has documented how women are not equal partners in decision making on climate change responses.

An African example of efforts of women's groups to mitigate climate change through controlling large polluters is found in the Niger Delta Region of Nigeria (Odigie-Emmanuel 2010b). Climate change is often related to increased political conflict among and within communities. Adaptation itself is a political process, as adjustments made to livelihoods have uneven outcomes (Eriksen and Lind 2009). Power relations and their related effects on social interactions within and between communities are often challenged by change (Allen 2003).

Collective decisions regarding local adjustments are hotly contested. In dryland Kenya, Eriksen and Lind (2003) document how alliances are formed between different groups of villagers and communities of pastoralists, politicians, clan elders, chiefs, and groups such as youth and women's community-based organizations. As they build these alliances, they rely on the values of the other capitals to demonstrate their spatial primacy when productive land and water is scarce due to drought. When pastoralists and villagers peacefully negotiate drought adjustments, these are often challenged by entrenched elites who see their control threatened by community-level action.

13.2.6 Financial Capital

Financial capital is often privileged in development schemes that are built around increasing market participation, including export. In many communities, there are men's crops and women's crops. And when they do grow the same crop, they have different value chains. For example, when women gather firewood, it is generally for domestic use. Men gather it to sell.

Because women have less access to financial capital, they are limited in the degree to which they can take advantage of the input-responsive crop varieties and animal species recommended by outside experts seeking to increase productivity (Uttaro 2002). Fortunately, there is research under way in Africa to enhance women's capacity to increase the productivity of traditional varieties and species through mixed cropping and fodder systems.

13.2.7 Built Capital

Built capital refers to technology, infrastructure, tools, and machinery. While an individual can accumulate tools and machinery, collective goods such as roads, water systems, school buildings, and community centers are generally best generated by a community working together. Often, built capital is biased toward the wealthy and toward males, with little investment in infrastructure that reduces women's work for the market or the home.

Of increasing concern are large projects such as monoculture plantations using large machinery and specialized seed, dams, mines, and petroleum exploration that drastically limit the adaptive possibilities for smallholders and herders. Furthermore, the land uses from these types of built capital provide few positive livelihood-strengthening impacts (De Schutter 2011).

13.3 ADAPTATION AND COPING

Globalization, changes in land use, including land grabs, changes in political regimes and increasing inequality in wealth and income, compound climate change. Predictability, never high, becomes even lower. Yet agriculturalists persist, often doing the same thing they have always done because they are not in a position to consider alternatives. There are two responses to climate change: coping and adaptation. Coping tends to mine the soil of carbon, while adaptation has the possibility of renewing it.

13.3.1 Coping

Coping approaches impose agricultural systems adapted to one context on a totally different context. Coping approaches clear land for industrial input crops: palm oil, soybeans, and corn, among others. Land grabs can be seen as a way of responding to declining soil quality and water availability in one area to capture that soil and water, often through crop production and export (Woodhouse 2012). Coping utilizes immediate market responses over long-term ecological responses. These coping approaches disadvantage smallholders, biodiversity, and soil quality.

Coping is generally individual and reactionary and does not involve planning, sometimes leading to actions that can exacerbate community vulnerability to climate change. That is called "erosive coping" (Van der Geest and Dietz 2004; Warner et al. 2012). Warner and colleagues illustrate this in the Budalangi Division in Kenya, where coping in response to a devastating flood caused temporary relocation, which took children out of schools, resulting in the loss of social capital. Households sold their traction animals to buy food, reducing the amount of land they could farm in subsequent years.

13.3.2 Adaptation

Communities and households that adapt see change as constant. They build on internal capitals first to change farming systems and diversify their livelihood strategies. They are mindful of the changes and what is happening with their soil, often by taste or smell or feel. And they seek alliances with market, state, and civil society actors to constantly innovate. Warner et al. (2013) see national governments and national market institutions joining forces to provide resilience-producing insurance schemes to share risk, particularly if they provide *ex ante* mechanisms to reduce vulnerability and prevent risk.

Communities and households that cope view extreme events and warmer weather as temporary. They look to infrastructure and technological solutions (built capital). They keep doing what they have been doing, only more so—use more pesticides, cultivate the land more often, and put marginal land into cash crops. Coping is often supported by state policies that subsidize inputs and offer a high level of crop insurance for a limited number of crops planted by certain times. Even those "copers" who acknowledge that climate change is occurring are convinced that technology will solve the problem—better engineered seeds, better "chemistry" to put on the crop, more extensive use of groundwater for irrigation. Sociologists refer to this as *ecological modernization* (Mol 2001; Horlings and Marsden 2011).

TABLE 13.1
Adaptation Practices Corresponding to Basic Types of Adaptation

	Community Level		Agency Level	
Type of Adaptation	**Individually Undertaken**	**Collectively Undertaken**	**Individual**	**Community**
Mobility	Wage labor migration, remittances (together with market exchange)	Agropastoral migration, wage labor migration, involuntary migration, remittances (together with market exchange)	Relocation expenses to less vulnerable area	Aid to a community to relocate together
Storage	Water storage, food storage, live animal storage	Water storage, food storage (crops, seeds, forest products), animal/live storage, dikes, embankments	Distribution to individuals of containers for water or for seed	Aid in building storage facilities and small reservoirs, crop intensification programs
Diversification	Asset portfolio diversification, skills and occupational training, diversify crop choices, production technologies, consumption choices, agroforestry	Community animal breeding programs, collective enterprises	Offer individual courses	Provide community capacity for new collective enterprises, including agroforestry
Communal pooling		Forestry, infrastructure development, information gathering, disaster preparation	Build water capturing infrastructure for individual households	Support building water-capturing infrastructure
Market exchange	Product sales, sell property, sell labor	Improved market access; seeds, animal, and other input purchases; remittances (in conjunction with mobility)	Insurance provision	Insurance provision, social safety nets, transfer payments
Mobilization	Refusal to move	Collective petition to government		Link to international movements with similar concerns

Source: Adapted from Agrawal, A. and N. Perrin. 2008. Climate adaptation, local institutions, and rural livelihoods. IFRI (International Forestry Resources and Institutions Program) working paper #W08I-6.

13.3.2.1 Types of Adaptation

Adaptation has several dimensions as illustrated in Table 13.1. It can take place at the individual level or the household level, or it can take place at a community level. Adaptation is not only undertaken by individuals, such as agriculturalists State agencies and institutions either cope or adapt as well. And when they adapt it can be through increased assistance to individuals or through working with communities to make innovative adaptation possible. Agrawal and Perrin (2008) have identified five types of adaptation: mobility, storage, diversification, communal pool, and market exchange. To these five, I would add a sixth: political mobilization. Work with agricultural communities suggests that these adaptations can be individual or collective. Furthermore, it is important to understand that outside entities, including governments and NGOs, facilitate or impede implementing a particular adaptation. Often such outside incentives and policy mechanisms are focused on individual producers (generally male) and not the community (Ashwill et al. 2011).

It is imperative that adaptation occurs at more than the individual and community level. Possible avenues for adaptation must include dealing with drought, floods, high temperatures, waterlogging, new and increasing incidences of plant pests and diseases, a shorter growing season, and associated human health concerns such as malaria and sleeping sickness in the Sahel due to wetter conditions favorable to mosquitoes and the tsetse fly (Jalloh et al. 2013).

13.4 MOBILITY

As in many cultures, around Lake Chad in Nigeria men migrate to the city to seek jobs as laborers in response to climate change disruption of their livelihoods. Women stay home because of lack of urban employment opportunities and the need to care for their children and elders (Odigie-Emmanuel 2010a). While this occurs in the community, these are individual decisions and no structures exist for the women who remain to deal with the increased work caused by fewer workers, disrupted agricultural cycles, and longer distances to go for wood and water. In other areas of Africa, young single women without children migrate to nearby countries such as Sudan, Dubai, and Saudi Arabia (Kassie et al. 2013).

Mobility is often forced by outside interventions such as land acquisition by mining, oil companies, and plantation agriculture investors. Even tourism and nature preservation can force communities to leave their ancestral lands. Although the causes cited for migration are not directly climate related, the impacts of migration are often the same. Migration can be hugely disruptive of cultural, human, and social capitals, even if there are short-term gains for a few in terms of financial capital (Chansinga et al. 2013). With some FDI, land is acquired and the people are removed, which is referred to as a new enclosure movement (McMichael 2012). In Tanzania, Kusiluka et al. (2011) found that migration resulted in loss of land, loss of means of livelihood, disruption of economic activities, persistent land-related conflicts, relocations to poorly developed areas, inadequate and late compensation, and environmental degradation. Geographic mobility is often exacerbated by climate change, particularly when the relocation is to arid areas.

13.5 STORAGE

Communities can work together to build stores of basic food to be shared for emergencies, although this more often takes place at a household level. Women's groups within communities often take on the responsibilities for storage of emergency supplies and distributing them when needed.

Water storage is often best done on a community level, as there has to be agreement on provision and access to the collective water source. Water can be stored in the soil when there is good organic matter available. Storage can occur by returning nutrients and organic matter to the soil through such activities as using mulch and compost (Goldman and Heldenbrand 2002), although often not at a level to improve soil fertility.

13.6 DIVERSIFICATION

Women commonly respond to climate change by adapting what is planted to precipitation and temperature changes. This can be done collectively at the community level, where seeds and knowledge is interchanged to increase everyone's potential return, or it can be done individually. Unfortunately, most institutional input focuses on monoculture, rather than diversification. There is often institutional assistance in getting irrigation through boreholes; however, often women have difficulty accessing these new sources of water (Vincent et al. 2010).

One of the goals of structural adjustment was to get smallholders to diversify from farming to wage labor and craft sales, through an emphasis on commodification, specialization, and standardization. This has led to the accumulation of resources in high potential areas with higher and more reliable rainfall on farmers with more resources vs. farmers with fewer resources, and on men rather than women, increasing inequalities (Bernstein 1990).

13.7 COMMUNAL POOLING

Women's groups have mobilized around advocacy and sharing of the domestic tasks of hauling water and wood (community response). However, more often, it is males who mechanize the process and sell the water and wood, with little alleviation of the women's tasks, turning traditional women's work into male market exchange.

13.8 MARKET EXCHANGE

Government transfers to households whose livelihood strategies have been disrupted represent a form of market adaptation. In Limpopo Province, South African women who live with elderly relatives with a monthly social pension manage the insecurities of climate change with cash. Remittances from relatives in the city also help (part of the mobility response). These are agency solutions for individuals (Vincent et al. 2010).

Sharing and transferring risk through insurance can be an adaptation, when it does not reward behavior that ignores climate change realities. Increasingly, insurance policies, some with state underwriting, are offered to individuals to reimburse them for losses due to extreme events such as floods, drought, hurricanes, and hail (Warner et al. 2013). These policies are normally offered to individuals and purchased, as in the Ghana Agricultural Insurance Program.

Community-level insurance schemes are possible but not common. In Nigeria, an index-based livestock insurance program was introduced. Payouts were based on areal statistical models, involving all community pastoralists that had purchased insurance. Most of the smallholders and pastoralists in Africa are uninsured. Ethiopia has instituted the HARITA program, which takes a holistic approach by linking insurance with risk reduction measures (Warner et al. 2013). Such schemes require public–private partnerships between insurance companies and governments. When there are multigovernment or regional agreements, multicountry programs such as African Risk Capacity further share the risks. Women farmers and pastoralists need access to such resources, as well as their male counterparts.

Often in the face of climate change or by pressure from large-scale land buyers, communities with customary land access sell their land—or it is sold for them by the government. The land, previously a crucial asset for the rural poor, is now lost to them. German et al. (2013) document such cases in Ghana, Mozambique, Tanzania, and Zambia. These break up communities and particularly disadvantage women.

13.9 MOBILIZATION

Responses to climate change are limited by access to all the capitals, including natural capital, such as land. A number of social movements have arisen in and across agricultural communities as access to land and water are increasingly constrained (McKeon 2013).

13.10 CONCLUSIONS

Many of the proposed adaptations from international agencies address developing new technologies, educating individuals and changing traditional farming patterns (Jalloh et al. 2013). These strategies would be strengthened by locating them within agricultural communities and taking into account gendered access and control of potential resources within communities. Women are particularly vulnerable to climate change, as their lack of access and control over the resources, particularly political, financial, and built capitals, and their responsibilities for the day-to-day upkeep of their families (human and cultural capital) give them fewer alternatives to adaptation (Arora-Jonsson 2011). Yet, when women organize within their communities and across communities, they can be powerful actors to influence adaptation to climate change (Odigie-Emmanuel 2010b).

Traditional communities, especially the women of these communities, are very vulnerable to climate change and globalization. Women's cultural capital, combined with bridging and bonding social capital, results in adaptation in areas where there has not been a strong presence of global resource extraction. The ability of women to adapt to climate change is greatly reduced where there are areas of large-scale land acquisition, including petroleum extraction, large-scale mining, and industrial agriculture. These activities destabilize communities and greatly marginalize women.

While it is easy to fault vulnerable communities for not adapting to climate change, government and donor programs, combined with market incentives that encourage risky livelihood strategies, provide perverse incentives to vulnerable communities not to adapt, further disadvantaging women. Enhancing and building on women's capitals in the context of their communities is an effective approach to help them adapt to climate change. It can result in increasing soil quality and enhanced agricultural intensification for food security.

REFERENCES

Agrawal, A., and N. Perrin. 2008. Climate adaptation, local institutions, and rural livelihoods. IFRI (International Forestry Resources and Institutions Program) Ann Arbor, MI: Working Paper #W08I-6.

Allen, J. 2003. *Lost Geographies of Power.* Malden, MA: Blackwell.

Arora-Jonsson, S. 2011. Virtue and vulnerability: Discourses on women, gender and climate change. *Global Environmental Change*, 21, 744–751.

Ashwill, M., C.B. Flora, and J.L. Flora. 2011. *Building Community Resilience to Climate Change: Testing the Adaptation Coalition Framework in Latin America.* Washington, DC: World Bank. Available at http://siteresources.worldbank.org/EXTSOCIALDEVELOPMENT/Resources/244362-1232059926563/5747581-1239131985528/Adaptation-Coalition-Framework-Latin-America_web.pdf. Accessed February 3, 2014.

Bernstein, H. 1990. Agricultural "modernization" and the era of structural adjustment: Observations on sub-Saharan Africa. *Journal of Peasant Studies*, 18, 3–35.

Cepeda Gómez, C., I. Gutiérrez-Montes, A. Imbach, F. Alpizar, and N. Windevoxhel. 2008. Tiburón ballena y bienestar comunitario en Holbox, Quintana Roo, México. *Recursos Naturales y Ambiente*, 55, 109–117.

Chansinga, B., M. Chasukwa, and S.P. Zuka. 2013. The political economy of land grabs in Malawi: Investigating the contribution of Limphasa Sugar Corporation to rural development. *Journal of Agricultural and Environmental Ethics*, 26, 1064–1084.

De Schutter, O. 2011. How not to think of land-grabbing: Three critiques of large-scale investments in farmland. *The Journal of Peasant Studies*, 38, 249–279.

Doss, C.R. 2002. Men's crops? Women's crops? The gender patterns in Ghana. *World Development*, 30, 1987–2000.

Due, J.M., and C.H. Gladwin. 1991. Impacts of structural adjustment programs on Africa women farmers and female-headed households. *American Journal of Agricultural Economics*, 73, 1431–1439.

Eriksen, S., and J. Lind. 2009. Adaptation as a political process: Adjusting to drought and conflict in Kenya's drylands. *Environmental Management*, 43, 817–835.

Flora, C.B. 1985. Women and agriculture. *Agriculture and Human Values*, 2, 5–12.

Flora, C.B., and C. Bregendahl. 2012. Collaborative community-supported agriculture: Balancing community capitals for producers and consumers. *International Journal of Sociology of Agriculture and Food*, 19, 329–346.

Flora, C.B., and M. Delaney. 2012. Mobilizing civil society for environmental protection in a context dominated by industrial agriculture: Avoiding projects. In I.S. Sjöblom, K. Andersson, T. Marsden, S. Skerratt (Eds.), *Sustainability and Short-Term Policies: Improving Governance in Spatial Policy Interventions* (pp. 201–213). Burlington, VT: Ashgate.

Flora, C.B., and J.L. Flora. 2013. *Rural Communities: Legacy and Change* (4th Edition). Boulder, CO: Westview Press.

Garrity, D. 2013. Transformational opportunities in regreening to create an evergreen agriculture. Presentation at Climate Change, Sustainable Intensification and Food Security in Sub-Saharan Africa iAGRI Conference, Morogoro, Tanzania, November 13–15.

Gasteyer, S., and T. Araj. 2009. Empowering Palestinian community water management capacity: Understanding the intersection of community cultural, political, social and natural capitals. *Community Development: Journal of the Community Development Society*, 40, 163–184.

German, L., G. Schoneveld, and E. Mwangi. 2013. Contemporary processes of large-scale land acquisition in Sub-Saharan Africa: Legal deficiency or elite capture of the rule of law? *World Development*, 48, 1–18.

Gladwin, C.H. 1991. *Structural Adjustment and African Women Farmers*. Gainesville, FL: University of Florida Press.

Gladwin, C.H. 2002. Introduction to gender and soil fertility in Africa. In C.H. Gladwin, L. Cassidy, P.N. Hoon (Eds.), *Gender and Soil Fertility in Africa*. Special issue of *African Studies Quarterly*, 6(1–2), 1–26. Gainesville, FL: Center for African Studies, University of Florida.

Gladwin, C.H., L. Cassidy, and P.N. Hoon (Eds.). 2002. *Gender and Soil Fertility in Africa*. Special issue of *African Studies Quarterly*, 6(1–2). Gainesville, FL: Center for African Studies, University of Florida.

Gladwin, C.H., A.M. Thomson, J.S. Peterson, and A.S. Anderson. 2001. Addressing food security in Africa via multiple livelihood strategies of women farmers. *Food Policy*, 26, 177–207.

Goldman, A., and K. Heldenbrand. 2002. Gender and soil fertility management in Mbale district, southeastern Uganda. In C.H. Gladwin, L. Cassidy, P.N. Hoon (Eds.), *Gender and Soil Fertility in Africa*. Special issue of *African Studies Quarterly*, 6(1–2), 45–75. Gainesville, FL: Center for African Studies, University of Florida.

Guyer, J.I. 1986. Intra-household processes and faming systems research: Perspectives from anthropology. In J.L. Moock (Ed.), *Understanding Africa's Rural Households and Farming Systems* (pp. 92–104). Boulder, CO: Westview Press.

Hassan, S.M., C.E. Ikuenobe, S. Jalloh, G.C. Nelson, and T.S. Thomas. 2013. Ch. 10. Nigeria. In A. Jalloh, G.C. Nelson, T.S. Thomas, R. Zougmoré, H. Roy-Macauley (Eds.), *West African Agriculture and Climate Change* (pp. 259–290). Washington, DC: International Food Policy Research Institute.

Horlings, L.G., and T.K. Marsden. 2011. Towards the real green revolution? Exploring the conceptual dimensions of a new ecological modernization of agriculture that could "feed the world." *Global Environmental Change*, 21, 441–452.

Intergovernmental Panel on Climate Change. 2012. *Managing the Risks of Extreme Events and Disasters to Advance Climate Change Adaptation*. Cambridge: Cambridge University Press.

Jalloh, A., G.C. Nelson, T.S. Thomas, R. Zougmoré, and H. Roy-Macauley. 2013. *West African Agriculture and Climate Change*. Washington, DC: International Food Policy Research Institute.

Kassie, B.T., H. Hengsdijk, R. Rötter, H. Kahiluotot, S. Asseng, and M. Van Ittersum. 2013. Adapting to climate variability and change: Experiences from cereal-based farming in the Central Rift and Kobo Valleys, Ethiopia. *Environmental Management*, 52, 1115–1131.

Kusiluka, M.M., S. Kongela, M.A. Kusiluka, E.D. Karimuribo, and L.J.M. Kusiluka. 2011. The negative impact of land acquisition on indigenous communities' livelihood and environment in Tanzania. *Habitat International*, 35, 66–73.

MacKenzie, F. 1993. Exploring the connections: Structural adjustment, gender and the environment. *Geoforum*, 24, 71–87.

McKeon, N. 2013. One does not sell the land upon which people walk: Land grabbing, transnational rural social movements, and global governance. *Globalizations*, 10, 105–122.

McMichael, P. 2012. The land grab and corporate food regime restructuring. *Journal of Peasant Studies*, 39, 681–701.

Meena, R. 1991. The impact of structural adjustment programs on rural women in Tanzania. In C.H. Gladwin (Ed.), *Structural Adjustment and African Women Farmers* (pp. 169–190). Gainesville, FL: University of Florida Press.

Mol, A. 2001. *Globalization and Environmental Reform: The Ecological Modernisation of the Global Economy*. London: MIT Press.

Nombo, C.I., S.J. Kabote, D.P. Mamiro et al. 2013. Adaptation to climate change effects: Changing gender relations. Presentation at Climate Change, Sustainable Intensification and Food Security in Sub-Saharan Africa iAGRI Conference, Morogoro, Tanzania, November 13–15.

Odigie-Emmanuel, O. 2010a. The gender impact of climate change in Nigeria. In I. Dankleman (Ed.), *Gender and Climate Change: An Introduction* (pp. 123–129). Sterling, VA: Earthscan.

Odigie-Emmanuel, O. 2010b. Women and the environmental justice movement in the Niger Delta region of Nigeria. In I. Dankleman (Ed.), *Gender and Climate Change: An Introduction* (pp. 247–248). Sterling, VA: Earthscan.

Organisation for Economic Co-operation and Development (OECD). 2012a. OECD Factbook 2012. Available at http://www.oecd-ilibrary.org/sites/factbook-2013-en/04/02/01/index.html?itemId=/content/chapter/factbook-2013-34-en. Accessed January 31, 2014.

OECD. 2012b. *Farmer Behaviour, Agricultural Management and Climate Change*. Paris, France: OECD Publishing. Available at http://dx.doi.org/10.1787/9789264167650-en. Accessed January 31, 2014.

Pigg, K., S. Gasteyer, K. Martin, G. Apaliya, and K. Keating. 2013. The community capitals framework: An empirical examination of internal relationships. *Community Development: Journal of the Community Development Society*, 44, 4, 492–502.

Siles, C.J., I. Gutiérrez, and C.B. Flora. 2013. Acción colectiva de consorcios locales para la gobernanza ambiental: Un análisis a partir de los capitales de la comunidad. *Biocenosis*, 27, 94–105.

Skinner, E. 2011. *Gender and Climate Change: Overview Report*. London: Institute for Development Studies.

Tchingulou, A., A. Jalloh, T.S. Thomas, and G.C. Nelson. 2013. Togo. In A. Jalloh, G.C. Nelson, T.S. Thomas, R. Zougmoré, H. Roy-Macauley (Eds.), *West African Agriculture and Climate Change* (pp. 353–382). Washington, DC: International Food Policy Research Institute.

United Nations Development Programme. 1998. Capacity assessment and development. Technical Advisory Paper No. 3. Washington, DC: United Nations Development Program.

Uttaro, R.P. 2002. Diminishing choices: Gender, small bags of fertilizer, and household food security decisions in Malawi. In C.H. Gladwin, L. Cassidy, P.N. Hoon (Eds.), *Gender and Soil Fertility in Africa*. Special issue of *African Studies Quarterly*, 6(1–2), 77–110. Gainesville, FL: Center for African Studies, University of Florida.

Van der Geest, K., and T. Dietz. 2004. A literature survey about risk and vulnerability in drylands, with a focus on the Sahel. In T. Dietz, R. Rueben, J. Verhagen (Eds.), *The Impact of Climate Change on Drylands* (pp. 117–146). Dordrecht: Kluwer.

Vincent, K., T. Cull, and E.R.M. Archer. 2010. Gendered vulnerability to climate change in Limpopo Province, South Africa. In I. Dankleman (Ed.), *Gender and Climate Change: An Introduction* (pp. 130–137). Sterling, VA: Earthscan.

Warner, K., K. Van de Geest, S. Kreft et al. 2012. Evidence from the frontlines of climate change: Loss and damage to communities despite coping and adaptation. Report No. 9. Bonn: United Nations University Institute for Environment and Human Security (UNU-EHS).

Warner, K., K. Yuzva, M. Zissener, S. Gille, J. Voss, and S. Wanczeck. 2013. Innovative insurance solutions for climate change: How to integrate climate risk insurance into a comprehensive climate risk management approach. Report No. 12. Bonn: United Nations University Institute for Environment and Human Security (UNU-EHS).

Woodhouse, P. 2012. New investment, old challenges: Land deals and the water constraint in African agriculture. *Journal of Peasant Studies*, 39, 777–794.

Xu, X., and Y. Sheng. 2011. Productivity spillovers from foreign direct investment: Firm level evidence from China. *World Development*, 40, 62–74.

14 Small-Farmer Choice and Decision Making for Sustainable Soil Management*

Keith M. Moore, Jerod Myers, and Corinna Clements

CONTENTS

14.1 INTRODUCTION

Agriculture in sub-Saharan Africa is constrained by degraded soils, limiting the productivity of small farmers who manage the land, whatever their tenure status may be. Despite the potential for soil management technologies to improve soil fertility, studies examining the uptake of these technologies have demonstrated no universally predictive factors influencing adoption (Knowler and Bradshaw 2007; Prager and Posthumus 2010; Knowler 2012). Knowler and Bradshaw's (2007) meta-analysis of 31 empirical studies found 170 significant variables to explain farm-level adoption of conservation agriculture practices by small farmers. However, detailed analysis determined that no single variable was universally critical to adoption. Such results suggest that single or combined variable guidance for the promotion of integrated soil

* This chapter was made possible by the US Agency for International Development and the generous support of the American People for the Sustainable Agriculture and Natural Resources Management Collaborative Research Support Program under terms of Cooperative Agreement No. EPP-A-00-04-00013-00 to the Office of International Research, Education, and Development at Virginia Polytechnic Institute and State University.

fertility management (ISFM) is minimal. Thus, Knowler and Bradshaw (2007) concluded that a new approach would need to be developed to conduct further research on ISFM technology adoption. Future research efforts should focus on obtaining locally meaningful results rather than striving for universal understanding of factors influencing adoption. This will require efforts tailored to local conditions to promote the diffusion of ISFM.

In this chapter, we consider these findings as a point of departure for an examination of the factors that contribute to the spread of sustainable soil management practices in networks of small farmers. We analytically distinguish between two phases in the process of agricultural innovation leading to changes in soil management practices. The first phase involves framing the problem, its context, and the consequent choices faced by small farmers (both men and women who make technological choices in the process of agricultural production). The second phase addresses farmer decision making with respect to those choices. Most previous work has emphasized this latter phase, leaving unquestioned how choices are framed. Here, we examine conditional factors shaping the overall process, paying particular attention to how technological problems and solutions are framed from the outset. We present the evolution of perceptions of innovations and their diffusion, and we conclude by suggesting that new approaches to systematic collaboration and stakeholder negotiation incorporating agricultural innovation systems, platforms, or networks provide a solid foundation for moving forward (Biggs 1989; Röling and Wagemakers 1998; Sayer and Campbell 2004; Leeuwis 2008; Ekboir et al. 2009; Moore 2009; Anandajayasekeram 2011; World Bank 2012).

Recognizing the significance of local context forces us to consider the differences between the universal knowledge and perceptions of agricultural scientists and the locally informed perspectives of small farmers with respect to soil management. The issue is not the technical feasibility of scientifically sound soil management practices, but their assumed universality. The universality of these choices does not hold in every farmer's world. Furthermore, given the knowledge-intensive and dynamic nature of ISFM, a unilinear extension approach does not provide a mechanism for the required learning and adaptation (Vanlauwe et al. 2006; Davis 2013). Farmer perception of choice is not a straightforward matter either. Farmer choices based on locally situated knowledge may appear irrational or contradictory to what is deemed by outsiders as the farmer's best interest (Enyong et al. 1999; Sanchez 2002). Yet farmers must choose their investments carefully to sustain livelihoods dependent on mixed income sources and market options (Ajayi 2007). This frame-of-reference approach allows us to consider how farmers view options in complex adaptive systems (Gunderson and Holling 2002; Colfer 2005; Moore 2009). Our objective is to determine the most successful ways to increase sustainable choices that are meaningful for farmers. This leads to a discussion of innovation platforms (IPs) as a promising method to increase chances for more positive soil management outcomes and enhance the adaptive capacity of small farmers.

In Section 14.2, we address the differences between the perspectives of farmers and those traditionally held by agricultural scientists and extension agents that largely inform current practice. We then consider the economic factors in Section 14.3, questioning whether farm- or field-level modeling provides the most appropriate framing for small-farmer choices. Section 14.4 elaborates on the role of faith-based framings of agricultural knowledge and the potential of this perspective, highlighting the significance of mindsets that shape the choices farmers perceive. Following this, we consider the literature on the diffusion of innovations in Section 14.5. Finally, in Section 14.6, we introduce the innovation systems perspective and address the growing recognition of the importance of stimulating and assisting local innovation networks and platforms.

14.2 PERCEPTIONS OF SOIL FERTILITY

The soils of sub-Saharan Africa are notorious for their levels of degradation, a major impediment to agricultural productivity in the region. Soil nutrients have been mined or are carried away by rain or wind, creating a soil system with scarce nutrients and little water-holding capacity. Consequently, declining soil fertility is a major concern across many parts of sub-Saharan Africa (Vanlauwe et al. 2006; Swift and Shepherd 2007; Giller et al. 2009). The agricultural scientist or extension agent often frames this issue from an agroecological standpoint, citing nutrient deficiency, low soil organic matter, water stress, and high erodibility as limiting factors in soil fertility and crop production. The current practices of the small farmer, such as overgrazing and continuous cropping without nutrient replacement, are targeted as improper soil management practices that further exacerbate these agroecological issues (Vanlauwe et al. 2006; Kolawole 2013). Mitigating the loss of nutrients has proved to be a significant challenge as the rate of input intensity for inorganic fertilizers is estimated to be between 8 and 12 kg ha^{-1} compared with an average of 83 kg ha^{-1} for all developing countries (Mwangi 1997). Owing to the expense and the inaccessibility of these inorganic inputs, alternative approaches that build on traditional low-cost soil fertility management are being promoted.

Despite the apparently obvious nature of these problems, farmer knowledge and perception of soil health and fertility often differs from that of research scientists and extension agents (Gray and Morant 2003). In a study to assess farmer perception and knowledge of soil fertility in Ethiopia, Corbeels et al. (2000) note that scientists often quantify soil fertility by nutrient chemical composition, without giving much consideration to physical properties. From a scientific perspective, soil fertility refers to the ability of the land to produce consistent, high crop yields that can be achieved through inputs of inorganic fertilizer or biological nitrogen fixation. In contrast, the small farmer defines fertility by a series of easily observable factors quite distinct from the technical focus imposed by the scientists and extension agents. From a farmer's perspective, the soil characteristics affecting plant growth of most interest are indicated by soil color, crop yield, water-holding capacity, foliar color, and the presence of indicator weeds. One of the most important factors for the Ethiopian farmers was soil color, which reflects otherwise hidden parent material and determines specific soil characteristics (Corbeels et al. 2000). These findings were repeated for studies of farmers in Zambia (Ajayi 2007) and Ghana (Dawoe et al. 2012).

Scientists and farmers have the potential to enrich one another's knowledge of soil fertility and improve on the capacity for innovation. Indeed, Schuler et al. (2006) found that collaboration between farmers and soil scientists significantly increased the quality and utility of soil mapping efforts in northern Thailand. However, this process of collaboration is often impeded by the tendency of agricultural scientists to perceive their knowledge as superior to that of the small farmer, leading to communication failures between the two perspectives (Leeuwis 2008). This finding is consistent with a broad range of literature that illustrates the principle of homophily: similar people tend to communicate better and think alike, whereas dissimilar people tend not to work so well together (Ruef et al. 2003; Crona and Bodin 2006; Isaac et al. 2007; Bodin and Crona 2009).

Agricultural scientists are driven by a narrow focus on resource efficiency and productivity and the desire to provide policy recommendations (Kolawole 2013). Consequently, many adoption studies led by agronomists and/or economists frame the choice between the "traditional" (perceived as inferior technology) and a new "innovation" (perceived as superior). These educated frames of reference shape the messages of agricultural scientists and extension agents as they seek to promote a specific innovation among a targeted group of farmers. The choices thus framed on the basis of such presumed farmer decision-making priorities (more on this in Section 14.3) become the dependent variable in the adoption studies to identify the factors that influence the innovation outcome. For the scientists, the choices are clear—either the old practice or the new and improved one.

Kolawole (2013) provides further insight into the contrasting perceptions of small farmers and scientists regarding soil management. A survey of 140 small farmers and 100 scientists in southwestern Nigeria addressed perceptions of collaboration in ISFM (Olorunfemi 2010; Quardre 2010). The study was conducted in two states, Osun and Ondo, in Nigeria. The results from Osun (paralleling those from Ondo) demonstrate that agricultural scientists largely ignore the knowledge and understanding of the smallholder. Despite the majority of small farmers having a favorable opinion of their local knowledge, 80% of agricultural scientists believe that "small farmers lack the requisite knowledge of soil fertility management" (Kolawole 2013). Around 92% of scientists believe that there are no reasons why they should learn from small farmers. In contrast, 68.5% of the smallholder farmers are of the opinion that "scientists capitalize on their Western knowledge to suppress our knowledge systems" (Kolawole 2013). Further supporting this finding, the study found that 78% of scientists strongly believed that "the Western scientists' soil management options are always the best. Hence, farmers need to accept them in good faith" (Kolawole 2013).

As Leeuwis (2008) points out, distinct "theories of knowing" emerge from different cultures and groups of people. Farmers' adoption of improved soil management practices is highly dependent on how the innovation is perceived within their sociocultural framework and consequently on the soil scientists' ability to communicate within that framework.

14.3 ECONOMIC FACTORS SHAPING FARMER CHOICE

In addition to sociocultural frames of reference, farmers' choices are also framed by their perception of financial outcome, income stability, and food security. Small farmers view agricultural outputs as only part of their economic concerns. They must choose carefully what investments they make to sustain a livelihood dependent on mixed income sources and market options. The pursuit of mixed income-generating activities is often linked to farmer perception of risk. Strategies for risk management include overlapping methods that provide protection against food deficits. While some methods seek to reduce yield risk directly, others focus on generating compensatory income to mitigate production losses (Malton 1991). The diversification of income sources is understood as a method of self-insurance in which the small-farmer exchanges foregone expected earnings for the reduced income variability that is realized through an assortment of assets and activities that have a low to negative correlation with incomes (Barrett et al. 2001; Babatunde et al. 2010; Bigsten and Tengstam 2011).

The issue is not simply a matter of risk. In a thought-provoking analysis on the profitability of rainfed crop production for small farmers, Harris (2012) found that even if the average-sized small farm household were to apply all recommended technologies in rainfed crop production, the household would not be able to achieve incomes sufficient to exceed the poverty level of a dollar a day per capita. The limitation Harris (2012) found is technical, but not a matter of the technology per se. The critical limitation to achieving sufficient income was the amount of land available to small-farmer households and the size of their families.

This insufficiency is compounded by the fact that small-farm households are mixed production and consumption units, consuming part of what is produced and selling the rest. Most households do not produce enough to feed themselves for the entire year. Sutherland et al. (1999) describe food shortages among smallholder households in semiarid Kenya throughout the year. Most households experience food shortages from late September through late December as early food crops are harvested. Less affluent households are forced to sell their crops shortly after harvest to generate income. All households, excluding the ones capable of purchasing cabbage at local markets, experience a lack of green vegetables in their diet during dry periods that typically last for 7 months out of the year. Furthermore, drought periods that occur every 3–5 years create severe food shortages, forcing households to rely on famine-relief supplies to meet dietary needs.

On the basis of a review of farm household income studies across Africa, Reardon (1997) noted that an average of 45% of small-farmer income was derived from nonfarm sources, with this percentage increasing over time. This increase is seen as a response to declining on-farm income and the need to insure against production and market risks (Babatunde et al. 2010). Off-farm activities have become an important element in maintaining the livelihood of the small farmer. Off-farm activities are diversification mechanisms driven by population growth and crop and market failures that "push" the small farmer into alternative income-generating activities and by the perception that nonfarm activities are less risky than agricultural production.

A recent analysis by Duflo et al. (2009) applies a behavioral economics approach to explain why farmers fail to apply adequate amounts of fertilizer at the appropriate moments in the production process. Continued farmer resistance to fertilizer use is seen as a function of poorly functioning formal and informal networks that influence production behavior. Here the term "formal networks" refers to those market relations formally regulated and promoted by the state, such as extension and fertilizer subsidy programs, and "informal networks" refers to nonmarket relations often based on kinship and/or social capital. On the one hand, behavioral economists suggest that the reason farmers often fail to use fertilizer is that they wait too long to purchase the fertilizer, and when they seek out fertilizer for purchase in the following spring, prices have risen too high for profitability. On the other, informal markets place social pressure on farmers to spend the income they earn from agricultural production activities on customary events such as weddings. These informal network pressures limit the ability of farmers to purchase fertilizer shortly after the growing season when prices are low.

Duflo et al. (2009) contrast these findings with surveys in which small farmers report on their intent to utilize more fertilizer in the following growing season, only to continue in the same cycle of insufficient and often late application. Interestingly, Duflo et al. (2009) attempt to step back from traditional structural approaches (that focus on either the structure of economic relations or characteristics of the adoption environment) by creatively applying behavioral economic theories to provide an alternative voice that may be more empowering for farmers. The analysis concludes that the extension service and the farmers themselves are to blame for their inability to adhere to a logic that aligns with the conventional agricultural production frame of reference. In theorizing about farmer ability to save, Duflo et al. (2009) assume the voice of the farmers without taking their values and priorities into account.

Farmers must choose their investments carefully to sustain such precarious livelihoods. In an article reviewing literature on the adoption of innovations, Feder and Umali (1993) suggest that decision making is based on a trade-off between risks and payoffs, with the number of adopters increasing over time. Adoption occurs when the additional benefits from investment outweighs the cost. Small farmers base their decision on the option that offers the highest net returns, whether it is in terms of capital generation or reduced risk. Investments in sustainable soil management rarely pay off in the short term. Within the framework of the farmer, foregoing soil conservation practices may be the highest (but short-term) net return, where the costs of adoption and implementation outweigh the potential long-term benefits. Thus, on face value, investing in soil management is unlikely to be among the choices that a small farmer considers (Shiferaw et al. 2009).

Feder and Umali's (1993) study demonstrates how adoption studies and diffusion models formulate this choice framework. Despite their recognition of the household nature of the decision-making unit, the adoption and diffusion models they describe apply the parameters of a farm business. Their analysis shows that such models are truncated and that below a certain farm size, superior technologies are not applicable. This conclusion led to further improvements in adoption diffusion model design. A model by Traxler and Byerlee (1992) includes the notion of farmer-inspired choices through the introduction of a fodder/grain economic trade-off. However, none of the models incorporate a combined consumption/production function that includes choices beyond agricultural production. Indeed, Feder and Umali (1993) note that these models are based on two underlying assumptions: that markets are perfectly competitive, and that production and consumption decisions are separable. Furthermore, they note that the household has no other production functions or role in determining socially optimal farmer behavior.

Despite this narrowed perception of economic relevance for small farmers, some additional findings of Feder and Umali's (1993) review of the adoption of innovations are worth noting. Learning processes are seen as critical to the diffusion process (although innovation is not analyzed in this regard). Decision-maker learning can make a difference even under highly risk-averse circumstances. They also determined that infrastructural and agroclimatic variables had substantial explanatory power. Following Knowler and Bradshaw (2007), we would consider these latter as local choice framing factors.

14.4 ROLE OF FAITH-BASED KNOWLEDGE

As suggested by Duflo et al. (2009) above, the adoption of innovations is more than simply a matter of how knowledge is produced and validated in the biophysical and economic dimensions. Innovation adoption also involves the farmer's broader attitudes, beliefs, and practices (Ajayi 2007; Kolawole 2013). Spiritual and religious beliefs within the sociocultural constructs of the community significantly shape how choices are framed. The framing of these choices goes beyond individual preference. Spirituality and religion have been used to mobilize farming practices both for the introduction of the moldboard plow in Zimbabwe as well as, a few decades later, the promotion of conservation agriculture (Baudron et al. 2011). Agricultural practices can be understood as components of collective consciousness shaping understandings between the individual and intangible elements in the social life of communities.

Technical approaches to the environmental challenges faced by the small farmer routinely overlook beliefs that motivate human behavior. To shape or guide choice, the religious or spiritual values that condition the decision-making process need to be taken into account. A person's religious beliefs help frame a worldview that influences farm management decisions and perceptions of agricultural problems (Corselius et al. 2003). ISFM messages should seek to build on this frame of reference; Andersson and Giller (2012) argue that faith, when framed properly within the context of conservation agriculture innovations, can supersede both agroecological and economic frameworks traditionally imposed by the agricultural scientist.

Building on these insights, in September 2008, a group of African biodiversity experts convened in Dar es Salaam, Tanzania, to develop a vision statement for the future of Africa's biodiversity. The statement, termed the "Dar Vision," stressed the inclusion of communities of faith as a driving force behind conservation. They concluded that faith-based communities are the largest social organizations in Africa and could be mobilized to promote and enhance value-based sustainable practices linking nature and human beings (quoted in Gambrill 2011).

Faith-based organizations, international donors, and nongovernmental organizations have been at the forefront of promoting conservation agriculture in sub-Saharan Africa utilizing the Judeo–Christian ethic of environmental stewardship, citing biblical texts that place emphasis on caring for God's creation (Baudron et al. 2011). One example of the use of faith to promote soil conservation initiatives has origins in the experiences of a commercial tobacco farmer in Zimbabwe (Andersson and Giller 2012). As a newly converted Christian, Brian Oldreive considered tobacco cultivation unethical and switched to maize. After 2 years of hardship from poor yields, the banks agreed to administer an agricultural loan if he switched back to tobacco farming. Unwilling to do so, he sold his farm and became manager of a farm located in northern Zimbabwe. It was there in the early 1980s that he developed minimum tillage technology in response to declining yields. This development was originally motivated by cost and conservationist concerns, but it quickly became shrouded in much deeper meaning. He had a revelation that there were no natural God-given mechanisms that inverted the soil; rather, soil was naturally protected by a blanket of leaves and grass (Oldreive 2009).

On the basis of this revelation, the promotion of conservation agriculture became an act of faith for some, where agronomic practice took on religious meaning, and the mulch cover consisting of a "thick blanket of fallen leaves and grass" became known as "God's blanket" (Andersson and Giller 2012). Oldreive was able to transform the dialogue and vocabulary surrounding the technological framework of the agricultural scientist (that of no-till farming) and create messages that draw on the religious background of the smallholder. The idea of faith-based agricultural extension has led several evangelical churches and faith-based organizations to promote conservation agriculture just as they do the gospel, with a missionary-like zeal that paints a picture of the misguided ways of the nonbelievers and drawing on the strengths of those who have "converted."

In addition to the Judeo–Christian approaches to the promotion of conservation agriculture, extension may seek to build on indigenous beliefs and traditional, spiritual values. A profound respect for ancestors and desire for their guidance when faced with choice is a common element within many sub-Saharan African cultures. An interviewer once asked a Ghanaian farmer to imagine that he had been given a millet variety that yielded three times as much as the current variety under cultivation. A hypothetical situation was created in which the millet crop fails and the farmer was asked what he would then do. In response, the farmer said that in accordance with his beliefs, he would consult the ancestors via the village soothsayer. If the ancestors were unhappy with the variety, he would abandon the plot immediately and not return to the area where the variety was planted (Millar 2007).

Falvey (2005) claims that today's paradigm of agricultural science is concerned with problem-solving technology, and not with the philosophy or ethics that go beyond the pursuit of overcoming constraints. Falvey (2000, 2004) refers to the nature of modern sustainable agriculture and the perpetuity of uncertainty that is associated with attempts to sustain outputs with constant technological innovation. He argues that the real insights of science and its resulting innovation occur only when spiritual dimensions are engaged. The Alliance of Religions and Conservation estimates that >90% of the African population self-identifies as Christian or Muslim, with nearly all of the population holding traditional, indigenous beliefs. Consequently, there is a strong foundation of faith on which to engage the smallholder in agricultural development.

In coming to understand how the choices of small farmers are framed, we must take into account their worldviews and how those perspectives provide meaning to perceived choices. Recognizing that what is regarded as sacred is more likely to be treated with care and respect, Paul Sagan noted that a much wider and deeper understanding of science and technology is needed that incorporates the vital role of religion (Sagan and 32 Nobel Laureates 1990). Consequently, the science of the agricultural scientist may find a natural compatibility with the deeply rooted beliefs of the smallholder.

14.5 TIME DYNAMICS, ACTOR IDENTITIES, AND INNOVATION

Up to this point, we have learned a great deal about the framing of small-farmer choices for ISFM. From this exploration of the first phase, four broad points can be retained: (i) farmers and scientists see the world differently; (ii) economic factors shaping ISFM choices extend beyond the field and farm to include complex farm household livelihood systems; (iii) dissemination processes occur over time and involve some sort of social learning; and (iv) ideologies and religions can be mobilized to help frame ISFM choices. In contrast to the limited study of the framing of decision making, the second phase, or that of the actual decision making, has been studied extensively. Rogers' *Diffusion of Innovations* (1962) is paradigmatic of that approach. Here we examine how this theory of technological change in agriculture characterizes the process of innovation by examining actor identification, the time dynamics, and the concept of innovation itself.

Rogers' *Diffusion of Innovations* (1962) was first to consider the adoption of innovations as process and identified a range of adopter categories that would successively take up an innovation. These adopters could be modeled as successive groups extending from one end of the bell curve to the other: innovators, early adopters, early majority, later majority, and laggards (Figure 14.1). The time of adoption was attributed to the individuals' degree of innovativeness. The accumulation of these adopters over time could be graphed as an S-curve (Figure 14.2). Here, the shape of the S-curve was explained by differences among adopter categories. The time dimension was measured by the rate of adoption, or the length of time before a larger percentage of the population adopted the innovation (Rogers 1983). This approach has been highly influential.

Originally the work focused on a rather static perspective coming out of the Green Revolution based on cross-sectional studies that did not allow for much emphasis of the time dimensions that are integral to diffusion as a process. More dynamic approaches to modeling adoption processes emerged (see Besley and Case 1993;

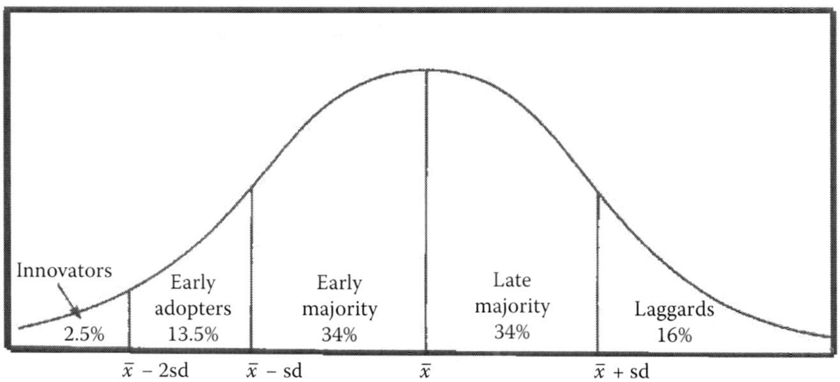

FIGURE 14.1 Adopter categorization on the basis of innovativeness (measured by time). (Adapted from Rogers, E.M. 1983. *The Diffusion of Innovations*. 3rd ed. New York: The Free Press. With permission.)

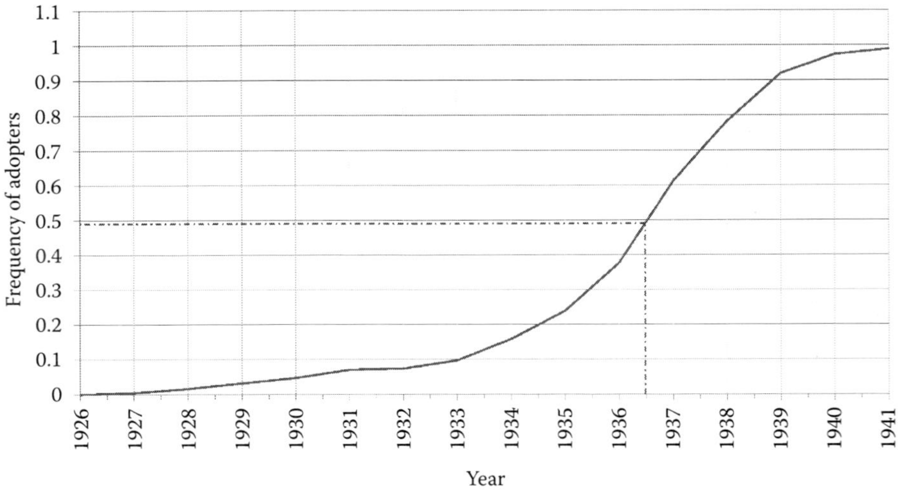

FIGURE 14.2 Long-tailed S-curve. (From Henrich, J., *American Anthropologist*, 103, 992, 2001.)

Feder and Umali 1993; Berger 2001). These approaches emphasized changes occurring over time in learning (increasing both knowledge of the technology and of other actors experiences and decisions) and in the market conditions (especially credit and prices).

Besley and Case (1993) highlighted the importance of marketing networks, the relative market position (power) of adopters *vis-à-vis* one another, and learning from the experiences of early adopters who change the landscape for late adopters. Attempting to reconstruct the actual dynamics of real-world decision making, they stressed the importance of modeling interdependent decision making. However, to emphasize the significance of time dimensions, they avoided the vagaries of complex technologies and focused on variety adoption among cotton growers (a simple technology and homogeneous sample allowed the time-variant dimensions to be accentuated). Drawing on agent-based models, Berger (2001) was able to simulate technology diffusion within a landscape over time, but retained the assumption that each actor was making the same innovation.

This work begins to take into account the social network findings of Granovetter (1974, 1985) who argued that economic behavior is embedded in a network of interpersonal relations. His work rejected the neoclassical economics assumption of atomized individuals whose decision making is completely autonomous and based on rational calculation of payoffs, but did not ascribe to the overly socialized view that people's behaviors are limited by the roles they play and the variation in their attributes. Instead, he argued that economic behavior and institutions are constrained by or embedded in social relations, but that individuals are capable of making independent decisions, thus allowing other factors, as we have seen above, to play a role.

From these agent-based theories and models emerged a perspective where networks of farmers and other stakeholders interact and technological innovation arises from that interaction (Röling and Wagemakers 1998; Biggs and Matsaert 2004; Sayer and Campbell 2004). Building on the work of Coughenour and Chamala (2000) and Ekboir et al. (2002), Swenson and Moore (2009) demonstrated the influence of networks and social relations on decision-making and technology adoption by reviewing case studies on the development of conservation agriculture, and found that field-specific conditions and unique farm and household characteristics are only part of the explanation for adoption of conservation agricultural practices. Successful adaptation of these practices appears to involve vast networks of relationships that reinforce certain sets of knowledge, beliefs, and behaviors (Davis 2013). Swenson and Moore (2009) focused on the critical relationship between actors and ecologically specific development problems within complex networks as they resolve systemic issues of input supply and delivery mechanisms, on-farm labor and biophysical adaptation, and reliable output markets.

Further analyzing the effect of social relations on decision making, Henrich (2001) suggested that payoffs are only part of what drives adoption; biased cultural transmission figures prominently in the process. People make investment decisions even in the face of evidence arguing against such behavior (e.g., Baudron et al. 2011; Andersson and Giller 2012). Henrich (2001) argued that Rogers' classification of adopters based on degrees of innovativeness does not result in the S-shaped curve (Figure 14.2) that defines the diffusion process. Instead, he noted that biased cultural transmission was the predominant force in the process of diffusion. Biased cultural transmission used information that may include but certainly goes beyond the innovation–evaluation information relevant to the payoffs of any particular technology adoption. Henrich (2001) builds on Boyd and Richerson's (1988) work, which distinguishes between "learners" who acquire behavior through a process of experimentation, and "imitators" whose behavior is acquired through social learning. Henrich attached this concept to Rogers' classification of adopters based on innovativeness by categorizing individuals as innovators and imitators. His analysis of the S-curve of adoption (see Figure 14.2) indicated that imitators are copying ideas and practices not directly related to an analysis of costs and benefits.

Henrich (2001) argues that purely environmental learning, where decisions are based on a cost–benefit analysis, creates R-shaped diffusion models (Figure 14.3), or graphs where the inflection point, the moment of maximum rate of growth, occurs at time zero. In other words, in most diffusion situations, which are best represented by S-curves, adoption is slow until a threshold level is achieved. At this inflection point, a threshold level of adoption has been achieved that results in a much higher rate of diffusion. Up to this point, the rate of adoption is slow. "Long-tailed" S-curves (Figure 14.2) demonstrate the slowness of initial adoption relative to the rate at which individuals imitate previous adopters. Rogers (1995) recognizes that a critical mass must be achieved to make individuals more open to adoption. Henrich explains this through Boyd and Richerson's (1985) description of conformist transmission, which indicates that the more frequently a trait appears in the population, the more valuable it seems.

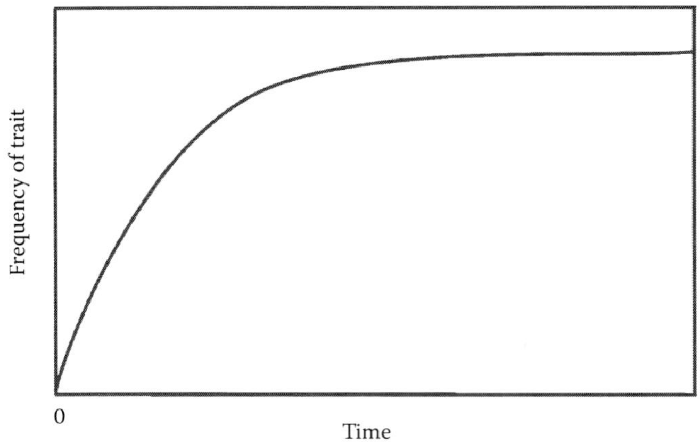

FIGURE 14.3 R-shaped diffusion curve. (Adapted from Rogers, E.M. 1983. *The Diffusion of Innovations*. 3rd ed. New York: The Free Press.)

In this way, innovations are adopted through a process of social learning in which actors imitate early adopters. Critical to this process are the networks through which actors are connected. While Rogers (1983) recognizes the importance of networks in disseminating information, Henrich (2001) identifies the key information flowing through networks and informing imitators as the prestige of an initial adopter and the popularity of the idea or practice. Henrich (2001) also demonstrated that farmers are more likely to also imitate particularly influential actors in the local network. For example, small farmers are more likely to adopt a technology when a person of power adopts that technology successfully (Kiptot et al. 2006). This process of biased cultural transmission helps explain the influence of religious groups on small farmers' adoption of conservation agriculture referenced earlier. Thus, adoption processes are not individualized rational choice decisions, but rather processes of imitation of other actors in the system. Rogers has also come to recognize the process of imitation and the importance of networks. In the 1995 edition of *Diffusion of Innovations*, Rogers introduced a discussion of innovation networks and the significance of time sequencing that reveals the important role of opinion leaders in a phased model of diffusion.

However, many of these diffusion studies do not study the innovation process itself (Thirtle and Ruttan 1987), continuing to view an innovation as a static entity to be communicated within a social system. This conforms to Rogers' (1967) diffusion of innovations perspective that differentiates between the generation of innovations and their diffusion. Biggs (1990) and Biggs and Clay (1981) rejected this belief, describing a cyclical process that invokes a broadened concept of innovation. Biggs (1990) argued that because technology evolves, reformulation itself is an innovation. Farmers engage in a continuous process of innovation to maintain equilibrium with their constantly changing environment (Biggs and Clay 1981). Innovation, then, is a reflexive, ongoing process that brings together various stakeholders for collaboration. Sayer and Campbell (2004) describe the process of innovation in terms of adaptive management and social learning, recognizing that knowledge is a fluid, constantly changing outcome of socio–material relations (Clark and Murdoch 1997). The concept of social learning moves beyond the idea that actors are influenced by one another to the notion of stakeholders coming together to collaborate for the purpose of technological or institutional innovation. Recent work has applied this revised conception of innovation that comprises institutional and organizational change (Tenywa et al. 2011; Nederlof and Pyborn 2012). An innovation is far from a monolithic technology to be diffused; it is as dynamic, context specific, and time responsive as any other actor.

This paradigm shift offers new conceptions of actor identities, time dynamics, innovation, and the relationships between them. In Rogers' (1962) original formulation of the diffusion of innovations, actors are identified by the time frame in which they make their decision. This decision is assumed to be a dualistic choice between an innovation and traditional knowledge and is largely autonomous, based only marginally on social influences. However, a greater recognition of the importance of actor interaction and social learning has altered the categorization of actors from various classes of adopters to innovators and imitators. A deeper analysis of time dynamics reveals that as time progresses, innovations themselves are adapted to fit changing needs and conditions. This has led to the shift in the perception of innovation as an unchanging, diffusible object to ongoing, dynamic processes. Within these processes, there is not a single moment of individual decision making; rather, a continuous process of group negotiation and adaptation takes place. This paradigm shift has significant implications for our approach to the promotion of ISFM.

14.6 AGRICULTURAL INNOVATION SYSTEMS, NETWORKS, AND PLATFORMS

Our examination of the first phase, in which small-farmer decisions were framed in terms of religious influences and economic quandaries, demonstrates that decision making cannot be isolated from its context. In exploring the second phase, we have come to recognize the systemic nature of innovation by acknowledging the variable influence of network connections and time dynamics on decision making, and revising our concept of innovation from that of an object to one of an ongoing process. This innovation systems perspective takes into account stakeholder priorities and builds shared understandings that frame farmers' choices, thus facilitating ISFM decision making (Davis et al. 2008; Buck and Scherr 2009; World Bank 2012). Undergirding the components of an innovation system is the practice of adaptive management, which integrates scientific and local knowledge while recognizing that an innovation must develop alongside changes in relationships among actors and the environment over time (Sayer and Campbell 2004; Moore 2009). Vital to this perspective is the concept of social learning: as stakeholders collaborate, new insights are generated and common understandings are fostered (Reij and Waters-Bayer 2001; Buck and Scherr 2009). This facilitates the process of innovation and leads to concerted action.

Essential to the process of social learning are innovation networks and platforms. The focus on social learning increase the range of choice for small farmers. Active participation in IPs improves farmers' knowledge and understanding of existing options, allowing them to become ISFM connoisseurs and recognize potential opportunities. This approach may not increase the likelihood that any particular small farmers will decide to adopt particular improved soil management practices, but the collaborative effort among the network partners ensures that a locally relevant range of options is stimulated and ready for application under conducive market and policy conditions stemming from the wider economy.

Research and agricultural extension have gradually shifted toward ISFM, which places more value on promoting options that are in tune with localized knowledge networks and perceptions (Corbeels et al. 2000; Vanlauwe et al. 2006). The World Bank, after many years of supporting the Training and Visit System, has also come to recognize the need for interaction among a diverse set of stakeholders to address complex problems holistically and contextually (World Bank 2012). Engel (1997) argues that, rather than seeking to unilinearily transfer science-based messages, approaches to integrate indigenous knowledge with scientific knowledge should create opportunities for sharing, joint learning, reflection, and mutual respect. This requires participatory and collaborative approaches that promote quality interaction between farmers and outside agents (Pretty 1995; Dawoe et al. 2012). Extension should focus on improving knowledge and capacity to observe and experiment while building on small-farmer knowledge as it relates to specific local conditions of production (Deugd et al. 1998). Doing so allows for improved understanding of the sociocultural conditions that shape choices and, consequently, farmer decisions. Engaging the small farmer in the process of soil improvement requires researchers and agricultural extension agents to recognize the importance of farmer perceptions and knowledge of soil fertility (Corbeels et al. 2000; Marenya et al. 2008).

An illuminating example comes from the Ethiopian experience of moving from an aggressive technology transfer approach that pushed a comprehensive package of hybrid seeds, fertilizer, and credit in the late 1980s and early 1990s (Borlaug and Dowswell 1994) toward the more recent, widespread, and gradual progress made by the Integrated Nutrient Management Program (Corbeels et al. 2000). The revised Integrated Nutrition Management Program and its Farmer Field School Network have become a platform for facilitating the process of finding a common ground for local and scientific knowledge. The farmer field school approach has been particularly helpful in allowing for local agroecological problem solving and legitimizing farmer knowledge of soils and their properties to enhance production of a range of agricultural commodities. Moreover, encouraging farmers to experiment and share their findings with others to make farmer-led choices about technology development and adoption decisions mobilizes farmers' own frames of reference. Farmers also observe and compare ecological outcomes of different methods to improve soil fertility. For example, farmer trials demonstrated that organic manure keeps the soil softer and is better for retaining moisture, while artificial fertilizer produces a darker-colored leaf and improved grain yield (Kebebe et al. 2007). With such results, many farmers opted for a 50/50 organic and artificial fertilizer regimen, an outcome that allowed farmers to improve their yield, manage their short term risk, and invest in the long-range productivity of their soils (Kebebe et al. 2007).

Essential to the incorporation and development of local knowledge is the network through which it is transferred. Composed of farmers, farmer organizations, extension, input suppliers, researchers, agencies, and policy makers, networks emerge when different actors realize a mutual desire to improve a product or process (Buck and Scherr 2009; Ekboir 2012). Innovation networks can form by the deliberate actions of actors (formation of IPs), or emerge organically as actors collaborate (Ekboir 2012). This collaboration facilitates social learning by fostering information sharing (Buck and Scherr 2009). Strong networks foster access to knowledge and physical inputs, increasing farmers' access to options (Swenson and Moore 2009). Building networks and strengthening existing ones is an investment in social capital necessary for fostering technological change.

Strengthening network ties improves the flow of information between local actors. This is supported by Conley and Udry's (2001, 2010) investigation into how farmers learn through strong social network ties. Working in Ghana with a project on pineapple production for European markets, Conley and Udry (2010) surveyed farmers to see how and why farmers were redesigning the technological package with specific regard to fertilizer use. It was discovered that farmers only communicated about fertilizer usage with a very small number of contacts as opposed to widespread sharing of experiences at the village level. For instance, when one farmer in the group applied significantly high amounts of fertilizer and experienced a vast yield gain, those individuals in that farmers "close" network were significantly more likely to increase their fertilizer application in the following growing season (Conely and Udry 2010). Thus, production behavior was highly influenced by their social ties in these small networks.

IPs create an environment that fosters the process of innovation by assembling a variety of stakeholders to identify and resolve systematically interdependent issues in a production network (Adekunle et al. 2010; Nederlof and Pyborn 2012). Researchers do not define a problem and recommend a solution; instead, various stakeholders identify issues and collaborate to resolve them. When scientific and local knowledge are equalized, an environment is created where stakeholders can contribute their own knowledge, skills, and perspective (Lamb and Moore 2010; Nederlof and Pyborn 2012). This establishes the foundation for negotiation between differing sources of knowledge (Moore 2011). As a viable solution is agreed on, the IP facilitates connections to other network actors, driving learning and adoption.

Tenywa et al. (2011) demonstrated that a market-based IP, in which participants innovate to take advantage of a market opportunity, results in a quicker formulation of a win–win situation than does a researcher-led IP. This is not to say that an IP must be centered on a market issue. Rather, the strength of an IP is in framing an issue in terms of market incentives (Posthumus et al. 2011; Nederlof and Pyborn 2012) and fostering connections with private sector stakeholders. In Zambia, IPs were developed at the local, district, and national levels to support the adoption of conservation agriculture and proved to be instrumental in forging relationships among stakeholders, developing connections with private sector actors such as seed companies, financial service providers, and traders (Nederlof et al. 2011).

Innovation networks and platforms enhance farmers' ability to adapt to constantly changing market and agroecological conditions, thus enriching a system's resilience by improving its capacity for self-organization, learning, and adaptation (Walker et al. 2002). A resilient community possesses the tools to respond to fluctuating conditions. Small farmers cannot wait for researchers to deliver answers as neatly packaged technological innovations. Rather, farmers are innovating constantly to stay in equilibrium with their environment (Biggs and Clay 1981). As this adaptive behavior is by its nature collaborative, innovation networks and platforms are essential to a system's adaptive capacity. A well-functioning network will facilitate the spread of market information to farmers that informs them of changing market conditions to which they must adapt. For adaptation at the community, regional, or national level, IPs provide the necessary forum to assemble stakeholders to negotiate the trajectory of response (Walker et al. 2002).

14.7 CONCLUSION

At the beginning of the chapter, we proposed that small-farmer decision making could be analytically differentiated into a two-phase process. The first phase frames the problem and specifies the critical decision-making parameters, and the second phase consists of the actual process of choosing. This distinction enables the recognition of how the local context matters. Local agroecology, knowledge, household livelihood strategies, belief systems, networks, and leadership each play a role. These factors, however, do not come together in any universal fashion that can be replicated from one situation to the next, as efforts to scale up technological fixes often propose.

To better understand framing of innovation decision making, we learned that (i) small farmers and scientists see the world differently; (ii) economic factors shaping ISFM choices extend beyond the field and farm to include complex farm household livelihood systems; (iii) dissemination processes occur over time and involve social learning; and (iv) ideologies and religions can be mobilized to help frame ISFM choices. We then reviewed the construction of a new paradigm to characterize the process of decision making to replace the unilinear framing of the diffusion of innovations model. Under such a paradigm, actors are seen as innovators and imitators rather than as various types of adopters. The duration of the innovation process presents opportunities for interaction among these classes of actors and introduces the potential for change in the innovation and evolution of the problem that it is intended to solve. This leads us to a renewed understanding of innovation and a transformation in the meaning of adoption to that of adaptation.

What we have learned is that innovation is not simply an individual act, but part of a process of social learning. This process is often characterized as adaptive management. It may involve a multitude of partners and the object of innovation may evolve as farmers and other value chain partners adapt to changing climate and market circumstances. Such an understanding forces us to transform the way we approach small-farmer ISFM decision making. Once we recognize that the framing of small-farmer ISFM choices is a collective endeavor, support for building IPs becomes self-evident. These local platforms provide a structured context for the introduction and development of improved practices that are adapted to the specificities of the network partners and their agroecological and socioeconomic circumstances. It is through participation in these networks that soil scientists can share their knowledge in a common development effort.

14.7.1 Where Do We Go from Here?

The difficulty we face in moving forward with implementing innovation networks and platforms more broadly is the lack of sufficient numbers of trained "innovation network facilitators and brokers" (Kibwika et al. 2009; World Bank 2012; Moore et al. 2014). An innovation broker can be an individual or an organization with a neutral role in the platform that invokes actors around a problem and facilitates collaboration throughout the innovation process (Nederlof et al. 2011). Klerkx and Gildemacher (2012) identify three main functions of innovation brokers: analyzing the context and articulating demand, bringing together stakeholders into networks, and facilitating interaction for innovation. In contrast to traditional extension agents, their job is not to deliver knowledge, but to catalyze the development of knowledge.

However, this role is challenged by formal education and training that reinforces unilinear approaches (Davis et al. 2008). Even when an innovation broker is sufficiently adept at facilitation, there are also difficulties in acquiring and maintaining funding for these champions whose success rests in the invisibility of their contributions (Klerkx et al. 2009). Another obstacle involves maintaining neutrality, especially when brokers have overlapping functions. Klerkx and Gildemacher outline investments to overcome these challenges in the 2012 World Bank investment sourcebook on agricultural innovation systems. They assert that a combination of formal training and practical experience is needed to develop the skills necessary for facilitating the innovation process: leadership, multistakeholder facilitation, trust building, and communication. Furthermore, they argue that the value of innovation brokers must also be recognized by national governments and the donor community.

REFERENCES

Adekunle, A.A., A.O. Fatunbi and M.P. Jones. 2010. *How to Set Up an Innovation Platform: A Concept Guide for the Sub-Saharan Africa Challenge Program (SSA CP)*. Accra, Ghana: Forum for Agricultural Research in Africa.

Ajayi, O.C. 2007. User acceptability of sustainable soil fertility technologies: Lessons from farmers' knowledge, attitude and practice in southern Africa. *Journal of Sustainable Agriculture* 30(3): 21–40.

Anandajayasekeram, P. 2011. The role of agricultural research and development within the agricultural innovation systems framework. ASTA/IFPRI. FARA Conference Working Paper 6, Accra, Ghana, December 5–7, 2011.

Andersson, J.A. and K.E. Giller. 2012. On heretics and God's blanket salesmen: Contested claims for Conservation Agriculture and the politics of its promotion in African smallholder farming. In: Sumberg, J. and Thompson, J. (eds.) *Contested Agronomy: Agricultural Research in a Changing World*. London: Earthscan.

Babatunde, R.O., F.I. Olagunju, S.B. Fakayode and A.O. Adejobi. 2010. Determinants of participation in off-farm employment among small-holder farming households in Kwara State, Nigeria. *Production Agriculture and Technology* 6(2): 1–14.

Barrett, C.B., T. Reardon and P. Webb. 2001. Nonfarm income diversification and household livelihood strategies in rural Africa: Concepts, dynamics, and policy implications. *Food Policy* 26(4): 315–331.

Baudron, F., J.A. Anderson, M. Corbeels and K.E. Giller. 2011. Failure to yield? Ploughs, conservation agriculture and the problem of agricultural intensification: An example from the Zambezi Valley, Zimbabwe. *Journal of Development Studies* 48(3): 393–412.

Berger, T. 2001. Agent-based spatial models applied to agriculture: A simulation tool for technology diffusion, resource use changes and policy analysis. *Agricultural Economics* 25: 245–260.

Besley, T. and A. Case. 1993. Modeling technology adoption in developing countries. *The American Economic Review* 83(2): 396–402.

Biggs, S. 1990. A multiple source of innovation model of agricultural research and technology promotion. *World Development* 18(11): 1481–1499.

Biggs, S. and H. Matsaert. 2004. Strengthening Poverty Reduction Programmes using an Actor-Oriented Approach: Examples from Natural Resources Innovation Systems. Agricultural Research and Extension Network Paper No. 134. UK Department for International Development (DFID).

Biggs, S.D. 1989. *Resource-Poor Farmer Participation in Research: A Synthesis of Experiences from Nine National Agricultural Research Systems*. The Hague, Netherlands: International Service for National Agricultural Research.

Biggs, S.D. and E.J. Clay. 1981. Sources of innovation in agricultural technology. *World Development* 9(4): 321–336.

Bigsten, A. and S. Tengstam. 2011. Smallholder diversification and income growth in Zambia. *Journal of African Economies* 20(5): 781–822.

Bodin, Ö. and B.I. Crona. 2009. The role of social networks in natural resource governance: What relational patterns make a difference? *Global Environmental Change* 19: 366–374.

Borlaug, N.E. and C.E. Dowswell. 1994. Feeding a human population that increasingly crowds a fragile planet. Keynote lecture, 15th World Congress of Soil Science, Acapulco, Mexico, Supplement to Transactions, 15 pp.

Boyd, R. and P.J. Richerson. 1985. *Culture and Evolutionary Process*. Chicago: University of Chicago Press.

Boyd, R. and P.J. Richerson. 1988. An evolutionary model of social learning: The effects of spatial and temporal variation. In: Zentall, T.R. and Galef, B.G. (eds.) *Social Learning: Psychological and Biological Perspectives*. Hillsdale, NJ: Lawrence Erlbaum Associates, pp. 29–48.

Buck, L.E. and S.J. Scherr. 2009. Building innovation systems for managing complex landscapes. In: Moore, K.M. (ed.) *The Sciences and Art of Adaptive Management: Innovating for Sustainable Agriculture and Natural Resource Management*. Ankeny, IA: The Soil and Water Conservation Society.

Clark, J. and J. Murdoch. 1997. Local knowledge and the precarious extension of scientific networks: A reflection on three case studies. *Sociologia Ruralis* 37: 38–60.

Colfer, C.J.P. 2005. *The Complex Forest: Communities, Uncertainty, and Adaptive Collaborative Management*. Washington, DC: Resources for the Future and the Center for International Forestry Research (CIFOR).

Conley, T. and C. Udry. 2001. Social learning through networks: The adoption of new agricultural technologies in Ghana. *American Journal of Agricultural Economics* 83(3): 668–673.

Conley, T. and C. Udry. 2010. Learning about a new technology: Pineapple in Ghana. *The American Economic Review* 100(1): 35–69.

Corbeels, M., A. Shiferaw and M. Haile. 2000. Farmers' knowledge of soil fertility and local management strategies in Tigray, Ethiopia. *Managing Africa's Soils* No. 10. International Institute for Environment and Development.

Coughenour, C.M. and S. Chamala. 2000. *Conservation Tillage and Cropping Innovation: Constructing the New Culture of Agriculture*. Ames, IA: Iowa State University Press.

Crona, B. and Ö. Bodin. 2006. What you know is who you know? Communication patterns among resource users as a prerequisite for co-management. *Ecology and Society* 11(2): 7

Davis, K. 2013. Keynote speech: Rural advisory services—A global status report and five key areas for mobilization. In: Binder, J. and Schöning, A. (eds.) *Modes of Sustainable Delivery: Documentation of the Symposium on Rural Advisory Services*. Bonn, Germany: Deutsche Gesellschaft für Internationale Zusammenarbeit (GIZ) GmbH.

Davis, K.E., J. Ekboir and D.J. Spielman. 2008. Strengthening agricultural education and training in sub-Saharan Africa from an innovation systems perspective: A case study of Mozambique. *The Journal of Agricultural Education and Extension* 14(2): 35–51.

Deugd, M., N. Röling and E.M.A. Smaling. 1998. A new praxeology for integrated nutrient management, facilitating innovation with and by farmers. *Agriculture, Ecosystems & Environment* 71: 271–285.

Duflo, E., M. Krenner and J. Robinson. 2009. Nudging Farmers to Use Fertilizer: Evidence from Kenya. NBER Working Paper No. 15131. Cambridge, MA: National Bureau of Economic Research.

Ekboir, J. 2012. How to build innovation networks. In: *The World Bank. Agricultural Innovation Systems: An Investment Sourcebook*. Washington, DC: The World Bank.

Ekboir, J., K. Boa and A. Dankyi. 2002. *Impact of No-Till Technologies in Ghana*. Mexico, DF: CIMMYT.

Engel, P.G. 1997. *The Social Organization of Innovation; A Focus on Stakeholder Interaction*. Amsterdam, The Netherlands: Royal Tropical Institute KIT Press.

Enyong, L.A., S.K. Debrah and A. Bationo. 1999. Farmers' perceptions and attitudes towards introduced soil fertility enhancing technologies in western Africa. *Nutrient Cycling in Agroecosystems* 53: 177–187.

Falvey, L. 2000. *Thai Agriculture: Golden Cradle of Millennia*. Bangkok: Kasetsart University Press, White Locus.

Falvey, L. 2004. *Sustainability: Elusive Or Illusion?: Wise Environmental Intervention*. Adelaide, Australia: Institute for International Development.

Falvey, L. 2005. *Religion and Agriculture: Sustainability in Christianity and Buddhism*. Adelaide, Australia: Institute for International Development.

Feder, G. and D. Umali. 1993. The adoption of agricultural innovation: A review. *Technological Forecasting and Social Change* 43: 215–239.

Gambrill, A. 2011. *From Practice to Policy to Practice: Connecting Faith and Conservation in Africa*. Washington, DC: International Resources Group for USAID Bureau for Africa.

Giller, K.E., E. Witter, M. Corbeels and P. Tittonel. 2009. Conservation agriculture and smallholder farming in Africa: The heretics' view. *Field Crops Research* 114(2009): 23–34.

Granovetter, M.S. 1974. *Getting a Job: A Study of Contacts and Careers*. Cambridge, MA: Harvard University Press.

Granovetter, M.S. 1985. Economic action and social structure: The problem of embeddedness. *The American Journal of Sociology* 91(3): 481–510.

Gray, L.C. and P. Morant. 2003. Reconciling indigenous knowledge with scientific assessment of soil fertility changes in southwestern Burkina Faso. *Geoderma* 111(3): 425–437.

Gunderson, L.H. and C.S. Holling (eds.). 2002. *Panarchy: Understanding Transformations in Human and Natural Systems*. Washington, DC: Island Press.

Harris, D. 2012. *Limits to Agricultural Development from a Smallholder Household Perspective: The Profitability of Rainfed Crop Production and How it Affects the Potential For Intensification*. Nairobi: ICRISAT. Available at www.ksiconnect.icrisat.org/wp-content/uploads/2012/12/Intensificationability-v5-ICRAF-color.pdf. Accessed on May 18, 2013.

Henrich, J. 2001. Cultural transmission and the diffusion of innovations: Adoption dynamics indicate that biased cultural transmission is the predominate force in behavioral change. *American Anthropologist* 103(4): 992–1013.

Isaac, M.E., B.H. Erickson, S. Quashie-Sam and V.R. Timmer. 2007. Transfer of knowledge on agroforestry management practices: The structure of farmer advice networks. *Ecology and Society* 12(2): 32. Available at http://www.ecologyandsociety.org/vol12/iss2/art32/.

Kebebe, E., E. Sheleme and W. Wondimu. 2007. Learning experiences of farmers field school on Integrated Nutrient Management: Evidence form Wolaita in southern Ethiopia. *Journal of Agronomy* 6(4): 992–1013.

Kibwika, P., A.E.J. Wals and M.G. Nassuna-Musoke. 2009. Competence challenges of demand-led agricultural research and extension in Uganda. *Journal of Agricultural Education and Extension* 15(1): 5–19.

Kiptot, E., S. Franzel, P. Hebinck and P. Richards. 2006. Sharing seed and knowledge: Farmer to farmer dissemination of agroforestry technologies in western Kenya. *Agroforestry Systems* 68(3): 167–179.

Klerkx, L. and P. Gildemacher. 2012. The role of innovation brokers in agricultural innovation systems. In: *The World Bank. Agricultural Innovation Systems: An Investment Sourcebook*. Washington, DC: The World Bank.

Klerkx, L., A. Hall and C. Leeuwis. 2009. Strengthening Agricultural Innovation Capacity: Are Innovation Brokers the answer? UNU-Merit Working Paper # 2009-019. Maastricht, the Netherlands: United Nations University–Maastricht Economic and Social Research and Training Centre on Innovation and Technology (UNU-MERIT).

Knowler, D. 2012. Progress on the economics and adoption of conservation agriculture by smallholders. Presentation at the SANREM CRSP/Conservation Agriculture Symposium, ASA/SSA/CSA Annual Meeting, Cincinnati, OH.

Knowler, D. and B. Bradshaw. 2007. Farmers' adoption of conservation agriculture: A review and synthesis of recent research. *Food Policy* 32: 25–48.

Kolawole, O.D. 2013. Soils, science and the politics of knowledge: How African smallholder farmers are framed and situated in the global debates on integrated soil fertility management. *Land Use Policy* 30(1): 470–484.

Lamb, J. and K.M. Moore. 2010. From technology transfer to adaptive management. Paper presented at the Rural Sociological Society Annual Meeting in Atlanta, GA, August 12–15, 2010.

Leeuwis, C. 2008. *Communication for Rural Innovation: Rethinking Agricultural Extension*. Wiley-Blackwell.

Malton, I. 1991. Efficient order picking—The need for it and possible solutions. Proceedings of the 11th International Conference on Automation in Warehousing, Helsinki, Finland, June, pp. 97–109.

Marenya, P., C.B. Barrett and T. Gulick. 2008. *Farmers' Perceptions of Soil Fertility and Fertilizer Yield Response in Kenya*. Department of Agricultural Economics Paper. Ithaca, NY: Cornell University, August 2008.

Millar, D. 2007. Ancestral guidance in Africa's everyday life. *Compas: Endogenous Development Magazine* No. 1, December 4–6.

Moore, K.M. (ed.). 2009. *The Sciences and Art of Adaptive Management: Innovating for Sustainable Agriculture and Natural Resource Management*. Ankeny, IA: The Soil and Water Conservation Society.

Moore, K.M. 2011. Global networks in local agriculture: A framework for negotiation. *Journal of Agricultural and Food Information* 12(1): 23–39.

Moore, K.M., J.N. Lamb, D.N. Sikuku, D.S. Ashilenje, R. Laker-Ojok and J. Norton. 2014. Multiple knowledges for agricultural production: Implications for the development of conservation agriculture in Kenya and Uganda. *Journal of Agricultural Education and Extension* 20(3): 291–307.

Mwangi, W.M. 1997. Low use of fertilizers and low productivity in sub-Saharan Africa. *Nutrient Cycling Agro-Forestry Systems* 47: 135–147.

Nederlof, S. and R. Pyborn (eds.). 2012. *One Finger Cannot Lift a Rock: Facilitating Innovation Platforms to Trigger Institutional change in West Africa*. Amsterdam, The Netherlands: Royal Tropical Institute.

Nederlof, S., M. Wongtschowski and F. van der Lee (eds.). 2011. *Putting Heads Together: Agricultural Innovation Systems in Practice*. Bulletin 396. Amsterdam, The Netherlands: KIT Publishers.

Oldreive, B. 2009. *Trainers Manual, Beginners Course*. Harare: Foundations for Farming.

Olorunfemi, B. 2010. *Comparative Analysis of Smallholder Farmers and Scientists' Perceptions of Integrated Soil Fertility Management in Ondo State, Nigeria*. B. Agric. Project. Ile-Ife, Nigeria: Obafemi Awolowo University.

Posthumus, H., B. Pound, N. Andrieu and B. Triomphe. 2011. Enhancing adoption of conservation agriculture practices through co-innovation platforms in sub-Saharan Africa. In: *Resilient Food Systems for a Changing World*. 5th World Congress Conservation Agriculture and 3rd Farming Systems Design Conference, Brisbane, Australia, September 26–29, 2011.

Prager, K. and H. Posthumus. 2010. Socio-economic factors influencing farmers' adoption of soil conservation practices in Europe, Chapter 12. In: Napier, T.L. (ed.) *Human Dimensions of Soil and Water Conservation*. Hauppauge, NY: Nova Science Publishers.

Pretty, J.N. 1995. Participatory learning for sustainable agriculture. *World Development* 23(8): 1247–1263.

Quardre, M.T. 2010. *Comparative Analysis of Smallholder Farmers and Scientists' Perceptions of Integrated Soil Fertility Management in Osun State, Nigeria*. B. Agric. Project. Ile-Ife, Nigeria: Obafemi Awolowo University.

Reardon, T. 1997. Using evidence of household income diversification to inform study of the rural non-farm labor market in Africa. *World Development* 25(5): 735–748.

Reij, C. and A. Waters-Bayer. 2001. An initial analysis of farmer innovators and their innovations. In: Reij, C. and Waters-Bayer, A. (eds.) *Farmer Innovation in Africa: A Source of Inspiration for Agricultural Development*. London, England: Routledge.

Rogers, E.M. 1962. *Diffusion of Innovations*, 1st ed. New York: The Free Press.

Rogers, E.M. 1983. *Diffusion of Innovations*, 3rd ed. New York: The Free Press.

Rogers, E.M. 1995. *Diffusion of Innovations*, 4th ed. New York: The Free Press.

Röling, N.G. and M.A.E. Wagemakers. 1998. *Facilitating Sustainable Agriculture: Participatory Learning and Adaptive Management in Times of Environmental Uncertainty*. Cambridge, UK: Cambridge University Press.

Ruef, M., H.E. Aldrich and N.M. Carter. 2003. The structure of founding teams: Homophily, strong ties, and isolation among U.S. entrepreneurs. *American Sociological Review* 68: 195–222.

Sagan, P. and 32 Nobel Laureates. 1990. Open Letter to the Global Forum of Spiritual and Parliamentary Leaders Conference. Moscow, Russia, January 1990. http://earthrenewal.org/Open_letter_to_the_religious_.htm.

Sanchez, P.A. 2002. Soil fertility and hunger in Africa. *Science* 295(5562): 2019–2020.

Sayer, J. and B. Campbell. 2004. *The Science of Sustainable Development: Local Livelihoods and the Global Environment*. Cambridge, UK: University Press.

Schuler, U., C. Choocharoen, P. Elstner et al. 2006. Soil mapping for land-use planning in a karst area of N Thailand with due consideration of local knowledge. *Journal of Plant Nutrition and Soil Science* 169(3): 444–452.

Shiferaw, B.A., J. Okello and R.V. Reddy. 2009. Adoption and adaptation of natural resource management innovations in smallholder agriculture: Reflections on key lessons and best practices. *Environment, Development and Sustainability* 11(3): 601–619.

Sutherland, A.J., J.W. Irungu, J. Kang'ara, J. Muthamia and J. Ouma. 1999. Household food security in semi-arid Africa—The contribution of participatory adaptive research and development to rural livelihoods in Eastern Kenya. *Food Policy* 24(4): 363–390.

Swenson, S. and K.M. Moore. 2009. Developing Conservation Agriculture Production Systems: An analysis of Local Networks. SANREM CRSP Working Paper 10-09. Blacksburg, VA: Office of International Research, Education, and Development (OIRED), Virginia Tech.

Swift, M.J. and K.D. Shepherd. (eds.). 2007. *Saving Africa's Soils: Science and Technology for Improved Soil Management in Africa*. Nairobi: World Agroforestry Centre.

Tenywa, M.M., K.P.C. Rao, J.B. Tukarhirwa et al. 2011. Agricultural innovation platforms as a tool for development oriented research: Lessons and challenges in the formation and operationalization. *Learning Publics Journal of Agriculture and Environmental Studies* 2(1): 117–146.

Thirtle, C.G. and V.W. Ruttan. 1987. *The Role of Supply and Demand in the Generation and Diffusion of Technical Change*. Newark, NJ: Harwood Academic Publishers.

Traxler, D. and G. Byerlee. 1992. *Crop Management Research and Extension: The Products and Their Impact on Productivity*.

Vanlauwe, B., J.J. Ramisch and N. Sanginga. 2006. Integrated soil fertility management in Africa: From knowledge to implementation. In: Uphoff, N., Bail, A.S., Fernandes, E., Herren, H., Husson, O., Laing, M., Palm, C., Pretty, J., Sanchez, P., Sanginga, N. and Thies, J. (eds.) *Biological Approaches to Sustainable Soil Systems*. New York: Taylor & Francis, pp. 257–272.

Walker, B., S. Carpenter, J. Anderies et al. 2002. Resilience management in social-ecological systems: A working hypothesis for a participatory approach. *Conservation Ecology* 6(1): 14.

World Bank. 2012. *Agricultural Innovation Systems: An Investment Sourcebook*. Washington, DC: The World Bank.

15 Factors Affecting Adoption of Soil and Water Conservation Production Systems in Lesser-Scale Societies

Ted L. Napier and Corey Henry Cockerill

CONTENTS

15.1 ADOPTION OF CONSERVATION PRODUCTION SYSTEMS: SOIL DEGRADATION—A GLOBAL PROBLEM

Erosion of agricultural land occurs in every country on this planet, to some greater or lesser extent. Some of the highest rates of soil loss occur on cropland in lesser-scale societies* where subsistence farming is most frequently employed and on soils that can least sustain such levels of soil loss without serious declines in productivity. Scherr (1999) discusses soil degradation on a global scale and observes that about 3.2 billion hectares† of land are potentially arable and about 1.5 billion hectares of the arable land are used to produce crops. The remaining 1.7 billion hectares and most of the nonarable lands are used for pasture and forests. She notes that soil quality has remained basically stable on about 75% of the world's agricultural lands during the past five decades. However, productivity has declined on about 16% of cropland in lesser-scale societies, especially in Africa and Central America. She observes that about 75% of the agricultural land in Central America has been seriously degraded and about 20% of the cropland in Africa have been negatively affected by soil erosion.

Scherr (1999) observes that continued degradation of soil resources at present rates will result in an increase in world food prices from 17% to 30% during the present decade. This estimate is based on the assumption that global population will increase about 33% between 1995 and 2020. While global agricultural supplies may not become a serious problem to the world's population owing to increased food and fiber production in countries not seriously affected by soil degradation, poor people living and operating subsistence farms in lesser-scale societies will suffer significantly because they will be unable to pay inflated prices for essential commodities.

Poor farmers also will not be able to accumulate sufficient economic resources to invest in conservation production systems that have the potential to increase farm output while simultaneously protecting soil and water resources. The ultimate outcome of increased commodity prices will be retarded adoption of soil and water conservation production systems in lesser-scale societies where such farming systems are sorely needed.

Scherr (1999) notes that between 5 and 12 million hectares of farmland are lost to soil degradation every year. Assuming that present trends continue, about 35 million to 84 million hectares of agricultural land will be lost by the year 2020. Such losses of cropland will further reduce the opportunities for poor farmers to establish viable farming operations and will reduce the availability of food and fiber. Flora (2010) even suggests that soil loss in lesser-scale societies and other regions of the planet will threaten food security in the future. Scherr's (1999) bleak overview of existing soil degradation on this planet strongly suggests that adoption of soil and water conservation production systems is essential if future productivity of agricultural land in lesser-scale societies is to be protected (Figure 15.1).

---

* Lesser-scale societies are defined as countries that are primarily agriculturally based with a major portion of the population engaged in subsistence agriculture. Industrial production is usually low in lesser-scale societies, and such societies usually do not have institutional structures to satisfy basic service needs of resident populations. Marketing and infrastructures of lesser-scale societies are usually inadequate for the needs of the populace. Production of goods and services tend to rely heavily on labor-intensive approaches.

† A hectare of land is equal to approximately 2.47 acres.

FIGURE 15.1 Landscape view of village in High Atlas Mountains, Morocco.

While identification of socioenvironmental problems associated with use of inappropriate production systems is relatively easy, motivating land managers to actually change production systems is extremely difficult. While many subsistence farmers in lesser-scale societies are aware that farm production systems presently in use do not produce food and fiber of sufficient quantities to feed and clothe their families, they continue to use the same farming systems because they have few or no alternatives to what they are presently doing. Unless barriers to adoption of conservation systems at the farm level can be identified and removed, subsistence farmers in lesser-scale societies will continue to exploit the land and environmental degradation will continue until cropland is no longer useful for any agricultural purpose. The primary objectives of this chapter are to identify factors that have been demonstrated to be significantly related with adoption of soil and water conservation production systems at the farm level in lesser-scale societies, and to propose a model designed to increase the probability that adoption of conservation systems will occur.

15.2 SOIL CONSERVATION PRODUCTION SYSTEMS IN LESSER-SCALE SOCIETIES

Anthony et al. (1998) argues that soil erosion will continue to exist in poor countries because subsistence farmers have no alternative to operating small farms on sloped land. They note that it is unrealistic to expect national governments to remove subsistence farmers from highly erodible land to protect natural resources from degradation because it is not politically acceptable to remove people who have no other alternative. Social upheaval would follow such a draconian approach to resolve environmental degradation issues. It is much more likely that governments in lesser-scale societies will continue to provide poor farmers with ownership of land resources on steep slopes as a means of reducing social unrest.

Providing subsistence farmers with ownership of land resources not only reduces political upheaval, but land tenure rights also encourage land managers to invest in conservation efforts. Farmers will be more likely to adopt conservation production systems if they are able to capture future income streams from adoption of conservation practices and/or technologies. However, it must be noted that ownership alone will not ensure adoption of conservation production systems because many factors influence the adoption decision.

One of the most important factors associated with adoption of soil and water conservation production systems is the existence of farming techniques and/or technologies to address erosion problems. Anthony et al. (1998) note that many farm production systems have been developed to address soil loss on sloped agricultural land in lesser-scale societies. They discuss several types of soil and water conservation techniques that have been effectively used to reduce soil loss on steep slopes within various geographic regions of the planet. They argue that no

FIGURE 15.2 Extreme terrain modified for agriculture, near Rimetea, Romania.

conservation production system is appropriate for every socioenvironmental situation. Slope of the land, climatic conditions, availability of resources, availability of labor, technical skills of the land manager, and a host of institutional factors influence the type(s) of conservation techniques that are appropriate for specific situations (Figure 15.2). The same position has been advanced by El-Swaify (1997) and Napier et al. (1994).

Anthony et al. (1998) posit that land managers must adopt production systems that reduce runoff and reduce splash impacts via use of ground cover and the establishment of barriers to surface water mobility. These authors suggest that soil erosion control approaches can be subsumed under three broad categories that they refer to as physical measures, vegetative measures, and integrated watershed management. Physical measures consist of bench terraces, hillside ditches, individual basins for tree planting, tree terraces, natural terraces composed of bunds that fill with sediment over time, hexagons composed of roads that provide multiple entry by machinery, vegetative barriers, grass waterways and barriers, alley cropping, surface drainage systems that divert runoff to reduce water velocity, water-harvesting systems consisting of small catchments, and holistic conservation systems using the watershed as the unit of planning.

Anthony et al. (1998) present a number of examples to show how each technique is implemented. Examples of rice terraces in northern Luzon demonstrate how terraces can be an effective means of conserving soil and water resources. Personal observation in Bali and other islands in the Pacific Ocean by the principal author of this chapter are indicative of the utility of rice terraces for controlling erosion on steep slopes (Figure 15.3).

Conservation of water resources via use of runoff farming in the Negev is discussed by Anthony et al. (1998), as well as tea and coffee plantations on bench

FIGURE 15.3 Terrace farming (right) as a strategy for controlling soil erosion and/or chemical or nutrient runoff. (Illustration by Corey Cockerill and Jamie Henry of ZigNine Design Studio.)

terraces in Sri Lanka, Kenya, Taiwan, Java, Colombia, and Jamaica. Personal observation of tea plantations in Taiwan, the People's Republic of China, and other regions of the planet by the principal author of this chapter provides further evidence of the success of bench terraces for controlling erosion. Examples of the use of rock walls in South and Central America by Anthony et al. (1998) show how physical barriers can be effectively employed to control runoff and to capture eroded soil from upslope. They note that permanent set-aside programs in lesser-scale societies have been implemented in some countries such as Costa Rica, where several national parks have been created to protect forests and wildlife. Such parks not only protect soil, water, forest, and wildlife from destruction associated with subsistence farming, but they also serve as potential sources of income in the form of ecotourism.

Vegetative approaches include planting various types of vegetation between rows of crops to retard water runoff. Small bushes and trees planted along hillslope contours can slow runoff and result in the development of small flat areas for planting crops. These efforts become even more effective when they are combined with the development of terraces. The principal author of this chapter has observed many examples of successful vegetative soil conservation techniques used in several countries in Africa and Asia. Integrated watershed management consists of holistic planning that incorporates multiple conservation approaches at the watershed level. Integrated approaches require massive quantities of human and economic resources to be effective. The only successful examples the principal author of this chapter has observed are in Taiwan where watershed-level conservation programs have been implemented using government resources to finance large-scale projects. While this type of conservation effort is highly desirable, it is extremely expensive to accomplish. Such approaches also require extensive human resources to effectively implement.

Anthony et al. (1998) discuss many conservation options that presently exist for consideration by potential adopters. They observe that land managers must become familiar with a multitude of relevant conservation production systems and choose the option that is most relevant to their specific farming operation. Unfortunately, information about many adoption options is not readily available owing to lack of institutional delivery systems.

15.2.1 Facilitators and Barriers to Adoption

During the past two decades, extensive research has been conducted in many societies to identify factors that influence the adoption of soil and water conservation production systems at the farm level. Considerable attention has been focused on adoption of conservation farming technologies and techniques among subsistence farmers in lesser-scale societies because soil and water degradation is most severe on land operated by these types of farmers. Many socioeconomic factors have been shown to influence adoption of soil and water conservation production systems in lesser-scale societies. These factors have been classified into two broad categories for discussion purposes, and are referred to as facilitators of adoption and barriers to adoption. The facilitator category is composed of factors that increase the probability of adoption, while the barriers category consists of factors that impede adoption.

Some of the factors that have been demonstrated to be useful for understanding adoption of soil and water conservation within lesser-scale societies are as follows: availability of information about soil and water conservation production systems, awareness that existing farm production systems contribute to degradation of soil and water resources, availability of economic resources that can be used to adopt conservation production systems at the farm level, availability of institutional structures to provide requisite resources to implement conservation production systems at the farm level, public policies that make it possible for land managers to benefit from investments made in conservation production systems, institutional means for controlling population expansion, availability of government programs designed to reduce poverty, availability of markets so that incentives exist to efficiently produce food and fiber, availability of conservation production technologies at the local level, government investment in conservation infrastructures, and characteristics of the farm manager and farm enterprise (El-Swaify 1997; Napier et al. 1994; Napier 2010).

Each of these factors will be discussed in the context of how they facilitate and/or impede the adoption of soil and water conservation production systems within lesser-scale societies. These factors are discussed in the context of the traditional diffusion model that has been shown to have utility for predicting adoption of innovations* in diverse socioenvironmental situations.

15.3 DIFFUSION MODEL FOR CONSERVATION DECISION MAKING

15.3.1 Steps in the Diffusion Model

One of the most significant paradigms for understanding the decision-making process associated with adoption of innovations is the Traditional Diffusion Model (Cockerill and Napier 2010; Napier 1991; Rogers 2003). This model has been employed extensively to examine the adoption process associated with a host of innovations within lesser-scale societies throughout the world. The traditional diffusion model basically argues that potential adopters of any innovation must become aware that a problem exists before they will consider changing existing behavioral patterns. The model asserts that potential adopters must be provided information about the existence of problems created by the use of existing behavioral practices and be provided information about potential solutions. Once potential adopters have been made aware of existing problems and possible solutions, many factors affect the ultimate decision to adopt specific action options.

One of the most important factors affecting adoption of a specific innovation is attitude toward the innovation being considered for adoption. Attitudes about specific innovations are formulated via access to information that demonstrates the potential utility of the innovation being considered. Information about how a specific innovation will affect identified problems will facilitate decision making because

* An innovation is defined as any technology, technique, object, or cultural component that is perceived to be new to a potential adopter. Although something may have been in existence for decades, it can be perceived as an innovation to those who have never encountered it before.

potential adopters will be better able to assess probable outcomes associated with adoption. If the innovation being assessed is perceived to reduce or eliminate identified problems, then it will tend to be perceived favorably and a positive attitude will begin to evolve.

Once a positive attitude toward a specific innovation has been developed, other factors become operative in the adoption decision-making process. Potential adopters begin collecting information about potential consequences of adoption at this stage of the decision-making process. If information about specific innovations suggests that benefits associated with adoption exceed costs, then the innovation being assessed will continue to be considered for adoption. If the assessment of potential outcomes suggests that the costs of adoption will be higher than the benefits, then it is highly probable that no further consideration will be given to that specific innovation.

Assuming the assessment of possible outcomes is favorable toward a specific innovation, the diffusion model argues that the adoption process will continue into the next stage of decision making. Factors that are most important in the determination of the outcome of the decision-making process at this stage are factors that influence the ability of the potential adopter to act on the attitudes and perceptions that have been developed about specific innovations. Adoption of innovations will not occur if the potential adopter cannot implement a decision to adopt regardless of the desire to do so. Many factors act as barriers and/or facilitators of adoption of innovations at this stage of the decision-making process. The types of factors that can facilitate or impede adoption of a specific innovation vary from individual to individual and from innovation to innovation. Adoption of some innovations requires very little in terms of material inputs, while others require extensive allocation of economic resources to effectively implement. The nature of the inputs required to adopt an innovation acts as a major consideration in the adoption decision-making process. Innovations that require the commitment of extensive economic resources to implement tend to be less-readily adopted.

The human skill level required to implement an innovation significantly affects the outcome of the adoption decision-making process. Some innovations require extensive human skills to adopt, while others can be effectively employed with little training. Innovations that require higher levels of human skills to be effectively implemented tend to be less extensively and more slowly adopted because potential adopters often do not possess such skills. This barrier is further compounded when institutional structures do not exist for potential adopters to obtain the needed skills via education and technical training.

The perceived short-term and long-term impacts of adopting a specific innovation are very important in the adoption decision-making process. The level of risk associated with adoption of specific innovations varies tremendously. Innovations that have the potential to threaten the socioeconomic well-being of adopters and their families tend to be adopted more slowly, and the duration of the evaluation stage of the decision-making process tends to be much longer.

Adoption of some innovations can result in extremely negative consequences for the adopter, if expected outcomes of adoption are not realized. Innovations that are more certain to produce beneficial outcomes will tend to be more quickly and more

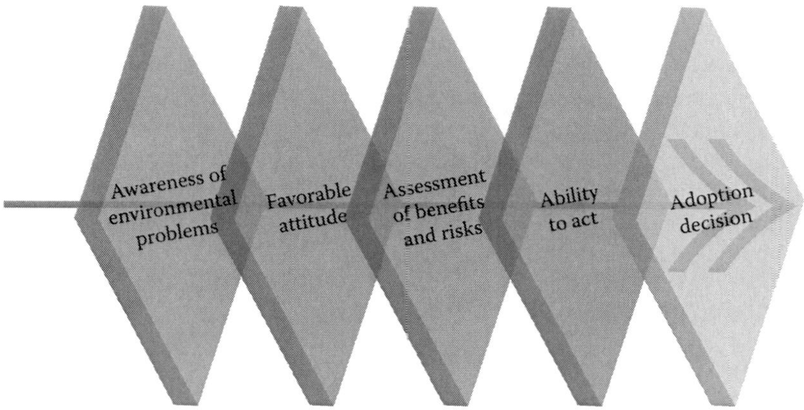

FIGURE 15.4 Stages of adoption: factors that impede and/or facilitate adoption decisions. (Illustration by Corey Cockerill and Jamie Henry of ZigNine Design Studio.)

extensively adopted. The complexity associated with implementation of specific innovations is an important consideration in the adoption decision-making process. Complexity refers to the level of difficulty associated with implementing a specific innovation. Some innovations are quite difficult to implement because they require use of complex technologies to achieve expected benefits. If potential adopters do not have the human and/or economic resources to implement such innovations, it is highly unlikely they will be adopted. Conversely, many innovations employ low-level technologies and/or techniques that are much easier to effectively implement even with low levels of human and economic resources.

Another consideration in the adoption decision-making process is whether potential adopters can adopt innovations on a piecemeal basis, or whether they must adopt all components of the innovation simultaneously. Some innovations require adoption of multiple components to effectively achieve anticipated outcomes. Other innovations may be implemented on a sequential component basis. Technology-intensive production systems frequently require adoption of multiple components simultaneously, while more labor-intensive production systems can often be implemented sequentially. Adoption of innovations that require implementation of multiple components simultaneously tends to be more slowly and less extensively adopted.

The traditional diffusion model as a decision-making process is conceptualized in Figure 15.4. The double arrow in the "Adoption decision" diamond indicates continued assessment of the decision made.

15.3.2 Diffusion Model Applied to Agricultural Pollution Abatement

It is argued that the diffusion model is appropriate for understanding the adoption of conservation production systems in lesser-scale societies because many conservation technologies and techniques are totally new to farmers in lower-scale societies although they are well known and frequently used in high-scale societies. The various elements of the traditional diffusion model can be combined to develop a decision-making framework to explain how adoption of soil and water conservation can be implemented within lesser-scale societies. While adoption of conservation production systems will never be accepted by all farm managers within any society, use of the diffusion model to introduce new conservation production systems will increase the probability that the outcome of the process will be greater adoption of conservation practices and/or technologies at the farm level.

The first stage in the traditional diffusion model is the creation of awareness among farmers about socioenvironmental problems associated with use of production systems that contribute to degradation of agricultural land. To accomplish this objective, education and exposure to environmental information must be provided to land managers. While provision of information and training may appear to be a very simple task to perform for people living in high-scale societies, such as the United States, this task is often extremely difficult to accomplish within lesser-scale societies. Within high-scale societies, government agencies have been developed to provide farmers with information in a prompt and efficient manner. Mass media systems have been created that instantaneously provide information to millions of people via multiple venues. While information and training institutions are quite common and easily accessible within high-scale societies, such institutions seldom exist in lesser-scale societies, especially within rural areas of poor countries. Without access to training and information-dissemination structures, the conservation decision-making process is often terminated at this point.

Most often, conservation training/information within lesser-scale societies are delivered by government agents or by change agents provided by nongovernment organizations. Local farmers who have experience with conservation production systems sometimes act as instructors and communicate their knowledge to other farmers through informal systems. The latter approach can prove to be problematic because farmers engaged in teaching may not be aware of some important elements of the conservation production system they are discussing. Failure to inform potential adopters about critical components may render what they communicate faulty or inadequate to achieve anticipated conservation/production goals. Failure to achieve anticipated outcomes may result in rejection of future conservation initiatives among potential adopters who have observed what has occurred.

Assuming potential adopters have been made aware of environmental problems associated with the agricultural production systems presently being employed, the second stage of the diffusion model becomes operative. At this stage of the decision-making process, land managers should be willing to consider possible solutions. Access to information about alternative action options becomes very important. Land managers begin the process of developing attitudes toward alternative action options, and this requires access to information about all action options being considered.

The same barriers exist at this stage of the decision-making process that existed in the initial stage of the diffusion model. If information about alterative action options is difficult to access, the options that can be assessed are very limited. Frequently, the most relevant action option for a specific land manager is not considered because information about that specific innovation is not available. If information provided about even a limited number of action options is inadequate to formulate definitive attitudes toward the options being assessed, then it is highly likely that adoption will be slow to occur because potential adopters may be uncertain about possible outcomes. Land managers will be reluctant to change existing production systems unless they are certain the change will result in a more viable farm operation. Inaccurate and/or inadequate information could also result in the development of negative attitudes about the action options being assessed because important components of the innovations being assessed are not provided. If some of the possible benefits of specific innovations are not provided, potential adopters may perceive the options being assessed less favorably and elect not to adopt. If adoption takes place using inadequate or inaccurate information, expected outcomes of adoption may not be realized because the perceptions formulated may be unrealistic relative to possible outcomes.

Assuming that adequate and accurate information about soil and water conservation has been provided to potential adopters so that they have been able to formulate valid attitudes about action options being considered, land managers must eventually make the decision to adopt or to reject specific conservation technologies and/or techniques. Major factors that influence the ultimate decision at this stage of the decision making process are primarily related to the ability of the land manager to effectively implement the adoption decision. Land managers must determine if they have relevant farm management skills and the economic resources to implement the adoption decision. Individual farmers must determine if the action option being assessed is relevant to their farm operation and if it is consistent with other components of the farm enterprise. There are many other factors farmers must consider when making the adoption decision, such as the level of technical skills required to effectively implement the conservation production system being evaluated, the potential impact of failure of the selected action option to produce expected outcomes on the socioeconomic viability of the farm enterprise, the short-term and long-term impacts of the adoption decision, the ability to reject the adoption decision at some future time without harming the future viability of the farm enterprise, expected return to investment made in the conservation action option being considered, and the potential sociocultural impacts of the adoption decision on the family.

Assuming subsistence farmers elect to adopt new conservation production systems, they must marshal economic resources and technical expertise to implement the adoption decision. In lesser-scale societies, this is often a very difficult task. Relatively few farmers in lesser-scale societies possess sufficient economic resources to implement conservation production systems at the farm level. If the conservation production system being considered requires the allocation of extensive economic resources, it is highly unlikely that implementation will occur unless some sponsor provides the economic resources to finance adoption. Most farmers in lesser-scale societies are subsistence farmers who are barely able to feed and clothe their family. Surplus income is a rarity among subsistence farmers, and such farmers have little or no access to credit institutions. Lack of capital to invest in conservation is a very significant barrier to adoption of conservation production systems among subsistence farmers.

Another consideration is the land tenure system that often operates in lesser-scale societies. Many poor people have little or no security in terms of ownership of the land resources they operate. Lack of land tenure rights makes land managers insecure relative to making long-term investments in the future productivity of the land resources they operate. If land managers have little or no assurance that they will be able capture future income streams from investments made in conservation, they will be very reluctant to invest in land protection technologies and/or techniques.

The final stage of the diffusion model is the evaluation/assessment of the decision made. Farmers continue to assess the outcomes of the adoption decision long after the initial decision has been made. Assessments of actual outcomes of the adoption decision are initiated as soon as the implementation stage has been completed. Land managers will compare the actual outcomes of adopting the new production system with expected outcomes and with previously used production systems. If the assessment demonstrates that adoption has produced valued outcomes in excess of costs, then the land manager will probably continue to use the new system. If the outcomes associated with adoption are negative in the near term, it is highly likely that the new production system will be rejected. It also must be recognized that the planning horizon for subsistence farmers is very short. Failure of one crop year could prove to be extremely problematic for subsistence farmers and their families. Newly adopted conservation production systems must produce immediate benefits for subsistence farmers, or the new systems will be quickly rejected.

15.3.3 Adoption Studies in Lesser-Scale Societies

El-Swaify and Marten (2010) and Oldeman et al. (1991) discuss the impact of soil degradation on poor people living in lesser-scale societies. They observe that continued degradation of soil and water resources will have significant adverse impacts on the future socioeconomic well-being of humanity. They note that land degradation associated with agricultural production is particularly problematic in lesser-scale societies owing to lack of institutional structures required to address such problems. In many economically poor countries, agricultural productivity is being threatened by soil loss so extensively that it is highly probable that future agricultural productivity will decline significantly. Environmentally sensitive land will continue to be brought into production using production systems that are highly degrading of land and water resources. Cropland presently in production will be farmed even more intensely as productivity declines and land degradation will increase. Unless conservation production systems are adopted and used, the future is very bleak for people living in lesser-scale countries. Poor countries do not have the economic resources to purchase food and fiber in the world market, which means that many people will suffer terribly if domestic agriculture is unable to feed and clothe resident populations.

El-Swaify and Marten (2010) argue that it is highly unlikely that conservation production systems will be widely adopted among subsistence farmers in lesser-scale societies. They note that soil and water conservation efforts in lesser-scale societies have much lower priority in terms of national development agendas because of the following reasons: level of poverty that exists, lack of institutional structures to provide educational opportunities, lack of educational infrastructure to provide needed technical skills to implement conservation production systems, lack of institutional structures to generate new knowledge about conservation issues, lack of institutional structures to deliver newly discovered scientific knowledge about conservation, lack of access to conservation technologies and techniques, lack of economic institutions to provide needed credit to low-income farmers, lack of access to foreign and domestic markets to motivate land managers to emphasize more efficient and productive agricultural systems, lack of transportation systems to access foreign and domestic markets, lack of dependable institutional structures to deliver production inputs at the local level, lack of access to newly created biotechnology, and lack of institutional structures to put public policies in place that will facilitate conservation efforts within the countries. Unless institutional structures are established in lesser-scale societies to address these issues, it is highly probable that subsistence farmers will continue using traditional production systems that will degrade land and water resources until future agricultural productivity is severely threatened.

Napier (2013) posits that the dire predictions made by El-Swaify and Marten (2010) and Oldeman et al. (1991) about the consequences of continued degradation of land and water resources in lesser-scale societies will be realized if expectations about the impacts of global warming on future agricultural productivity are shown to be valid. While farmers within high-scale societies will probably be able to adapt production systems to significantly nullify the effects of increasing global temperature owing to their access to technology-intensive production systems and educational structures, farm managers in lesser-scale societies will not have access to such resources and will not be able to adapt to changing socioenvironmental conditions. It is also highly likely that global warming will result in more frequent and intense storm events. If this prediction is shown to be true, land operators in lesser-scale societies who have not incorporated erosion control techniques into their farm production systems will observe much greater erosion and subsequent degradation of their land resources over time.

15.3.4 Soil and Water Conservation Studies by Geographic Region

Adoption of soil and water conservation production systems has been examined in a host of lesser-scale societies. This literature is examined in the context of the geographical location of the studies.

15.3.4.1 Soil Conservation in the Tropics

El-Swaify and Marten (2010) discuss the nature of soil erosion and land degradation within the tropics. The authors note that soil erosion is particularly problematic between the tropics of Cancer and Capricorn due to climatic conditions of the region and the high incidence of poverty that adversely affects resident populations. Many of the 160 countries within this geographic region of the planet are populated by subsistence farmers who operate very small land holdings. Most of the land operators in the tropics are very poor and have few resources to invest in alternative production systems. This is especially true for production systems that require access to capital and require technical skills to effectively implement at the farm level. Unemployment and disease further complicate the situation and impede the adoption of conservation production systems.

Many community and natural resources development scholars have advocated the introduction of complex agricultural technologies as a means of addressing environmental degradation of land and water resources within the tropics. El-Swaify and Marten (2010) argue that such approaches will not be successful because farmers do not possess requisite human skills or the capital to invest in such production systems. They also note that suitable land is not available for use of technology-intensive production approaches and that land tenure is not assured. They suggest that lack of institutional structures to effectively address the barriers to adoption will ultimately prevent widespread adoption of soil and water conservation production systems within the tropics (Figure 15.5).

Each of the variables that El-Swaify and Marten (2010) have identified as being barriers to adoption of soil and water conservation production systems in the tropics is consistent with the traditional adoption diffusion model. Each of the factors they observe as contributing to environmental degradation within the tropics is closely aligned to those discussed in the diffusion literature as influencing adoption of innovations. Lack of human skills, lack of institutions to generate new agricultural knowledge, lack of communication systems, inability of subsistence farmers to assume risk, the culture of poverty that permeates the region, and a host of other socioeconomic variables significantly influence the ability of farmers to adopt conservation systems. These barriers make it very difficult to motivate subsistence farmers to adopt soil conservation production systems even when it would be in the best interests of individual farmers and society for them to do so.

FIGURE 15.5 Steep slope farming near Pinantura, Ecuador. (Courtesy of Dr. Tom Stilwell.)

15.3.4.2 Soil Conservation in Africa

Wangia and Prato (1994) and Lovejoy and Sanders (1994) synthesized a host of studies focused on the adoption of conservation production systems in Africa and made many of the same observations as those noted by El-Swaify and Marten (2010). They observed that studies conducted within all regions of Africa demonstrate that the major causes of soil erosion and subsequent degradation of soil resources are the result of adoption of poor cropping and land management systems, overgrazing of pastures, deforestation, failure to properly terrace sloped land, and the removal of crop cover and residue (Figure 15.6).

Lovejoy and Sanders (1994) suggest that erosion problems in Africa are the result of overpopulation that has placed high pressure on farmland and has resulted in the fragmentation of land holdings into small segments that are too small to sustain resident populations. They also argue that technology-intensive production systems that have been introduced as a means of overcoming low productivity of land resources have failed because subsistence farmers do not possess skills and economic resources to effectively implement such development strategies.

Lovejoy and Sanders (1994) also posit that property rights compound the issue of soil degradation because land operators are not certain that they will receive the benefits of investments in land improvement. Public policies that favor urban residents over rural populations, such as cheap food policies, impede adoption of production systems that will increase productivity. Farmers will not produce larger quantities of food and fiber when government policies keep commodity prices low. Wangia and Prato (1994) support the position of Lovejoy and Sanders (1994) when they state that erosion of soil and water conservation will never be adequately addressed on the continent of Africa until public conservation policies are implemented and enforced. They submit that African countries must establish institutional structures to address such issues as ignorance among farmers concerning the causes and consequences

FIGURE 15.6 Steep slope shanty farm on the hills of Rwanda.

of soil erosion, as well as the lack of knowledge about potential solutions to erosion problems; create regulations regarding access to and use of commons*; formulate tenure arrangements that provide land managers rights to the land they operate; establish institutional structures that will elevate conservation initiatives to a much higher level in the hierarchy of development priorities of the society; create funding systems to ensure that conservation institutions will be adequately financed and staffed by capable agents for change; and create national conservation institutions that will complement existing cultural traditions that are highly valued by potential adopters.

Lovejoy and Sanders (1994) also submit that rejection of the use of agricultural technologies is a poor strategy for agricultural development within many African countries. Unlike other conservationists who advance more indigenous approaches to conservation (Shaxson 1993; Shaxson et al. 1989), Lovejoy and Sanders argue that increased productivity, which is essential to the development of African nations, is only possible via adoption and effective use of more technology-intensive farming systems. While these authors recognize the need to continue use of some indigenous production practices, they also note that sole reliance on such approaches will not produce greater output nor will use of such practices result in lower rates of erosion. Unless higher levels of technologies are effectively employed, low levels of output will mire farmers in poverty and degradation of soil resources will continue until the land can no longer produce enough output to sustain resident populations.

Kraybill (2010) notes that sub-Saharan countries in Africa are primarily populated by subsistence farmers and that approximately 60% of the population within sub-Saharan Africa depend on farming to survive. Unfortunately, agricultural production is very low and it is highly unlikely that greater production will be achieved in the future. Southgate and Graham (2010) observe that more than half of sub-Saharan African land is not suitable for agricultural production, and rapidly increasing populations further increase the demand for food and fiber. Climatic conditions keep agricultural production low. Drought combined with high temperatures destroys soil organic matter, which adversely affects water retention in most sub-Saharan countries.

Kraybill (2010) and Southgate and Graham (2010) have synthesized existing literature focused on the adoption of soil and water conservation production systems within sub-Saharan Africa, and conclude that widespread adoption of conservation production technologies and techniques will probably not occur within the region. They note that the primary reasons adoption of conservation production systems has not, and probably will not, occur within the region are due to socioeconomic

* "Commons" means land that is collectively owned and/or managed by a specific group (Crowe 1969; Hardin 1968). Access and use of common land are usually determined by rules established by the group that owns/manages the land. Unless the commons are protected by explicit rules and the rules strictly enforced, the commons will be degraded because there is no motivation to conserve any portion of the commons by any individual user. Individual users will seldom invest in conservation efforts because there is no means of securing benefit streams from investments made in the commons. Users are often motivated to mine the commons. Individual users are aware that the commons can sustain only a limited amount of use and if individual farmers do not maximize their use of common resources, other users will do so. Individual farmers will maximize their short-term use of the commons so that others will not have access to the resources within the commons to do so.

factors that impede adoption. Some of the most important factors impeding adoption are as follows: lack of capital accumulation among poor farmers, incompetent government structures and leadership, exploitation of natural resources by external groups, lack of free market policies, little investment in conservation infrastructures, lack of enforcement of existing conservation policies, uncertain land tenure rights, cheap food policies by government to keep food and fiber prices low for urban elites, little investment in infrastructures and communication systems, and a host of other needed public investments to reduce the level of poverty. Kraybill (2010) and Southgate and Graham (2010) note that very few subsistence farmers have any formal education and/or technical training, which means human skills required to adopt and effectively use new conservation production systems do not exist among potential adopters. Subsistence farmers do not possess requisite economic resources to adopt conservation production systems. This means poor farmers will continue to exploit farmland because they have no other options.

Subsistence farmers "mine"* the land (Henao and Baanante 2006) because they do not have access to resources to make investments in soil and water conservation initiatives on their farms. Baidu-Forson (1994) discusses adoption of soil and water conservation in the Sudano–Sahelian zone of Africa, and notes that socioeconomic factors are very important factors in the decision to adopt and/or reject conservation production systems. While wind and water erosion are very severe within the study region and adoption of soil and water conservation production systems is essential to long-term productivity of land resources, adoption of conservation production systems is very low. The author notes that lack of access to machinery, fertilizer, and other farm inputs impede the adoption of conservation production systems. He also notes that lack of access to capital to invest in conservation, lack of access to labor due to migration of males for work in other areas, low and slow return to investments made in conservation, lack of awareness of benefits associated with adoption of conservation practices, inability of subsistence farmers to assume risk associated with adoption, lack of knowledge about how to implement conservation production systems, lack of awareness of the need for specific soil and water conservation practices, insecure land tenure rights to land operated by subsistence farmers, and a host of other factors are the major reasons that adoption has been slow within the study region.

Pagiola (1994) examined soil conservation issues in Kenya in the context of the economic return to investment in conservation structures. His research demonstrated that creation of certain types of physical structures resulted in improved income and enhanced environmental quality. He also noted that all farmers did not adopt the new terracing system although it was demonstrated that adoption often produced net benefits for those who adopted the new production techniques. He observed that the two most important barriers to adoption of the new conservation terraces were lack of human skills and the labor required to construct them. The terrace building technique being diffused consisted of removing soil along the contour of sloped land

* "Mine the land" means that land managers extract everything possible from land resources without making investments in the improvements of soil quality. Output is maximized without regard for future consequences of land degradation. Exploitation of land resources occurs most often when land managers have no other option in terms of survival.

and placing the displaced soil above the excavated trench. The ridge of excavated soil acts as a barrier to silt-laden water from upslope, which results in sedimentation behind the barrier, forming a small terrace that is used for planting. The economic costs associated with the development of this type of terrace are practically zero. However, the labor costs associated with the implementation of this terracing approach is high. Assuming that labor exists to do the initial excavation, many subsistence farmers could implement such a terracing system over time.

Pagiola (1994) notes that one of the major factors that impeded the adoption of the new conservation technique was the perception of return to investment of time and labor associated with building the terraces. When it was demonstrated that risk associated with adoption was low and the benefits high, farmers had a higher propensity to adopt the new terracing technique. He also observed that farmers wanted information about the risk level and the expected return to investment before they would consider adopting the new technique. Within the Kenyan study region, communication systems were practically nonexistent and the means of providing information to farmers was lacking. The lack of access to relevant information about the innovation acted as a barrier to adoption. Pagiola (1994) also submits that land tenure was an important consideration in the adoption decision-making process. If land managers did not have property rights to the land being operated, they tended to reject the conservation practice being assessed because they would not benefit from investments in building the terraces.

Giller et al. (2009) assessed the probabilities of adoption of soil and water conservation production systems in Africa and suggest that most conservation production systems will never be adopted. They assert that subsistence farmers in Africa do not have access to technologies, and do not possess required inputs and labor to implement such production systems. They further suggest that many subsistence African farmers adopt some components of conservation production systems but not all parts of the production systems. This observation strongly suggests that many production systems that are being diffused are not relevant to the needs of African farmers. They conclude that development and diffusion of more relevant conservation systems may result in higher levels of adoption.

Research reported by Adeola (2010, 2012) strongly suggests that exposure to information and access to technical support systems are significant in the adoption of soil conservation practices among subsistence farmers in Africa. Perceptions of declines in soil fertility associated with existing farm production systems and adoption of seven conservation farming practices were examined using data collected from subsistence farmers in the Ibadan/Ibarapa region of Nigeria. Study findings revealed that level of education, contact with extension agents, farming experience, and farm size contributed to the development of awareness of soil fertility problems associated with erosion of farm land and the adoption of the conservation measures assessed.

Adeloa (2010, 2012) argues that socioeconomic factors are important in the adoption decision-making process when land operators within the study region are considering adopting conservation production systems. He notes that exposure to information about soil erosion and environmental degradation is useful in the development of perceptions about declining fertility of land resources and contributes to a willingness to consider adopting conservation practices to reduce degradation of soil resources.

Faltermeier (2007) reports research findings from a study focused on the adoption of water conservation practices in northern Ghana. The study region is characterized as being seasonably dry, which necessitates water conservation during the early portion of the growing season to produce sufficient levels of output to sustain resident populations. The author observes that northern Ghana is primarily populated by small-scale subsistence farmers who produce cereals on small land holdings. Most farmers operate about 1 ha of land and do so with little technological knowledge and with very little mechanization. She observed that rice production was increasing over time owing to subsidies provided by external sponsors for purchase of technological inputs. After subsidies were withdrawn, access to technological inputs at the local level was significantly reduced. The ultimate outcome of the elimination of external subsidies was a substantial decline in rice output within the region.

A project was organized to encourage the adoption of conservation structures to reduce water loss and reduce the erosion of land resource. The project was also designed to encourage the adoption of innovative planting techniques, provide access to farm inputs, provide access to credit, provide information about conservation production systems, provide farm management technical support, and provide collective structures to capture water. The study focused on the assessment of adoption of physical structures that consisted of small dams built along the contour of the slopes and the use of dibbling* techniques for planting rice fields. The contour dams were designed to capture excess water during seasons when rain was more abundant to be used later to provide water to rice fields. Faltermeier (2007) observed that more farmers adopted dribbling than dam construction because the costs of dam construction were much higher, especially labor costs. Access to economic resources to implement adoption of the farming innovations and perception that the innovations would result in increased profitability of the farm enterprise were important determinants of adoption. Formal education was not significant as a predictive variable because there was no variance. Practically all of the study participants had no formal education.

Swenson and Moore (2009) examined existing literature focused on adoption of no-till conservation production systems within several regions of the planet that are primarily populated by subsistence farmers. The authors state that adoption and use of no till in lesser-scale societies in Africa is consistent with established farming traditions. While adoption of no-till production systems has been very slow in nearly all African countries and almost nonexistent among small-scale farms, the potential exists for no-till farming systems to be accepted because subsistence farmers have used no-till approaches in their farming systems for centuries. Subsistence farmers have used a stick to open holes in the earth for seeds. More modern no-till approaches employ the same principles but on a much larger and more complex scale.

* Dibbling is a technique of manual seed planting on hillslopes using a pole with a scoop attached. The farmer lifts the soil with the scoop and plants the seed in the hole. The seed is covered with the soil displaced by the scoop. This planting technique is appropriate in areas where access to machine planters is limited or on land that is not accessible for planting by mechanical drills.

Swenson and Moore (2009) attribute the lack of adoption of no-till farming systems in many African countries to the inappropriateness of no-till production systems for small-scale farm operations, lack of access to chemical inputs to control weeds, lack of access to economic resources to implement no-till systems, lack of access to information and technical assistance to effectively use no-till technologies and techniques, lack of access to institutional structures to provide prompt and appropriate information about local socioenvironmental conditions, and lack of a host of institutions to combat poverty. They note that adoption of no till has occurred to some extent in government-sponsored programs in Ghana and Zambia.

The most frequently adopted no-till conservation practice in Zambia is "planting basins" that are constructed by digging small shallow pits by hand with a hoe and filling the hole with soil, compost, fertilizer, manure, and any other organic matter available. Seeds are planted with little disturbance of the soil. This practice has been shown to be successful in maintaining and improving production even in very dry areas. The primary reason this technique has been adopted is the very low economic requirements for its adoption. When labor exists to construct the basins, this is an inexpensive means of building crop-producing areas.

Swenson and Moore (2009) conclude their assessment of the situation in Africa by noting that adoption of soil and water conservation production with few exceptions has been unsuccessful. Adoption has been slow because subsistence farmers do not possess requisite resources to adopt and the institutional structures do not exist within most countries to effectively address the barriers to adoption. While some large-scale farm operators may adopt some appropriate conservation production systems, it is highly unlikely that widespread adoption will occur in small-scale farm operations where such systems are sorely needed.

Research conducted within communal areas of Zimbabwe (Musara et al. 2012) revealed that land resources were being exploited owing to use of inappropriate agricultural practices. It was observed that continued use of traditional production systems resulted in severe degradation of soil and water resources. The Zimbabwe government, in conjunction with several nongovernment organizations, developed a project to diffuse conservation production systems to small-scale farmers who were operating land in communal areas of the country. The project employed an approach that emphasized exposure to conservation information via workshops, seminars, and extension personnel. Potential adopters were shown the benefits of adopting conservation practices and were offered seeds as an incentive to participate in the project.

Adopters were compared with nonadopters, and the findings revealed that younger farmers who were operating larger farms tended to adopt more often. Adopters also tended to report larger family size and had been involved in farming for longer periods of time. Older farmers tended not to adopt conservation production systems.

The authors note that farmers with these characteristics were more innovative and were more willing and able to assume the risks associated with the adoption of new production systems. It was observed that older farmers tended to prefer use of traditional production techniques owing to social tradition and the certainty of output associated with systems presently being employed. It was also noted that practically all of the adopters continued to use traditional production systems on a large portion of their cropped land to ensure some minimal level of output. While adopters were willing to assume some risk, they were not willing to completely embrace the new production systems. The ability to experiment with the new production systems on a portion of their land was a significant factor in the adoption decision.

15.3.4.3 Soil Conservation in Asia

Huszar et al. (1994) examined adoption of soil and water conservation production systems within the uplands of Indonesia. The objective of the project assessed was to introduce new farm production systems that would increase farm income while simultaneously reducing degradation of soil and water resources. The Indonesian government offered economic subsidies for construction of terraces and for purchase of annual inputs such as seed, fertilizer, and pesticides. It was expected that removal of the economic barriers to adoption would result in wholesale adoption of the new production system. It was hypothesized that, over time, subsistence farmers would no longer require subsidies to operate their farms using the new conservation production system because farm income would increase and farmers would be financially able to operate the farm without government assistance.

Study findings revealed that removal of the economic barriers to initial adoption facilitated the extensive adoption of the new production system in the short term. However, when the government subsidies were withdrawn, farmers who had adopted the new conservation production system quickly ceased to use the production systems being assessed. It was noted that farm income increased significantly for adopters compared with nonadopters within the study area when the project was initiated. However, differences in farm income between adopters and nonadopters eroded quickly after the subsidies were withdrawn. It was discovered that farmers who had been receiving subsidies from the government used the supplemental income to increase their lifestyles rather than reinvesting additional income in the farm operation. When the subsidies were withdrawn, the use of inputs to increase yields ceased and productivity reverted back to the preproject levels. The positive conservation and productivity impacts of the structures created using government subsidies were shown to be longer lived; however, over time, it is expected that the terraces and other conservation structures would be destroyed because of lack of maintenance.

Huszar et al. (1994) submit that alternative approaches need to be explored rather than direct subsidies of the conservation production systems being diffused. They suggest that a credit institution be created to aid farmers in need of capital to invest in conservation rather than using subsidies. They argue that subsistence farmers will be more likely to reinvest additional income produced by conservation production systems in the farm enterprise if they are aware that they must repay a loan. When the government was subsidizing the adoption of conservation production systems, most farmers perceived the subsidy as being a grant to be used in any manner the farmer wished, and they spent the subsidies on lifestyle enhancement. The authors further argue that the government should develop and fund an extension service to deliver technical information to potential adopters and to develop local input suppliers so that farmers will have access to farm inputs in the local community.

Ruaysoongnern (1999) and McDonald and Brown (2000) discuss soil conservation adoption research produced in northeast Thailand. An agricultural extension project was initiated in 1995 with the expressed goal of diffusing soil conservation practices among local farmers. The authors suggest that lack of flexibility in project implementation resulted in rejection of conservation practices in some areas. They note that when farmers were included in the planning process and when flexibility was included in the implementation procedures, adoption was more frequently observed. Factors that were shown to be most important in facilitating adoption at the village level were the following: conservation problems were clearly demonstrated, farmers groups were organized to address conservation problems, the conservation practices being diffused were shown to be effective, and education/training was available to potential adopters. Factors shown to influence the adoption decision at the farm level were as follows: farm size, availability of labor, access to economic resources, off-farm employment, availability of water resources, and the congruence of the practices with the existing farming systems. All of these factors are consistent with the factors discussed in the traditional diffusion model.

Saguiguit et al. (1999) and McDonald and Brown (2000) discuss the Integrated Social Forestry Program introduced in the Nueva Vizcaya Province in northern Philippines. The program was designed to bring about adoption of agroforestry and soil conservation production systems among highland farmers. Conservation efforts were focused on motivating land managers to adopt and use conservation practices by providing subsidies to access production inputs, making training and technical assistance available to potential adopters, enhancing the community support systems, and making tenure rights available to program participants. The use of tenure rights as an incentive to participate in the project was somewhat unusual. The opportunity to gain tenure rights to the land being operated as a reward for participating in the project acted as a very powerful incentive for subsistence farmers to participate in the conservation program.

While all project participants adopted some aspect of the program, most program participants adopted contour hedgerows rather than the more complex alternatives. The primary reasons for adoption were the economic subsidies provided by the program, the land tenure reward for program participation, and the encouragement of respected community leaders. Saguiguit et al. (1999) note that awareness of the benefits associated with adoption of soil conservation practices was not a significant issue for program participants. They suggest that the reason for this finding is that farmers were motivated to adopt by economic and land tenure factors and not by environmental concerns. The authors also posit that it is highly doubtful that adopters will continue using the new conservation production systems once the program resources are withdrawn because farmers have not been informed about the benefits of the new practices. The authors observed that some conservation practices employed before the project were still being used and diffused among other farmers as the project progressed. This finding suggests that the type of conservation practices being diffused by the project were probably not the most relevant to the needs of local farmers.

Sureshwaran et al. (1996) discuss research focused on the adoption of Sloping Agricultural Land Technology (SALT) among subsistence farmers operating farms on Leyte Island in the Philippines. The project sponsors used an approach very similar to the traditional diffusion model to implement the conservation program.

Subsistence farmers were informed about the conservation effort and provided economic support, training, and technical assistance to adopt the conservation production system being advanced by the change agents. These efforts are consistent with the information and decision-making stages of the diffusion model. The barriers to adoption associated with lack of access to information and technical assistance within the study region were negated by the contributions of the project sponsors. Access to economic resources to adopt new production systems was also eliminated as a barrier by sponsors.

The conservation program being advanced by project sponsors consisted of contour planting of hedgerows composed of fast-growing legumes across sloping land. Cuttings from the hedgerows were applied to the land, which increased soil fertility. Legumes used to develop the hedgerows fixed nitrogen in the soil, which further improved the soil. Over time, the areas upslope from the hedgerows became more level because of sediments being precipitated from runoff water. The upslope land leveled by sedimentation made it possible to introduce additional crops that resulted in a more diversified and more economically viable farm operation. Soil erosion from cultivated land upslope from the hedgerows was significantly reduced owing to the slowing of runoff by the hedgerows acting as barriers to water flow. Reduced erosion and increased fertility translated into economic benefits to the adopters, since improved soil quality increased property values.

The cost of implementing the new conservation production system for participating farmers was minimal given the contributions of the project sponsors. Increased demand for labor for manual trimming of the hedgerows and production losses at the site of hedgerow development were the primary costs associated with adoption for the farmer. The study reported by Sureshwaran et al. (1996) compared SALT adopters and nonadopters. Study findings demonstrated that younger farmers who were operating larger farms tended to adopt the SALT production system more frequently. These findings are consistent with other diffusion studies because younger people tend to be more willing to assume the risks associated with the adoption of innovations. While the costs associated with adopting SALT were low, there was some risk associated with changing existing production practices. The authors suggest that younger people were more interested in long-term impacts of change than older people. The farm size variable is also consistent with the diffusion model because operators of larger farms are better able to assume any economic losses if the conservation production system failed to produce the expected outcomes.

The income and education of study participants were not significant in explaining adoption, as the diffusion model suggests. The reason for these findings is that little variance existed between adopters and nonadopters in terms of income and education. The study subjects were subsistence farmers who possessed little income and practically no education. Land tenure was shown to facilitate adoption. This is also consistent with the diffusion model because farmers must be able to capture the benefit streams of their investments in conservation. The authors conclude that conservation efforts such as SALT should focus their efforts on younger farmers with larger land acreage who have land tenure rights to the land they are operating. They also note that assistance in the form of education, training, technical aid, and economic subsidies will be required when attempting to motivate subsistence farmers to adopt and use new conservation production systems. These conclusions strongly support the diffusion model as developed for this discussion because all of these factors affect the ability of the potential adopter to implement a decision to adopt new conservation production systems.

Research reported by Fujisaka (1999) and McDonald and Brown (2000) focused on conservation adoption among farmers in the Misamis Oriental Province of the Philippines. Study findings revealed that adoption occurred for some types of conservation practices; however, many conservation production systems were not adopted. Contour farming was adopted and used but trenching and bunds were never adopted because of the labor required to develop them. Green manures were not adopted because farmers did not have an interest in using this approach. A-frames were adopted and used to construct contours because the frames were easy to build and, once constructed, required little human skill to effectively use. Most plant species that were recommended for hedgerow construction were rejected. Preference was given to local weeds to form the hedgerows. The reasons given for this decision were that local weeds retarded runoff better and did not require as much land to be permanently retired from production. Farmer-to-farmer communication and training was the method used to continue the conservation program after the project was discontinued.

Adoption research findings produced in the Philippines support the traditional diffusion model. Access to information, technical assistance, and economic resources were shown to be significant factors in bringing about adoption of new conservation production systems among small-scale farmers. Adoption of conservation production systems was facilitated by access to land tenure and by access to extension service personnel. Lack of awareness of the need for conservation and the environmental consequences of land degradation acted as barriers to adoption and/or continued use of soil and water conservation production systems at the farm level.

Another study focused on evaluations of SALT was conducted in the Hindu–Kush–Himalayan region of Asia (McDonald and Brown 2000; Tang 1999) and revealed that SALT production systems were only adopted in two countries in the region. Some adoption occurred in Bangladesh and China. However, adoption was slow and the rate of adoption was low. Factors shown to influence adoption were lack of understanding of how the SALT system functioned and lack of awareness of environmental problems resulting from soil erosion of cropland. Tang (1999) argues that lack of government investment in communication systems and the development of an extension service for people within the region are the primary reasons why adoption is so low. Another factor influencing adoption was lack of demonstrations of the positive and/or negative impacts of SALT for individual farmers. It is argued that lack of awareness of possible outcomes and uncertainties attached to the outcomes act as barriers to adoption. Farmers are unwilling to adopt any new production system when they are not certain about what will be the advantages and/or disadvantages associated with adoption.

Research focused on the adoption of conservation tillage systems in Mongolia was conducted by Lafond et al. (2009), and the findings are quite similar to those associated with the adoption of all technology-intensive production systems in lesser-scale societies. The authors discovered that the primary factors influencing adoption of conservation tillage systems were as follows: lack of access to inputs including herbicides and application technologies, lack of mechanization for seeding and harvest, uncontrolled grazing, lack of forage in cropping systems, and lack of access to capital to invest in the new tillage system. The authors conclude that adoption of conservation tillage systems will be very slow in Mongolia because of the lack of access to knowledge and the technologies to implement such tillage systems.

Hannam (2010) examined soil and water conservation issues in Mongolia and argues that the country should engage in long-term planning to introduce national soil and water conservation programs at the farm level. He asserts that soil and water conservation is sorely needed in Mongolia because land resources have been in a constant state of degradation since the 1950s when the country embraced an industrial model for development. While Mongolia is the seventh largest country in terms of land area in the world, most of the land is not appropriate for agricultural purposes. Land that is productive is rapidly being developed, and soil resources are being degraded. Most of Mongolia's exports are raw or partially processed natural resources.

Mongolia's climate is slowly becoming drier, which contributes to land degradation from overgrazing. Forests are being removed, which subjects deforested land to severe erosion. Wheat production was expanded tremendously from the 1960s through the 1990s. At the present time, about 60% of the crop land previously devoted to wheat production has been abandoned and subject to wind and water erosion. About 50% of the remaining cultivated cropland has been severely eroded, and >95% of the total land area of the country is susceptible to becoming desert.

Existing farming systems are slowly destroying soil fertility and food, and fiber production is declining. Only 1.5% of the land area is devoted to crop production and 1% for production of hay. More than 95% of the land is devoted to pasture. Overgrazing is widespread, and herds have been expanded without consideration of the environmental impacts. Given the high level of soil degradation that has been observed, the country must develop a means of reducing soil loss quickly before soil degradation reaches such a low level that future productivity of land resources is destroyed.

Hannam (2010) has examined multiple action options to control erosion and concluded that institutional structures must be developed and implemented within the country. One of the first problems the government needs to address from his perspective is land tenure rights. Land is presently state owned, and land managers pay rent to the government to access land for agricultural purposes. Land managers do not invest resources in land that the government owns because they may not benefit from their investments. Problems created by existing land tenure policies are complicated by the fact that most land managers do not possess management skills necessary to operate farms nor do they have access to economic resources to purchase needed farm inputs.

Farm technologies are not accessible to improve productivity. Poverty is increasing in rural areas, and little is being done to correct the problem. Institutional structures do not exist to deliver environmental information or technical assistance. No institutional structures exist to conduct scientific research on problems unique to Mongolian agriculture, and no means exists to have new knowledge delivered to land managers. Institutional structures to provide education to rural residents have not been developed. Mongolian leaders have recognized the seriousness of the environmental problems that have emerged from mismanagement of natural resources within the country, and have pledged to address social problems and invest in environmental action. Since Mongolia was under the rule of Soviets for an extended period of time, it is not surprising that a national plan was developed to correct the problems identified as being associated with degradation of soil and water resources. Unfortunately, the grand plan has not achieved much since 2008 because the government does not possess adequate economic resources or human skills to influence any of the major environmental problems associated with natural resources degradation within the country.

The issues identified in Mongolia are the same as those observed in other lesser-scale societies. The development of macro-level plans will achieve little without first addressing the factors that contribute to environmental degradation in lesser-scale societies. Unlike many other countries that have attempted to implement conservation programs via the provision of information, technical assistance, and economic subsidies, Mongolia has relied heavily on a top–down program implementation with national planning combined with local implementation. The approach has not been successful because the model failed to address the basic causes of resource exploitation. In Mongolia, poor people are operating farms with few relevant skills while attempting to implement a national plan that calls for huge commitments of economic resources and human expertise that do not exist.

15.3.4.4 Soil Conservation in South and Central America

Alwang and Sowell (2010) synthesized many studies focused on adoption of soil and water resources conservation in the Andean Mountains of South America. Factors shown to be significant in the adoption decision-making process within the region are consistent with the traditional diffusion model. Unfortunately, most of the factors have been shown to impede adoption.

Economic assessments of the impacts of erosion throughout the Andean Mountains show that significant costs are associated with erosion of cultivated land. On-site damage due to erosion of cultivated farmland has reduced the productivity of land resources and has resulted in extensive off-site damage in the form of siltation of roadways, ditches, and impoundments constructed to produce hydroelectricity. Water quality has also been degraded. Concern has increased among local and national organizations to reduce the degradation caused by use of inappropriate agricultural production systems. While many conservation technologies and techniques presently exist to address environmental problems associated with the use of inappropriate production systems, adoption of conservation production systems among subsistence farmers within the Andean Mountains has been very slow. Subsistence farmers have refused to change production systems, and the reasons expressed for doing so are very similar to those noted in the discussion of the traditional diffusion model.

Alwang and Sowell (2010) and Alwang et al. (2009) note that many erosion control structures have been in existence within the Andean Mountains since the pre-Colombian days. However, most of these conservation structures are in disrepair and the terraces are no longer functional. Increasing population has resulted in the subdivision of cropland and has resulted in marginal lands being brought into production. Food and fiber production is now being introduced into fragile ecosystems that should never have been exposed to such farming activity. Farms are small in the Andean Mountains and are managed by economically poor and uneducated farmers. Most of the farm managers operate their farms at barely subsistence level with little or no capital accumulation. Without surplus food production to sell, there is little hope of accumulating sufficient economic resources to invest in conservation production systems. Without education and/or technical assistance to implement new production systems, the ability to actually use conservation practices and/or technologies is very low (Figure 15.7). While incentives have been offered to poor farmers to adopt and use conservation production systems, these efforts have been unsuccessful in resolving problems associated with erosion of farmland. Efforts to increase the long-term profitability of agriculture have not been successful.

While many local governments have acquired the authority to address conservation issues, they do not have the capacity to do so. It will take many years for local governments to achieve the level of institutional development required to address the needs of poor farmers. Schools, roads, technical assistance agencies, local marketing systems linked to national markets, communication systems, and a host of other institutional structures will have to be created, staffed, and funded before the institutional structures of local government can begin to address soil and water conservation issues on a scale that will be required to reduce the incidence of existing problems.

One of the major factors that act as barriers to adoption of conservation technologies and techniques within the Andean Mountains is the ability to assume risks associated with adoption. Subsistence farmers live on the edge of economic disaster.

FIGURE 15.7 Poza Rica, Veracruz State, Mexico, 1978. International Maize and Wheat Improvement Center (CIMMYT) training of extension agents in maize, zero tillage. (Courtesy of Dr. Tom Stilwell.)

Failure of one crop year may result in the family starving. Living under such a serious threat of destruction most often produces a very cautious decision maker. Unless new production systems can produce short-term production benefits in the form of increased productivity, subsistence farmers will not adopt conservation production systems.

Subsistence farmers will adopt labor-intensive production systems if labor is available. Unfortunately, many areas of the Andean Mountains no longer have a surplus of labor because many young people have migrated to the cities, seeking better employment opportunities. Conservation programs that rely heavily on low-technology and high-labor approaches will find it more difficult to motivate land managers to adopt because local labor no longer exists to implement such conservation production systems.

Land tenure rights alone will not strongly influence the adoption of conservation technologies and/or techniques within the Andean Mountains. Subsistence farmers may possess property rights and still not have sufficient human and economic resources to implement new conservation production systems. Land tenure will affect the orientation of potential adopters if they have the resources to invest in conservation efforts because such rights guarantee that the owner can claim the long-term benefits associated with making conservation investments in their land.

Downstream costs associated with erosion of cropland are ignored in the Andean Mountains. Off-site damages associated with food and fiber production are basically ignored because institutional structures do not exist to address the issue. Conservation policies are seldom in existence, and when they do exist, there are no institutional structures in place to enforce such policies. Conservation policies without enforcement capabilities are basically useless tools for implementing soil and water conservation efforts.

Alwang and Sowell (2010) note that conservation change agents must take into consideration several things when developing a strategy for bringing about adoption of soil and water conservation among subsistence farmers. They assert that conservation efforts must produce positive economic results quickly and that attention should be focused on conservation systems that are appropriate for small-scale agriculture. Large-scale/technology-intensive farm operations do not exist in subsistence farm production areas in the Andean Mountains. Soil and water conservation production systems appropriate for large farm operations will not be relevant to small farming units. The authors strongly suggest that short-term subsidies to encourage adoption of conservation production systems should be de-emphasized when working with subsistence farmers in favor of investments in long-term solutions to existing poverty conditions. Unless poverty conditions are reduced, soil conservation will continue unabated.

Considerable research attention has been focused on the adoption of soil and water conservation production systems within Central America because soil erosion has severely degraded soil and water resources. The primary cause of erosion is the extensive use of traditional agricultural techniques by subsistence farmers. As subsistence agriculture was extended into higher elevations, soil loss due to erosion increased. Soil erosion has reduced agricultural productivity and has created extensive environmental problems so significant that several national governments within the region have initiated conservation efforts to reduce the environmental damages.

Solis et al. (2007) examined two nationally sponsored conservation projects to determine what factors were significant in predicting the adoption of conservation production systems at the farm level. One project was assessed in El Salvador and one in Honduras. Farmers were asked to indicate if they operated >50% of their cultivated land using some type of conservation production practice. Respondents with positive responses were defined as adopters, and those with <50% were defined as nonadopters. The socioeconomic factors of the two groups were compared.

Study findings demonstrated that younger farmers, better-educated farmers, and farmers with higher farm incomes tended to adopt conservation production systems more frequently than farmers operating larger farms and those reporting more off-farm income. Managers who owned the land they were operating and had participated in the government-sponsored conservation program tended to adopt conservation practices more frequently. Farmers who reported greater contact with extension agents and had developed a positive attitude toward conservation tended to adopt conservation practices more frequently.

Findings reported by Solis et al. (2007) clearly support the traditional diffusion model approach for introducing new conservation production systems. Access to information, formation of positive attitudes, ability to assume risk, and ability to implement conservation practices via government aid are important factors in the diffusion model, and they were shown to be important predictors of adoption among study participants. Study findings for managers of larger land holdings are interesting because such farmers tended not to adopt the innovative conservation farming practices. This suggests that land managers of larger farms tend to remain committed to traditional subsistence farming that employ fewer inputs. It may also mean they have previously adopted a production system that meets their desired level of output. It should be noted that operators of larger farms also report a larger portion of their family income from off-farm employment. Part-time farmers would not be as motivated to protect future land productivity as farmers with no off-farm employment opportunities.

Bunch (1999) and McDonald and Brown (2000) discuss adoption of conservation practices in Honduras and Guatemala. They observed that use of soil and water conservation technologies and techniques often were discontinued over time. However, they observed that project participants had learned how to experiment with innovative conservation practices and had discovered new approaches that were adopted and used.

Study findings were used to develop specific recommendations for program implementation in the future. It was recommended that soil and water conservation production systems be combined with other systems to increase productivity so that farmers would have more income to invest in conservation. The authors argued that selection of conservation innovations for diffusion should consider the following: require low labor input to implement, produce benefits quickly, produce observable outcomes that are expected by the adopter, employ low-level technologies, and contribute to adopters learning how to develop their own conservation innovations.

15.4 CONCLUSIONS

Research focused on factors affecting the adoption of soil and water conservation production systems among subsistence farmers throughout the world consistently indicates that the traditional diffusion model is a useful approach for understanding why potential adopters elect to adopt or reject conservation production systems. Existing literature revealed that the most important factors shown to influence the adoption of soil and water conservation production systems among subsistence farmers in lesser-scale societies are as follows: access to information about soil and water conservation problems, access to information about possible solutions to environmental problems, access to technical training to develop requisite skills to effectively implement conservation production systems, access to information delivery systems that disseminate relevant agroecological information about local socioenvironmental situations, access to economic resources to adopt requisite conservation farming inputs, access to institutions that make local and national markets available to subsistence farmers, creation of national economic policies that encourage the production of food and fiber in an efficient manner, creation and implementation of national land tenure policies that ensure subsistence farmers will be able to capture the benefits from investments in conservation, creation of institutions to make credit available to subsistence farmers so they may purchase requisite inputs to implement farming conservation systems, access to local and national infrastructures to provide opportunity for social mobility of individual subsistence farmers within the community and society, creation and enforcement of national soil and water conservation policies to protect soil and water resources, and creation of national policies to address socioeconomic problems associated with acute poverty so that subsistence farmers will be better able to assume risks associated with the adoption of conservation farming systems at the farm level. Each of these factors is closely aligned with several components of the traditional diffusion model.

The diffusion model strongly suggests that subsistence farmers in lesser-scale societies must be provided detailed information about the nature of environmental problems that result from use of inappropriate agricultural production systems. They also must be provided information that demonstrates how various conservation production systems will resolve the identified problems. The model states that farmers must have access to learning opportunities to develop the human skills to implement the conservation production systems they are considering. Potential adopters must also have access to institutional structures that can provide them access to economic resources to purchase requisite inputs to adopt the conservation production systems being considered. The model posits that institutional structures must be put into place to provide relevant and reliable information about agricultural issues that are appropriate for the region into which conservation innovations are being introduced. Governments must introduce antipoverty programs that will provide subsistence farmers with education and training to become integrated members of society, and with the infrastructures to remove the many barriers to adoption of conservation production systems.

While the traditional diffusion model can be extremely useful in the identification of required actions to promote adoption of conservation production systems within lesser-scale societies, the reality of the situation is that it will be almost impossible for lesser-scale societies to address the factors impeding adoption. To adequately address soil and water conservation issues within lesser-scale societies populated by subsistence farmers requires massive human and financial resources that these types of countries simply do not possess. Most lesser-scale societies would have to practically restructure their total institutional system and empower masses of lower-class people. These actions would be strongly resisted by those who presently possess the political, social, and economic power within lesser-scale societies.

It is highly unlikely that lesser-scale societies will ever be able to successfully address the environmental problems associated with degradation of agricultural land without extensive support from high-scale societies. Support in the form of technical assistance, provision of economic resources to purchase requisite inputs, aid in the development of market and educational infrastructures, access to farm technologies and biotechnological innovations that are appropriate for subsistence farmers, and development of information diffusion systems are the most important contributions that high-scale societies can make available to lesser-scale societies.

Gould (1994) notes that economic and technical assistance will be required from high-scale societies, or subsistence farmers in lesser-scale societies have no hope of adopting soil and water conservation production systems at the farm level. Gould observes that high-scale societies must be prepared to provide these types of support for extended periods of time because it will take many years for lesser-scale societies to achieve the level of institutional development to effectively address environmental problems associated with soil erosion of agricultural lands.

While Gould and others who advance these types of development approaches are probably correct that such actions are sorely needed by lesser-scale societies, it is highly unlikely that high-scale societies will be able to provide such aid in the near term. Given the severe economic difficulties presently being experienced by many high-scale societies, it is highly unlikely that surplus socioeconomic resources will be available to aid other countries. The United States is certainly not in any position at the present time to offer large sums of money and technical assistance to lesser-scale societies because the country has experienced severe deficit spending and rapidly increasing national debt during the last decade. Priority will have to be given to domestic conservation issues before serious consideration can be given to international conservation problems at the level required to have an effect. Since the present economic situation in the United States is problematic and will probably remain so for years, combined with the fact that domestic conservation problems are becoming more severe over time, it will probably be decades before domestic conservation problems will be adequately addressed to the point that resources can be reallocated to provide significant aid to other countries.

Napier (2009) notes that domestic and world demand for food and fiber has increased commodity prices to the point that most US farmers are now strongly motivated to maximize production. Marginal land has been brought back into production, and the costs associated with implementing set-aside programs are rapidly increasing. Practically all domestic soil and water conservation programs in the United States are no longer being adequately funded owing to declining federal, state, and local economic resources. The cost of maintaining domestic conservation programs will continue to increase, while budgets will continue to decline. The result of all of these interconnected factors will be a substantial increase in environmental problems in the United States. It is highly doubtful that the United States will be able to adequately address domestic conservation problems in the future unless significant economic growth is quickly achieved. That is highly unlikely to occur with existing domestic economic policies.

It also must be acknowledged that it is not possible for the US government to provide biotechnological innovations to lesser-scale societies as suggested by some writers. Biotechnology is often privately owned in the United States, and no entity can make those technologies available to anyone without the consent of the owners. It is highly unlikely that owners of agricultural biotechnology will be willing to make their products available to anyone without being adequately compensated.

One of the most significant threats to adoption of conservation production systems at the farm level in lesser-scale societies is global warming. If predictions about the adverse agricultural impacts of global warming are realized, then it is highly likely that subsistence agriculture in many lesser-scale societies will be devastated (Napier 2013). While farmers in most high-scale societies will probably be able to mitigate many of the adverse consequences of increasing temperature and intensity of weather events by rapid changes in farm production systems, subsistence farmers in lesser-scale societies will not have access to human skills nor the technologies/techniques to address such modifications in weather. Although adoption of soil and water conservation farming systems would aid in the adjustment to agroecological problems generated by global warming, the inability of subsistence farmers to adopt such systems will result in wholesale destruction of land resources in many lesser-scale societies.

Soil and water conservation efforts will undoubtedly continue in lesser-scale societies with some small measure of success. Unfortunately, the types of conservation efforts that will be introduced will address none of the root causes of environmental degradation associated with agricultural production. The principal factors contributing to soil and water degradation in lesser-scale societies are socioeconomic in nature, and these types of factors cannot be addressed by provision of a stick, hoe, seed, and fertilizer by some government agency or nongovernment organization. Unless the institutional factors impeding adoption are adequately addressed in the near term, the future productivity of soil resources will be destroyed and subsistence farmers will be further stressed.

REFERENCES

Adeola, R.G. 2010. Influence of socio-economic factors on the adoption of soil conservation measures in Ibadan/Ibarapa agricultural zone of Oyo State. *Report* 2(7): 42–45.

Adeola, R.G. 2012. Farmers' perception of soil fertility decline under yam production in OYO State, Nigeria. *Continental Journal of Agricultural Science* 6: 38–47.

Alwang, J., V. Barrera, R. Andrade, S. Hamilton, and G.W. Norton. 2009. Adaptive watershed management in the South American highlands: Learning and teaching on the fly. In *The Sciences and Art of Adaptive Management: Innovating for Sustainable Agriculture and Natural Resources*, ed. K.M. Moore, 209–227. Ankeny, IA: Soil and Water Conservation Society Press.

Alwang, J. and A. Sowell. 2010. Social factors, soil and water conservation, and vulnerable populations in the Andean Mountains of South America: What do we know? In *Human Dimensions of Soil and Water Conservation: A Global Perspective*, ed. T.L. Napier, 305–319. New York: Nova Science Publishers Inc.

Anthony, S., R. Jou, and T.L. Thurow. 1998. *Sustainable Technologies for Use and Conservation of Steeplands*, 1–13. Food and Fertilizer Technology Center Publication. Available at http://www.agnet.org/library.php?func=view&id=20110804181442.pdf.

Baidu-Forson, J. 1994. Adoption of soil and water conservation practices in the Sudano–Sahelian zone: Constraints and incentives. In *Adopting Conservation on the Farm: An International Perspective on the Socioeconomics of Soil and Water Conservation*, eds. T.L. Napier, S.M. Camboni and S. El-Swaify, 157–169. Ankeny, IA: Soil and Water Conservation Society Press.

Bunch, R. 1999. Learning how to "make the soil grow": Three case studies on soil recuperation adoption and adaptation from Honduras and Guatemala. In *Issues and Options in the Design of Soil and Water Conservation Projects*, eds. M.A. McDonald and K. Brown, 23–38. University of Wales. School of Agricultural and Forest Sciences Publication no. 17.

Cockerill, C.H. and T.L. Napier. 2010. The impact of information on the adoption of soil and water conservation practices at the farm level. In *Human Dimensions of Soil and Water Conservation: A Global Perspective*, ed. T.L. Napier, 47–61. New York: Nova Science Publishers Inc.

Crowe, B. 1969. The tragedy of the commons revisited. *Science* 166: 1103–1107.

El-Swaify, S.A. 1997. *Multiple Decision Making for Land, Water, and Environmental Management*. New York: Lewis Publishing Company, Inc.

El-Swaify, S.A. and G.G. Marten. 2010. Human dimensions for conservation-effective natural resource management in the tropics. In *Human Dimensions of Soil and Water Conservation: A Global Perspective*, ed. T.L. Napier, 321–339. New York: Nova Science Publishers Inc.

Faltermeier, L. 2007. *Adoption of Water Conservation and Intensification Technologies in the Lowland Rice Production Systems of Northern Ghana*. Keil, Germany: Department of Food Economics and Consumption Studies, University of Keil.

Fujisaka, S. 1999. A retrospection of soil conservation in Claveria, Philippines. In *Issues and Options in the Design of Soil and Water Conservation Projects*, eds. M.A. McDonald and K. Brown, 39–46. University of Wales. School of Agricultural and Forest Sciences Publication No. 17.

Giller, K.E., E. Witter, M. Corbeels, and P. Tittonell. 2009. Conservation agriculture and smallholder farming in Africa: The heretics' view. *Field Crops Research* 114: 23–34.

Gould, T. 1994. Socioeconomic barriers to adoption of conservation practices in developing countries. In *Adopting Conservation on the Farm: An International Perspective on the Socioeconomics of Soil and Water Conservation*, eds. T.L. Napier, S.M. Camboni and S. El-Swaify, 111–116. Ankeny, IA: Soil and Water Conservation Society Press.

Hannam, I. 2010. Human dimensions of soil and water conservation: Mongolia. In *Adopting Conservation on the Farm: An International Perspective on the Socioeconomics of Soil and Water Conservation*, eds. T.L. Napier, S.M. Camboni, and S. El-Swaify, 341–356 Ankeny, IA: Soil and Water Conservation Society Press.

Hardin, G. 1968. The tragedy of the commons. *Science* 162: 1243–1248.

Henao, J. and C. Baanante. 2006. *Agricultural Production and Soil Nutrient Mining in Africa*. Muscle Shoals, AL: International Fertilizer Development Center.

Huszar, P.C., H.S. Pasaribu, and S.P. Ginting. 1994. Sustainability of Indonesia's upland conservation projects. In *Adopting Conservation on the Farm: An International Perspective on the Socioeconomics of Soil and Water Conservation*, eds. T.L. Napier, S.M. Camboni, and S. El-Swaify, 199–211. Ankeny, IA: Soil and Water Conservation Society Press.

Kraybill, D.S. 2010. Climate change and natural resource management in sub-Saharan Africa. In *Human Dimensions of Soil and Water Conservation: A Global Perspective*, ed. T.L. Napier, 275–286. New York: Nova Science Publishers Inc.

Lafond, G., B.G. McConkey, and M. Stumborg. 2009. Conservation tillage models for small-scale farming: Linking the Canadian experience to the small farmers of Inner Mongolia Autonomous region in China. *Soil and Tillage Research* 104: 150–155.

Lovejoy, S.B. and J.H. Sanders. 1994. Resource degradation and new technologies: Success in Africa. In *Adopting Conservation on the Farm: An International Perspective on the Socioeconomics of Soil and Water Conservation*, eds. T.L. Napier, S.M. Camboni and S. El-Swaify, 133–142. Ankeny, IA: Soil and Water Conservation Society.

McDonald, M. and K. Brown. 2000. Soil and water conservation projects and rural livelihoods: Options for design and research to enhance adoption and adaptation. *Land Degradation and Development* 11: 343–361.

Musara, J.P., J. Chimvurahwe, and R. Borerwe. 2012. Adoption and efficiency of selected conservation farming technologies in Madziva communal area, Zimbabwe: A transcendental production function approach. *Bulletin of Environment, Pharmacology and Life Sciences* 1(4): 27–38.

Napier, T.L. 1991. Factors affecting acceptance and continued use of soil conservation practices in developing societies: A diffusion perspective. *Agriculture, Ecosystems and Environment* 36: 127–140.

Napier, T.L. 2009. Grain scarcity: A new era for conservation policies and programs. *Journal of Soil and Water Conservation* 64(1): 7A–10A.

Napier, T.L. 2010. Conclusion. In *Human Dimensions of Soil and Water Conservation: A Global* Perspective, ed. T.L. Napier, 373–379. New York: Nova Science Publishers Inc.

Napier, T.L. 2013. Socio-Economic Consequences of Global Warming on US Agriculture and Rural Communities. Working manuscript for the Soil and Water Conservation Society task force on global warming. Ankeny, IA.

Napier, T.L., S.M. Camboni, and S.A. El-Swaify. eds. 1994. *Adopting Conservation on the Farm: An International Perspective on the Socioeconomics of Soil and Water Conservation*. Ankeny, IA: Soil and Water Conservation Society Press.

Oldeman, L.R., R.T.A. Hakkeling, and W.G. Sombroek. 1991. *World Map of the Status of Human Induced Soil Degradation*. Rome: Food and Agricultural Organization of the United Nations; Wageningen: International Institute for Aerospace Survey and Earth Sciences; Nairobi: United Nations Environmental Programme.

Pagiola, S. 1994. Soil conservation in a semi-arid region of Kenya: Rates of return and adoption of farmers. In *Adopting Conservation on the Farm: An International Perspective on the Socioeconomics of Soil and Water Conservation*, eds. T.L. Napier, S.M. Camboni, and S. El-Swaify, 171–188. Ankeny, IA: Soil and Water Conservation Society Press.

Rogers, E.M. 2003. *Diffusion of Innovations* (5th edition). Glencoe: Free Press.

Ruaysoongnern, S. 1999. Adoption of sloping land conservation practices by farmers in north-east of Thailand. In *Issues and Options in the Design of Soil and Water Conservation Projects*, eds. M.A. McDonald and K. Brown. 47–56. University of Wales, School of Agricultural and Forest Sciences Publication No. 17.

Saguiguit, G., J.N. Garcia, and R. Cramb. 1999. Promotion of soil conservation measures in the Philippines: The ISF project in Sitio Domang. In *Issues and Options in the Design of Soil and Water Conservation Projects*, eds. M.A. McDonald and K. Brown. University of Wales, School of Agricultural and Forest Sciences Publication No. 17.

Scherr, S.J. 1999. *Soil Degradation: A Threat to Developing-Country Food Security by 2020. A 2020 Vision for Food, Agriculture and the Environment*. International Food Policy Research Institute Publication No. 58.

Shaxson, T.F. 1993. Conservation-effectiveness of farmers' actions: A criterion of good land husbandry. *Topics in Applied Resource Management* 3: 103–128.

Shaxson, T.F., N.W. Hudson, D.W. Sanders, E. Roose, and W.C. Moldenhauer. 1989. *Land Husbandry: A Framework for Soil and Water Conservation*. Ankeny, IA: Soil and Water Conservation Society Press.

Solis, D., B.E. Bravo-Ureta, and R.E. Quiroga. 2007. Soil conservation and technical efficiency among hillside farmers in Central America: A switching regression model. *Australian Journal of Agricultural and Resource Economics* 51: 491–510. doi 10.1111/j.1467-8489.2007.00394.x.

Southgate, D. and D. Graham. 2010. Economic progress, agricultural development and environmental conservation: The challenges in sub-Saharan Africa. In *Human Dimensions of Soil and Water Conservation: A Global* Perspective, ed. T.L. Napier, 263–273. New York: Nova Science Publishers Inc.

Sureshwaran, S., S.R. Londhe, and P. Frazier. 1996. Factors influencing soil conservation in developing countries: A case study of upland farmers in the Philippines. *Journal of Agribusiness* 14(1): 83–94.

Swenson, S. and K.M. Moore. 2009. Developing Conservation Agricultural Production Systems: An Analysis of Local Networks. Working Paper Number 10-09. Office of International Research, Education, and Development, Virginia Tech.

Tang, Y. 1999. Factors influencing farmers' adoption of soil conservation programmes in the Hindu Kush-Himalaya region. In *Issues and Options in the Design of Soil and Water Conservation Projects*, eds. M.A. McDonald and K. Brown. University of Wales, School of Agricultural and Forest Sciences Publication No. 17.

Wangia, C. and T. Prato. 1994. A review of socioeconomic and policy aspects of soil conservation in Africa. In *Adopting Conservation on the Farm: An International Perspective on the Socioeconomics of Soil and Water Conservation*, eds. T.L. Napier, S.M. Camboni and S. El-Swaify, 119–132. Ankeny, IA: Soil and Water Conservation Society.

16 Sustainable Intensification of Smallholder Agriculture

Rattan Lal and Bobby A. Stewart

CONTENTS

16.1 INTRODUCTION

There are about 500 million small farms worldwide, and an estimated 2 billion people depend on them for their livelihood and well-being. Furthermore, 80% of the food consumed in Asia and sub-Saharan Africa are produced on small farms (International Fund for Agricultural Development [IFAD] 2011; United Nations Environment Programme [UNEP]/IFAD 2013). Alleviating poverty of smallholder farmers is the key to lifting more than 1 billion people out of poverty (UNEP 2013). Therefore, the Millennium Development Goals (MDG), of reducing poverty and food insecurity (MDG 2000) and sustainable development, cannot be realized without improving the productivity and sustainability of smallholder farms. However, agronomic productivity and the well-being of smallholder farms are being undermined by soil degradation, decline in ecosystem functions and services, poor infrastructure, and weak institutional support. While, collectively, smallholder farmers have wealth of traditional knowledge and practical solutions for addressing site-specific constraints, most have neither the access nor the capacity to avail the benefits of modern technologies. For example, adoption of no-till (NT) farming and conservation agriculture (CA) is most relevant to improving productivity of smallholder farms. Despite the strong adoption of NT farming to the extent of 125 million hectare (Mha) worldwide (Friedrich et al. 2012), the adoption by smallholders has lagged behind because of several constraints (Lal 2007). Yet the strategy is to promote adoption of best management practices (BMPs) and strategies of "sustainable intensification." Thus, the objective of this concluding chapter of *Advances in Soil Science* is to identify appropriate technological options for "sustainable intensification" of agriculture practiced by smallholder farmers to advance MDGs with specific reference to achieving food security, alleviating poverty, and adapting to and mitigating climate change.

16.2 SUSTAINABLE INTENSIFICATION

Sustainable intensification of agriculture implies adoption of BMPs that enhance the natural resource base (e.g., soil, water, biodiversity, microclimate) while improving agronomic productivity, increasing farm income and alleviating poverty, and advancing food security. Processes and practices of sustainable intensification of smallholder farmers are outlined in Table 16.1 and Figure 16.1. The strategy is to reduce risks of new/additional degradation, restore degraded soils, and protect natural ecosystems for nature conservancy. Achieving these goals implies producing more from less. In other words, it means producing more agricultural output from less land area, lower water input, and lesser use of chemical fertilizers and pesticides. Therefore, farm practices that reduce input, reuse whatever can be salvaged, and recycle all by-products must be used. Soil erosion must be controlled, water must be conserved, nutrients must not be leached or volatilized, and all by-products must be recycled. Appropriate technologies recommended as BMPs must be those that are resilient against changing and harsh climate (e.g., drought stress, heat wave, and incidence of pests and pathogens), produce minimal assured agronomic yields even under the worst-case scenario, reduce adverse impacts on the environment, minimize drudgery especially of the women farmers, enhance farm income, and ensure food security.

TABLE 16.1
Processes and Practices of Sustainable Intensification of Smallholder Farms

Processes	Practices
1. Soil and water conservation	No-till farming, retention of crop residue mulch, cover cropping, contour hedges, buffer strips, continuous ground cover
2. Enhancing soil fertility	Integrated nutrient management, judicious use of chemical fertilizers, biological N fixation, mycorrhizal inoculation, nutrient recycling, compost and manure use, integration of crops and livestock
3. Sequestering soil C and improving soil organic matter	Creating positive C budget, mulch farming, conservation-effective measures, deep-rooted plants, recycling crop and animal residues, conservation agriculture, balanced application of plant nutrients
4. Climate-resilient agriculture	Conservation agriculture, mulch farming, complex cropping/farming systems, flexible farm operations, drought-resistant varieties and species, integrated crops/trees/livestock systems, recycling of crops and animal residues

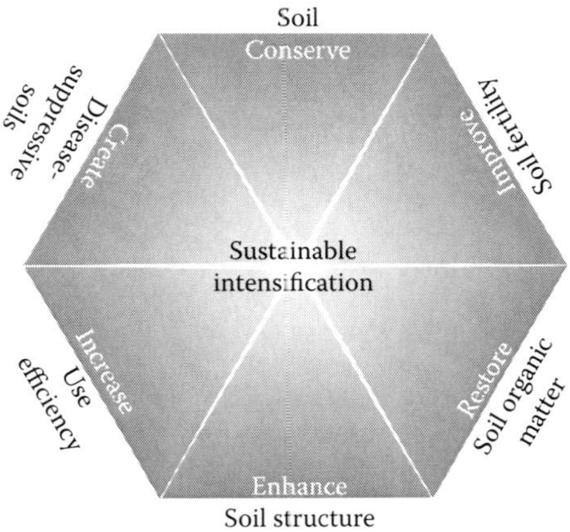

FIGURE 16.1 Strategies of sustainable intensification of smallholder agriculture.

16.3 ISSUES OF SMALLHOLDER FARMING

Despite its global significance and the impact on lives of >2 billion people, smallholder agriculture has numerous issues that need to be addressed through sustainable intensification. Low agronomic productivity has created a poverty trap for its practitioners, which can only be broken by enhancing crop yields and profitability through sustainable intensification. Increasing agronomic production is also essential to ensuring food and nutritional security of the resource-poor households.

Over and above the issues pertaining to the human dimensions (e.g., alleviating poverty and advancing food/nutritional security), there are numerous environmental issues. Indeed, the extractive agriculture practiced by resource-poor small landholders is a major determinant/control of soil/environmental degradation. Deforestation, biomass burning, and cultivation of agriculturally marginal lands by the small landholders lead to loss of biodiversity, accelerated erosion, transport of pollutants and sediments into natural water, and transforming state of the agroecosystems to beyond the safe planetary boundaries (Röckstrom et al. 2009). Any new expansion of agriculture that may be occurring in ecologically sensitive ecoregions, *viz* the tropical rainforest (Foley et al. 2011), must be avoided.

Thus, an appropriate strategy is to enhance productivity, bridge the yield gap, and improve use efficiency of inputs through adoption of the BMPs for sustainable intensification. Increasing productivity per unit of land area and inputs (e.g., water, fertilizers, energy) would also reduce and avoid gaseous emissions (Burney et al. 2010). Thus, the strategies of promoting adoption of the BMPs of sustainable intensification by resource-poor farmers are to (i) reduce soil erosion; (ii) conserve and recycle water and nutrients; (iii) enhance soil fertility through integrated nutrient management involving a judicious combination of organic amendments, inorganic fertilizers, biological N fixation, and mycorrhizal inoculation; (iv) improve soil quality through C sequestration as humus and secondary carbonates while increasing the activity and species diversity of soil fauna and flora, and strengthening the disease-suppressive properties of soils; (v) reduce emissions of N_2O (NO_x) and increase CH_4

uptake in soils by improving its structure and aeration; (vi) restore soils degraded by erosion, salinization, nutrient depletion, and other degradation processes; (vii) minimize water and air pollution by agricultural operations; (viii) reduce water use by improving water productivity; (ix) enhance resilience of the agroecosystems to harsh, variable, and uncertain climate; and (x) use decision support systems for identifying management decisions to enhance productivity and environmental stewardship.

16.4 PAYMENTS FOR ECOSYSTEM SERVICES

Indiscriminate promotion of the so-called sustainable agricultural practices for reducing the environmental footprint can reduce the flexibility of smallholder farmers, minimize opportunities for growth, and create additional barriers to alleviating poverty. While linking smallholder to market and strengthening institutional support (e.g., extension services, access to credit, availability of input) are important, creating another income stream is essential to alleviating poverty and creating opportunities for investments in soil restoration. Thus, payments for ecosystem services provisioned by farmers would promote adoption of BMPs that are otherwise inaccessible. Principal among ecosystem services are C sequestration, water quality, biodiversity, net primary productivity, etc. The use of payments through a just and transparent system, such as trading of C credits, is a win–win strategy. Sequestration of C in soils and biota is a conservation-effective strategy and has numerous co-benefits. However, it requires soil and crop management options that create positive ecosystem C and nutrient (N, P, S) budgets. In this context, some inputs are either expensive or have competing uses (e.g., crop residues). Thus, for provisioning of ecosystem services of societal or global relevance, farmers must be appropriately compensated. Conversion of biomass-C into humus requires incorporation of nutrients (e.g., N, P, S) that might otherwise be used by subsequent crops because humus has a higher concentration of these elements than crop residues. The societal value of C sequestered must be assessed on the basis of services provisioned, and on inherent characteristics of humus that stabilize the soil C and enhance its mean residence time. For example, the societal value of C in soil humus may be computed on the basis of its constituents (e.g., concentration of plant nutrients and water retention capacity). Alternatively, the societal value of soil C may be computed on the basis of the cost of other methods of C sequestration. The cost of geological sequestration of C by injection into a stable geological formation is estimated at \$600 to \$800 Mg^{-1} CO_2 (Economist 2012). Farmers and land managers may be compensated on the basis of the cost of geological sequestration adjusted for some co-benefits on soil quality, agronomic productivity, and other factors.

16.5 TECHNOLOGICAL OPTIONS FOR SUSTAINABLE INTENSIFICATION OF SMALLHOLDER AGRICULTURE

There exists a wide range of technological options for sustainable intensification of smallholder agriculture. Promising technologies for soil management include NT farming or CA, mulching techniques including organic and inorganic mulches, cover cropping, and those systems that create a positive soil/ecosystem C budget. Recommended technologies for water management include options that enhance the green water by reducing losses through runoff and evaporation and enhance water productivity. Soil fertility enhancement and nutrient management options must be those that replace what is removed so that risks of nutrient mining are minimal, the balanced use of plant nutrients is achieved through a judicious combination of inorganic fertilizers and organic amendments, and modern/innovative fertilizers are developed on the basis of the principles of nanotechnologies and slow-release formulations. The energy use efficiency must be increased by choice of appropriate farm operations (e.g., tillage, grain drying, transport, harvesting) and judicious use of energy-based inputs (e.g., fertilizers, pesticides, irrigation). Similarly, the management of growing season duration is relevant for adaptation to the changing and uncertain climate through choice of crops and cropping systems and timing of farm operations to make adjustments for any alteration in the duration of the growing season. In this context, the importance of crops (animals and trees) is important, including those of genetically modified organisms, which may reduce the inputs of pesticides.

Agricultural technologies for sustainable intensification must be scale neutral but gender sensitive. There is neither a panacea nor a single approach, and the choice of a technology depends on site-specific situations—biophysical, social, economic, political, and cultural. All technological processes and practices outlined in Figure 16.2 have pros and cons, merits and demerits, costs and benefits. Nothing is free, given, or appropriated. Thus, trade-offs must be carefully evaluated. Furthermore, technologies must also be appropriate to the women farmers of Africa and Asia.

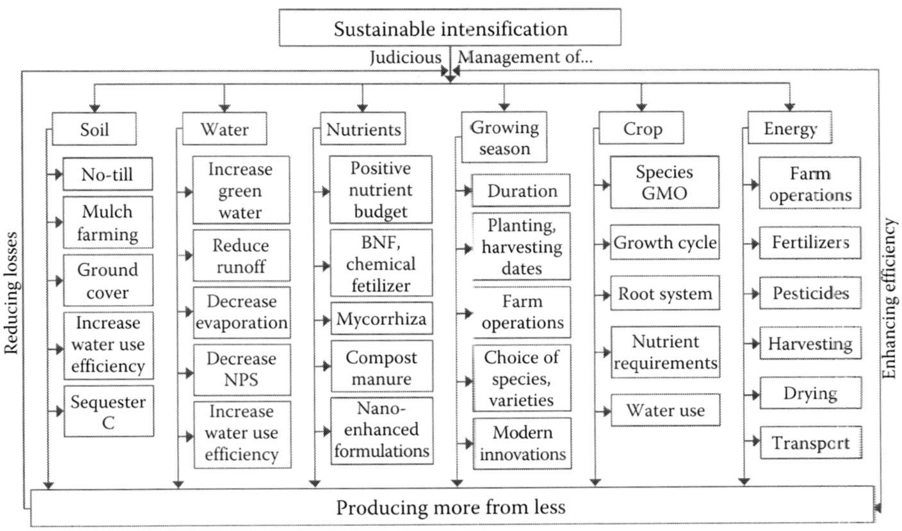

FIGURE 16.2 Processes and practices for sustainable intensification of agriculture practiced by smallholders. NPSP, nonpoint source pollution; C, carbon; GMO, genetically modified organisms.

16.6 RESEARCH AND DEVELOPMENT PRIORITIES

Transforming smallholder farming from subsistence to productive and environmentally restorative agriculture is a high priority in the context of (i) increasing population, growing affluent lifestyle, increasing consumption, and preferences toward an animal-based diet; (ii) a changing and harsh climate with increasing frequency of extreme events and growing variability and uncertainty; (iii) reducing availability of renewable freshwater supply and increasing risks of pollution, eutrophication, and contamination; (iv) degrading soils and desertifying lands, and increasing conversion of agricultural lands to industrial, urban, and recreational uses; and (v) decreasing per capita arable land area; and the ever increasing demand for food production while reducing the environmental footprint of agroecosystems. There are long-term climate implications of increasing concentrations of CO_2 and greenhouse gases (Friedlingstein et al. 2011) to small landholders. Thus, it is pertinent to consider historical and future perspectives of global soil (and carbon) response to climate and land use changes (Eglin et al. 2010). Thus, there are research, outreach, education, and communication priorities that must be addressed (Figure 16.3).

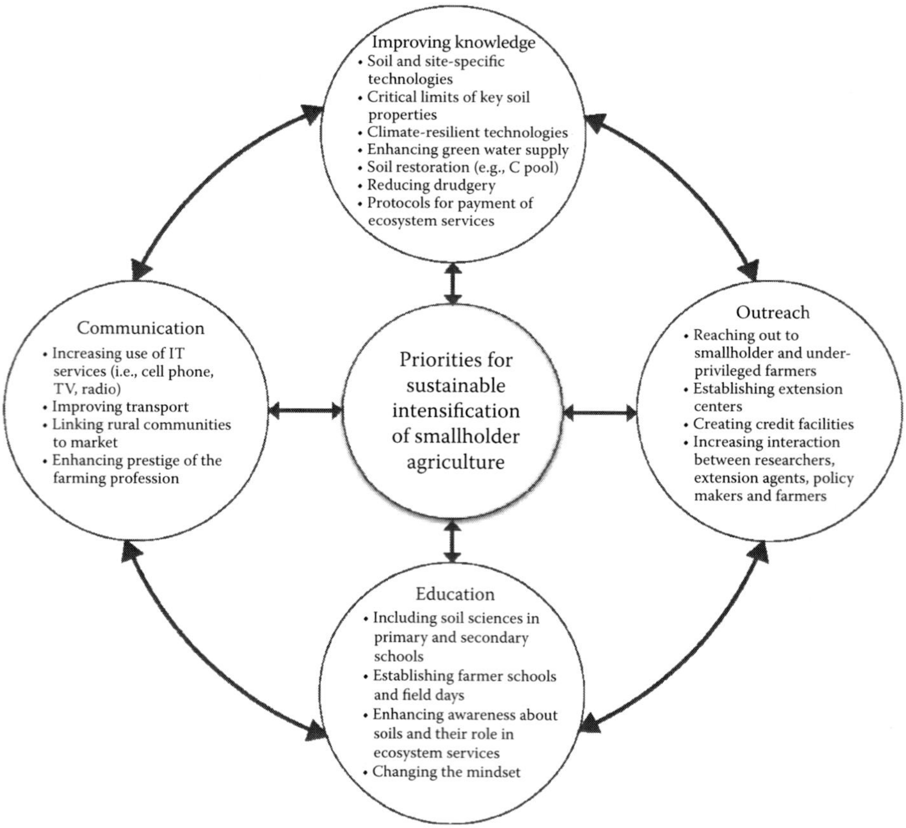

FIGURE 16.3 Priorities for promoting adoption of sustainable intensification.

A high priority must be given to improving knowledge about the soil-related constraints to adopting sustainable intensification. Site-specific and credible research information is needed for the critical limits of key soil properties, technological options for climate-strategic practices, options of improving the green water supply in the root zone and reducing risks of drought, techniques to restore soil quality and improve fertility, and tools and practices that reduce drudgery and enhance the respectability of the farming profession (Figure 16.3). The scientific data must be translated into practical tools that farmers can use and policy makers can understand.

Outreach and economic development are closely interlinked. A high priority must be given to reaching out to the underprivileged and poorest among the resource-poor farmers. Extension centers must be established in the rural communities, and credit facilities must be made available. It is important to increase interactions among researchers, extension agents, policy makers, and the farming communities.

Teaching soil science must be integral to curricula in primary and secondary schools (K to 12). Farmer schools must be established through extension centers for enhancing awareness about new and emerging technologies. The mindset about traditional versus new options must be changed through dialogue and demonstration under on-farm conditions by participatory approaches.

Lack of proper communication with the farming communities in remote areas is a major constraint to the adoption of BMPs. It is critical to enhance the use of information technologies services such as cell phones, TV programs, radio talk shows, and bulletins or fact sheets. Improving transport that links farmers to market (to purchase seed, fertilizers, pesticides, tools, and sell farm produce) is critical to adoption of BMPs for sustainable intensification (Figure 16.3).

16.7 CONCLUSIONS

Increasing food production by 70% between 2010 and 2050 to meet the demands of the expected increase in population from 7.3 billion in 2014 to 9.5 billion by 2050 (UN 2013) is the biggest challenge facing humanity since the dawn of settled agriculture. The challenge is even more daunting because of the warming Earth, dwindling and polluting water resources, degrading soils, reducing per capita arable land area, and changing dietary preferences toward animal-based diets.

The strategy is to produce more from less through promoting the adoption of the BMPs of sustainable intensification. The goal is to minimize expansion of agricultural land area, avoid deforestation of tropical rainforest, reduce use of surface water and groundwater, minimize the use of fertilizers and pesticides, and recycle all by-products. Therefore, losses must be reduced and the use efficiency of all inputs drastically increased.

There exists a large gap between the actual and the attainable crop yield. The gap can be narrowed or eliminated through adoption of BMPs. Adoption of these technologies can be promoted through payments to land managers for provisioning of ecosystem services.

The challenges facing smallholder or any other agriculture are unlike any experienced by humanity since the dawn of settled agriculture. Thus, addressing these challenges necessitates adoption of innovative and revolutionary (rather than gradual or evolutionary) approaches. It is in this context that sustainable intensification holds a vast but unexploited potential.

REFERENCES

Burney, J.A., Steven, J.D., Lobell, D.B. 2010. Greenhouse gas mitigation by agricultural intensification. *PNAS* 107: 12052–12057.

Economist. 2012. Combating climate change: Net benefits. *Economist* March 17, 2012, p. 89.

Eglin, T., Ciais, P., Piao, S.L. et al. 2010. Historical and future perspectives of global soil carbon response to climate and land use changes. *Tellus* 62B: 700–718.

Foley, J.A., Ramankutty, N., Brauman, K.A. et al. 2011. Solutions for a cultivated planet. *Nature* 448: 337–342.

Friedlingstein, P., Solomon, S., Plattner, G.-K., Knutti, R., Ciais, P., Raupach, M.R. 2011. Long-term climate implications of twenty-first century options for carbon dioxide emission mitigation. *Nat. Climate Change* 1: 457–461.

Friedrich, T., Derpsch, R., Kassam, A. 2012. Overview of the global spread of conservation agriculture. *J. Field Actions*. Special Issue 6: 1–33. Available at http://factsreports.revues.org/1941.

IFAD. 2011. *Viewpoint: Smallholders Can Feed the World*. International Fund for Agricultural Development, Rome, Italy. Available at http://www.ifad.org/pub/viewpoint/smallholder.pdf.

Lal, R. 2007. Constraints to adopting no-till farming in developing countries. *Soil Tillage Res.* 94: 1–3.

MDG. 2000. Millennium Development Goals. Available at http://www.un.org/millenniumgoals/.

Röckstrom, J., Steffen, W., Noone, K. et al. 2009. A safe operating space for humanity. *Nature* 461: 472–475.

UN. 2013. World Population Prospects: The 2012 Revision, Key Finding and Advance Tables. Working Paper #ESA/P/WP 227.

UNEP. 2013. *Smallholder Farmers Key to Lifting Over One Billion People Out of Poverty*. United Nations Environment Programme. Nairobi, Kenya, June 4. Available at http://www.unep.org/NEWSCENTRE/default.aspx?DocumentId=2718&ArticleId=9524.

UNEP/IFAD. 2013. *Smallholders, Food Security and the Environment*. United Nations Environment Programme. Nairobi, Kenya. International Fund for Agricultural Development, Rome, Italy. Available at http://www.unep.org/pdf/SmallholderReport_WEB.pdf#sthash.869xkD1N.dpuf.

Index

Page numbers followed by f and t indicate figures and tables, respectively.

D

G

M

N

T

U

V

W